S. Haykin

Theory of Optimal Control
and Mathematical Programming

McGRAW-HILL SERIES IN SYSTEMS SCIENCE

Editorial Consultants

A. V. Balakrishnan
George Dantzig
Lotfi Zadeh

Berlekamp *Algebraic Coding Theory*
Canon, Cullum, and Polak *Theory of Optimal Control and Mathematical Programming*
Gill *Linear Sequential Circuits*
Harrison *Introduction to Switching and Automata Theory*
Jelinek *Probabilistic Information Theory: Discrete and Memoryless Models*
Mangasarian *Nonlinear Programming*
Nilsson *Learning Machines: Foundations of Trainable Pattern-classifying Systems*
Papoulis *Probability, Random Variables, and Stochastic Processes*
Papoulis *Systems and Transforms with Applications in Optics*
Viterbi *Principles of Coherent Communication*
Weber *Elements of Detection and Signal Design*
Zadeh and Desoer *Linear System Theory: The State Space Approach*

Theory of Optimal Control and Mathematical Programming

MICHAEL D. CANON
Research Staff
San Jose Research Laboratory
IBM Corporation
San Jose, California

CLIFTON D. CULLUM, JR.
Research Staff
Thomas J. Watson Research Center
IBM Corporation
Yorktown Heights, New York

ELIJAH POLAK
Professor of Electrical Engineering and Computer Sciences
University of California, Berkeley

McGRAW-HILL BOOK COMPANY
New York   San Francisco   St. Louis   Toronto
London   Sydney   Mexico   Panama

Theory of Optimal Control and Mathematical Programming

Copyright © 1970 by McGraw-Hill, Inc. All rights reserved.
No part of this publication may be reproduced, stored in a
retrieval system, or transmitted, in any form or by any means,
electronic, mechanical, photocopying, recording, or otherwise,
without the prior written permission of the publisher.

Printed in the United States of America.

Library of Congress catalog card number: 73-78954

34567890 KPKP 7987654

07-009760-7

Preface

Until recently, optimal control and nonlinear programming were considered to be distinct disciplines. In the late 1960s, however, it became apparent that a unified approach was not only feasible, but also highly desirable. This volume started out as a research monograph, presenting the authors' original work on a unified theory of constrained optimization in finite-dimensional spaces. This theory deals with conditions of optimality and with their utilization in algorithms. While this volume was being developed, the material to be used was presented in graduate courses on optimization in the department of Electrical Engineering and Computer Sciences, University of California, Berkeley, and in the department of Industrial Engineering, Columbia University. As a result, the scope of the book was enlarged, and a number of examples and exercises were added to provide the reader with a means for testing his understanding of the material. Also, two appendixes were added: one to provide the necessary mathematical background in convexity and one to derive the Pontryagin maximum principle as an illustration of how results are extended from finite-dimensional spaces to infinite-dimensional spaces.

This book has three basic aims: to present a unified theory of optimization, to introduce nonlinear programming algorithms to the control engineer, and to introduce the nonlinear programming expert to optimal control. This volume can be used either as a graduate text or as a reference text. As a basis for a graduate course, the book can be used on several levels. An advanced two-quarter sequence should cover the entire material. Alternatively, to lower the level of the course and reduce the time required, particularly difficult material may be omitted without loss of continuity: Sections 2.3, 2.4, 4.1, and some of the convergence proofs in Chapters 6 and 7, and Appendix B. Furthermore, the course may be biased toward optimal control by omitting Chapter 3, or toward nonlinear programming by omitting Chapter 4 and the optimal control applications in the other chapters. Thus in effect, this book offers several options.

At the end of each chapter is a short list of references, chosen either because they serve as a source for the material of that chapter or because they present an alternative, or complementary, point of view. Thus the references listed constitute a representative sample and not a comprehensive bibliography.

Basically, this book consists of two parts. In the first part, optimal control and nonlinear programming problems are shown to be equivalent to a simple canonical form of a mathematical programming problem. Necessary and sufficient conditions of optimality are derived for this canonical problem and are then specialized to obtain a number of specific results for nonlinear programming and optimal control problems. Much of this material is original. In the second part, a selection of linear and nonlinear programming algorithms is presented, and it is shown how these can be used for the solution of discrete optimal control problems. These algorithms were also chosen to illustrate how optimality conditions are used either as stopping rules in an algorithm, or conceivably, to suggest certain major features of an algorithm. To highlight these applications of optimality conditions, it is necessary to present the algorithms in a somewhat unorthodox manner. This will cause no difficulty to the person unfamiliar with the algorithms, while to the expert it will offer an alternative point of view.

The reader will observe that dynamic programming, stochastic optimal control, and stochastic mathematical programming have been omitted. Dynamic programming is omitted because the theory presented here adds little to its understanding and because it is covered extensively in other books. Stochastic problems are omitted chiefly because their inclusion would have increased this volume considerably beyond its projected size.

The authors are grateful to Dr. E. J. Messerli for a number of examples, exercises, and specialized optimality conditions; to Drs. J. Cullum, H. Halkin, S. Winograd, and Mr. L. P. Kalfon for their critical comments and suggestions; and to the graduate students who took the course EECS 226 at the University of California, Berkeley for their assistance in eliminating errors and ambiguities from the manuscript. Last, but not least, we wish to thank Mrs. Billie Vrtiak for the great care she took in typing the manuscript.

The preparation of this volume involved a great amount of research which would have been impossible without the generous support received from the International Business Machine Corporation; from the National Aeronautics and Space Administration, under grant NsG 354 and supplements 1, 2, 3, 4; from the National Science Foundation under grant GK 716; and from the University of California. This support is gratefully acknowledged.

Michael D. Canon
Clifton D. Cullum, Jr.
Elijah Polak

NOTE TO THE READER

The system of numbering and cross referencing used in this book is described as follows. The top of the right-hand page carries the section number and an abbreviated title of the section to which the page belongs. The top of the left-hand page carries the title of the chapter to which the page belongs. For example, SEC. 5.2 LINEAR CONTROL PROBLEMS appears on the right-hand page in Section 2 of Chapter 5, whose title OPTIMAL CONTROL AND LINEAR PROGRAMMING appears on the left-hand pages. Within each section, definitions, theorems, equations, remarks, and so forth, are numbered consecutively by means of boldface numerals appearing in the left-hand margin. In reference to a section within the same chapter, the section number only is used; in reference to a section in another chapter, both the chapter number and the section number are used. For example, "it was shown in Section 3" refers to Section 3 of the same chapter, while "it was shown in Section 2.3" refers to Section 3 of Chapter 2. Similarly, "substituting from (3)" refers to item 3 in the same section, "substituting from (2.3)" refers to item 3 in Section 2 of the same chapter, and "substituting from (3.2.3)" refers to item 3 in Section 2 of Chapter 3. The two appendixes are lettered A and B, and in references the letters A and B are used in the position of chapter numbers.

Preceding the index is a glossary of symbols which describes notational conventions and contains brief definitions and references to the principal symbols used in this book. The reader is advised to examine this glossary before reading the book.

Contents

Preface   v

1   PROBLEM FORMULATION   1

   1.1   Introduction   1
   1.2   Statement of the discrete optimal control problem   4
   1.3   A Canonical form of the discrete optimal control problem   7
   1.4   The mathematical programming problem   9
   1.5   Equivalence of the optimization problems   10

2   CONDITIONS OF OPTIMALITY FOR THE BASIC PROBLEM   13

   2.1   Introduction   13
   2.2   A first approach to necessary conditions   14
   2.3   The fundamental theorem   23
         A few second-order conditions of optimality   35
         Conditions for $\psi^0 < 0$ in theorem (12)   37
         Problems without differentiability assumptions   38
   2.4   The two-point-boundary-value optimal control problem   39

3 SOME NECESSARY AND SOME SUFFICIENT CONDITIONS FOR NONLINEAR PROGRAMMING PROBLEMS — 49

- 3.1 Introduction — 49
- 3.2 Theory of LaGrange multipliers — 50
- 3.3 The Kuhn-Tucker theory — 54
- 3.4 General nonlinear programming problems — 60
- 3.5 A further generalization — 67
- 3.6 A sufficient condition — 70
  - *An extension to quasi-convex and pseudoconvex functions* — 72

4 DISCRETE OPTIMAL CONTROL PROBLEMS — 75

- 4.1 The general case — 75
  - *Conditions for* $p^0 < 0$ — 82
- 4.2 A maximum principle — 84

5 OPTIMAL CONTROL AND LINEAR PROGRAMMING — 98

- 5.1 Introduction — 98
- 5.2 Linear control problems — 99
- 5.3 Control problems with linear dynamics, linear constraints, and piecewise-linear cost — 101
- 5.4 The Canonical linear programming problem — 105
- 5.5 Characterization of an optimal solution to the Canonical linear programming problem — 106
- 5.6 Some preliminary remarks on the Simplex algorithm — 109
- 5.7 Procedure for determining an initial extreme point of Ω' — 111
- 5.8 Generating improved adjacent extreme points — 112
- 5.9 The Simplex algorithm — 117
- 5.10 Resolution of degeneracy — 119
- 5.11 Bounded-variable linear programming — 125

6 OPTIMAL CONTROL AND QUADRATIC PROGRAMMING — 130

- 6.1 Formulation of the general control problem with quadratic cost and transformation to a quadratic programming problem — 130
- 6.2 The existence of an optimal solution to the quadratic programming problem — 132
- 6.3 A sufficient condition for a unique optimal solution to the quadratic programming problem — 137

6.4	Necessary and sufficient conditions of optimality for the quadratic programming problem	139
6.5	Applications to unbounded control problems	140
	Solution of (1) by direct transcription	140
	Solution of (1) by alternative transcription	144
	The control approach to the minimum-energy problem	145
	Solution of (53) by direct transcription	150
	Solution of (53) by alternative transcription	151
6.6	Quadratic Control problems with inequality constraints	154
6.7	Optimality conditions for the Canonical quadratic programming problem	156
6.8	The derived minimization problem	157
6.9	The simplex algorithm for quadratic programming	159
6.10	A further generalization	163
6.11	Convergence	168

7   CONVEX PROGRAMMING ALGORITHMS   177

7.1	Introduction	177
7.2	Methods of feasible directions	179
7.3	Steepest descent and gradient projection	196
	Unconstrained minimization problems	197
	Gradient projection: a special case	199
	Gradient projection: the general case	201
7.4	Penalty functions	205
	Exterior penalty functions	206
	Interior penalty functions	211

8   FREE-END-TIME OPTIMAL CONTROL PROBLEMS   214

8.1	Description of the free-end-time problem	214
8.2	The free-end-time problem as a sequence of fixed-time problems	215
8.3	The first time for which a free-end-time problem has a feasible solution: the minimum-time problem	218
8.4	A linear minimum-time problem	221
8.5	A geometric approach to the linear minimum-time problem	224

Appendixes   231

A   CONVEXITY   231

A.1	Introduction	231
A.2	Lines and hyperplanes	231

A.3	Convex sets	233
A.4	Convex cones	240
A.5	Separation of sets: supporting hyperplanes	245
A.6	Convex functions	252

B   **CONSTRAINED MINIMIZATION PROBLEMS IN INFINITE-DIMENSIONAL SPACES**   **259**

B.1	An extension of the fundamental theorem (2.3.12)	259
B.2	The maximum principle	262

Glossary of Symbols   277

Index   281

1
Problem formulation

1.1 INTRODUCTION

With the exception of Appendix B, the sections of this book which deal with optimal control consider discrete optimal control problems only. In addition, optimal control problems are treated as special cases of mathematical programming problems.

There are essentially two reasons for stressing discrete optimal control. The first reason is pedagogical; the second, and more important one is technical. Discrete optimal control problems are optimization problems in finite-dimensional spaces and, as such, require considerably less mathematical sophistication in their treatment than continuous optimal control problems. In addition, as we shall see in Appendix B, the extension of results from finite-dimensional spaces to infinite-dimensional spaces, and hence to continuous optimal control problems, is conceptually straightforward. It therefore seems to be pedagogically efficient to study discrete optimal control in depth and then to learn about continuous optimal control through natural extensions.

Our main reason for attaching so much importance to discrete optimal control is technical and stems from the constantly increasing use of digital computers in the control of dynamical systems. In any

2 PROBLEM FORMULATION

computation carried out on a digital computer, we can do no better than to obtain a finite set of real numbers. Thus in solving a continuous optimal control problem of the form:

1 $\qquad$ Minimize $\int_0^{t_f} f^0(x(t),u(t),t)\,dt$ subject to the constraints that for $t \in [0,t_f]$

2 $\qquad \dfrac{d}{dt} x(t) = f(x(t),u(t),t) \qquad x(t) \in E^n,\ u(t) \in E^m$

3 $\qquad x(0) \in X_0 \qquad x(t_f) \in X_f$

4 $\qquad u(t) \in U,$

we are forced to resort to some form of discretization.

If the discretization is governed only by the nature of the integration formulas to be used in solving (2) and in computing (1), then, in any iterative scheme, we must compute and store a very large number of points $x(t_i)$ and $u(t_i)$, with $t_i \in [0,t_f]$, at each iteration. Thus a straightforward discretization requires large memory capacity and usually results in long computation times, both of which are unacceptable if relatively small computers are to be used to control relatively fast dynamics.

The discretization favored by the authors is the one resulting from the restriction of the inputs $u(\cdot)$ to a class of functions representable by a finite set of parameters. The choice of the class of functions to which $u(\cdot)$ is to belong and of the number of parameters to be computed can be used by the designer to gain a great amount of freedom in controlling the dimension, and hence also the computational complexity of the resulting optimization problem. The price of such a simplification is a reduction in performance. However, without additional restrictions, problem (1) may not be solvable at all (within a prescribed time on the computer we must use), and there may be no choice but to further restrict the problem to make it tractable.

We shall now consider a few simple examples of commonly used discretizations of the inputs $u(\cdot)$. Suppose that in addition to (4) we require that the inputs be piecewise constant. In particular, we may require that

5 $\qquad u(t) = u_i \qquad t \in [iT, (i+1)T),\ u_i \in U,\ i = 0, 1, 2 \ldots, k-1,$

where $t_f = kT$ and k is an integer design parameter. The resulting discrete-time optimal control problem has the form:

6 $\qquad$ Minimize $\displaystyle\sum_{i=0}^{k-1} f_i^0(x_i,u_i)$ subject to the constraints

7 $$x_{i+1} - x_i = f_i(x_i,u_i) \qquad i = 0, 1, 2, \ldots, k-1$$
8 $$x_0 \in X_0 \qquad x_k \in X_f \qquad u_i \in U,$$

where $f_i(x_i,u_i)$ and $f_i^0(x_i,u_i)$ are computed as follows. For $t \in [iT, (i+1)T)$ let $x_i(t)$ be the solution of (2) corresponding to $u(t) \equiv u_i$ for $t \in [iT, (i+1)T)$ and satisfying $x_i(iT) = x_i$. In addition, we must have that for $i = 0, 1, 2, \ldots, k-1$, $x_{i+1} = x_i((i+1)T)$. Consequently,

9 $$x_{i+1} - x_i = \int_{iT}^{(i+1)T} f(x_i(t),u_i,t)\, dt,$$

and since $x_i(t)$ is uniquely determined by x_i and u_i, the functions in (6) and (7) are properly defined, as follows:

10 $$f_i^0(x_i,u_i) = \int_{iT}^{(i+1)T} f^0(x_i(t),u_i,t)\, dt.$$
11 $$f_i(x_i,u_i) = \int_{iT}^{(i+1)T} f(x_i(t),u_i,t)\, dt.$$

Other useful discretizations of the input that might be considered are

12 $$u(t) = \sum_{j=0}^{s} u_i^j t^j \qquad t \in I_i,\, u_i^j \in A_j \subset E^m,$$

where $\bigcup_i I_i = [0,t_f]$ and the A_j are suitably related to the original constraint set U. Note that when this particular discretization is interpreted in terms of a discrete dynamical system such as (7), we we find that the vector $u_i = (u_i^0, u_i^1, \ldots, u_i^s) \in E^m \times E^m \times E^m \times \cdots \times E^m$ and therefore has different dimension from the vector $u(t)$.

Once a discretization of the above type has been performed, the original optimal control problem becomes a finite-dimensional mathematical programming problem. This will be demonstrated in detail in the following sections of this chapter. It should be noted at this point, however, that a continuous optimal control problem may generate a finite-dimensional mathematical programming problem without simultaneously giving rise to a discrete optimal control problem. For example, this would be the case if we found it necessary to restrict the inputs $u(\cdot)$ as follows: for $t \in [0,t_f]$

13 $$u(t) = \sum_{i=1}^{s} u_i \varphi_i(t) \qquad u_i \in U_i \subset E^1,$$

where $\varphi_i: E^1 \to E^m$.

The realization that optimal control and mathematical programming problems are essentially one and the same thing, has led the authors [1] (as well as Neustadt [2], and Halkin and Neustadt [3]) to the construction of a unified theory of optimization. As we shall see later, this theory results in substantial conceptual simplifications and facilitates the transcription of highly sophisticated nonlinear programming algorithms for use in discrete optimal control.

1.2 STATEMENT OF THE DISCRETE OPTIMAL CONTROL PROBLEM

To define an optimal control problem we must specify the dynamics of the system, the constraints on the controls and on the trajectories, and in addition, we must specify a cost function. In this book we shall consider finite-dimensional systems whose dynamics satisfy a difference equation of the form

$$1 \qquad x_{i+1} - x_i = f_i(x_i, u_i) \qquad i = 0, 1, 2, \ldots,$$

where $x_i \in E^n$ is the system state at time $i = 0, 1, 2, \ldots$; $u_i \in E^m$ is the system input, or control, at time $i = 0, 1, 2, \ldots$; and the $f_i(\cdot,\cdot)$ for $i = 0, 1, 2, \ldots$ are functions mapping $E^n \times E^m$ into E^n, assumed to be continuously differentiable always in x, but not always in u.

The duration of an optimal control process may be either preassigned or not; i.e., it is either fixed or free. It is necessary to emphasize this distinction at the very outset, because considerably more results, both qualitative and algorithmic, are available for fixed-time problems than for free-time problems. Indeed, we shall confine ourselves almost exclusively to fixed-time optimal control problems, and we shall later see how free-time problems can often be solved by solving a sequence of fixed-time problems of increasing duration. In any event, we shall always assume that the duration of the optimal process is finite, since otherwise the optimization might have to be carried out in an infinite-dimensional space, which is outside the scope of this text.

Now let us turn our attention to the type of constraints we shall encounter in fixed-time processes, *which we assume to be of k steps duration*. We assume that we are given k subsets of E^m, which we shall designate by U_i for $i = 0, 1, \ldots, k-1$; $k+1$ subsets of E^n, which we shall designate by X_i for $i = 0, 1, \ldots, k$; a function $h(\cdot,\cdot)$ from $E^{n(k+1)} \times E^{km}$ into E^s, where s is some positive integer; and a subset $D \subset E^s$. Depending on the result desired, we shall later impose various conditions on the sets U_i, X_i, and D and the function $h(\cdot,\cdot)$.

Now let $\mathfrak{U} = (u_0, u_1, \ldots, u_{k-1})$ be a control sequence, and let $\mathfrak{X} = (x_0, x_1, \ldots, x_k)$ be a corresponding trajectory determined by system (1).† We shall say a control sequence $\mathfrak{U}$ and a corresponding

† Note that the trajectory $\mathfrak{X}$ is not defined uniquely by the control sequence $\mathfrak{U}$, since it also depends on the initial state x_0.

SEC. 1.2 STATEMENT OF THE DISCRETE OPTIMAL CONTROL PROBLEM

trajectory $\mathfrak{X}$ are *admissible*[1] if they satisfy the *control constraints*

(2) $u_i \in U_i \quad i = 0, 1, \ldots, k - 1,$

the *state-space constraints*

(3) $x_i \in X_i \quad i = 0, 1, \ldots, k,$

and the *trajectory constraints*

(4) $h(\mathfrak{X}, \mathfrak{U}) \in D.$

In a typical boundary-value problem, the sets U_i may be intervals of the form $U_i = \{u : |u| \leq 1\}$ for $i = 0, 1, \ldots, k - 1$; the sets X_0 and X_k may be manifolds of the form $X_i = \{x : g_i(x) = 0\}$ for $i = 0, k$ where $g_i : E^n \to E^{l_i}$ for $i = 0, k$; and the other constraints may be nonexistent; that is, $X_i = E^n$ for $i = 1, 2, \ldots, k - 1$ and $h \equiv 0$. The function h is commonly used to express limitations imposed by the total quantity of available resources, such as energy or fuel, in which case (4) may assume the form

(5) $\sum_{i=0}^{k-1} \|u_i\|^2 \leq d \quad \text{or} \quad \sum_{i=0}^{k-1} \sum_{j=1}^{m} |u_i^j| \leq d,$

where $\|\cdot\|$ is the euclidean norm.

Finally, we assume that we are given a *real-valued cost function*

(6) $f(\cdot, \cdot)$

defined on $E^{n(k+1)} \times E^{mk}$.† As a typical example of the function $f(\cdot, \cdot)$, consider the one defined by

(7) $f(\mathfrak{X}, \mathfrak{U}) = \|x_k - x_d\|^2 + \sum_{i=0}^{k-1} \|u_i\|^2,$

which expresses the deviation of the terminal state x_k from a given point $x_d \notin X_k$, summed with the energy expended by taking system (1) from an initial state $x_0 \in X_0$ to the terminal state $x_k \in X_k$.

We can now combine the dynamical system (1) with the constraints (2) to (4) and the cost function (6) to obtain a precise formulation of the most general fixed-time optimal control problem that we can consider in this framework.

(8) **The fixed-time optimal control problem.** Given a dynamical system

(9) $x_{i+1} - x_i = f_i(x_i, u_i) \quad i = 0, 1, \ldots, k - 1$

(where the states $x_i \in E^n$ for $i = 0, 1, \ldots, k$, the controls $u_i \in E^m$

---

[1] Later on, the words *admissible* and *feasible* will be used synonymously.
† We shall always consider a control sequence $\mathfrak{U} = (u_0, u_1, \ldots, u_{k-1})$ to be a vector in E^{mk} and a trajectory $\mathfrak{X} = (x_0, x_1, \ldots, x_k)$ to be a vector in $E^{n(k+1)}$.

6 PROBLEM FORMULATION

for $i = 0, 1, \ldots, k - 1$, and k is the specified duration for the process), together with subsets $X_i \subset E^n$ for $i = 0, 1, \ldots, k$, subsets $U_i \subset E^m$ for $i = 0, 1, \ldots, k - 1$, a subset $D \subset E^s$, a constraint function $h(\cdot,\cdot)$ from $E^{n(k+1)} \times E^{mk}$ into E^s, and a real-valued cost function $f(\cdot,\cdot)$ defined on $E^{n(k+1)} \times E^{mk}$, find a control sequence $\hat{\mathfrak{U}} = (\hat{u}_0, \hat{u}_1, \ldots, \hat{u}_{k-1})$ and a corresponding trajectory $\hat{\mathfrak{X}} = (\hat{x}_0, \hat{x}_1, \ldots, \hat{x}_k)$ satisfying (9), with

10      $\hat{u}_i \in U_i \qquad i = 0, 1, \ldots, k - 1$
11      $\hat{x}_i \in X_i \qquad i = 0, 1, \ldots, k$
12      $h(\hat{\mathfrak{X}}, \hat{\mathfrak{U}}) \in D,$

such that for every control sequence $\mathfrak{U} = (u_0, u_1, \ldots, u_{k-1})$ and every corresponding trajectory $\mathfrak{X} = (x_0, x_1, \ldots, x_k)$ satisfying (9) to (12)

13      $f(\hat{\mathfrak{X}}, \hat{\mathfrak{U}}) \leq f(\mathfrak{X}, \mathfrak{U}). \quad \square$

A free-time problem differs from a fixed-time problem in one important detail: the duration of the process, k, is not specified in advance. The easiest way to extend the definition of the fixed-time optimal control problem to free-time problems is by assuming that instead of having a fixed cost function f and constraint function h, we have a sequence of such functions, $f_{(k)}$ and $h_{(k)}$, parametrized by the duration k. Similarly, we have to assume that we have a sequence of subsets D_k. Thus we shall consider the free-time problem as a sequence of fixed-time problems.

14   **The free-time optimal control problem.** Given a dynamical system

15      $x_{i+1} - x_i = f_i(x_i, u_i) \qquad i = 0, 1, 2, \ldots,$

together with subsets $X_i \subset E^n$ for $i = 0, 1, 2, \ldots$, subsets $U_i \subset E^m$ for $i = 0, 1, 2, \ldots$, subsets $D_k \subset E^s$ for $k = 0, 1, 2, \ldots$, a sequence of constraint functions $h_{(k)}(\cdot,\cdot)$ mapping $E^{n(k+1)} \times E^{mk}$ into E^s for $k = 0, 1, 2, \ldots$, and a sequence of real-valued cost functions $f_{(k)}(\cdot,\cdot)$ defined on $E^{n(k+1)} \times E^{mk}$ for $k = 0, 1, 2, \ldots$, find an integer $\hat{k}$, a control sequence $\hat{\mathfrak{U}}_{\hat{k}} = (\hat{u}_0, \hat{u}_1, \ldots, \hat{u}_{\hat{k}-1})$, and a corresponding trajectory $\hat{\mathfrak{X}}_{\hat{k}} = (\hat{x}_0, \hat{x}_1, \ldots, \hat{x}_{\hat{k}})$ satisfying (15), with

16      $\hat{u}_i \in U_i \qquad i = 0, 1, \ldots, \hat{k} - 1$
17      $\hat{x}_i \in X_i \qquad i = 0, 1, \ldots, \hat{k}$
18      $h_{(\hat{k})}(\hat{\mathfrak{X}}_{\hat{k}}, \hat{\mathfrak{U}}_{\hat{k}}) \in D_{(\hat{k})},$

such that for every $k = 0, 1, \ldots$, for every control sequence $\mathfrak{U}_k = (u_0, u_1, \ldots, u_{k-1})$, and for every corresponding trajectory $\mathfrak{X}_k = (x_0, x_1, \ldots, x_{k-1})$ satisfying (15) to (18), with k taking the place of $\hat{k}$,

(19) $$f_{(k)}(\hat{\mathfrak{X}}_k, \hat{\mathfrak{U}}_k) \leq f_{(k)}(\mathfrak{X}_k, \mathfrak{U}_k). \quad \square$$

1.3 A CANONICAL FORM OF THE DISCRETE OPTIMAL CONTROL PROBLEM

There are various ways of associating a cost with a fixed-time control process $(\mathfrak{X}, \mathfrak{U})$, that is, a control sequence $\mathfrak{U}$ and a corresponding trajectory $\mathfrak{X}$. The most common one is to assign a cost to each state transition, in which case the total cost is the sum of these individual costs, and we therefore have

(1) $$f(\mathfrak{X}, \mathfrak{U}) = \sum_{i=0}^{k-1} f_i^0(x_i, u_i),$$

where the $f_i^0(\cdot, \cdot)$ for $i = 0, 1, \ldots, k-1$ are real-valued functions defined on $E^n \times E^m$. As an example, consider a minimum-energy problem for which

(2) $$f(\mathfrak{X}, \mathfrak{U}) = \sum_{i=0}^{k-1} \|u_i\|^2,$$

where $\|\cdot\|$ denotes the euclidean norm. As another example, suppose that we wish to minimize the value $\varphi(x_k)$ of a real-valued function φ of the terminal state, that is,

(3) $$f(\mathfrak{X}, \mathfrak{U}) = \varphi(x_k).$$

If we now examine (3) and (1), we find that by letting $f_i^0(\cdot, \cdot) \equiv 0$ for $i = 0, 1, \ldots, k-2$ and $f_{k-1}^0(x_{k-1}, u_{k-1}) = \varphi(x_{k-1} + f_{k-1}(x_{k-1}, u_{k-1}))$, we can convert (3) to the form of (1).

It should be clear from these examples that many optimal control problems have cost functions which can be written in form (1). However, there are also a number of optimal control problems for which the cost associated with a control process $(\mathfrak{X}, \mathfrak{U})$ cannot be decomposed into a sum. A simple example of this occurs when the cost function $f(\cdot, \cdot)$ has the form

(4) $$f(\mathfrak{X}, \mathfrak{U}) = \max_{i=0,1,\ldots,k-1} \varphi^i(x_i, u_i),$$

that is, when we wish to minimize the maximum "deviation" from a desired control process.

Since fixed-time optimal control problems with costs of form (1) are by far the most common, they have received the lion's share of the attention. It is sometimes convenient to combine the dynamic equations (2.1) with the cost expression (1) into a single augmented dynamical system as follows. Let the scalars x_i^0 for $i = 0, 1, \ldots, k$ be determined by the difference equation

(5) $$x_{i+1}^0 - x_i^0 = f_i^0(x_i, u_i) \qquad i = 0, 1, \ldots, k-1,$$

with $x_0{}^0 = 0$, where $((x_0, x_1, \ldots, x_k), (u_0, u_1, \ldots, u_{k-1}))$ is any control process of the system (2.1). We now let $\mathbf{x} = (x^0, x) \in E^{n+1}$, where $x^0 \in E^1$ and $x \in E^n$ [that is, $\mathbf{x} = (x^0, x^1, \ldots, x^n)$], and for $i = 0, 1, \ldots, k-1$ we define the functions $\mathbf{f}_i \colon E^{n+1} \times E^m \to E^{n+1}$ by $\mathbf{f}_i(\mathbf{x}, u) = (f_i{}^0(x, u), f_i(x, u))$. Finally, we combine equations (2.1) and (5) into the augmented system

6 $\qquad \mathbf{x}_{i+1} - \mathbf{x}_i = \mathbf{f}_i(\mathbf{x}_i, u_i) \qquad i = 0, 1, \ldots, k-1.$

With the introduction of the augmented system (6), we are led to the following important special case of the fixed-time optimal control problem (2.8).

7 **The canonical optimal control problem (fixed-time).** Given a dynamical system

8 $\qquad \mathbf{x}_{i+1} - \mathbf{x}_i = \mathbf{f}_i(\mathbf{x}_i, u_i) \qquad i = 0, 1, \ldots, k-1$

(where the states $\mathbf{x}_i \in E^{n+1}$ for $i = 0, 1, \ldots, k$, the controls $u_i \in E^m$ for $i = 0, 1, \ldots, k-1$, and k is the specified duration of the process), together with subsets $\mathbf{X}_i \subset E^{n+1}$ for $i = 0, 1, \ldots, k$, subsets $U_i \subset E^m$ for $i = 0, 1, \ldots, k-1$, a subset $\mathbf{D} \subset E^s$, and a constraint function $\mathbf{h}(\cdot, \cdot)$ mapping $E^{(n+1)(k+1)} \times E^{km} \to E^s$, find a control sequence $\hat{\mathfrak{U}} = (\hat{u}_0, \hat{u}_1, \ldots, \hat{u}_{k-1})$ and a corresponding trajectory $\hat{\mathfrak{X}} = (\hat{\mathbf{x}}_0, \hat{\mathbf{x}}_1, \ldots, \hat{\mathbf{x}}_k)$ satisfying (8), with

9 $\qquad \hat{u}_i \in U_i \qquad i = 0, 1, \ldots, k-1$
10 $\qquad \hat{\mathbf{x}}_i \in \mathbf{X}_i \qquad i = 0, 1, \ldots, k$
11 $\qquad \mathbf{h}(\hat{\mathfrak{X}}, \hat{\mathfrak{U}}) \in \mathbf{D},$

such that for every control sequence $\mathfrak{U}$ and every corresponding trajectory $\mathfrak{X}$ satisfying (8) to (11)

12 $\qquad \hat{x}_k{}^0 \leq x_k{}^0,$

where $x_k{}^0$ is the first component of $\mathbf{x}_k$. □

13 **Remark.** Observe that in the above definition we have *not* made the assumption implied by (6) that the functions $\mathbf{f}_i$ for $i = 0, 1, \ldots, k-1$ and $\mathbf{h}$ do not depend on the cost variable x^0. As will be seen from example (16) below, the removal of this assumption enables us to treat an important class of problems without increasing the dimension of the dynamical system. □

When the system (8) is indeed an augmented system, then the functions $\mathbf{f}_i$ and $\mathbf{h}$ do not depend on x^0, that is, $\mathbf{f}_i \colon E^n \times E^m \to E^{n+1}$ and $\mathbf{h} \colon E^{n(k+1)} \times E^m \to E^s$, and the sets $\mathbf{X}_i$ assume the form

14 $\qquad \begin{aligned} \mathbf{X}_0 &= \{0\} \times X_0 \\ \mathbf{X}_i &= E^1 \times X_i \qquad i = 1, 2, \ldots, k. \end{aligned}$

The effect of transcribing an optimal control problem into the canonical form is to make the problem completely geometric, as will be seen in Section 2.4.

15 **Exercise.** Obtain a canonical formulation for free-time optimal control problems analogous to (7) above. □

16 **Example.** It is not always necessary to augment the dynamic equations (2.1) in order to cast the problem into canonical form. This is clearly the case when the cost is of the form

17 $$f(\mathfrak{X}, \mathfrak{U}) = x_k^1.$$

Thus consider the case when we wish to take an object constrained to move on a line from some initial position x_0 to a position as far as possible from the initial point (for example, a ballistics problem). In this case we may care little about the terminal values of the other state variables. Then if for $i = 0, 1, \ldots, k$ we let $x_i^0 = -(x_i^1 - x_0^1)$ and renumber the other state variables $x_i^2, x_i^3, \ldots, x_i^n$ as $x_i^1, x_i^2, \ldots, x_i^{n-1}$, we find that the original problem has been transcribed into canonical form without our having augmented the state space. □

1.4 THE MATHEMATICAL PROGRAMMING PROBLEM

For the purpose of this book it will be convenient to adopt a canonical form for the mathematical programming problem. We shall call this form the *Basic Problem* because all the optimization problems which we shall consider can be transcribed into this form and it will play a central role in the following chapters.

1 **The Basic Problem.** Given a real-valued continuously differentiable function $f(\cdot)$ defined on E^n, a continuously differentiable function $r(\cdot)$ from E^n into E^m, and a subset $\Omega \subset E^n$, find a vector $\hat{z} \in E^n$ satisfying

2 $\qquad \hat{z} \in \Omega$
3 $\qquad r(\hat{z}) = 0$

such that

4 $\qquad f(\hat{z}) \leq f(z)$

for all $z \in E^n$ satisfying (2) and (3). □

Note that we have refrained from saying how the set Ω is characterized and that we have chosen to highlight the presence of equality constraints on the minimization of $f(z)$. The reason for breaking up the constraints on the minimization of $f(z)$ into equality constraints and other constraints is that it is often impossible to obtain meaningful necessary conditions of optimality without imposing this much additional structure. This fact will become increasingly clear in Chapters 2 and 3.

5 Remark. The formulation of the Basic Problem does *not* imply that equality constraints *cannot* enter into the characterization of the set Ω. The set Ω may well be of the form $\Omega = \{z : s(z) = 0, q(z) \leqq 0\}$, where $s: E^n \to E^l$ and $q: E^n \to E^m$. We simply choose to highlight *some* of the equality constraints. □

Finally, in the chapters to follow it will often be necessary to refer to vectors z which satisfy (2) and (3) as well as to vectors $\hat{z}$ which are solutions to the Basic Problem. We shall therefore give these vectors special names. These names are commonly used in mathematical programming literature.

6 Definition. A vector $z \in E^n$ will be said to be *feasible* (or a *feasible solution*) for the Basic Problem if it satisfies (2) and (3). A vector $\hat{z} \in E^n$ will be said to be *optimal* (or an *optimal solution*) for the Basic Problem if it is feasible and satisfies (4) for all feasible vectors $z \in E^n$. □

1.5 EQUIVALENCE OF THE OPTIMIZATION PROBLEMS

We shall now show that the fixed-time optimal control problems (2.8)† and (3.7) can be transcribed into the form of the Basic Problem (4.1), and similarly, that the Basic Problem (4.1) can be transcribed into a one-step two-point-boundary-value optimal control problem of form (2.8) or (3.7). The fact that these transformations are possible is important, since it enables us to interpret results obtained for either one of these problems in the light of the other.

Thus consider again the fixed-time optimal control problem (2.8), and let $z = (x_0, x_1, \ldots, x_k, u_0, u_1, \ldots, u_{k-1})$, where $\mathfrak{U} = (u_0, u_1, \ldots, u_{k-1})$ and $\mathfrak{X} = (x_0, x_1, \ldots, x_k)$ are a control sequence and a corresponding trajectory of the system (2.9).

† When the cost function $f(\mathfrak{X}, \mathfrak{U}) \triangleq \max \varphi^i(x_i, u_i)$ for $i = 0, 1, \ldots, k-1$, as in (3.4), it becomes necessary to introduce a new variable, ζ, and to consider instead the equivalent problem: minimize ζ subject to $\zeta - \varphi^i(x_i, u_i) \geq 0$ for $i = 0, 1, \ldots, k-1$ and the constraints in (2.8). This transcription is necessary to make the cost function differentiable in the final form (4.1) of the problem.

SEC. 1.5 EQUIVALENCE OF THE OPTIMIZATION PROBLEMS

Suppose that for $i = 0, 1, \ldots, k$ the subsets X_i of E^n [see (2.3)] are of the form

$$X_i = \{x: g_i(x) = 0, x \in X'_i\},$$

where $g_i(\cdot)$ is a continuously differentiable function mapping E^n into E^{l_i} and X'_i is a subset of E^n which cannot be described by a system of equations. Also suppose that for $i = 0, 1, \ldots, k$ the subsets U_i [see (2.2)] of E^m are of the form

$$U_i = \{u_i: \psi_i(u_i) = 0, u_i \in U'_i\},$$

where ψ_i is a continuously differentiable function mapping E^m into E^{s_i} and U'_i is a subset of E^m which, again, cannot be described by a system of equations. Finally, let $V = h^{-1}(D)$ [see (2.4)]. Then, to convert the fixed-time optimal control problems (2.8) or (3.7) into the Basic Problem (4.1), we may let[1]

$$\Omega = (X'_0 \times X'_1 \times \cdots \times X'_k \times U'_0 \times U'_1 \times \cdots \times U'_{k-1}) \cap V$$

and define the function $r(\cdot)$ by

$$r(z) = \begin{bmatrix} x_1 - x_0 - f_0(x_0, u_0) \\ x_2 - x_1 - f_1(x_1, u_1) \\ \vdots \\ x_k - x_{k-1} - f_{k-1}(x_{k-1}, u_{k-1}) \\ g_0(x_0) \\ g_1(x_1) \\ \vdots \\ g_k(x_k) \\ \psi_0(u_0) \\ \psi_1(u_1) \\ \vdots \\ \psi_{k-1}(u_{k-1}) \end{bmatrix}.$$

The cost function, of course, remains the same, $f(z) = f(\mathfrak{X}, \mathfrak{U})$. □

---

[1] As we shall see in Section 2.4, the transcription of problem (2.8) or (3.7) into the form of the Basic Problem (4.1) can be carried out in more ways than one. By and large, the transcription given here is more convenient for the purpose of analysis, while the transcription given in Section 2.4 is more suitable for computation when the sets $X_i = E^n$ for $i = 1, 2, \ldots, n - 1$, since it results in a lower-dimensional function $r(\cdot)$.

We transcribe the Basic Problem (4.1) into a one-step optimal control problem of form (2.8) by treating the variable z as a control and by constructing the fictitious dynamical system given by the difference equation

5 $$x_{i+1} - x_i = f_i(x_i, z) \qquad i = 0,$$

where $x_i = (x_i^1, x_i^2, \ldots, x_i^m) \in E^m$ for $i = 0, 1$ and $f_0(x_0, z) \triangleq r(z)$.

The constraints now become $X_0 = \{0\}$, $X_1 = \{0\}$, and $U_0 = \Omega$. The cost of this one-step transition is given by $f_0^0(x_0, z) \triangleq f(z)$. To obtain (3.7), we simply proceed as in Section 3. □

We have thus seen that for the purpose of analysis the fixed-time optimal control problem (2.8) and the Basic Problem (4.1) are equivalent, and we shall therefore choose the form to be used in the analysis of any particular problem simply on the basis of convenience.

REFERENCES

1. M. D. Canon, C. D. Cullum, and E. Polak: Constrained Minimization Problems in Finite Dimensional Spaces, *SIAM J. Control*, **4**:528–547 (1966).
2. L. W. Neustadt: An Abstract Variational Theory with Applications to a Broad Class of Optimization Problems, I. General Theory, II. Applications, *SIAM J. Control*, **4**:505–527 (1966), **5**:90–137 (1967).
3. H. Halkin and L. W. Neustadt: General Necessary Conditions for Optimization Problems, *Proc. Natl. Acad. Sci.*, **56**(4):1066–1071 (1966).

2
Conditions of optimality for the basic problem

2.1 INTRODUCTION

We have already seen in Section 1.5 that fixed-time discrete optimal control problems, with very general constraints, are transcribable into the form of the Basic Problem (1.4.1). We observe at this point that the standard nonlinear programming problem is simply the Basic Problem with the set Ω described by a system of inequalities. We also saw in Section 1.5 that the Basic Problem is transcribable into a one-step two-point-boundary-value optimal control problem.[1] Thus we could adopt either the Basic Problem or the two-point-boundary-value discrete optimal control problem as our canonical constrained optimization problem. However, since it contains no dynamic equations, the Basic Problem has less structure, and it is therefore the simpler, and in a sense the more general, of the two candidates for a canonical form. We shall therefore adopt it as our standard problem from now on.

[1] A two-point-boundary-value optimal control problem is the particular case of the control problem (1.2.7) in which $X_0 = \{\hat{x}_0\}$ and $X_k = \{\hat{x}_k\}$, that is, the initial and terminal states are given, and there are no other trajectory constraints, that is, $h \equiv 0$ and $X_i = E^n$ for $i = 2, 3, \ldots, k-1$.

One of our major goals is a unified theory of constrained optimization. We propose to achieve this goal in two steps. The first step will consist of obtaining conditions of optimality for the Basic Problem. The second step, which will be undertaken in the following chapters, will be to interpret these conditions of optimality in terms of specific optimization problems.

The reader might be perturbed by the fact that the unified approach we are about to present does not seem to relate to the perturbed-trajectory approach used in the past by Pontryagin et al. [7], Jordan and Polak [8], Halkin [9], and many others. We shall therefore demonstrate in Section 4 that the perturbed-trajectory approach actually can be recovered by transcribing optimal control problems into the form of the Basic Problem (1.4.1) in a manner somewhat more complicated than the one given in Section 1.5.

2.2 A FIRST APPROACH TO NECESSARY CONDITIONS

We propose to accomplish two objectives in this section. The first is to obtain a necessary condition of optimality for the Basic Problem (1.4.1) without differentiating between equality and other constraints. Such an approach readily yields a necessary condition, but unfortunately, as will be seen in Chapter 3, it cannot be applied meaningfully to a large class of nonlinear programming and optimal control problems. Since the difficulties are usually caused by the equality constraints, our next objective is a heuristic development of an "approximate" necessary condition of optimality which does take into account the fact that some of the constraints are equality constraints. This approximate condition will be refined into a rigorous necessary condition of optimality in the next section.

To avoid the need for excessive leafing back and forth, the Basic Problem (1.4.1) is restated here.

1 The Basic Problem. Given a real-valued continuously differentiable function $f(\cdot)$ defined on E^n, a continuously differentiable function $r(\cdot)$† mapping E^n into E^m, and a subset Ω of E^n, find a vector $\hat{z} \in E^n$ satisfying

2 $\quad\quad\hat{z} \in \Omega$
3 $\quad\quad r(\hat{z}) = 0$

such that for all $z \in E^n$ satisfying (2) and (3)

4 $\quad\quad f(\hat{z}) \leq f(z)$. $\square$

† The vector $r(z) \in E^m$ will be assumed to be a column vector, so that $r(z) = (r^1(z), r^2(z), \ldots, r^m(z))^T$, where T denotes transposition.

Since for the time being we do not propose to differentiate between equality and other constraints, we combine them into a set Ω', defined by

(5) $\quad \Omega' = \{z : z \in \Omega, r(z) = 0\}.$

The Basic Problem now becomes:

(6) **Basic Problem.** Find a $\hat{z} \in \Omega'$ so that $f(\hat{z}) \leq f(z)$ for all $z \in \Omega'$. □

Also, in terms of (5), definition (1.4.6) can be rephrased as follows.

(7) **Definition.** We shall say that a vector $\hat{z} \in E^n$ is a *feasible solution* to the Basic Problem if $\hat{z} \in \Omega'$, and we shall say that $\hat{z}$ is an *optimal solution* to the Basic Problem if $\hat{z}$ solves problem (6). □

Suppose that $\hat{z}$ is an optimal solution. We now define a cone of directions which are tangent to all the "smooth" paths in Ω' which originate at $\hat{z}$.

(8) **Definition.** The *tangent cone* to the set Ω' at $\hat{z} \in \Omega'$, denoted by $TC(\hat{z},\Omega')$, is the set of all vectors $\delta z \in E^n$ with the property that for every $\delta z \in TC(\hat{z},\Omega')$ there exists a $t_1 > 0$ and a continuous function $o(\cdot)$ from the reals into E^n, satisfying $\|o(t)\|/t \to 0$ as $t \to 0$, such that

(9) $\quad z(t) \triangleq [\hat{z} + t\,\delta z + o(t)] \in \Omega' \qquad 0 \leq t \leq t_1,$

where t_1 and $o(\cdot)$ may depend on $\hat{z}$ and δz. □

(10) **Remark.** Thus if $\delta z \in TC(\hat{z},\Omega')$, then there is in Ω' a finite arc, emanating from $\hat{z}$ and defined by (9), which is differentiable at $\hat{z}$ and to which δz is a tangent at $\hat{z}$. □

(11) **Exercise.** Use scaling to show from (9) that if $\delta z \in TC(\hat{z},\Omega')$, then so does $\lambda\,\delta z$ for all $\lambda \geq 0$, thus establishing that $TC(\hat{z},\Omega')$ *is indeed a cone*. □

We are now ready to prove our first necessary condition of optimality.

(12) **Theorem.**[1] If $\hat{z}$ is an optimal solution to the Basic Problem (1) and $TC(\hat{z},\Omega')$ is the tangent cone to Ω' at $\hat{z}$, where Ω' is as defined in (5), then

[1] A form of this theorem was first presented by Kuhn and Tucker [5].

16 CONDITIONS OF OPTIMALITY FOR THE BASIC PROBLEM

(13) $$\langle -\nabla f(\hat{z}), \delta z \rangle \leq 0 \qquad \text{for all } \delta z \in \overline{\text{co}} \ TC(\hat{z}, \Omega').\dagger$$

Proof. Suppose that there is a $\delta z^* \in TC(\hat{z}, \Omega')$ which does not satisfy (13), that is,

(14) $$\langle -\nabla f(\hat{z}), \delta z^* \rangle > 0.$$

Then, by definition (8), there is a $t_1^* > 0$ and a continuous function $o^*(\cdot)$ such that

(15) $$[\hat{z} + t \ \delta z^* + o^*(t)] \in \Omega' \qquad 0 \leq t \leq t_1^*.$$

Expanding the cost function $f(\cdot)$ along the arc defined by (15), we get

(16) $$f(\hat{z} + t \ \delta z^* + o^*(t)) = f(\hat{z}) + t \langle \nabla f(\hat{z}), \delta z^* \rangle + o(t)$$
$$0 \leq t \leq t_1^*,$$

where $o(\cdot)$ is a continuous function such that $o(t)/t \to 0$ as $t \to 0$. Since $\langle \nabla f(\hat{z}), \delta z^* \rangle < 0$, by (14), and in (16) the term linear in t dominates $o(t)$ for t sufficiently small, we conclude that there exists a t^* in the interval $(0, t_1^*]$ such that

(17) $$f(\hat{z} + t^* \ \delta z^* + o^*(t)) < f(\hat{z}),$$

which contradicts the optimality of $\hat{z}$, since $z^* = \hat{z} + t^* \ \delta z^* + o(t^*)$ is a feasible solution; i.e., it is in Ω'. Consequently, for every $\delta z \in TC(\hat{z}, \Omega')$

(18) $$\langle -\nabla f(\hat{z}), \delta z \rangle \leq 0.$$

Now let δz be an arbitrary vector in co $TC(\hat{z}, \Omega')$; then for some integer s

(19) $$\delta z = \sum_{i=1}^{s} \mu^i \ \delta z_i,$$

where $\delta z_i \in TC(\hat{z}, \Omega')$, $\sum_{i=1}^{s} \mu^i = 1$, and $\mu^i \geq 0$ for $i = 1, 2, \ldots, s$. But since each $\delta z_i \in TC(\hat{z}, \Omega')$ satisfies (18) and the $\mu^i \geq 0$, it is clear that every $\delta z \in$ co $TC(\hat{z}, \Omega')$ must satisfy (18). Now, the function $\langle \nabla f(\hat{z}), \cdot \rangle$ is obviously continuous, and hence if (18) holds for all $\delta z \in$ co $TC(\hat{z}, \Omega')$, it must also hold for all $\delta z \in \overline{\text{co}} \ TC(\hat{z}, \Omega')$, the closure of this set. $\square$

(20) **Remark.** In anticipation of the fact that we shall have to rely heavily in the next section on the Brouwer fixed-point theorem, let us see how the first part of the proof of theorem (12) can be accom-

† The symbol co $TC(\hat{z}, \Omega')$ denotes the convex hull of $TC(\hat{z}, \Omega')$, while $\overline{\text{co}} \ TC(\hat{z}, \Omega')$ denotes the closure of the convex hull of $TC(\hat{z}, \Omega')$ [see (A.3.32)].

plished by showing that a certain map has a fixed point. Suppose that $\delta z^* \in TC(\hat{z}, \Omega')$ satisfies (14). Let $t_0 = \frac{1}{2}t_1^*$, where t_1^* is as in (15), and let σ be a real variable satisfying $|\sigma| \leq \rho \triangleq t_0$. Then for $t = \alpha(t_0 + \sigma)$, with $\alpha \in (0,1]$, (16) becomes

21
$$f(\hat{z} + \alpha(t_0 + \sigma)\, \delta z^* + o^*(\alpha(t_0 + \sigma))$$
$$= f(\hat{z}) + \alpha t_0 \langle \nabla f(\hat{z}), \delta z^* \rangle + \alpha \sigma \langle \nabla f(\hat{z}), \delta z^* \rangle + o(\alpha(t_0 + \sigma)),$$

where $o(\alpha(t_0 + \sigma))/\alpha \to 0$ as $\alpha \to 0$ uniformly in all $\sigma \in [-\rho, +\rho]$. Now, $\alpha t_0 \langle \nabla f(\hat{z}), \delta z^* \rangle < 0$ for all $\alpha > 0$; hence to establish (17) we need simply find an $\alpha^* \in (0,1]$ and a $\sigma^* \in [-\rho, +\rho]$ which satisfy the equation

22
$$\alpha \sigma \langle \nabla f(\hat{z}), \delta z^* \rangle + o(\alpha(t_0 + \sigma)) = 0.$$

We let $x = \sigma \langle \nabla f(\hat{z}), \delta z^* \rangle$, $y_0 = t_0 \langle \nabla f(\hat{z}), \delta z^* \rangle$, and $r = \rho |\langle \nabla f(\hat{z}), \delta z^* \rangle|$. If we substitute for σ and t_0 in (22), then (22) becomes

23
$$G_\alpha(\alpha x) \triangleq \alpha x + o\left(\frac{\alpha}{\langle \nabla f(\hat{z}), \delta z^* \rangle}(y_0 + x)\right) = 0.$$

Now let $\bar{o}(z) = o(z/\langle \nabla f(\hat{z}), \delta z^* \rangle)$, with $o(\cdot)$ as in (23). Then (23) becomes

24
$$G_\alpha(\alpha x) = \alpha x + \bar{o}(\alpha(y_0 + x)) = 0,$$

where $\bar{o}(\alpha(y_0 + x))/\alpha \to 0$ as $\alpha \to 0$ uniformly in $x \in [-r, +r]$. Hence there exists an $\alpha^* \in (0,1]$ such that

25
$$\alpha^* r > |\bar{o}(\alpha^* y_0 \pm r)|.$$

Referring to Figure 1, we see immediately that there must be a point $\alpha^* x^*$ in the interval $[-\alpha^* r, +\alpha^* r]$ such that $G_{\alpha^*}(\alpha^* x^*) = 0$, that is,

26
$$-\bar{o}(\alpha^*(y_0 + x^*)) = \alpha^* x^*,$$

which shows that $\alpha^* x^*$ is a fixed point of the map $-\bar{o}(\alpha^*(y_0 + \cdot))$ from the interval $[-\alpha^* r, +\alpha^* r]$ into the reals, and at the same time that (22) [and hence (17)] has a solution. □

Obviously, as far as theorem (12) is concerned, this is a long-winded approach to take, but we shall see in the next section that it cannot always be avoided in conjunction with other theorems.

27 **Exercise.** Let $\hat{z}$ be an optimal solution to the Basic Problem (1). Show that theorem (12) [that is, inequality (13)] remains valid for all $\delta z \in \overline{co}\, TC_s(\hat{z}, \Omega')$, where $TC_s(\hat{z}, \Omega')$ is as defined in (28). □

28 **Definition.** The *sequential tangent cone* to the set Ω' at $\hat{z} \in \Omega'$, denoted by $TC_s(\hat{z}, \Omega')$, is the set consisting of the origin together with

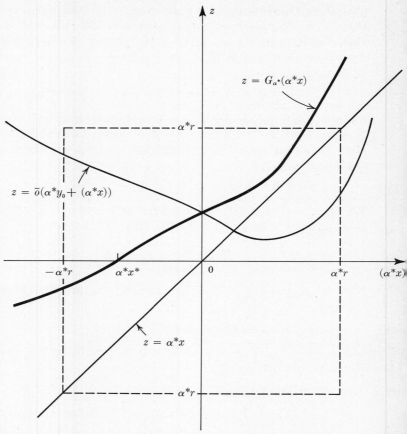

Figure 1

all nonzero vectors δz which have the property that there exists a sequence of vectors $\{\delta z_i\}$ such that $(\hat{z} + \delta z_i) \in \Omega'$ for $i = 1, 2, 3, \ldots$, $\delta z_i \to 0$ as $i \to \infty$ and $\delta z_i/\|\delta z_i\| \to \alpha\, \delta z$ as $i \to \infty$ for some $\alpha > 0$. The sequence $\{\delta z_i\}$ may depend on $\hat{z}$ and δz. ☐

29 Exercise. Show that $TC_s(\hat{z},\Omega')$ is a closed cone. Show that $TC_s(\hat{z},\Omega') \supset TC(\hat{z},\Omega')$. Show that $TC_s(\hat{z},\Omega') = TC(\hat{z},\Omega')$ when Ω' is convex. ☐

Let us now return to theorem (12). Statement (13) is very simple, but it cannot be applied meaningfully unless we have the tangent cone $TC(\hat{z},\Omega')$, or at least a subset of $TC(\hat{z},\Omega')$. Thus the utility of theorem (12) depends entirely on our ability to obtain a characterization for the tangent cone. A detailed discussion of

SEC. 2.2 A FIRST APPROACH TO NECESSARY CONDITIONS

tangent cones is postponed until Chapter 3, and we shall see there that when the jacobian matrix $\partial r(\hat{z})/\partial z = \partial r^i(\hat{z})/\partial z^j$ has rank less than m, the tangent cone to Ω' at $\hat{z}$ cannot be constructed by standard methods. However, we shall also see that it is not exceptionally difficult to construct either tangent cones or subsets of tangent cones to the set Ω in a large number of interesting cases. We therefore proceed to construct a necessary condition of optimality which depends only on tangent cones or subsets of tangent cones to the set Ω.

Our first step is to combine the *cost function* $f(\cdot)$ with the *equality-constraint function* $r(\cdot)$, appearing in the statement of the Basic Problem (1), to define a continuously differentiable function $F(\cdot)$ mapping E^n into E^{m+1}:

$$30 \qquad F(z) = \begin{bmatrix} f(z) \\ r(z) \end{bmatrix}.$$

For the sake of convenience, we shall always write $F(z) = (f(z), r(z))$, although we mean a column vector, as in (30). We shall also number the components of E^{m+1} from 0 to m, rather than the usual 1 to $m+1$, so that $y \in E^{m+1}$ has the form $y = (y^0, y^1, \ldots, y^m)$; again we mean a column vector, but we omit the transposition sign. Consequently, $F(z) = (F^0(z), F^1(z), \ldots, F^m(z))$, where, by (30), $F^0(z) = f(z), F^1(z) = r^1(z), \ldots, F^m(z) = r^m(z)$. The jacobian matrix of $F(\cdot)$ evaluated at z will be denoted by $\partial F(z)/\partial z$; its first row is $\nabla F^0(z) = \nabla f(z)$, its second row is $\nabla F^1(z) = \nabla r^1(z)$, and so forth. Thus it is an $(m+1) \times n$ rectangular matrix whose ijth element is $\partial F^i(z)/\partial z^j$.

Now, suppose that $\hat{z}$ is an optimal solution to the Basic Problem (1) and that δz is some small perturbation vector. Then, since $F(\cdot)$ is assumed to be continuously differentiable, we can expand it about $\hat{z}$ to obtain

$$31 \qquad F(\hat{z} + \delta z) = F(\hat{z}) + \frac{\partial F(\hat{z})}{\partial z} \delta z + o(\delta z),$$

where $o(\cdot)$ is a continuous function from E^n into E^{m+1} with the property that $\|o(\delta z)\|/\|\delta z\| \to 0$ as $\|\delta z\| \to 0$.

To avoid having to evaluate (31) with δz moving along a curve, as was done in (16), we assume that the entire tangent cone to Ω at $\hat{z}$, or at least a large subset of it, is made up of tangents to linear segments l of the form $l = \{z: z = \hat{z} + t\,\delta z, 0 \leq t \leq 1\}$ contained in Ω. We shall call such a subset a *radial cone*, which we now define formally.

32 Definition. The *radial cone* to Ω at $\hat{z} \in \Omega$, denoted by $RC(\hat{z},\Omega)$, is the set of all vectors δz for which there exists an $\epsilon > 0$ such that $(\hat{z} + \alpha\, \delta z) \in \Omega$ for all $\alpha \in [0,\epsilon]$. □

Thus the radial cone $RC(\hat{z},\Omega)$ is the cone generated by all the straight-line segments which emanate from $\hat{z}$ and are contained in Ω. When the set Ω is convex, the closure of the cone $\{\hat{z}\} + RC(\hat{z},\Omega)$, whose vertex is at $\hat{z}$, is the smallest closed cone with vertex at $\hat{z}$ which contains the set Ω.

33 Exercise. Let Ω be a convex subset of E^n. Show that the radial cone to Ω at $\hat{z} \in \Omega$ is the cone generated by the set $\Omega - \{\hat{z}\}$ [see (A.4.7).] □

Example. Referring to Figure 2a, we see that when the point $\hat{z}$ is interior to the set Ω, the radial cone $RC(\hat{z},\Omega)$ is the whole space. When the set Ω is convex and $\hat{z}$ is a boundary point of Ω, $RC(\hat{z},\Omega)$ is as shown in Figure 2b, and when Ω is not convex but $\hat{z}$ is on the boundary of Ω, then $RC(\hat{z},\Omega)$ is as shown in Figure 2c. It is not difficult to show that when Ω is a closed polyhedral set, $RC(\hat{z},\Omega) = TC(\hat{z},\Omega) = TC_s(\hat{z},\Omega)$. □

It is rather clear that the major source of difficulty in deriving necessary conditions of optimality will be the term $o(\delta z)$ in (31). To reduce its effect, we begin by assuming that only the cost function $f(\cdot)$ is nonlinear. This enables us to obtain the following simple result.

34 Theorem. Let $\hat{z}$ be an optimal solution to the Basic Problem (1). If the function $r(\cdot)$ is affine, then the *negative cost ray*

$$35 \quad R = \{y: y = \beta(-1,0,0,\ldots,0) \in E^{m+1}, \beta > 0\}$$

cannot belong to the cone

$$36 \quad K(\hat{z}) = \frac{\partial F(\hat{z})}{\partial z} RC(\hat{z},\Omega) = \left\{y: y = \frac{\partial F(\hat{z})}{\partial z} \delta z,\ \delta z \in RC(\hat{z},\Omega)\right\}.^\dagger$$

Proof. Suppose that R belongs to the cone $K(\hat{z})$; then there is a $\delta z^* \in RC(\hat{z},\Omega)$ such that

$$37 \quad \frac{\partial F(\hat{z})}{\partial z} \delta z^* \in R.$$

† Additional geometric insight into the nature of this theorem as well as theorem (3.12) can be obtained from the observation that $\hat{z}$ is an optimal solution to the Basic Problem (1) *if and only if* the ray R is disjoint from the set $F(\Omega) - \{\hat{y}\}$, where $\hat{y} = (f(\hat{z}),0,0,\ldots,0)$. Theorem (34) states that if $\hat{z}$ is optimal, then the ray R must be disjoint from $K(\hat{z})$, which is a first-order approximation to $F(\Omega) - \{\hat{y}\}$; that is, (34) "approximates" the statement that R and $F(\Omega) - \hat{y}$ must be disjoint.

SEC. 2.2 A FIRST APPROACH TO NECESSARY CONDITIONS

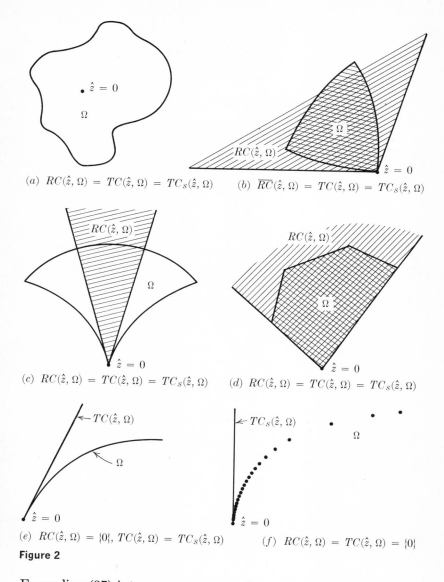

(a) $RC(\hat{z}, \Omega) = TC(\hat{z}, \Omega) = TC_S(\hat{z}, \Omega)$
(b) $\overline{RC}(\hat{z}, \Omega) = TC(\hat{z}, \Omega) = TC_S(\hat{z}, \Omega)$
(c) $RC(\hat{z}, \Omega) = TC(\hat{z}, \Omega) = TC_S(\hat{z}, \Omega)$
(d) $RC(\hat{z}, \Omega) = TC(\hat{z}, \Omega) = TC_S(\hat{z}, \Omega)$
(e) $RC(\hat{z}, \Omega) = \{0\}, TC(\hat{z}, \Omega) = TC_S(\hat{z}, \Omega)$
(f) $RC(\hat{z}, \Omega) = TC(\hat{z}, \Omega) = \{0\}$

Figure 2

Expanding (37) into component equations, we get

$$\langle \nabla f(\hat{z}), \delta z^* \rangle < 0 \tag{38}$$
$$\langle \nabla r^i(\hat{z}), \delta z^* \rangle = 0 \qquad i = 1, 2, \ldots, m. \tag{39}$$

But $r(\cdot)$, by assumption, is affine, and hence [from (39)] we have

$$r^i(\hat{z} + \delta z^*) = r^i(\hat{z}) + \langle \nabla r^i(\hat{z}), \delta z^* \rangle = 0 \qquad i = 1, 2, \ldots, m. \tag{40}$$

Now,

(41) $$f(\hat{z} + \alpha\, \delta z^*) = f(\hat{z}) + \alpha\langle \nabla f(\hat{z}), \delta z^*\rangle + o(\alpha),$$

where $o(\alpha)/\alpha \to 0$ as $\alpha \to 0$, and it follows from (38) and (41) that there exists an $\epsilon' > 0$ such that

(42) $$f(\hat{z} + \alpha\, \delta z^*) < f(\hat{z}) \qquad \text{for all } \alpha \in (0, \epsilon'].$$

Furthermore, by definition (32), there is an $\epsilon(\hat{z},\delta z^*) > 0$ such that $\hat{z} + \alpha \delta z^* \in \Omega$ for all $\alpha \in [0, \epsilon(\hat{z},\delta z^*)]$. Let $\epsilon^* = \min\,[\epsilon', \epsilon(\hat{z},\delta z^*)]$; then $\hat{z} + \epsilon^*\,\delta z^*$ is a feasible solution; that is,

(43) $$r(\hat{z} + \epsilon^*\,\delta z^*) = 0 \quad \text{and} \quad (\hat{z} + \epsilon^*\,\delta z^*) \in \Omega,$$

and

(44) $$f(\hat{z} + \epsilon^*\,\delta z^*) < f(\hat{z}),$$

which contradicts the optimality of $\hat{z}$. $\square$

(45) **Corollary.** Suppose that the assumptions of theorem (34) are satisfied and, in addition, the cone $RC(\hat{z},\Omega)$ is convex. Then the cone $K(\hat{z})$ and the ray R can be separated, that is, there exists a nonzero vector $\psi \in E^{m+1}$, with $\psi^0 \leq 0$, such that

(46) $$\left\langle \psi, \frac{\partial F(\hat{z})}{\partial z}\,\delta z \right\rangle \leq 0 \qquad \text{for all } \delta z \in RC(\hat{z},\Omega)$$

and

(47) $$\langle \psi, y\rangle \geq 0 \qquad \text{for all } y \in R.$$

Proof. When $RC(\hat{z},\Omega)$ is convex, the cone $K(\hat{z})$ is also convex. Since $K(\hat{z})$ does not contain the ray R, it follows from theorem (A.5.5) that there exists a vector $\psi \in E^{m+1}$ such that (46) and (47) are satisfied and hence $\psi^0 \leq 0$. $\square$

(48) **Corollary.** In addition to the hypothesis of corollary (45), suppose that the cone $K(\hat{z})$ is closed. Then there is a nonzero vector $\psi = (\psi^0, \psi^1, \ldots, \psi^m)$, with $\psi^0 < 0$, such that

(49) $$\left\langle \psi, \frac{\partial F(\hat{z})}{\partial z}\,\delta z \right\rangle \leq 0 \qquad \text{for all } \delta z \in RC(\hat{z},\Omega).$$

Proof. Since $R \cap K(\hat{z}) = \phi$, the point $l = (-1,0,0,\ldots,0) \in E^{m+1}$ does not belong to the closed convex cone $K(\hat{z})$. It follows from corollary (A.5.28) and exercise (A.5.27) that there exists a vector $\psi \in E^{m+1}$ satisfying (49) and such that $\langle \psi, l\rangle > 0$; that is, $-\psi^0 > 0$.

50 Example. In many instances the radial cone $RC(\hat{z},\Omega)$ can be given a very simple description. Consider the case when the set Ω is a polyhedron of the form

$$\Omega = \{z : \langle q^i, z \rangle \leq d^i,\ i = 1, 2, \ldots, k\},$$

where $q^i \in E^n$ for $i = 1, 2, \ldots, k$. Let $I(\hat{z})$ be the index set identifying the *active constraints;* that is,

$$I(\hat{z}) = \{i : \langle q^i, \hat{z} \rangle = d^i,\ i \in \{1, 2, \ldots, k\}\}.$$

If $I(\hat{z})$ is the empty set, then $\hat{z}$ is in the interior of Ω, and hence $RC(\hat{z},\Omega)$ is the whole space. If $I(\hat{z})$ is not empty, then it is easy to see that

51 $$RC(\hat{z},\Omega) = \{\delta z : \langle q^i, \delta z \rangle \leq 0,\ i \in I(\hat{z})\}.\quad\square$$

We can now easily guess the extent to which theorem (34) will have to be modified to include the case where $r(\cdot)$ is not necessarily affine.

52 **A heuristic condition of optimality.** Suppose that R is not just any ray of the cone $K(\hat{z})$, but that it is an interior ray of $K(\hat{z})$. Then the second-order term $o(\cdot)$ in (31) is not likely to dislodge all the points of $F(\hat{z}) + R$ from the set $F(\{\hat{z}\} + RC(\hat{z},\Omega))$ [the image of $\{\hat{z}\} + RC(\hat{z},\Omega)$ under F], and hence the set $F(\Omega)$ will contain admissible points $y \in F(\hat{z}) + R$ of lower cost $f(\hat{z})$. Thus we would expect that the statement of theorem (34) need only be modified to read " . . . then the negative cost ray R cannot be an *interior* ray of the cone $K(\hat{z})$ . . . ." $\square$

We shall see in the next section that this guess is correct, and we shall also see what further modifications are necessary to accommodate the case when the radial cone $RC(\hat{z},\Omega)$ contains only the zero vector.

2.3 THE FUNDAMENTAL THEOREM

With the preliminary analysis of the Basic Problem completed, we are now ready to approach the question of necessary conditions in earnest. We begin by introducing two types of conical approximations to the set Ω. These conical approximations will define the class of *perturbations* about $\hat{z}$, an optimal solution to the Basic Problem, in terms of which we shall state our necessary condition of optimality for $\hat{z}$. Thus a salient feature of the theorem to be derived is that it is not valid with respect to arbitrary perturbations about an optimal point, but is valid only with respect to a specific subclass, which may or may not exhaust all the possible perturbations about the optimal solution in question.

1 Definition. A convex cone $C(\hat{z},\Omega) \subset E^n$ will be called a *conical approximation of the first kind* to the constraint set Ω at the point $\hat{z} \in \Omega$ if for any collection $\{\delta z_1, \delta z_2, \ldots, \delta z_k\}$ of linearly independent vectors in $C(\hat{z},\Omega)$ there exists an $\epsilon > 0$, possibly depending on $\hat{z}$ and $\delta z_1, \delta z_2, \ldots, \delta z_k$, such that co $\{\hat{z}, \hat{z} + \epsilon\, \delta z_1, \hat{z} + \epsilon\, \delta z_2, \ldots, \hat{z} + \epsilon\, \delta z_k\} \subset \Omega$. □

If a cone $C(\hat{z},\Omega)$ is a conical approximation of the first kind to Ω at $\hat{z} \in \Omega$, then for every $\delta z \in C(\hat{z},\Omega)$ there exists, by definition, an $\epsilon > 0$ such that $(\hat{z} + \alpha\, \delta z) \in \Omega$ for all α satisfying $0 \leq \alpha \leq \epsilon$. Thus every conical approximation of the first kind to the set Ω at $\hat{z}$ is contained in the radial cone to Ω at $\hat{z}$, which was defined in (2.32).

2 Lemma. Suppose that Ω is a convex subset of E^n. Then the radial cone to Ω at any point $\hat{z} \in \Omega$ is a conical approximation of the first kind. Furthermore,

$$RC(\hat{z},\Omega) = \{\delta z : \delta z = \lambda(z - \hat{z}),\ \lambda \geq 0,\ z \in \Omega\}.$$

Proof. From exercise (2.33), the radial cone to Ω at $\hat{z}$ is the cone generated by $\Omega - \{\hat{z}\}$; that is, $RC(\hat{z},\Omega)$ is as given above. Now, $RC(\hat{z},\Omega)$ is convex, since Ω is convex, and the cone generated by a convex set is a convex cone [see (A.4.8)]. Let $\{\delta z_1, \delta z_2, \ldots, \delta z_k\}$ be any set of vectors in $RC(\hat{z},\Omega)$. Then there are scalars $\lambda_i \geq 0$ and vectors $z_i \in \Omega$ such that $\delta z_i = \lambda_i(z_i - \hat{z})$ for $i = 1, 2, \ldots, k$. Let $\epsilon > 0$ be chosen such that $0 \leq \epsilon \lambda_i \leq 1$ for every $i = 1, 2, \ldots, k$, and let $\xi_i = \epsilon \lambda_i$. Then $\hat{z} + \epsilon\, \delta z_i = \xi_i z_i + (1 - \xi_i)\hat{z} \in \Omega$, since Ω is convex. Therefore, by theorem (A.3.16), co $\{\hat{z}, \hat{z} + \epsilon\, \delta z_1, \ldots, \hat{z} + \epsilon\, \delta z_k\} \subset \Omega$. □

For most problems of interest we can derive necessary conditions of optimality in terms of perturbations δz belonging to a conical approximation of the first kind. However, there are a few important exceptions for which conical approximations of the first kind are either unavailable or else unsuitable. To provide for these situations we now introduce a somewhat more complex conical approximation to the set Ω at a point $\hat{z} \in \Omega$.

3 Definition. A convex cone $C(\hat{z},\Omega) \subset E^n$ will be called a *conical approximation of the second kind* to the set Ω at $\hat{z} \in \Omega$ if for any collection $\{\delta z_1, \delta z_2, \ldots, \delta z_k\}$ of linearly independent vectors in $C(\hat{z},\Omega)$ there exist an $\epsilon > 0$, possibly depending on $\hat{z}$, $\delta z_1, \delta z_2, \ldots, \delta z_k$, and a continuous map $\zeta(\cdot)$ from co $\{\hat{z}, \hat{z} + \epsilon\, \delta z_1, \hat{z} + \epsilon\, \delta z_2, \ldots,$

$\hat{z} + \epsilon \, \delta z_k\}$ into Ω such that $\zeta(\hat{z} + \delta z) = \hat{z} + \delta z + o(\delta z)$, where $\lim_{\|\delta z\| \to 0} \|o(\delta z)\|/\|\delta z\| = 0.$† ☐

4 Remark. Observe that if $C(\hat{z},\Omega)$ is a conical approximation of the first kind to Ω at $\hat{z} \in \Omega$, then it is also a conical approximation of the second kind to Ω at $\hat{z}$, with the map ζ being the identity map. However, while we can usually tell by inspection whether a set Ω has conical approximations of the first kind, it is often quite difficult to establish the existence of approximations of the second kind. This will become very clear in the next chapter. Consequently, the main reason for introducing conical approximations of the first kind is that they constitute a very important special class. *When we have no specific reason for indicating whether a cone $C(\hat{z},\Omega)$ is a conical approximation of the first kind or of the second kind, we shall refer to it simply as a conical approximation to Ω at $\hat{z} \in \Omega$, and we shall understand that it is a conical approximation of the second kind.* ☐

Let us consider a very simple example of a set Ω which has nontrivial conical approximations of the second kind but not of the first.

5 Example. Let $\Omega = \{z: z = z(\theta) = (r \cos \theta, r \sin \theta), 0 \leq \theta \leq \pi/2\}$; that is, Ω is a circular arc in the plane E^2. By inspection, the radial cone $RC(\hat{z},\Omega)$ to Ω at any point $\hat{z} \in \Omega$ is the singleton $\{0\}$, and hence there are no nontrivial conical approximations of the first kind to Ω. Now let $\hat{z} = (r \cos \hat{\theta}, r \sin \hat{\theta})$ be any point in Ω; we shall show that when $\hat{\theta} < \pi/2$, the tangent ray $C(\hat{z},\Omega) = \{\delta z: \delta z = \alpha \, dz(\hat{\theta})/d\theta = \alpha(-r \sin \hat{\theta}, r \cos \hat{\theta}), \alpha \geq 0\}$ is a conical approximation of the second kind to Ω at $\hat{z}$.

Let $\delta z_1 = dz(\hat{\theta})/d\theta$; then, expanding the expression for $z(\theta)$ in the definition of Ω about $\hat{\theta}$, we get

6 $$z(\hat{\theta} + \alpha) = \hat{z} + \alpha \, \delta z_1 + o(\alpha) \qquad 0 \leq \hat{\theta} + \alpha \leq \frac{\pi}{2},$$

where o is a continuous function such that $\|o(\alpha)\|/\alpha \to 0$ as $\alpha \to 0$. Now let $\delta \hat{z} \in C(\hat{z},\Omega)$ be arbitrary, but not zero. Then $\delta \hat{z} = \bar{\alpha} \, \delta z_1$ for some $\bar{\alpha} > 0$. We now choose $\epsilon = [(\pi/2) - \hat{\theta}]/\bar{\alpha}$ and define the map ζ

---

† When it is specifically known that the set Ω will be used with a *given* constraint function $r: E^n \to E^m$, it is convenient to make the definition of $C(\hat{z},\Omega)$ dependent on r in the following way: In definition (3), replace the phrase "any collection of linearly independent vectors" by the phrase "any collection of $(m + 1)$ linearly independent vectors." Furthermore, when $m = 0$, delete the requirement that $C(\hat{z},\Omega)$ be convex.

from co $\{\hat{z}, \hat{z} + [(\pi/2) - \hat{\theta}] \delta z_1\}$ into Ω by

7
$$\zeta(\hat{z} + \alpha\, \delta z_1) = \hat{z} + \alpha\, \delta z_1 + o(\alpha) \qquad 0 \leq \alpha \leq \frac{\pi}{2} - \hat{\theta}.$$

A comparison with (6) shows that ζ maps co $\{\hat{z},\hat{z} + [(\pi/2) - \hat{\theta}]\delta z_1\}$ into Ω, and consequently, the ray $C(\hat{z},\Omega)$ is a conical approximation of the second kind to Ω at $\hat{z}$. □

8 **Remark.** Comparing the definition of a conical approximation $C(\hat{z},\Omega)$ given in (3) with that of a tangent cone given in (2.8), we see that every vector in a conical approximation to a set Ω at a point $\hat{z} \in \Omega$ is also contained in the tangent cone $TC(\hat{z},\Omega)$. This is so because of the form of the map ζ. However, the tangent cone is not always convex, nor is it always possible to satisfy the condition of existence of a suitable map ζ. Hence the tangent cone to an arbitrary constraint set Ω will not always be a conical approximation. As we shall now see, an important exception arises when we know that the function $r(\cdot) \equiv 0$, that is, when we know that $m = 0$. This corresponds to the case of lumping all the constraints together into a single set Ω', as was done in (2.5). □

9 **Lemma.** Let $TC(\hat{z},\Omega')$ be the tangent cone to Ω' at $\hat{z} \in \Omega'$, with Ω' as defined in (2.5), containing all the constraints of the Basic Problem (2.1). Then $TC(\hat{z},\Omega')$ is a conical approximation to Ω' at $\hat{z}$ (in the sense indicated in the footnote on page 25).

Proof. Referring to definition (3) of a conical approximation we see that since $m = 0$, we need only show that for any $\delta z_1 \in TC(\hat{z},\Omega')$ there exist an $\epsilon > 0$ and a continuous map ζ from co $\{\hat{z}, \hat{z} + \epsilon\, \delta z_1\}$ into Ω' of the form $\zeta(\hat{z} + \delta z) = \hat{z} + \delta z + o(\delta z)$, where $\|o(\delta z)\|/\|\delta z\| \to 0$ as $\|\delta z\| \to 0$. Now, since $\delta z_1 \in TC(\hat{z},\Omega')$, there is in Ω' an arc of the form $c_1 = \{z: z = \hat{z} + t\, \delta z_1 + o_1(t),\ 0 \leq t \leq t_1,\ t_1 > 0\}$, where $o_1(\cdot)$ is a continuous function such that $\|o_1(t)\|/t \to 0$ as $t \to 0$. Let $\epsilon = t_1$; then for any $z = \hat{z} + t\, \delta z_1$ in co $\{\hat{z}, \hat{z} + \epsilon\, \delta z_1\}$, with $0 \leq t \leq \epsilon$, we define

10
$$\zeta(\hat{z} + t\, \delta z_1) = \hat{z} + t\, \delta z_1 + o_1(t).$$

Obviously, since $\delta z = t\, \delta z_1$ and δz_1 is a fixed vector, $\|o_1(t)\|/\|\delta z\| \to 0$ as $\|\delta z\| \to 0$. □

SEC. 2.3 THE FUNDAMENTAL THEOREM

We now return to the Basic Problem (2.1). Recall that the continuously differentiable function F from E^n into E^{m+1} was defined in (2.30) as

11      $$F(z) = (f(z), r(z)),$$

where f is the cost function and r is the equality-constraint function of the Basic Problem (2.1). Recall also that we had agreed to denote the jacobian matrix $\partial F^i(z)/\partial z^j$ by $\partial F(z)/\partial z$ and to number the components of vectors in E^{m+1} from 0 to m; that is, $y \in E^{m+1}$ is given by $y = (y^0, y^1, \ldots, y^m)$.

We now state and prove the most important result of the first half of the book: a necessary condition of optimality for the Basic Problem (2.1), from which all the necessary conditions given for specific problems in the following chapters will be derived.

12   **The fundamental theorem.** If $\hat{z}$ is an optimal solution to the Basic Problem (2.1) and $C(\hat{z}, \Omega)$ is a conical approximation to Ω at $\hat{z}$, then there exists a nonzero vector $\psi = (\psi^0, \psi^1, \ldots, \psi^m) \in E^{m+1}$, with $\psi^0 \leq 0$, such that for all $\delta z \in \bar{C}(\hat{z}, \Omega)$ [the closure of $C(\hat{z}, \Omega)$ in E^n]

13      $$\left\langle \psi, \frac{\partial F(\hat{z})}{\partial z} \delta z \right\rangle \leq 0.$$

Proof. Let $K(\hat{z}) \subset E^{m+1}$ be the cone defined by

14      $$K(\hat{z}) = \frac{\partial F(\hat{z})}{\partial z} C(\hat{z}, \Omega) = \{ \delta y : \delta y = \frac{\partial F(\hat{z})}{\partial z} \delta z, \delta z \in C(\hat{z}, \Omega) \}.$$

The cone $K(\hat{z})$ is convex because the cone $C(\hat{z}, \Omega)$ is convex, and the map $\partial F(\hat{z})/\partial z$ is linear (see Figure 1).

We shall now show that the optimality of $\hat{z}$ implies that the cone $K(\hat{z})$ must be separated from the ray

15      $$R = \{ \delta y : \delta y = \beta(-1, 0, 0, \ldots, 0), \beta > 0 \};$$

that is, there must exist a nonzero vector $\psi \in E^{m+1}$ such that (see Figure 1)

16      $$\langle \psi, \delta y \rangle \leq 0 \quad \text{for every } \delta y \in K(\hat{z})$$

and

17      $$\langle \psi, \delta y \rangle \geq 0 \quad \text{for every } \delta y \in R.$$

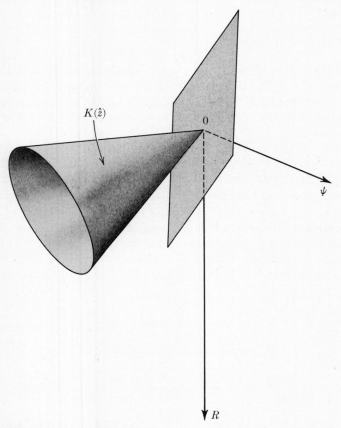

Figure 1

Note that (16) is only a slightly weaker statement than (13), and if (16) is true, then it must also be true that $\langle \psi, \delta y \rangle \leq 0$ for every $y \in \bar{K}(\hat{z})$, the closure of $K(\hat{z})$, which is a slightly stronger statement than (13). Also note that (17) implies that $\psi^0 \leq 0$. Thus to prove the theorem it is enough to establish the separation of $K(\hat{z})$ and R.

Suppose that the cone $K(\hat{z})$ and the ray R are not separated (see Figure 2). It then follows from theorem (A.5.5) that:

a. $K(\hat{z}) \cup R$ cannot be contained in an m-dimensional hyperplane.
b. $R \cap \text{rel int } K(\hat{z}) \neq \phi$.

Clearly, if one point of R belongs to the relative interior of $K(\hat{z})$, then every point of R belongs to the relative interior of $K(\hat{z})$. Thus, since $R \subset \text{rel int } K(\hat{z})$, it follows from (*a*) that $K(\hat{z})$ must be of dimension $m + 1$,† and hence that R is an interior ray of $K(\hat{z})$ [that is, all points of R are interior points of $K(\hat{z})$].

To obtain a contradiction we shall exhibit in $K(\hat{z})$ a vector δy^* which has in $C(\hat{z}, \Omega)$ a corresponding vector δz^* with the following properties:

18 $$\frac{\partial F(\hat{z})}{\partial z} \delta z^* = \delta y^*.$$
19 $$\zeta(\hat{z} + \delta z^*) \in \Omega.$$
20 $$[F(\zeta(\hat{z} + \delta z^*)) - F(\hat{z})] \in R,$$

where ζ is a continuous map of the form $\zeta(\hat{z} + \delta z) = \hat{z} + \delta z + o(\delta z)$ [$\|o(\delta z)\|/\|\delta z\| \to 0$ as $\|\delta z\| \to 0$]. Obviously, if (18) to (20) hold, then the vector $z^* = \zeta(\hat{z} + \delta z^*)$ is in Ω, and since $F = (f, r)$, it follows

† Note that for the cone $K(\hat{z})$ to be of dimension $m + 1$, that is, for the cone $K(\hat{z})$ to have an interior, it is necessary that the columns of the matrix $\partial F(\hat{z})/\partial z$ span E^{m+1}, that is, that $\partial F(\hat{z})/\partial z$ have rank $m + 1$.

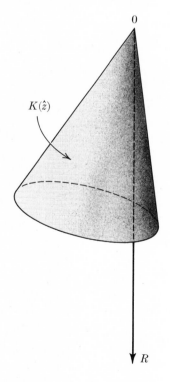

Figure 2

from (20) that $f(z^*) - f(\hat{z}) < 0$ and $r(z^*) - r(\hat{z}) = 0$. But $r(\hat{z}) = 0$, and therefore $z^* \in \Omega$, $r(z^*) = 0$ and $f(z^*) < f(\hat{z})$, contradicting the optimality of $\hat{z}$.

Our construction of the vectors δy^* and δz^* requires the use of the Brouwer fixed-point theorem[1] (see references [3, 4]). We begin by constructing in the cone $K(\hat{z})$ a simplex S (see Figure 3), with vertices 0 and $\delta y_1, \delta y_2, \ldots, \delta y_{m+1}$ and containing a segment $\Gamma = \{y: y = \beta(-1,0,0, \ldots ,0), 0 < \beta \leq \nu\}$ of the ray R in its interior. Since R is assumed to be in the interior of $K(\hat{z})$, we can always construct such a simplex. We assume that we have made the lengths of the vectors δy_i for $i = 1, 2, \ldots, m + 1$ sufficiently small to ensure that for a set of vectors δz_1 $\delta z_2, \ldots, \delta z_{m+1}$ in $C(\hat{z},\Omega)$ satisfying

21
$$\frac{\partial F(\hat{z})}{\partial z} \delta z_i = \delta y_i \qquad i = 1, 2, \ldots, m + 1$$

there exists a continuous map ζ from co $\{\hat{z}, \hat{z} + \delta z_1, \ldots, \hat{z} + \delta z_{m+1}\}$ into Ω of the form $\zeta(\hat{z} + \delta z) = \hat{z} + \delta z + o(\delta z)$, where $\|o(\delta z)\|/\|\delta z\| \to 0$ as $\|\delta z\| \to 0$.†

[1] The usual form of the Brouwer fixed-point theorem is as follows: *Brouwer fixed-point theorem*. If f is a continuous map from the closed unit ball $B = \{x \in E^n: \|x\| \leq 1\}$, then f has a fixed point.

However, for the purpose of our proof we find it more convenient to work with the following alternative form.

Theorem. If f is a continuous map from the closed unit ball $B = \{x \in E^n: \|x\| \leq 1\}$ into E^n, with $f(x) = x + g(x)$, where $\|g(x)\| \leq 1$ for all x with $\|x\| = 1$, then the origin is contained in the range of f.

Proof. To say that the origin is contained in the range of f is equivalent to saying that the function $h(x) = -g(x)$ has a fixed point. Let us define the function h_1 by

$$h_1(x) = \begin{cases} -g(x) & \text{if } \|g(x)\| \leq 1 \\ -g(x)/\|g(x)\| & \text{if } \|g(x)\| > 1. \end{cases}$$

Clearly, h_1 is a continuous function from the closed unit ball in E^n into the closed unit ball in E^n. Therefore, by the Brouwer fixed-point theorem, h_1 has a fixed point, say, x_1. If $\|h_1(x_1)\| < 1$, then $h_1(x_1) = -g(x_1)$, and x_1 is a fixed point of $-g$. Suppose $\|h_1(x_1)\| = 1$. Then $\|x_1\| = 1$, and consequently, $\|g(x_1)\| \leq 1$. Again $h_1(x_1) = -g(x_1)$ and x_1 is a fixed point of $-g$. (Note that this result is true for any closed ball centered at the origin, not just for the unit ball.)

The Brouwer fixed-point theorem follows immediately from this theorem, so they are in fact equivalent.

† Since R has points in the interior of $K(\hat{z})$, there must exist some simplex S' in $K(\hat{z})$ with vertices 0, $\delta y_1', \ldots, \delta y_{m+1}'$ containing a segment Γ' of R in its interior. By definition of a simplex, the vectors $\delta y_1', \ldots, \delta y_{m+1}'$ are linearly

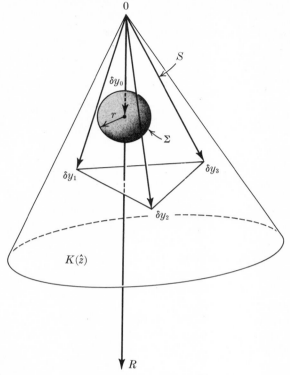

Figure 3

Let δy_0 be a point in the relative interior of $S \cap R$, and let $\Sigma \subset S$ be a closed ball with center δy_0 and radius r. For $0 < \alpha \leq 1$, let $\Sigma_\alpha = \alpha \Sigma = \{\alpha \, \delta y : \delta y \in \Sigma\}$; that is, Σ_α is a closed ball with center at $\alpha \, \delta y_0$ and radius αr. Since for every vector $\delta y \in S$, $\alpha \, \delta y \in S$ for $0 \leq \alpha \leq 1$, it is clear that Σ_α is contained in the simplex S for $0 < \alpha \leq 1$.

For a fixed α satisfying $0 < \alpha \leq 1$ we now construct a map G_α from the ball $\Sigma_\alpha - \{\alpha \, \delta y_0\}$ (with center at the origin and radius αr)

---

independent, and so, as a result, are any $\delta z_i' \in C(\hat{z}, \Omega)$ which satisfy

$$\delta y_i' = \frac{\partial F(\hat{z})}{\partial z} \delta z_i' \quad \text{for} \quad i = 1, 2, \ldots, m+1.$$

By definition (3) there exist an $\epsilon > 0$ and a continuous map $\zeta(\cdot)$ from co $\{\hat{z}, \hat{z} + \epsilon \delta z_1', \ldots, \hat{z} + \epsilon \delta z_{m+1}'\}$ into Ω. Setting $S = \epsilon S'$, so that $\delta y_i = \epsilon \delta y_i'$ and $\delta z_i = \epsilon \delta z_i'$ for $i = 1, 2, \ldots, m+1$, we obtain a simplex S with the desired properties.

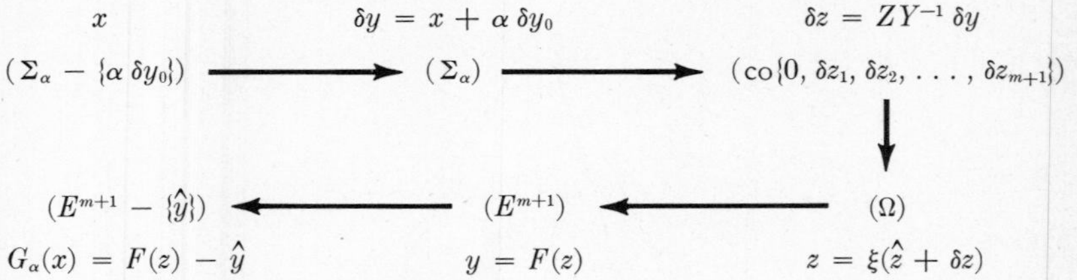

$$
\begin{array}{ccc}
x & \delta y = x + \alpha\, \delta y_0 & \delta z = ZY^{-1}\, \delta y \\
(\Sigma_\alpha - \{\alpha\, \delta y_0\}) \longrightarrow & (\Sigma_\alpha) \longrightarrow & (\mathrm{co}\{0, \delta z_1, \delta z_2, \ldots, \delta z_{m+1}\}) \\
& & \downarrow \\
(E^{m+1} - \{\hat{y}\}) \longleftarrow & (E^{m+1}) \longleftarrow & (\Omega) \\
G_\alpha(x) = F(z) - \hat{y} & y = F(z) & z = \zeta(\hat{z} + \delta z)
\end{array}
$$

Figure 4 Construction of $G_\alpha(x)$.

into E^{m+1} as follows (see Figure 4). For any $x \in (\Sigma_\alpha - \{\alpha\, \delta y_0\})$ let

$$
G_\alpha(x) = F(\zeta(\hat{z} + ZY^{-1}(\alpha\, \delta y_0 + x))) - (\hat{y} + \alpha\, \delta y_0), \tag{22}
$$

where $\hat{y} = F(\hat{z})$, ζ is our map from co $\{\hat{z}, \hat{z} + \delta z_1, \ldots, \hat{z} + \delta z_{m+1}\}$ into Ω, Y is an $(m + 1) \times (m + 1)$ matrix whose ith column is δy_i, and Z is an $n \times (m + 1)$ matrix whose ith column is δz_i. Since S is a simplex, the vectors δy_i are linearly independent, and therefore Y is nonsingular.

The composite map G_α can be seen to have been constructed for the purpose of finding a $\delta y^* \in K(\hat{z})$ and a $\delta z^* \in C(\hat{z}, \Omega)$ satisfying (18) to (20) as follows. First, for any $x \in (\Sigma_\alpha - \{\alpha\, \delta y_0\})$ the vector $\delta y = (x + \alpha\, \delta y_0) \in \Sigma_\alpha \subset S$. Next, since $\delta y \in S$, $\delta y = \sum_{i=1}^{m+1} \mu^i\, \delta y_i = Y\mu$, where $\mu^i \geqq 0$ and $\sum_{i=1}^{m+1} \mu^i \leqq 1$, and therefore

$$
z = \hat{z} + ZY^{-1}\, \delta y = \hat{z} + \sum_{i=1}^{m+1} \mu^i\, \delta z_i = (1 - \sum_{i=1}^{m+1} \mu^i)\, \hat{z} \\
+ \sum_{i=1}^{m+1} \mu^i\, (\hat{z} + \delta z_i)
$$

is a point in co $\{\hat{z}, \hat{z} + \delta z_1, \ldots, \hat{z} + \delta z_{m+1}\}$. Consequently, since $(\partial F(\hat{z})/\partial z)\, Z = Y$, the vector $\delta z = z - \hat{z}$ satisfies $(\partial F(\hat{z})/\partial z)\, \delta z = \delta y$ [cf. (18)] and the vector z satisfies $\zeta(z) \in \Omega$ [cf. (19)]. If we can now show that for some α^* satisfying $0 < \alpha^* \leqq 1$ there is an x^* in $\Sigma_\alpha^* - \{\alpha^*\, \delta y_0\}$ such that $G_{\alpha^*}(x^*) = 0$, then the vector $\delta y^* = x^* + \alpha^*\, \delta y_0$ will satisfy not only (18) and (19), but also (20).

Recalling that

$$
\zeta(\hat{z} + ZY^{-1}(\alpha\, \delta y_0 + x)) = \hat{z} + ZY^{-1}(\alpha\, \delta y_0 + x) \\
+ o(ZY^{-1}(\alpha\, \delta y_0 + x)),
$$

we can expand the right-hand side of (22) about $\hat{z}$ as follows:

$$
G_\alpha(x) = \hat{y} + \frac{\partial F(\hat{z})}{\partial z} ZY^{-1}(\alpha\, \delta y_0 + x) - (\hat{y} + \alpha\, \delta y_0) \\
+ o_G(ZY^{-1}(\alpha\, \delta y_0 + x)), \tag{23}
$$

where o_G is a continuous function such that $\|o_G(y)\|/\|y\| \to 0$ as $\|y\| \to 0$. Since $Y = (\partial F(\hat{z})/\partial z)Z$, equation (23) simplifies out to

(24) $$G_\alpha(x) = x + o_G(ZY^{-1}(\alpha\, \delta y_0 + x)).$$

Now, for $x \in \partial(\Sigma_\alpha - \{\alpha\, \delta y_0\})$, the boundary of the ball, $\|x\| = \alpha r$, and we may therefore write $x = \alpha\rho$, where ρ is a vector in E^{m+1} such that $\|\rho\| = r$. Hence for any $x \in \partial(\Sigma_\alpha - \{\alpha\, \delta y_0\})$

(25) $$G_\alpha(x) = G_\alpha(\alpha\rho) = \alpha\rho + o_G(\alpha ZY^{-1}(\delta y_0 + \rho)),$$

and since $\|o_G(\alpha ZY^{-1}(\delta y_0 + \rho))\|/\alpha \to 0$ as $\alpha \to 0$ uniformly in ρ, with $\|\rho\| = r$, there exists an α^* satisfying $0 < \alpha^* \leq 1$ such that

(26) $$\|o_G(\alpha^*\, ZY^{-1}(\delta y_0 + \rho))\| < \alpha^* r$$

for all ρ satisfying $\|\rho\| = r$. Referring to Brouwer's fixed-point theorem (see the footnote on page 30), we find that the map G_{α^*} satisfies its assumptions, and hence there is an x^* in $\Sigma_{\alpha^*} - \{\alpha^*\delta y_0\}$ such that

(27) $$G_{\alpha^*}(x^*) = 0.$$

Now, $\delta y^* = \alpha^*\, \delta y_0 + x^*$ is in $\Sigma_{\alpha^*} \subset K(\hat{z})$, $\delta z^* = ZY^{-1}\, \delta y^*$ is in co $\{\delta z_1, \delta z_2, \ldots, \delta z_{m+1}\} \subset C(\hat{z},\Omega)$, $z^* = \zeta(\hat{z} + \delta z^*)$ is in Ω, and, by (22) and (27),

(28) $$F(z^*) - F(\hat{z}) - \alpha^*\, \delta y_0 = 0.$$

But $\delta y_0 = (-\nu,0,0,\ldots,0)$, with $\nu > 0$, so that $f(z^*) - f(\hat{z}) = -\alpha^*\nu < 0$ and $r(z^*) = 0$. Thus z^* satisfies the constraints of the Basic Problem and also results in a lower cost than the optimal solution $\hat{z}$, which is a contradiction.

We therefore conclude that $K(\hat{z})$ and R must be separated, which establishes (16) and (17). But then (16) must also be true for all $y \in \bar{K}(\hat{z})$, and since $[\partial F(z)/\partial z]\, \bar{C}(\hat{z},\Omega) \subset \bar{K}(\hat{z})$, the theorem is proved. □

The significance of the assumption of convexity on a conical approximation to the set Ω about a point $\hat{z} \in \Omega$ is quite clear, and the reader should have no difficulty finding cases for which the Fundamental Theorem (12) does not hold with respect to the radial cone $RC(\hat{z},\Omega)$ when $RC(\hat{z},\Omega)$ is not convex. We shall now see that the assumption concerning the existence of a map ζ is quite subtle and cannot be eliminated, even when the radial cone $RC(\hat{z},\Omega)$ is convex. Let us examine a case in which theorem (12) breaks down when this assumption is not satisfied.

(29) **Example.** Suppose that for $z \in E^3$ we wish to minimize $f(z) = z^1$ subject to the constraints $r(z) = 0$ and $z \in \Omega$, where

CONDITIONS OF OPTIMALITY FOR THE BASIC PROBLEM

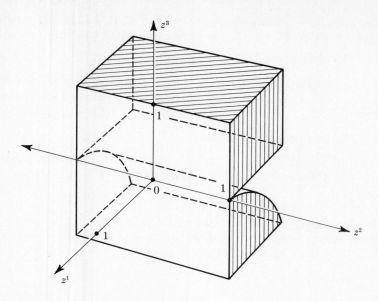

Figure 5

30
$$r(z) = \begin{bmatrix} z^2 \\ z^3 - \tfrac{1}{2}(z^1)^3 \end{bmatrix}$$

31
$$\Omega = \{z: -1 \leq z^1 \leq 0, -1 \leq z^2 \leq 1, 0 \leq z^3 \leq 1\}$$
$$\cup \{z: -1 \leq z^1 \leq 0, -1 \leq z^2 \leq 1, -1 \leq z^3 \leq -(z^1)^2\}.$$

From (30) and (31) we see that $\hat{z} = (0,0,0)$ is the only vector which satisfies our constraints, and hence it is the optimal solution.

Now let us examine the radial cone $RC(\hat{z},\Omega)$, which is seen to be the set $\{z: z^1 \leq 0\}$; that is, it is a half space. The reader may amuse himself by showing that although this cone is convex, there is no suitable map ζ for it. Hence it is not a conical approximation to the set Ω at $\hat{z}$, according to our definitions (1) and (3).

The jacobian matrix $\partial F(\hat{z})/\partial z$ is easily seen to be

$$\frac{\partial F(\hat{z})}{\partial z} = \begin{bmatrix} 1 & 0 & 0 \\ 0 & 1 & 0 \\ 0 & 0 & 1 \end{bmatrix},$$

and therefore the cone $K(\hat{z}) = [\partial F(z)/\partial z]\, RC(\hat{z},\Omega)$ is seen to be the set

$$K(\hat{z}) = \{y = (y^0,y^1,y^2): y^0 \leq 0\}.$$

By inspection, the ray $R = \{y: y = \beta(-1,0,0), \beta \geq 0\}$ belongs to the interior of the full-dimensional cone $K(\hat{z})$, in contradiction to the Fundamental Theorem (12). We therefore see that the requirement that there exist a suitable map ζ cannot be eliminated from definitions (1) and (3) of a conical approximation without invalidating our Fundamental Theorem (12). □

A Few Second-order Conditions of Optimality

Referring to the proof of the Fundamental Theorem (12), we see that this theorem degenerates, in the sense that it can be satisfied trivially, when the dimension of the cone $K(\hat{z})$ is less than $m + 1$. This is obviously the case when the vectors $\nabla f(\hat{z})$ and $\nabla r^i(\hat{z})$ for $i = 1, 2, \ldots, m$ are linearly dependent. To elaborate on this point, let us expand (13) into a sum of components: for all $\delta z \in \bar{C}(\hat{z},\Omega)$ we get

32
$$\left\langle \psi, \frac{\partial F(\hat{z})}{\partial z} \delta z \right\rangle = \left\langle \frac{\partial F(\hat{z})^T}{\partial z} \psi, \delta z \right\rangle$$
$$= \left\langle \psi^0 \nabla f(\hat{z}) + \sum_{i=1}^m \psi^i \nabla r^i(\hat{z}), \delta z \right\rangle \leq 0.$$

Now, if some of the vectors $\nabla f(\hat{z})$ and $\nabla r^i(\hat{z})$ for $i = 1, 2, \ldots, m$ are linearly dependent, then there exist multipliers $\psi^0, \psi^1, \ldots, \psi^m$, not all zero and with $\psi^0 \leq 0$, such that

33
$$\psi^0 \nabla f(\hat{z}) + \sum_{i=1}^m \psi^i \nabla r^i(\hat{z}) = 0.$$

Clearly, these multipliers satisfy the conditions of the Fundamental Theorem (12), but apparently, their existence does not depend on, and therefore does not reflect, the optimality of the point $\hat{z}$.

Nevertheless, we cannot conclude automatically that whenever the vectors $\nabla f(\hat{z})$ and $\nabla r^i(\hat{z})$ for $i = 1, 2, \ldots, m$ are linearly dependent the Fundamental Theorem is useless, because a degeneracy of this type may occur without making (33) a trivial statement. For example, suppose that $\Omega = E^n$; then (32) must hold for every $\delta z \in E^n$, implying (33); that is, that these vectors must be linearly dependent at every optimal $\hat{z}$. Similarly, suppose the point $\hat{z}$ is in the interior of Ω; then again (33) is a meaningful condition of optimality [note that when $\hat{z}$ is in the interior of Ω, then $C(\hat{z},\Omega)$ is E^n]. Also, while it may be possible to satisfy (32) trivially by choosing multipliers which satisfy (33), it may also be possible to find a set of multipliers which satisfy (32) but not (33). This nontrivial set of multipliers may, possibly, give a nontrivial result.

When (33) can be satisfied trivially with $\psi^0 = -1$, $\psi^i = 0$ for $i = 1, 2, \ldots, m$ because $\nabla f(\hat{z}) = 0$, and the vectors $\nabla r^i(\hat{z})$ are lin-

36 CONDITIONS OF OPTIMALITY FOR THE BASIC PROBLEM

early independent, we can strengthen the Fundamental Theorem (12) to some extent by introducing second-order effects, provided the function $f(\cdot)$ has continuous second-order partial derivatives.

34 Definition. Let the cost function $f(\cdot)$ be twice continuously differentiable. We define the *hessian* $H_{\hat{z}}(\cdot)$ mapping E^n into the reals to be the function

35
$$H_{\hat{z}}(\delta z) = \left\langle \delta z, \frac{\partial^2 f(\hat{z})}{\partial z^2} \delta z \right\rangle,$$

where $\partial^2 f(\hat{z})/\partial z^2$ is an $n \times n$ matrix whose ijth elements are $\partial^2 f(\hat{z})/\partial z^i \partial z^j$. □

Thus $\tfrac{1}{2} H_{\hat{z}}(\delta z)$ is the second term of the expansion of $f(\hat{z} + \delta z)$ about $\hat{z}$; that is,

36
$$f(\hat{z} + \delta z) = f(\hat{z}) + \langle \nabla f(\hat{z}), \delta z \rangle + \tfrac{1}{2} H_{\hat{z}}(\delta z) + \bar{o}(\delta z),$$

where $\bar{o}(\delta z)/\|\delta z\|^2 \to 0$ as $\|\delta z\| \to 0$.

37 Exercise. Prove the following theorems.

38 Theorem. If $\hat{z}$ is an optimal solution to the Basic Problem (2.1), with $\nabla f(\hat{z}) = 0$, and f is twice continuously differentiable, then

39
$$-H_{\hat{z}}(\delta z) \leqq 0 \qquad \text{for all } \delta z \in TC(\hat{z}, \Omega'),$$

where $TC(\hat{z}, \Omega')$ is the tangent cone to the constraint set Ω' defined in (2.5) and $H_{\hat{z}}(\cdot)$ is as defined in (35). □

Show that (39) also holds for all $\delta z \in TC_s(\hat{z}, \Omega')$, the sequential tangent cone to Ω' at $\hat{z}$, defined in (2.28). [*Hint:* Proceed as for theorem (2.12)].

40 Theorem. If $\hat{z}$ is an optimal solution to the Basic Problem (2.1), with $\nabla f(\hat{z}) = 0$, and f is twice continuously differentiable, then the ray

41
$$R = \{y \in E^{m+1}: y = \beta(-1, 0, 0, \ldots, 0), \beta > 0\}$$

is not an interior ray of the set

42
$$L(\hat{z}) = \{y = (y^0, x) \in E^{m+1}: y^0 = H_{\hat{z}}(\delta z), x = \frac{\partial r(\hat{z})}{\partial z} \delta z,$$
$$\delta z \in C(\hat{z}, \Omega)\},$$

where $H_{\hat{z}}$ is as defined in (35), r and Ω are as in (2.1), and $C(\hat{z}, \Omega)$ is any conical approximation to Ω at $\hat{z}$.

[*Hint:* This theorem is fairly difficult to prove. Show that if the theorem is false, then there must exist a simplex S in E^m, with vertices $x_1, x_2, \ldots, x_{m+1}$ and *containing the origin in its interior,* such that for $i = 1, 2, \ldots, m+1$

43 $$x_i = \frac{\partial r(\hat{z})}{\partial z} \delta z_i \quad \text{for some } \delta z_i \in C(\hat{z},\Omega),$$

with

44 $$H_{\hat{z}}(\delta z) < 0 \quad \text{for all } \delta z \in \text{co } \{\delta z_1, \delta z_2, \ldots, \delta z_{m+1}\}$$

and

45 $$\zeta(\text{co } \{\hat{z} + \delta z_1, \hat{z} + \delta z_2, \ldots, \hat{z} + \delta z_{m+1}\}) \subset \Omega,$$

where ζ is the map in the definition of the conical approximation (3). Now construct a map G_α, as in the proof of the Fundamental Theorem (12), and complete the present proof.] □

A complete study of second-order conditions of optimality can be found in reference [6].

Conditions for $\psi^0 < 0$ in Theorem (12)

As will be seen in the next section and in subsequent chapters, it is much easier to make use of the Fundamental Theorem (12) when we are certain that ψ^0 may be taken to be nonzero, that is, $\psi^0 < 0$. The reader is now invited to prove some sufficient conditions which ensure that $\psi^0 < 0$.

46 **Exercise.** Consider the statement of theorem (12). Show that if $\langle \nabla f(\hat{z}), \delta z \rangle \geqq 0$ for every $\delta z \in \bar{C}(\hat{z},\Omega)$ such that $\langle \nabla r^i(\hat{z}), \delta z \rangle = 0$ for $i = 1, \ldots, m$, then there exists a vector ψ satisfying the conditions of theorem (12), with $\psi^0 < 0$. □

47 **Exercise.** Consider the statement of theorem (12) and suppose that

48 $$\overline{\text{co }} TC_s(\hat{z},\Omega') = \overline{\text{co }} TC_s(\hat{z},\Omega) \cap \{\delta z : \langle \nabla r^i(\hat{z}), \delta z \rangle = 0,$$
$$i = 1, \ldots, m\},$$

where $\Omega' = \{z : r(z) = 0, z \in \Omega\}$. Show that under this assumption there exists a vector ψ satisfying the conditions of theorem (12), with $\psi^0 < 0$. [*Hint:* Recall that $\bar{C}(\hat{z},\Omega) \subset TC_s(\hat{z},\Omega)$; use exercises (46) and (2.27).] □

49 **Remark.** Assumption (48) is a condition involving *only* the constraints, which guarantees the existence of a vector ψ satisfying

theorem (12), with $\psi^0 < 0$. Conditions of this type are usually referred to as *constraint qualifications*, and it will be seen in Section 3.3 that both the *Kuhn-Tucker constraint qualification* and its weakened forms are equivalent to (48). □

50 Exercise. Consider the statement of theorem (12) and suppose that the origin in E^m is an interior point of the set

51
$$\tilde{K} = \{y : y^i = \langle \nabla r^i(\hat{z}), \delta z \rangle, \, i = 1, \ldots, m, \, \delta z \in C(\hat{z}, \Omega)\}.$$

Show that under this assumption if ψ is a vector satisfying the conditions of theorem (12), then $\psi^0 < 0$. □

52 Exercise. Show that the origin in E^m is an interior point of the set $\tilde{K}$ [see (51)] if and only if the vectors $\nabla r^i(\hat{z})$ for $i = 1, \ldots, m$ are linearly independent and the subspace

53
$$P = \{\delta z : \langle \nabla r^i(\hat{z}), \delta z \rangle = 0, \, i = 1, \ldots, m\}$$

is not contained in a support hyperplane to the conical approximation $C(\hat{z}, \Omega)$. □

54 Exercise. Show that the subspace P defined by (53) is not contained in a support hyperplane to the conical approximation $C(\hat{z}, \Omega)$ if there is a δz^* in the *interior* of $C(\hat{z}, \Omega)$ satisfying $\langle \nabla r^i(\hat{z}), \delta z^* \rangle = 0$ for $i = 1, \ldots, m$. □

Problems without Differentiability Assumptions

A question which may have crossed the mind of a reader familiar with such problems in optimal control as the minimum-fuel problem is what should be done when the cost function $f(\cdot)$ is not continuously differentiable. The following exercises should help clear up this point.

55 Exercise. Show that the original Basic Problem (2.1) is equivalent to the following augmented Basic Problem whose cost function is continuously differentiable:

56   Minimize $\langle e_1, \mathbf{z} \rangle$ subject to $\mathbf{r}(\mathbf{z}) = 0$ and $\mathbf{z} \in \Omega^*$, where $e_1 = (1,0,0,\ldots,0) \in E^{n+1}$, $\mathbf{z} = (z^0, z) \in E^{n+1}$, with $z^0 \in E^1$ and $z \in E^n$, $\mathbf{r}(\mathbf{z}) = r(z)$, and $\Omega^* = \{\mathbf{z} = (z^0, z) : z \in \Omega, \, z^0 - f(z) \geqq 0\}$.

Thus the nondifferentiable cost can be shifted into the constraint set by adding one component to z (we can obviously shift the nondifferentiable components of the function r into Ω without adding any new components to z). □

SEC. 2.4 THE TWO-POINT-BOUNDARY-VALUE OPTIMAL CONTROL PROBLEM

57 **Exercise.** Consider form (56) of the Basic Problem (2.1). Suppose that $f(\cdot)$ is a convex function and that $C(\hat{z},\Omega)$ is a conical approximation of the first kind to Ω at $\hat{z} \in \Omega$. Show that for $\hat{\mathbf{z}} = (f(\hat{z}),\hat{z})$

58
$$C(\hat{\mathbf{z}},\Omega^*) = \{\delta \mathbf{z} = \lambda(\delta z^0, \delta z): \delta z \in C(\hat{z},\Omega),$$
$$\lambda \geq 0,\ \delta z^0 \geq f(\hat{z} + \delta z) - f(\hat{z})\}$$

is a conical approximation of the first kind to Ω^* at $\hat{\mathbf{z}}$. ☐

59 **Exercise.** Suppose that the set Ω is convex and that the cost function $f(\cdot)$ is convex. Use the representation (56) of the Basic Problem and lemma (12) to show that condition (13) now takes on the form

60
$$\psi^0(f(z) - f(\hat{z})) + \sum_{i=1}^{m} \psi^i \nabla r^i(\hat{z})(z - \hat{z}) \leq 0 \quad \text{for all } z \in \Omega,$$

and that, in addition, when the functions r^i are affine, (60) becomes

61
$$\psi^0 f(\hat{z}) + \sum_{i=1}^{m} \psi^i r^i(\hat{z}) \geq \psi^0 f(z) + \sum_{i=1}^{m} \psi^i r^i(z) \quad \text{for all } z \in \Omega.$$

[This is a maximum-type condition and is true whether $f(\cdot)$ is differentiable or not.] ☐

62 **Exercise.** Let $F = (f,r)$, as before. Show that if the set $F(\Omega)$ is convex and $\hat{z}$ is an optimal solution to the Basic Problem, then there exist multipliers $\psi^0, \psi^1, \ldots, \psi^m$, not all zero, with $\psi^0 \leq 0$, such that

63
$$\psi^0 f(\hat{z}) + \sum_{i=1}^{m} \psi^i r^i(\hat{z}) \geq \psi^0 f(z) + \sum_{i=1}^{m} \psi^i r^i(z) \quad \text{for all } z \in \Omega.$$

[This result is independent of whether $f(\cdot)$ and $r(\cdot)$ are differentiable or not.] ☐

2.4 THE TWO-POINT-BOUNDARY-VALUE OPTIMAL CONTROL PROBLEM

In the next chapter we shall use very efficient methods to deduce from Fundamental Theorem (3.12) almost all the presently known necessary conditions of optimality for a solution to an optimal control problem. Since most control engineers are accustomed to methods based on perturbations about optimal trajectories, which, incidentally, are less efficient than the techniques we use in Chapter 3, we shall now consider a treatment of the two-point-boundary-value optimal control problem in terms of perturbation techniques. This will serve the purpose of clarifying the similarities and the differences in the two approaches. The reader will have no difficulty in establishing the fact that the differences increase as the constraints on the optimal trajectory become more complicated.

In dealing with optimal control problems it is traditional to augment the system of dynamic equations as in (1.3.6) and to state the probems in the canonical form (1.3.7). We now proceed in this manner.

1 The two-point-boundary-value optimal control problem. We are given a fixed integer k and an augmented system of dynamic equations

2 $$\mathbf{x}_{i+1} - \mathbf{x}_i = \mathbf{f}_i(x_i, u_i) \qquad i = 0, 1, \ldots, k-1,$$

where $\mathbf{x}_i = (x_i^0, x_i^1, \ldots, x_i^n)$ is the *augmented state*, or *phase*, of the system, $u_i = (u_i^1, u_i^2, \ldots, u_i^m)$ is the *control*, or *input*, applied to the system, and the $\mathbf{f}_i(\cdot, \cdot)$ are differentiable functions from $E^{n+1} \times E^m$ into E^{n+1}. We are also given the boundary conditions

3 $$\mathbf{x}_0 = (0, \hat{x}_0^1, \hat{x}_0^2, \ldots, \hat{x}_0^n)$$
4 $$\mathbf{x}_k \in \mathbf{X}_k = \{\mathbf{x} : \mathbf{x} = (x^0, \hat{x}_k^1, \hat{x}_k^2, \ldots, \hat{x}_k^n)\},$$

where $\hat{x}_0^1, \hat{x}_0^2, \ldots, \hat{x}_0^n$ and $\hat{x}_k^1, \hat{x}_k^2, \ldots, \hat{x}_k^n$ are given and x^0 is an arbitrary scalar. Thus the initial phase is a fixed zero cost point, while the terminal phase lies on a fixed line which is parallel to the cost axis. Finally, we are given k sets $U_i \subset E^m$ for $i = 0, 1, \ldots, k-1$, to which the controls u_i are restricted. The radial cones $RC(u, U_i)$, with $u \in U_i$, arbitrary, are assumed to be conical approximations of the first kind for $i = 0, 1, \ldots, k-1$ [see (3.1)].

We are required to find a control sequence $\hat{\mathbf{u}} = (\hat{u}_0, \hat{u}_1, \ldots, \hat{u}_{k-1})$, with $\hat{u}_i \in U_i$, whose corresponding phase trajectory $\hat{\mathfrak{X}} = (\hat{\mathbf{x}}_0, \hat{\mathbf{x}}_1, \ldots, \hat{\mathbf{x}}_k)$ satisfies the given boundary conditions (3) and (4) such that for every other control sequence $\mathbf{u} = (u_0, u_1, \ldots, u_{k-1})$, with $u_i \in U_i$, whose corresponding phase trajectory $\mathfrak{X} = (\mathbf{x}_0, \mathbf{x}_1, \ldots, \mathbf{x}_k)$ also satisfies (3) and (4) we have

5 $$\hat{x}_k^0 \leq x_k^0. \qquad \square$$

We shall call such a $\hat{\mathbf{u}}$ an *optimal control sequence*.

6 Remark. Fortunately, the usual constraint sets U_i, such as $U_i = \{u : \max_j |u^j| \leq \alpha^i\}$ or $U_i = \{u : \|u\| \leq \beta^i\}$ are convex and their radial cones are conical approximations of the first kind, by lemma (3.2). $\square$

For the problem under consideration all the trajectories start at the same point $\hat{\mathbf{x}}_0 = (0, \hat{x}_0^1, \hat{x}_0^2, \ldots, \hat{x}_0^n)$. Thus the solution of system (2) at time k, denoted by $\mathbf{x}_k(\hat{\mathbf{x}}_0, \mathbf{u})$, which corresponds to an

SEC. 2.4 THE TWO-POINT-BOUNDARY-VALUE OPTIMAL CONTROL PROBLEM

admissible control sequence $\mathfrak{U}$, defines a map $\mathbf{F}$ with domain $U = U_0 \times U_1 \times \cdots \times U_{k-1} \subset E^{mk}$ and range in E^{n+1}, that is,

7
$$\mathbf{F}(\mathfrak{U}) = (F^0(\mathfrak{U}), F^1(\mathfrak{U}), \ldots, F^n(\mathfrak{U})) = \mathbf{x}_k(\hat{\mathbf{x}}_\iota, \mathfrak{U}).$$

The map $\mathbf{F}$ is a composition map made up of the functions $\mathbf{f}_i(\cdot, \cdot)$ appearing in the difference equation (2). Since these functions are continuously differentiable, the map $\mathbf{F}$ is also continuously differentiable.

8 Remark. With the introduction of the map $\mathbf{F}$, the two-point-boundary-value optimal control problem (1) automatically assumes the form of the Basic Problem (2.1): minimize $F^0(\mathfrak{U})$, subject to $F(\mathfrak{U}) - \hat{x}_k = 0$ and $\mathfrak{U} \in U$, where we define $F = (F^1, F^2, \ldots, F^n)$; that is, $\mathbf{F} = (F^0, F)$. □

The above rephrasing of the optimal control problem is implied in the proofs given by various writers and will therefore be familiar to the reader of control literature. Note that the function F above, which takes the place of the function r in the Basic Problem (2.1), is completely different in form from the function r constructed in Chapter 1 [see (1.5.4)]. For the purpose of obtaining necessary conditions form (1.5.4) is usually more convenient; however, there are many cases involving the construction of optimal control algorithms when the form F defined by (7) is preferred. These points will be clarified in later chapters.

We can now restate the Fundamental Theorem (3.12) in terms of control problem (1).

9 Separation theorem. If $\hat{\mathfrak{U}}$ is an optimal control sequence for control problem (1) and the radial cone $RC(\hat{\mathfrak{U}}, U)$ is a conical approximation of the first kind to U at $\hat{\mathfrak{U}}$, then there exists a nonzero vector $\psi = (\psi^0, \psi^1, \ldots, \psi^n) \in E^{n+1}$, with $\psi^0 \leq 0$, such that for all $\delta \mathfrak{U} \in \overline{RC}(\hat{\mathfrak{U}}, U)$, the closure of $RC(\hat{\mathfrak{U}}, U)$ in E^{mk},

10
$$\left\langle \psi, \frac{\partial F(\hat{\mathfrak{U}})}{\partial \mathfrak{U}} \delta \mathfrak{U} \right\rangle \leq 0.$$

11 Remark. The radial cone to U at $\hat{\mathfrak{U}}$ is the cross product of the radial cones to the sets U_i at $\hat{u}_i$; that is,

12
$$RC(\hat{\mathfrak{U}}, U) = RC(\hat{u}_0, U_0) \\ \times RC(\hat{u}_1, U_1) \times \cdots \times RC(\hat{u}_{k-1}, U_{k-1}).$$

It may therefore be easily verified that if the radial cones $RC(\hat{u}_i, U_i)$ are conical approximations of the first kind, as we have assumed

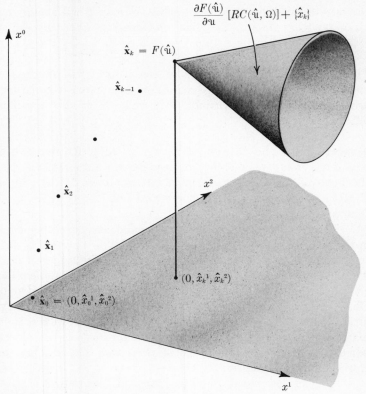

Figure 1

them to be, then their cross product $RC(\hat{\mathfrak{u}},U)$ is also a conical approximation the first kind. □

We must now interpret inequality (10) (see Figure 1). Let us begin by computing the jacobian matrix $\partial F(\hat{\mathfrak{u}})/\partial \mathfrak{u}$. Let $\delta \mathfrak{u} = (\delta u_0, \delta u_1, \ldots, \delta u_{k-1})$ be any perturbation about the optimal control sequence $\hat{\mathfrak{u}}$; then, since $\mathbf{F}$ is continuously differentiable, the perturbed terminal phase is given by

13 $$\mathbf{F}(\hat{\mathfrak{u}} + \delta \mathfrak{u}) = \mathbf{F}(\hat{\mathfrak{u}}) + \frac{\partial \mathbf{F}(\hat{\mathfrak{u}})}{\partial \mathfrak{u}} \delta \mathfrak{u} + \mathbf{o}(\delta \mathfrak{u}),$$

where the $(n + 1) \times km$ jacobian matrix $\partial \mathbf{F}(\hat{\mathfrak{u}})/\partial \mathfrak{u}$ has as its $\alpha\beta$ element $\partial \mathbf{F}^\alpha(\hat{\mathfrak{u}})/\partial u_0^\beta$ for $\alpha = 0, 1, \ldots, n$, and $\beta = 1, 2, \ldots, m$, $\partial \mathbf{F}^\alpha(\hat{\mathfrak{u}})/\partial u_1^{\beta-m}$ for $\alpha = 0, 1, \ldots, n$ and $\beta = m + 1, m + 2, \ldots,$ $2m$, etc., and $\mathbf{o}(\delta \mathfrak{u})$ is a continuous function of $\delta \mathfrak{u}$ such that

$$\lim_{\|\delta \mathfrak{u}\| \to 0} \|\mathbf{o}(\delta \mathfrak{u})\|/\|\delta \mathfrak{u}\| = 0.$$

SEC. 2.4 THE TWO-POINT-BOUNDARY-VALUE OPTIMAL CONTROL PROBLEM

The most direct way to compute the linear term in (13) is to go back to the difference equation (2), which defines the map $\mathbf{F}$. Since the functions $\mathbf{f}_i(\cdot,\cdot)$ are continuously differentiable for $i = 0, 1, \ldots, k-1$, we may expand the right-hand side of (2) about the optimal control sequence $\hat{\mathfrak{U}} = (\hat{u}_0, \hat{u}_1, \ldots, \hat{u}_{k-1})$ and its corresponding optimal trajectory $\hat{\mathfrak{X}} = (\hat{\mathbf{x}}_0, \hat{\mathbf{x}}_1, \ldots, \hat{\mathbf{x}}_k)$ to obtain

14
$$(\hat{\mathbf{x}}_{i+1} + \delta\mathbf{x}_{i+1}) - (\hat{\mathbf{x}}_i + \delta\mathbf{x}_i) = \mathbf{f}_i(\hat{\mathbf{x}}_i, \hat{u}_i) + \frac{\partial}{\partial \mathbf{x}} \mathbf{f}_i(\hat{\mathbf{x}}_i, \hat{u}_i) \, \delta\mathbf{x}_i + \frac{\partial}{\partial u} \mathbf{f}_i(\hat{\mathbf{x}}_i, \hat{u}_i) \, \delta u_i + \mathbf{o}(\delta\mathbf{x}_i, \delta u_i) \qquad i = 0, 1, \ldots, k-1,$$

where $\delta\mathbf{x}_i \in E^{n+1}$ for $i = 0, 1, \ldots, k$, $\delta u \in_i E^m$ for $i = 0, 1, \ldots, k-1$ and we have used the obvious substitutions in (2)

15
$$\mathbf{x}_i = \hat{\mathbf{x}}_i + \delta\mathbf{x}_i \qquad \text{and} \qquad u_i = \hat{u}_i + \delta u_i.$$

The element in the α row and β column of the jacobian matrices $(\partial/\partial\mathbf{x})\mathbf{f}_i(\hat{\mathbf{x}}_i, \hat{u}_i)$ and $(\partial/\partial u)\mathbf{f}_i(\hat{\mathbf{x}}_i, \hat{u}_i)$ is $[(\partial/\partial \mathbf{x}^\beta)f_i^\alpha(\mathbf{x}, \hat{u}_i)]_{\mathbf{x}=\hat{\mathbf{x}}_i}$, with $\alpha, \beta = 0, 1, \ldots, n$, and $[(\partial/\partial u^\beta)f_i^\alpha(\hat{\mathbf{x}}_i, u)]_{u=\hat{u}_i}$, with $\alpha = 0, 1, \ldots, n$ and $\beta = 1, 2, \ldots, m$, respectively; that is, $(\partial/\partial \mathbf{x})f_i(\hat{x}_i, \hat{u}_i)$ is an $(n+1) \times (n+1)$ matrix and $(\partial/\partial u)f_i(\hat{x}_i, \hat{u}_i)$ is an $(n+1) \times m$ matrix. The function $\mathbf{o}(\delta\mathbf{x}_i, \delta u_i)$ is continuous and has the property that $\lim_{(\delta\mathbf{x}_i,\delta u_i)\to 0} \|\mathbf{o}(\delta\mathbf{x}_i, \delta u_i)\|/\|(\delta\mathbf{x}_i, \delta u_i)\| = 0$, where $(\delta\mathbf{x}_i, \delta u_i)$ is a combined vector with $n + 1 + m$ components.

It is reasonably clear that the linear part of (13), $[\partial \mathbf{F}(\hat{\mathfrak{U}})/\partial \mathfrak{U}] \, \partial \mathfrak{U}$, can only be the result of the linear part of (14); i.e., it is the solution of the linear difference equation

16
$$\delta\mathbf{x}_{i+1} - \delta\mathbf{x}_i = \frac{\partial}{\partial \mathbf{x}} \mathbf{f}_i(\hat{\mathbf{x}}_i, \hat{u}_i) \, \delta\mathbf{x}_i + \frac{\partial}{\partial u} \mathbf{f}_i(\hat{\mathbf{x}}_i, \hat{u}_i) \, \delta u_i$$
$$i = 0, 1, \ldots, k-1$$

with $\delta\mathbf{x}_0 = 0$ and $\delta\mathfrak{U} = (\delta u_0, \delta u_1, \ldots, \delta u_{k-1})$. Thus

17
$$\frac{\partial \mathbf{F}(\hat{\mathfrak{U}})}{\partial \mathfrak{U}} \delta\mathfrak{U} = \delta\mathbf{x}_k(0, \delta\mathfrak{U}).$$

Let $G_{i,j}$, where $i = j, j+1, \ldots, k$ and $j \in \{0, 1, \ldots, k\}$, be the $(n+1) \times (n+1)$ matrix solution of

18
$$G_{i+1,j} - G_{i,j} = \frac{\partial}{\partial \mathbf{x}} \mathbf{f}_i(\hat{\mathbf{x}}_i, \hat{u}_i) \, G_{i,j} \qquad i = j, j+1, \ldots, k-1,$$

from the initial value $G_{j,j} = I$, the identity matrix; that is, $G_{i,j}$ is a state transition matrix. Hence $G_{j,j} = I$ and

19
$$G_{i,j} = \left[I + \frac{\partial}{\partial \mathbf{x}} \mathbf{f}_{i-1}(\hat{\mathbf{x}}_{i-1}, \hat{u}_{i-1})\right]\left[I + \frac{\partial}{\partial \mathbf{x}} \mathbf{f}_{i-2}(\hat{x}_{i-2}, \hat{u}_{i-2})\right]$$
$$\cdots \left[I + \frac{\partial}{\partial \mathbf{x}} \mathbf{f}_j(\hat{\mathbf{x}}_j, \hat{u}_j)\right] \quad i = j+1, j+2, \ldots, k$$

Now suppose that $\delta u_j \neq 0$ for some $j \in \{0, 1, \ldots, k-1\}$ and that $\delta u_i = 0$ for all other $i \in \{0, 1, \ldots, k-1\}$. Then $\delta \mathbf{x}_k$, the resulting solution of (16) at time k, is given by

20
$$\delta \mathbf{x}_k = G_{k,j+1} \frac{\partial}{\partial u} \mathbf{f}_j(\hat{\mathbf{x}}_j, \hat{u}_j) \, \delta u_j.$$

Since $\delta \mathbf{x}_0 = 0$, by assumption, we conclude from (17) and (20) that for an arbitrary $\delta \mathcal{U} = (\delta u_0, \delta u_1, \ldots, \delta u_{k-1}) \in E^{km}$

21
$$\delta \mathbf{x}_k = \frac{\partial F(\hat{\mathcal{U}})}{\partial \mathcal{U}} \delta \mathcal{U} = \sum_{j=0}^{k-1} G_{k,j+1} \frac{\partial}{\partial u} \mathbf{f}_j(\hat{\mathbf{x}}_j, \hat{u}_j) \, \delta u_j.$$

Thus we find that in partitioned form $\partial F(\hat{\mathcal{U}})/\partial \mathcal{U}$ can be expressed as

22
$$\frac{\partial F(\hat{\mathcal{U}})}{\partial \mathcal{U}} = \left(G_{k,1} \frac{\partial}{\partial u} f_0(\hat{\mathbf{x}}_0, \hat{u}_0) \middle| G_{k,2} \frac{\partial}{\partial u} f_1(\hat{\mathbf{x}}_1, \hat{u}_1) \middle| \right.$$
$$\left. \cdots \middle| G_{kk} \frac{\partial}{\partial u} f_{k-1}(\hat{\mathbf{x}}_{k-1}, \hat{u}_{k-1})\right).$$

If we substitute from (21) into (10), our necessary condition becomes

23
$$\left\langle \psi, \sum_{j=1}^{k-1} G_{k,j+1} \frac{\partial}{\partial u} \mathbf{f}_j(\hat{\mathbf{x}}_j \hat{u}_j) \, \delta u_j \right\rangle \leq 0 \quad \text{for all } \delta \mathcal{U} \in \overline{RC}(\hat{\mathcal{U}}, U).$$

Now let us choose $\delta u_i \neq 0$ and $\delta u_j = 0$ for all $j \neq i$. Then (23) becomes

24
$$\left\langle \psi, G_{k,i+1} \frac{\partial}{\partial u} \mathbf{f}_i(\hat{\mathbf{x}}_i, u_i) \, \delta u_i \right\rangle \leq 0 \quad \text{for all } \delta u_i \in \overline{RC}(\hat{u}_i, U_i).$$

Finally, we introduce the $(n+1)$-dimensional *cophase vectors* $\mathbf{p}_0$, $\mathbf{p}_1, \ldots, \mathbf{p}_k$, defined by

25
$$\mathbf{p}_i = G_{k,i}^T \psi \quad i = 0, 1, \ldots, k.$$

SEC. 2.4 THE TWO-POINT-BOUNDARY-VALUE OPTIMAL CONTROL PROBLEM

It can be verified by direct use of definition (19) of the matrices $G_{k,i}$ that the cophase vectors satisfy the adjoint difference equation

(26) $$\mathbf{p}_i - \mathbf{p}_{i+1} = \left[\frac{\partial}{\partial \mathbf{x}} \mathbf{f}_i(\hat{\mathbf{x}}_i, \hat{u}_i)\right]^T \mathbf{p}_{i+1}$$

$$i = 0, 1, \ldots, k-1, \quad \mathbf{p}_k = \boldsymbol{\psi}.$$

In terms of the cophase variables, (24) becomes

(27) $$\left\langle \mathbf{p}_{i+1}, \frac{\partial}{\partial u} \mathbf{f}_i(x_i, u_i) \, \delta u_i \right\rangle \leq 0 \qquad \text{for all } \delta u_i \in \overline{RC}(u_i, U_i),$$

$$i = 0, 1, \ldots, k-1.$$

We can now summarize the above findings.

(28) **Theorem.** If $\hat{\mathfrak{u}}$ is an optimal control sequence and $\hat{\mathfrak{X}}$ is the corresponding optimal trajectory for problem (1), and if for $i = 0, 1, \ldots, k-1$ the radial cones $RC(\hat{u}_i, U_i)$ are conical approximations to the sets U_i at $\hat{u}_i \in U_i$, then there exist vectors $\mathbf{p}_0, \mathbf{p}_1, \ldots, \mathbf{p}_k$ in E^{n+1}, not all zero, such that

(29) $$\mathbf{p}_i - \mathbf{p}_{i+1} = \left[\frac{\partial}{\partial \mathbf{x}} \mathbf{f}_i(\hat{\mathbf{x}}_i, \hat{u}_i)\right]^T \mathbf{p}_{i+1} \qquad i = 0, 1, \ldots, k-1$$

(30) $$p_k^0 \leq 0$$

and for all $\delta u_i \in RC(\hat{u}_i, U_i)$, with $i = 0, 1, \ldots, k-1$,

(31) $$\left\langle \frac{\partial}{\partial u} H(\hat{\mathbf{x}}_i, \hat{u}_i, \mathbf{p}_{i+1}, i), \delta u_i \right\rangle \leq 0,$$

where the hamiltonian $H(\cdot, \cdot, \cdot, \cdot)$ which maps $E^{n+1} \times E^m \times E^{n+1} \times \{0, 1, \ldots, k\}$ into the real line is defined by

(32) $$H(\mathbf{x}, u, \mathbf{p}, i) = \langle \mathbf{p}, \mathbf{f}_i(\mathbf{x}, u) \rangle. \quad \square$$

(33) **Corollary.** If the functions $\mathbf{f}_i(\mathbf{x}, u)$ for $i = 0, 1, \ldots, k-1$ do not depend on the cost variable x^0, then the vectors $\mathbf{p}_0, \mathbf{p}_1, \ldots, \mathbf{p}_k$ satisfying (29) to (31) also satisfy $p_0^0 = p_1^0 = \cdots = p_k^0 \leq 0$.

Proof. Since the functions $\mathbf{f}_i(\mathbf{x}, u)$ do not depend on the cost variable x^0, the first column of the matrices $[(\partial/\partial \mathbf{x})\mathbf{f}_i(\hat{\mathbf{x}}_i, \hat{u}_i)]^T$ is zero, and hence, from (29),

(34) $$p_i^0 - p_{i+1}^0 = 0 \qquad i = 0, 1, \ldots, k-1. \quad \square$$

35 **Remark.** It is customary to exhibit the relationship among the hamiltonian H, the system equations (2), and the adjoint equations (26). These are as follows:

36 $$\hat{\mathbf{x}}_{i+1} - \hat{\mathbf{x}}_i = \frac{\partial}{\partial \mathbf{p}} H(\hat{\mathbf{x}}_i, \hat{u}_i, \mathbf{p}_{i+1}, i) \qquad i = 0, 1, \ldots, k-1,$$

and

37 $$\mathbf{p}_i - \mathbf{p}_{i+1} = \frac{\partial}{\partial \mathbf{x}} H(\hat{\mathbf{x}}_i, \hat{u}_i, \mathbf{p}_{i+1}, i) \qquad i = 0, 1, \ldots, k-1. \quad \square$$

Observe that condition (31) of the necessary-conditions theorem (28) does not express a true maximum principle. At most, it may be interpreted to read: "If the gradient of the hamiltonian H with respect to u, evaluated on the optimal trajectory, is not zero and not orthogonal to all the admissible perturbations $\delta u \in RC(\hat{u}_i, U_i)$, then H has a local maximum with respect to u at $u = \hat{u}_i$." However, the fact that (28) is not a maximum principle does not necessarily decrease its utility as a necessary condition.

Let us now consider an example of how the conditions of theorem (28) can be used to reduce the optimization problem (1) to a two-point-boundary-value problem for a system of difference equations.

38 **A minimum-energy problem.** Consider the particular case of problem (1) when the augmented difference equation (2) has the form

39 $$\mathbf{x}_{i+1} - \mathbf{x}_i = A\mathbf{x}_i + \mathbf{g}(u_i) \qquad i = 0, 1, \ldots, k-1,$$

where A is an $(n+1) \times (n+1)$ constant matrix of the form

40 $$A = \begin{bmatrix} 0 & 0 & \cdots & \cdots & 0 \\ \hline 0 & & & & \\ \cdot & & & & \\ \cdot & & B & & \\ \cdot & & & & \\ 0 & & & & \end{bmatrix}.$$

The controls u_i are assumed to be real valued (that is, $m = 1$), and for $i = 0, 1, \ldots, k$, $U_i = \{u : |u| \leq 1\}$. The function $\mathbf{g}(\cdot)$ is defined to be

41 $$\mathbf{g}(u_i) = ((u_i)^2, u_i d^1, \ldots, u_i d^n).$$

SEC. 2.4 THE TWO-POINT-BOUNDARY-VALUE OPTIMAL CONTROL PROBLEM

That is, referring to Section 1.3, we observe that we have here a linear plant with a quadratic cost functional. The hamiltonian H for this problem has the form

(42) $$H(\mathbf{x}_i, u_i, \mathbf{p}_{i+1}, i) = \langle \mathbf{p}_{i+1}, A\mathbf{x}_i + \mathbf{g}(u_i) \rangle \qquad i = 0, 1, \ldots, k-1,$$

and condition (31) becomes

(43) $$\left(2\hat{u}_i p_{i+1}^0 + \sum_{j=1}^n p_{i+1}^j d^j\right) \delta u_i \leq 0 \qquad \text{for all } \delta u_i \in \overline{RC}(\hat{u}_i, U_i).$$

Examining (43) above, we conclude that if $(2\hat{u}_i p_{i+1}^0 + \Sigma_{j=1}^n p_{i+1}^j d^j) > 0$, then $RC(\hat{u}_i, U_i)$ must be the set $\{\delta u_i : \delta u_i \leq 0\}$, and hence $\hat{u}_i = +1$. Similarly, if this expression is negative, then $\hat{u}_i = -1$. Otherwise $2\hat{u}_i p_{i+1}^0 + \Sigma_{j=1}^n p_{i+1}^j d^j = 0$. We can now combine the above three statements to express the optimal control sequence in terms of the vectors $p_0, p_1, \ldots, p_k$ as follows:[1]

(44) $$\hat{u}_i = \text{sat}\left(-\sum_{j=1}^n \frac{p_{i+1}^j d^j}{2p_{i+1}^0}\right) \qquad i = 0, 1, 2, \ldots, k-1,$$

where

(45) $$\text{sat}(y) = \begin{cases} y & \text{if } |y| \leq 1 \\ \dfrac{y}{|y|} & \text{if } |y| > 1. \end{cases}$$

As a result, the problem is reduced to an ordinary two-point-boundary-value problem: solve the system of equations

(46) $$\hat{\mathbf{x}}_{i+1} - \hat{\mathbf{x}}_i = A\mathbf{x}_i + \mathbf{g}\left(\text{sat}\left(\sum_{j=1}^n \frac{p_{i+1}^j d^j}{2p_{i+1}^0}\right)\right)$$

(47) $$\mathbf{p}_i - \mathbf{p}_{i+1} = A^T \mathbf{p}_{i+1}$$

with the boundary conditions

(48) $$\hat{\mathbf{x}}_0 = (0, \hat{x}_0^1, \hat{x}_0^2, \ldots, \hat{x}_0^n) \qquad \hat{\mathbf{x}}_k \in \{\mathbf{x} : \mathbf{x} = (x^0, \hat{x}_k^1, \hat{x}_k^2, \ldots, \hat{x}_k^n)\},$$

where $\hat{x}_0^1, \hat{x}_0^2, \ldots, \hat{x}_0^n$ and $\hat{x}_k^1, \hat{x}_k^2, \ldots, \hat{x}_k^n$ are given. $\square$

REFERENCES

1. M. D. Canon, C. D. Cullum, and E. Polak: Constrained Minimization Problems in Finite Dimensional Spaces, *SIAM J. Control*, **4**:528–547 (1966).

---

[1] We anticipate here some of the results which will be developed in Chapter 4, where it will be shown that for minimum-energy problems such as the one described, $p_i^0 = $ constant < 0 (not zero) and condition (31) is not only necessary, but also sufficient.

2. L. W. Neustadt: An Abstract Variational Theory with Applications to a Broad Class of Optimization Problems, I. General Theory, II. Applications, *SIAM J. Control*, **4**:505–527 (1966), **5**:90–137 (1967).
3. L. M. Graves: "Theory of Functions of Real Variables," p. 149, McGraw-Hill Book Company, New York, 1946.
4. W. Hurewicz and H. Wallman: "Dimension Theory," pp. 40–41, Princeton University Press, Princeton, N.J., 1960.
5. H. W. Kuhn and A. W. Tucker: Nonlinear Programming, *Proc. Second Berkeley Symp. Math. Stat. Probability*, pp. 481–492, University of California Press, Berkeley, Calif., 1951.
6. E. J. Messerli and E. Polak: On Second Order Conditions of Optimality, *SIAM J. Control*, **7** (1969), in press.
7. L. S. Pontryagin, V. G. Boltyanskii, R. V. Gamkrelizde, and E. F. Mishchenko: "The Mathematical Theory of Optimal Processes," Interscience Publishers, Inc., New York, 1962.
8. B. W. Jordan and E. Polak: Theory of a Class of Discrete Optimal Control Systems, *J. Electron. Control*, **17**:697–713 (1964).
9. H. Halkin: A Maximum Principle of the Pontryagin Type for Systems Described by Nonlinear Difference Equations, *SIAM J. Control*, **4**:90–111 (1966).

3
Some necessary and some sufficient conditions for nonlinear programming problems

3.1 INTRODUCTION

The Fundamental Theorem (2.3.12) is remarkable in that virtually all the known necessary conditions of optimality for nonlinear programming and discrete optimal control problems can be deduced from it. However, the manner in which these necessary conditions are obtained is not always straightforward. This is because to obtain a meaningful condition we must first obtain a nontrivial conical approximation to the constraint set [see (2.3.1) and (2.3.3)]. This chapter is therefore devoted to the application of the Fundamental Theorem (2.3.12) to nonlinear programming problems and to establishing certain cases for which the resulting necessary conditions are also sufficient.

1 The nonlinear programming problem. Given a real-valued continuously differentiable function $f(\cdot)$ defined on E^n, a continuously differentiable function $r(\cdot)$ from E^n into E^m, and a continuously differentiable function $q(\cdot)$ from E^n into E^k, find a vector $\hat{z} \in E^n$ satisfying

2 $\qquad \hat{z} \in \Omega \triangleq \{z: q(z) \leq 0\}$
3 $\qquad r(\hat{z}) = 0$

such that

4 $\qquad f(\hat{z}) \leq f(z)$

for all $z \in E^n$ satisfying (2) and (3). □

Thus the nonlinear programming problem is simply a form of the Basic Problem (1.4.1) in which the constraint set Ω is given the specific description $\Omega = \{z: q(z) \leq 0\}$, as in (2). In the following sections we shall consider a number of versions of the nonlinear programming problem (1) which will be found to be of increasing complexity as far as application of the Fundamental Theorem (2.3.12) is concerned. We begin with the simplest case possible, $\Omega = E^n$, that is, $q(\cdot) \equiv 0$.

3.2 THEORY OF LAGRANGE MULTIPLIERS

The simplest case of the nonlinear programming problem (1.1) is the one in which $\Omega = E^n$, that is, where the inequality-constraint function $q(\cdot)$ is identically zero. Thus we are given the continuously differentiable functions $f(\cdot)$ and $r^1(\cdot), r^2(\cdot), \ldots, r^m(\cdot)$ [that is, $r(\cdot) = (r^1(\cdot), r^2(\cdot), \ldots, r^m(\cdot))$] from E^n into the reals, and we are required to find a $\hat{z} \in E^n$ satisfying $r^i(\hat{z}) = 0$ for $i = 1, 2, \ldots, m$ such that $f(\hat{z}) \leq f(z)$ for all $z \in E^n$ also satisfying $r^i(z) = 0$ for $i = 1, 2, \ldots, m$.

Obviously, this is a special case of the Basic Problem (1.4.1), with the set $\Omega = E^n$. Now, E^n is a conical approximation to E^n at any point $\hat{z} \in E^n$, and hence we conclude from the Fundamental Theorem (2.3.12) that if $\hat{z}$ is an optimal solution to the Basic Problem and $\Omega = E^n$, then there exists a nonzero vector $\psi \in E^{m+1}$ such that

1 $\qquad \left\langle \psi, \dfrac{\partial F(\hat{z})}{\partial z} \delta z \right\rangle \leq 0 \qquad \text{for all } \delta z \in E^n,$

where the function $F(\cdot)$ from E^n into E^{m+1} is as defined in (2.2.30): $F(z) = (f(z), r^1(z), r^2(z), \ldots, r^m(z))$; that is, $F^0(\cdot) = f(\cdot)$, $F^1(\cdot) = r^1(\cdot), \ldots, F^m(\cdot) = r^m$. Obviously, (1) may be rewritten as

2 $\qquad \left\langle \left[\dfrac{\partial F(\hat{z})}{\partial z}\right]^T \psi, \delta z \right\rangle \leq 0 \qquad \text{for all } \delta z \in E^n.$

SEC. 3.2 THEORY OF LAGRANGE MULTIPLIERS

Since for any $\delta z \in E^n$, $-\delta z$ is also in E^n, we conclude from (2) that

3
$$\left\langle \left[\frac{\partial F(\hat{z})}{\partial z}\right]^T \psi, \delta z \right\rangle = 0 \quad \text{for all } \delta z \in E^n,$$

and hence

4
$$\left[\frac{\partial F(\hat{z})}{\partial z}\right]^T \psi = 0.$$

Now, $\partial F(\hat{z})/\partial z$ is an $(m+1) \times n$ matrix with rows, from top to bottom, $\nabla f(\hat{z})$, $\nabla r^1(\hat{z})$, ..., $\nabla r^m(\hat{z})$, where $\nabla f(\hat{z}) = (\partial f(\hat{z})/\partial z^1, \ldots, \partial f(\hat{z})/\partial z^n)$ and $\nabla r^i(\hat{z}) = (\partial r^i(\hat{z})/\partial z^1, \ldots, \partial r^i(\hat{z})/\partial z^n)$ for $i = 1, 2, \ldots, m$. We may therefore expand (4) into the form

5
$$\psi^0 \nabla f(\hat{z}) + \sum_{i=1}^{m} \psi^i \nabla r^i(\hat{z}) = 0.$$

We have thus proved the following classical Lagrange multiplier rule.

6 **Theorem: Lagrange multiplier rule.** Let f and $r^1, r^2, \ldots, r^m$ be real-valued continuously differentiable functions on E^n. If $\hat{z} \in E^n$ minimizes $f(z)$ subject to the constraints $r^i(z) = 0$ for $i = 1, 2, \ldots, m$, then there exist scalar multipliers $\psi^0, \psi^1, \ldots, \psi^m$, not all zero, such that the real-valued function $L(\cdot)$ defined on E^n by

7
$$L(z) = \psi^0 f(z) + \sum_{i=1}^{m} \psi^i r^i(z)$$

has a stationary point at $z = \hat{z}$; that is, $\partial L(\hat{z})/\partial z = 0$. □

Clearly, this merely says that (5) is satisfied.

8 **Corollary.** If the functions r^i for $i = 1, 2, \ldots, m$ are affine, then the conditions of theorem (6) can be satisfied with $\psi^0 < 0$.

Proof. We shall show that the assumption of corollary (2.2.48) is satisfied, that is, that the cone $K(\hat{z})$ is closed. In the particular case where $\Omega = E^n$, $RC(\hat{z}, \Omega) = E^n$, and hence

$$K(\hat{z}) = \left\{\frac{\partial F(\hat{z})}{\partial z} \delta z : \delta z \in E^n\right\}.$$

Obviously, $K(\hat{z})$ is closed. □

9 **Corollary.** If the vectors $\nabla r^i(\hat{z})$ for $i = 1, 2, \ldots, m$ are linearly independent, then any vector ψ satisfying the conditions of theorem (6) also satisfies $\psi^0 < 0$. □

It is usual to assume that every point of the set $\{z: r^i(z) = 0, i = 1, 2, \ldots, m\}$ is a *normal point* i.e., that the gradient vectors $\nabla r^i(z)$ for $i = 1, 2, \ldots, m$ are linearly independent for all z satisfying $r^i(z) = 0$ for $i = 1, 2, \ldots, m$. When the gradient vectors $\nabla r^i(z)$ for $i = 1, 2, \ldots, m$ are not linearly independent, it is always possible to find multipliers $\psi^1, \psi^2, \ldots, \psi^m$, not all zeros, such that $\sum_{i=1}^{m} \psi^i \nabla r^i(z) = 0$. Hence by setting $\psi^0 = 0$ we can make $\psi^0 \nabla f(z) + \sum_{i=1}^{m} \psi^i \nabla r^i(z) = 0$ whether z is an optimal solution or not. Obviously, in this case theorem (6) is trivial. However, when the vectors $\nabla r^i(\hat{z})$ for $i = 1, 2, \ldots, m$ are linearly independent, it is not possible to satisfy (5) with $\psi^0 = 0$.

When $\psi^0 \neq 0$, we can multiply the ψ^i in (7) by $1/\psi^0$ and let $\hat{\lambda}_i = \psi^i/\psi^0$ for $i = 1, 2, \ldots, m$. This results in the following somewhat better-known condition.

10 **Theorem.** Let f and $r^1, r^2, \ldots, r^m$ be real-valued continuously differentiable functions on E^n. If $\hat{z}$ minimizes $f(z)$ subject to $r^i(z) = 0$ for $i = 1, 2, \ldots, m$ and either the $r^i(\cdot)$ are affine or the gradient vectors $\nabla r^i(\hat{z})$ for $i = 1, 2, \ldots, m$ are linearly independent, then there exists a vector $\hat{\lambda} \in E^m$ such that the real-valued function $L(\cdot,\cdot)$ defined on $E^n \times E^m$ by

11 $$L(z,\lambda) = f(z) + \sum_{i=1}^{m} \lambda^i r^i(z)$$

has a stationary point at $(\hat{z},\hat{\lambda})$.

Proof. Referring to (5), we note that $\partial L(\hat{z},\hat{\lambda})/\partial z = 0$, while $\partial L(\hat{z},\hat{\lambda})/\partial \lambda = r(\hat{z}) = 0$, by assumption. □

The reader should be careful not to read more into theorem (10) than it actually states. In particular, it is not true that every $\hat{z}$ which maximizes or minimizes, locally or globally, the function $f(\cdot)$ subject to the constraints $r^i(z) = 0$ for $i = 1, 2, \ldots, m$ also maximizes or minimizes the lagrangian $L(\cdot,\hat{\lambda})$ subject to no constraints on z, as will be seen from Example (14). The relationship between the extrema of these two functions is given in the next theorem, which is added as a digression to correct an occasionally encountered misconception.

12 **Theorem.** Let f and $r^1, r^2, \ldots, r^m$ be functions mapping E^n into the reals, and let $\hat{\lambda}_1, \hat{\lambda}_2, \ldots, \hat{\lambda}_m$ be given real numbers. If $\hat{z} \in E^n$

satisfies $r^i(\hat{z}) = 0$ for $i = 1, 2, \ldots, m$, and if for all $z \in N(\hat{z},\epsilon)$, an ϵ-ball about $\hat{z}$,

13
$$f(\hat{z}) + \sum_{i=1}^{m} \hat{\lambda}^i r^i(\hat{z}) \leq f(z) + \sum_{i=1}^{m} \hat{\lambda}^i r^i(z),$$

then for all $z \in N(\hat{z},\epsilon) \cap \{z: r^i(z) = 0, i = 1, 2, \ldots, m\}$

$$f(\hat{z}) \leq f(z).$$

(To paraphrase this theorem, if $\hat{z}$ satisfies $r^i(\hat{z}) = 0$ and if it is a point of unconstrained local minimum for the lagrangian $L(\cdot,\hat{\lambda})$, then it is also a point of local minimum for the function $f(\cdot)$ subject to the constraints $r^i(z) = 0$ for $i = 1, 2, \ldots, m$. Clearly, we could replace the word *minimum* by *maximum* and the word *local* by *global* in the statement of the theorem. Example (14) will show that the converse of this theorem is not true.)

Proof. Obviously, if $\hat{z}$ satisfies (13) for all $z \in N(\hat{z},\epsilon)$, it also satisfies (13) on the smaller set $N(\hat{z},\epsilon) \cap \{z: r^i(z) = 0, i = 1, 2, \ldots, m\}$ which is contained in $N(\hat{z},\epsilon)$. But for all z satisfying $r^i(z) = 0$ for $i = 1, 2, \ldots, m$ the lagrangian $L(z,\hat{\lambda}) = f(z) + \Sigma_{i=1}^{m}\hat{\lambda}^i r^i(z)$ is identically equal to $f(z)$. □

We conclude from the above theorem that if $\hat{z}$ is a point of constrained minimum (maximum) for a continuously differentiable function $f(\cdot)$ subject to continuously differentiable constraints $r^i(z) = 0$ for $i = 1, 2, \ldots, m$, and if $\hat{\lambda}^1, \hat{\lambda}^2, \ldots, \hat{\lambda}^m$ are Lagrange multipliers satisfying the conditions of theorem (10), then $\hat{z}$ is either a point of minimum (maximum) or a stationary point for the lagrangian $L(\cdot,\hat{\lambda})$, but it is never a point of maximum (minimum).

14 Example. Suppose that $z \in E^2$ and we wish to maximize the function

15
$$f(z) = (z^1)^4 + (z^2)^4$$

subject to the constraint

16
$$r^1(z) = (z^1)^2 + (z^2 - 1)^2 - 1 = 0.$$

Examining Figure 1, we see that $\hat{z} = (0,2)$ is the point of constrained maximum. We now form the lagrangian $L(\cdot,\cdot)$ on $E^2 \times E^1$ defined by

17
$$L(z,\lambda) = (z^1)^4 + (z^2)^4 + \lambda[(z^1)^2 + (z^2 - 1)^2 - 1]$$

and compute its stationary points $\hat{z}$ and $\hat{\lambda}$ by solving the equations $\partial L(z,\lambda)/\partial z = 0$ and $\partial L(z,\lambda)/\partial \lambda = 0$. One of the pairs that this yields

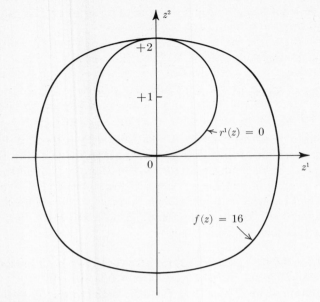

Figure 1

is ($\hat{z} = (0,2)$, $\hat{\lambda} = -16$). A simple calculation now shows that $\hat{z}$ is not a point of maximum for the function

18
$$L(z,-16) = (z^1)^4 + (z^2)^4 - 16[(z^1)^2 + (z^2 - 1)^2 - 1],$$

illustrating the fact that a point $\hat{z}$ which maximizes $f(z)$ subject to $r^i(z) = 0$ for $i = 1, 2, \ldots, m$ *does not necessarily* also maximize the lagrangian $L(z,\hat{\lambda})$, which is subject to no constraints. □

19 **Exercise.** Let f and $r^1, r^2, \ldots, r^m$ be twice–continuously differentiable functions defined on E^n, and let $\hat{z}$ and $\hat{\lambda}$ be such that $\partial L(\hat{z},\hat{\lambda})/\partial z = 0$ and $\partial L(\hat{z},\hat{\lambda})/\partial \lambda = 0$; that is, $\hat{z}$ and $\hat{\lambda}$ satisfy the necessary conditions of optimality in theorem (10). Show that if the $n \times n$ hessian matrix $\partial^2 L(\hat{z},\hat{\lambda})/\partial z^2$ of second-order partial derivatives is positive definite, then $\hat{z}$ is a local minimum of $f(\cdot)$ on the set $\{z: r^i(z) = 0, i = 1, 2, \ldots, m\}$. [*Hint:* Expand $L(\cdot,\hat{\lambda})$ about $\hat{z}$, using Taylor's theorem.] □

We now proceed to more complex optimization problems, those with inequality constraints.

3.3 THE KUHN–TUCKER THEORY

We shall examine in this section a particularly well-behaved class of nonlinear programming problems which were originally considered by Kuhn and Tucker [2]. Several extensions of their work can also be found in a monograph by Arrow, Hurwicz, and Uzawa [4].

This material is included partly for historical reasons and partly to exhibit the relation between some of the earlier work and the approach we are taking. Consequently, there is some duplication of results between this section and Sections 3.4 and 3.5. The reader will observe that with the Fundamental Theorem (2.3.12) established, it is often easier to obtain necessary conditions of optimality by proceeding as in Sections 3.5 and 3.6, rather than as in this section.

Kuhn and Tucker made the very interesting observation that in a number of cases the tangent cone [see (2.2.8)] to the set $\Omega' = \{z: r(z) = 0, q(z) \leq 0\}$ can be given a very simple characterization. They formalized this observation into an assumption, which they called a *constraint qualification*. A paraphrase of their statement is as follows.

1 Constraint qualification. Let $r(\cdot)$ be a continuously differentiable function from E^n into E^m, let $q(\cdot)$ be a continuously differentiable function from E^n into E^k, and let $\Omega' = \{z: r(z) = 0, q(z) \leq 0\}$. Let $\hat{z} \in \Omega'$, and let the index set $I(\hat{z}) = \{i: q^i(\hat{z}) = 0, i \in \{1, 2, \ldots, k\}\}$.† We say that the set Ω' satisfies the (Kuhn–Tucker) constraint qualification at $\hat{z}$ if

2
$$TC(\hat{z},\Omega') = \left\{\delta z: \frac{\partial r(\hat{z})}{\partial z}\, \delta z = 0,\ \langle \nabla q^i(\hat{z}), \delta z\rangle \leq 0 \text{ for all } i \in I(\hat{z})\right\},$$

where $TC(\hat{z},\Omega')$ is the tangent cone to Ω' at $\hat{z}$ [see (2.2.8)]. □

This assumption has subsequently been relaxed (see Arrow, Hurwicz, and Uzawa [3]) to the following form.

3 Weakened constraint qualification. We say that the set Ω', defined as in (1), satisfies the weakened constraint qualification at $\hat{z} \in \Omega'$ if

4
$$\overline{\text{co}}\ TC(\hat{z},\Omega') = \left\{\delta z: \frac{\partial r(\hat{z})}{\partial z}\, \delta z = 0,\ \langle \nabla q^i(\hat{z}), \delta z\rangle \leq 0 \right.$$
$$\left. \text{for all } i \in I(\hat{z})\right\}.\ \ \square$$

The effect of the constraint qualifications is twofold: first, they identify an important class of problems to which theorem (2.2.12) can be applied to yield nontrivial conditions of optimality; second, they result in a description of the tangent cone which enables us to

---

† The constraint functions $q^i(\cdot)$ satisfying $q^i(\hat{z}) = 0$ [that is, $i \in I(\hat{z})$] are usually called the *active constraints* at $\hat{z}$.

combine theorem (2.2.12) with Farkas' lemma (A.5.34) to obtain the following important result.[1]

5   **Theorem.** Let $\hat{z}$ be an optimal solution to the nonlinear programming problem (1.1). If the set $\Omega' = \{z: r(z) = 0, q(z) \leq 0\}$ satisfies at $\hat{z}$ either the constraint qualification (1) or the weakened constraint qualification (3), then there exist scalar multipliers $\psi^1, \psi^2, \ldots, \psi^m$, undefined in sign, and nonpositive scalar multipliers $\mu^1 \leq 0, \mu^2 \leq 0, \ldots, \mu^k \leq 0$ such that

6   $$-\nabla f(\hat{z}) + \sum_{i=1}^{m} \psi^i \nabla r^i(\hat{z}) + \sum_{i=1}^{k} \mu^i \nabla q^i(\hat{z}) = 0$$

and

7   $$\mu^i q^i(\hat{z}) = 0 \quad i = 1, 2, \ldots, k.$$

Proof. First we observe that if the set Ω' satisfies the constraint qualification (1) at $\hat{z}$, then $\overline{\text{co}}\, TC(\hat{z},\Omega') = TC(\hat{z},\Omega')$, and therefore Ω' also satisfies (3). Thus we need prove the theorem only under the assumption that (3) is satisfied.

From Theorem (2.2.12), if $\hat{z}$ is an optimal solution, then

8   $$\langle -\nabla f(\hat{z}), \delta z \rangle \leq 0 \quad \text{for all } \delta z \in \overline{\text{co}}\, TC(\hat{z},\Omega').$$

We now write (4) as

9   $$\overline{\text{co}}\, TC(\hat{z},\Omega') = \left\{ \delta z : \frac{\partial r(\hat{z})}{\partial z} \delta z \leq 0, -\frac{\partial r(\hat{z})}{\partial z} \delta z \leq 0, \right.$$
$$\left. \langle \nabla q^i(\hat{z}), \delta z \rangle \leq 0, i \in I(\hat{z}) \right\},$$

where $I(\hat{z}) = \{i: q^i(\hat{z}) = 0, i \in \{1,2, \ldots, k\}\}$, and use Farkas' lemma (A.5.34) to combine (8) and (9) to yield (6), with $\mu^i = 0$ for all i not in $I(\hat{z})$. Equation (7) then follows immediately, since for each $i \in \{1,2, \ldots, k\}$ either μ^i or $q^i(\hat{z})$ is zero. From Farkas' lemma, $\mu^i \leq 0$ for $i = 1, 2, \ldots, k$, but the sign of the ψ^i for $i = 1, 2, \ldots, m$ is indefinite, since each ψ^i can be seen to have arisen as the sum of a negative and a positive number. □

10 **Remark.** Probably the most important feature of theorem (5) is that it gives a condition for the multiplier associated with $\nabla f(\hat{z})$ in (6) to be -1. It is common experience that computational methods derived

---

[1] The following theorems remain true when the sequential tangent cone $TC_s(\hat{z},\Omega')$ [defined in (2.2.28)] is substituted for the tangent cone $TC(\hat{z},\Omega')$ in the constraint qualifications.

SEC. 3.3 THE KUHN-TUCKER THEORY 57

from necessary conditions become considerably simplified whenever the cost-gradient multiplier in (6) is -1. □

Generally speaking, it is not possible to establish by a direct approach whether a particular set Ω' does or does not satisfy condition (1) or (3). However, there are several sufficient conditions which can be checked numerically by pivotal methods and which ensure that a set Ω' satisfies (1) or (3). Let us examine some of them.

Our first observation is that $\overline{\text{co }} TC_s(\hat{z},\Omega')$ is always contained in the right-hand side of (4) and hence that we need only prove that the right-hand side of (4) is contained in $\overline{\text{co }} TC(\hat{z},\Omega')$ or in $TC(\hat{z},\Omega')$. We now prove this observation.

11 **Lemma.** Consider the constraint set $\Omega' = \{z : r(z) = 0,\ q(z) \leqq 0\}$, defined as in (1). Then

12 $$\overline{\text{co }} TC_s(\hat{z},\Omega') \subset \left\{ \delta z : \frac{\partial r(\hat{z})}{\partial z} \delta z = 0,\ \langle \nabla q^i(\hat{z}), \delta z \rangle \leqq 0 \right.$$
$$\left. \text{for all } i \in I(\hat{z}) \right\},$$

where the sequential tangent cone $TC_s(\hat{z},\Omega')$ is as defined in (2.2.28) and the active-constraint index set $I(\hat{z})$ is as defined in (1).

Proof. Let $\delta z \in TC_s(\hat{z},\Omega')$ be arbitrary. Then, by definition (2.2.28), there exists a sequence of vectors δz_i for $i = 1, 2, 3, \ldots$ such that $(\hat{z} + \delta z_i) \in \Omega'$, $\delta z_i \to 0$ as $i \to \infty$, and $\delta z_i / \|\delta z_i\| \to \alpha\, \delta z$ as $i \to \infty$ for some $\alpha > 0$. Hence

13 $$r(\hat{z} + \delta z_i) = 0 + \frac{\partial r(\hat{z})}{\partial z} \delta z_i + o(\delta z_i) = 0, \qquad i = 1, 2, 3, \ldots.$$

Dividing (13) by $\|\delta z_i\|$, we obtain

14 $$\frac{\partial r(\hat{z})}{\partial z} \frac{\delta z_i}{\|\delta z_i\|} + \frac{o(\delta z_i)}{\|\delta z_i\|} = 0.$$

But $\delta z_i \to 0$, and $\|o(\delta z)\|/\|\delta z\| \to 0$ as $\|\delta z\| \to 0$; hence, in the limit, (14) results in

$$\frac{\partial r(\hat{z})}{\partial z} \delta z = 0.$$

Now let $j \in I(\hat{z})$ be arbitrary; then $q^j(\hat{z}) = 0$ and

15 $$q^j(\hat{z} + \delta z_i) = 0 + \langle \nabla q^j(\hat{z}), \delta z_i \rangle + o(\delta z_i) \leqq 0.$$

Dividing (15) by $\|\delta z_i\|$ and letting $i \to \infty$, we obtain

$$\langle \nabla q^j(\hat{z}), \delta z \rangle \leq 0 \quad \text{for all } j \in I(\hat{z}),$$

since $o(\delta z)/\|\delta z\| \to 0$ as $\|\delta z\| \to 0$. This proves that

16 $$TC_s(\hat{z}, \Omega') \subset \left\{ \delta z : \frac{\partial r(\hat{z})}{\partial z} \delta z = 0, \langle \nabla q^j(\hat{z}), \delta z \rangle \leq 0 \text{ for all } j \in I(\hat{z}) \right\}.$$

The set $\{\delta z : [\partial r(\hat{z})/\partial z] \delta z = 0, \langle \nabla q^j(\hat{z}), \delta z \rangle \leq 0 \text{ for all } j \in I(\hat{z})\}$ is convex and closed [the set $TC_s(\hat{z}, \Omega')$ is closed, by (2.2.29)]; hence (12) follows. □

We now proceed with our theorems, with some of the proofs left as exercises.

17 **Theorem.** Consider the constraint set $\Omega' = \{z : r(z) = 0, q(z) \leq 0\}$, let $\hat{z}$ be any point in Ω', and let the index set $I(\hat{z}) = \{i : q^i(\hat{z}) = 0, i \in \{1, 2, \ldots, k\}\}$. If the vectors $\nabla r^i(\hat{z})$ for $i = 1, 2, \ldots, m$, together with the vectors $\nabla q^i(\hat{z})$ for $i \in I(\hat{z})$, are fewer than n in number and are linearly independent,[1] then the set Ω' satisfies the constraint qualification (1) at $\hat{z}$. □

18 **Remark.** This theorem is almost as long and difficult to prove as the Fundamental Theorem (2.3.12). Since the fact that $\psi^0 \neq 0$ in (6) follows trivially under the above assumptions from corollary (4.28) or theorem (5.11), the proof of theorem (17) is omitted here. The interested reader may wish to prove this theorem without resorting to the Fundamental Theorem. □

19 **Theorem.** Consider the constraint set $\Omega' = \{z : r(z) = 0, q(z) \leq 0\}$, and let $\hat{z}$ be any point in Ω'. If the functions r and q are affine, then the set Ω' satisfies the constraint qualification (1) at $\hat{z}$.

Proof. Let $\delta \hat{z} \in E^n$ be any vector such that

20
$$\frac{\partial r(\hat{z})}{\partial z} \delta \hat{z} = 0$$

$$\langle \nabla q^i(\hat{z}), \delta \hat{z} \rangle \leq 0 \quad \text{for } i \in I(\hat{z}).$$

[1] This assumption guarantees that the set $\{z : r(z) = 0, q^i(z) = 0, i \in I(\hat{z})\}$ is a manifold whose tangent plane is the set $\{\delta z : [\partial r(\hat{z})/\partial z] \delta z = 0, \langle \nabla q^i(\hat{z}), \delta z \rangle = 0, i \in I(\hat{z})\}$. When the functions r^i and q^i are linear or affine, this assumption is no longer necessary, since the set $\{z : r(z) = 0, q^i(z) = 0, i \in I(\hat{z})\}$ is then always a linear manifold [see (19)]. For a discussion of manifolds and tangent subspaces see reference [7].

Since r and q are affine, it follows that for any $t \geq 0$

$$r(\hat{z} + t\,\delta\hat{z}) = r(\hat{z}) + t\frac{\partial r(\hat{z})}{\partial z}\,\delta\hat{z} = 0$$

and there must exist a $t_1 > 0$ such that

$$q(\hat{z} + t\,\delta\hat{z}) = q(\hat{z}) + t\frac{\partial q(\hat{z})}{\partial z}\,\delta\hat{z} \leq 0 \qquad 0 \leq t \leq t_1.$$

Hence $\{z: z = \hat{z} + t\,\delta\hat{z}, 0 \leq t \leq t_1\}$ is the arc required by the definition of the constraint qualification. □

21 **Theorem.** Consider the constraint set $\Omega' = \{z: r(z) = 0, q(z) \leq 0\}$. If there exists a vector $\delta\hat{z} \in E^n$ such that $[\partial r(\hat{z})/\partial z]\,\delta\hat{z} = 0$ and $\langle \nabla q^i(\hat{z}), \delta\hat{z} \rangle < 0$ for $i \in I(\hat{z}) = \{i: q^i(\hat{z}) = 0, i \in \{1,2,\ldots,k\}\}$, and either the vectors $\nabla r^1(\hat{z}), \nabla r^2(\hat{z}), \ldots, \nabla r^m(\hat{z})$ are linearly independent or else the function $r(\cdot)$ is affine, then the set Ω' satisfies the constraint qualification (1) at $\hat{z}$ when it is stated in terms of the sequential tangent cone $TC_s(\hat{z},\Omega')$ to Ω' at $\hat{z}$. □

22 **Remark.** Again, the lengthy proof of theorem (21) is omitted, since under the above assumptions $\psi^0 \neq 0$ in (6) follows trivially from corollary (4.31). □

23 **Theorem.** Consider the constraint set $\Omega' = \{z: r(z) = 0, q(z) \leq 0\}$. Suppose that the function r is affine and for $i = 1, 2, \ldots, k$ the components q^i, of q are convex. If there is a vector z^* in Ω' satisfying $q^i(z^*) < 0$ for $i = 1, 2, \ldots, k$, then Ω' satisfies the constraint qualification (1) at any point $\hat{z}$ in Ω' when it is stated in terms of the sequential cone tangent $TC_s(\hat{z},\Omega')$.

Proof. Let $\hat{z} \in \Omega'$ be arbitrary, and let $\delta z = z^* - \hat{z}$. Then, since r is affine, $[\partial r(\hat{z})/\partial z]\,\delta z = r(z^*) - r(\hat{z}) = 0$. If $I(\hat{z})$ is empty, then the conditions of theorem (21) are satisfied automatically, and theorem (23) follows directly. Therefore suppose that $I(\hat{z})$ is not empty. Then, since the functions q^i for $i = 1, 2, \ldots, k$ are convex, for $\delta z = z^* - \hat{z}$, and $i = 1, 2, \ldots, k$ we obtain [see (A.6.25)]

24 $$0 > q^i(\hat{z} + \lambda\,\delta z) \geq q^i(\hat{z}) + \lambda\langle \nabla q^i(\hat{z}), \delta z \rangle, \quad 0 < \lambda \leq 1.$$

But for $i \in I(\hat{z})$, $q^i(\hat{z}) = 0$, and hence for all $i \in I(\hat{z})$

25 $$\langle \nabla q^i(\hat{z}), \delta z \rangle < 0.$$

Since this shows that the conditions of theorem (21) are satisfied, it follows that the set Ω' satisfies the constraint qualification (1). □

26 **Exercise.** Show that the following sets Ω' in E^2 (the plane) satisfy the constraint qualification (1) at all points in these sets:

27 $\quad \Omega' = \{z: -(z^1)^2 - (z^2)^2 + 1 \leq 0, (z^1)^2 + 2(z^2)^2 - 2 \leq 0\}.$

28 $\quad \Omega' = \{z: z^1 - (z^2)^3 = 0, (z^1 - 1)^2 + (z^2)^2 - 2 \leq 0,$
$\qquad\qquad\qquad\qquad\qquad (z^1 + 1)^2 + (z^2)^2 - 2 \leq 0\}.$ □

29 **Exercise.** Show that the set

30 $\quad \Omega' = \{z: z^1 z^2 = 0\}$

satisfies the weakened constraint qualification (3) but does not satisfy the constraint qualification (1) at $z = 0$. □

31 **Exercise.** Show that the set $\Omega_4 \subset E^2$ defined by

$$\Omega' = \{z: -(1 - z^1)^3 + z^2 \leq 0\} \cap \{z: -(1 - z_1)^3 - z^2 \leq 0\}$$

satisfies neither the constraint qualification (1) nor the weakened constraint qualification (3) at the point $\hat{z} = (1,0)$. Show, however, that there are nontrivial conical approximations of the first kind to Ω' at $\hat{z}$. □

We now turn to problems whose constraint sets do not necessarily satisfy the constraint qualification (1) or the weakened qualification (3).

3.4 GENERAL NONLINEAR PROGRAMMING PROBLEMS

In this section we relax the requirement that the weakened constraint qualification (3.3) be satisfied by the set $\Omega' = \{z: r(z) = 0, q(z) \leq 0\}$ to the requirement that it be satisfied by the set $\Omega = \{z: q(z) \leq 0\}$. Consequently, we shall consider a broader class of nonlinear programming problems than we did in the last section.

Since the problem no longer falls within the assumptions of theorem (2.2.12), our first task is to try to find a satisfactory convex approximation to the set $\Omega = \{z: q(z) \leq 0\}$. In connection with this, as in the preceding section, given a particular point $z \in \Omega$, we shall often need to divide the components q^i for $i = 1, 2, \ldots, k$ of the inequality constraint function q into two sets: the *active* constraints satisfying $q^i(z) = 0$ and the *inactive* constraints satisfying $q^i(z) < 0$. Recall that in (3.1) we introduced the index set $I(\hat{z})$ for identifying the active constraints. Its definition is repeated here for convenience.

SEC. 3.4 GENERAL NONLINEAR PROGRAMMING PROBLEMS

1 Definition. For any $\hat{z} \in \Omega = \{z: q^i(z) \leq 0,\ i = 1, 2, \ldots, k\}$ the *index set* $I(\hat{z})$ is defined by

$$I(\hat{z}) = \{i: q^i(\hat{z}) = 0,\ i \in \{1, 2, \ldots, k\}\}.$$

We shall denote by $\bar{I}(\hat{z})$ the complement of $I(\hat{z})$ in $\{1, 2, \ldots, k\}$. ☐

We can now define a rather natural conical approximation to the set Ω.

2 Definition. Let $\Omega = \{z: q(z) \leq 0\}$, where $q(\cdot)$ is a continuously differentiable function from E^n into E^k. For any $\hat{z} \in \Omega$ the *internal cone* to Ω at $\hat{z}$, denoted by $IC(\hat{z},\Omega)$, is defined by

3 $$IC(\hat{z},\Omega) = \{\delta z: \langle \nabla q^i(\hat{z}), \delta z \rangle < 0 \text{ for all } i \in I(\hat{z})\} \cup \{0\}.$$

4 Theorem. Let $\Omega = \{z: q(z) \leq 0\}$, where $q(\cdot)$ is a continuously differentiable function from E^n into E^k. If the internal cone to Ω, $IC(\hat{z},\Omega)$, at a point $\hat{z} \in \Omega$ is not the origin, then:

a. The set Ω satisfies the constraint qualification (3.1) [with the sequential tangent cone $TC_s(\hat{z},\Omega)$ in (3.3)].
b. The internal cone $IC(\hat{z},\Omega)$ is a conical approximation of the first kind [see (2.3.1)].
c. The closure of the internal cone $IC(\hat{z},\Omega)$ is given by

5 $$\overline{IC(\hat{z},\Omega)} = \{\delta z: \langle \nabla q^i(\hat{z}), \delta z \rangle \leq 0 \text{ for all } i \in I(\hat{z})\}.$$

Proof. Part (a) is trivial, since it follows directly from theorem (3.21).

To prove (b) we begin by showing that $IC(\hat{z},\Omega)$ is a convex cone. It is a cone, by inspection, so if we let δz_1 and δz_2 be any two vectors in $IC(\hat{z},\Omega)$, then

6 $$\langle \nabla q^i(\hat{z}), \delta z_1 + \delta z_2 \rangle \leq 0 \quad \text{for all } i \in I(\hat{z}),$$

with equality holding if and only if $\delta z_1 = \delta z_2 = 0$. Hence $IC(\hat{z},\Omega)$ is a convex cone [see (A.4.6)], since $(\delta z_1 + \delta z_2) \in IC(\hat{z},\Omega)$.

Now let us exhibit an $\epsilon > 0$ satisfying the assumptions of definition (2.3.1). Let $\delta z_1, \delta z_2, \ldots, \delta z_p$ be any set of linearly independent (and hence nonzero) vectors in $IC(\hat{z},\Omega)$, and let λ be a positive number. Then any vector $z \in \text{co }\{\hat{z}, \hat{z} + \lambda\,\delta z_1, \hat{z} + \lambda\,\delta z_2, \ldots, \hat{z} + \lambda\,\delta z_p\}$ can be expressed in the form

7 $$z = \hat{z} + \lambda \sum_{j=1}^{p} \mu^j\, \delta z_j,$$

where $\mu^j \geqq 0$ and $\sum_{j=1}^{p} \mu^j \leqq 1$, and since the functions $q^i(\cdot)$ for $i = 1, 2, \ldots, k$ are continuously differentiable, we obtain for $i = 1, 2, \ldots, k$

8
$$q^i(z) = q^i\left(\hat{z} + \lambda \sum_{j=1}^{p} \mu^j \, \delta z_j\right)$$
$$= q^i(\hat{z}) + \lambda \sum_{j=1}^{p} \mu^j \langle \nabla q^i(\hat{z}), \delta z_j \rangle + o^i\left(\lambda \sum_{j=1}^{p} \mu^j \, \delta z_j\right),$$

where $o^i(\lambda \Sigma_{j=1}^{p} \mu^j \, \delta z_j)/\lambda \to 0$ as $\lambda \to 0$ uniformly in $\delta z = \Sigma_{j=1}^{p} \mu^j \, \delta z_j$, with $\mu^j \geqq 0$ and $\Sigma_{j=1}^{p} \mu^j \leqq 1$ (that is, $\delta z \in$ co $\{\delta z_1, \delta z_2, \ldots, \delta z_p\}$). Now let

9
$$\beta = \max_{i,j} \langle \nabla q^i(\hat{z}), \delta z_j \rangle \qquad j \in \{1, 2, \ldots, p\}, i \in I(\hat{z}).$$

Clearly, since $\delta z_j \neq 0$, by assumption, $\beta < 0$ for $i \in I(\hat{z})$. Hence for all $i \in I(\hat{z})$ (8) and (9) yield

10
$$q^i\left(\hat{z} + \lambda \sum_{j=1}^{p} \mu^j \, \delta z_j\right) \leqq \lambda \beta + o^i\left(\lambda \sum_{j=1}^{p} \mu^j \, \delta z_j\right).$$

Since $o^i(\lambda \Sigma_{j=1}^{p} \mu^j \, \delta z_j)/\lambda \to 0$ as $\lambda \to 0$ uniformly in $\delta z = \Sigma_{j=1}^{p} \mu^j \, \delta z_j \in$ co $\{\delta z_1, \delta z_2, \ldots, \delta z_p\}$, and $\beta < 0$, there exists an $\epsilon' > 0$ such that

11
$$q^i\left(\hat{z} + \lambda \sum_{j=1}^{p} \mu^j \, \delta z_j\right) \leqq \lambda \beta + o^i\left(\lambda \sum_{j=1}^{p} \mu^j \, \delta z_j\right) \leqq 0$$
$$\text{for all } \lambda \in [0, \epsilon'], i \in I(\hat{z}).$$

The functions $q^i(\cdot)$ are continuous and $q^i(\hat{z}) < 0$ for all $i \in \bar{I}(\hat{z})$, the complement of $I(\hat{z})$ in $\{1, 2, \ldots, k\}$. Consequently, since co $\{\delta z_1, \delta z_2, \ldots, \delta z_p\}$ is compact, there exists an $\epsilon'' > 0$ such that for all $\sum_{j=1}^{p} \mu^i \, \delta z_i \in$ co $\{\delta z_1, \delta z_2, \ldots, \delta z_p\}$

12
$$q^i\left(\hat{z} + \lambda \sum_{j=1}^{p} \mu^j \, \delta z_j\right) \leqq 0 \qquad \text{for all } \lambda \in [0, \epsilon''], i \in \bar{I}(\hat{z}).$$

Let $\epsilon = \min \{\epsilon', \epsilon''\}$. Then, from (11) and (12), co $\{\hat{z}, \hat{z} + \epsilon \, \delta z_1, \hat{z} + \epsilon \, \delta z_2, \ldots, \hat{z} + \epsilon \, \delta z_p\}$ is in Ω, which proves that $IC(\hat{z}, \Omega)$ is a conical approximation of the first kind.

We shall now prove part (c). Let

13
$$C = \{\delta z \colon \langle \nabla q^i(\hat{z}), \delta z \rangle \leqq 0 \text{ for all } i \in I(\hat{z})\}.$$

Clearly, C is closed and $C \supset IC(\hat{z}, \Omega)$; hence $C \supset \overline{IC}(\hat{z}, \Omega)$. Now let δz be any vector in C. Since $IC(\hat{z}, \Omega) \neq \{0\}$, by assumption, there is a

SEC. 3.4 GENERAL NONLINEAR PROGRAMMING PROBLEMS 63

nonzero vector $\delta z^* \in IC(\hat{z},\Omega)$. Then for $j = 1, 2, 3, \ldots$ the vectors $\delta z_j = 1/j\, \delta z^* + \delta z$ are in $IC(\hat{z},\Omega)$, and $\delta z_j \to \delta z$ as $j \to \infty$. Hence $C \subset \overline{IC}(\hat{z},\Omega)$. □

14 **Exercise.** Suppose that the functions $q^i(\cdot)$ for $i = 1, 2, \ldots, l$, with $l \leq k$, are affine, and let

15
$$IC'(\hat{z},\Omega) = \{\delta z \colon \langle \nabla q^i(\hat{z}), \delta z \rangle \leq 0 \text{ for all } i \in I(\hat{z})$$
$$\cap \{1, 2, \ldots, l\}\} \cap \{\delta z \colon \langle \nabla q^i(\hat{z}), \delta z \rangle < 0 \text{ for all } i \in I(\hat{z})$$
$$\cap \{l+1, \ldots, k\}\} \cup \{0\}.$$

Show that if $IC'(\hat{z},\Omega) \neq \{0\}$, then conditions (a), (b), and (c), of theorem (4) hold. Note that if $l = k$, that is, if all the functions $q^i(\cdot)$ are affine, then the cone defined in (15) is the radial cone. □

Obviously, given a specific constraint set $\Omega = \{z \colon q(z) \leq 0\}$ and a point $\hat{z} \in \Omega$, $IC(\hat{z},\Omega)$ either contains some point other than the origin or it does not. We shall now concentrate on the more interesting case when the internal cone $IC(\hat{z},\Omega)$ is not the origin, and we shall consider the other possibility in the next section. Unfortunately, when $IC(\hat{z},\Omega) = \{0\}$, we can obtain only very weak necessary conditions of optimality.

For the remainder of this section we shall make the following assumption.

16 **Assumption.**[1] If $\hat{z} \in \Omega$ is an optimal solution to the nonlinear programming problem (1.1), then there exists at least one vector $v \in E^n$ such that

17
$$\langle \nabla q^i(\hat{z}), v \rangle < 0 \qquad \text{for all } i \in I(\hat{z});$$

that is, we shall assume that $IC(\hat{z},\Omega) \neq \{0\}$. □

It will be shown in corollary (5.10) that this condition is always satisfied if the vectors $\nabla q^i(\hat{z})$, with $i \in I(\hat{z})$, are linearly independent.

18 **Remark.** If the set $\Omega' = \{z \colon r(z) = 0, q(z) \leq 0\}$ satisfies the assumptions of theorem (3.21), then the set $\Omega = \{z \colon q(z) \leq 0\}$ obviously satisfies assumption (16). The converse, of course, is not true. This indicates the extent to which the assumptions in this section are weaker than those in the preceding section. □

When specialized to the nonlinear programming problem (1.1), the Fundamental Theorem (2.3.12) yields the following necessary

---

[1] When some of the functions q^i, with $i \in I(\hat{z})$, are affine, it suffices to require that there exist a vector $h \in E^n$ such that $\langle \nabla q^i(\hat{z}), h \rangle \leq 0$ for these functions and that $\langle \nabla q^i(\hat{z}), h \rangle < 0$ for the remaining functions q^i, with $i \in I(\hat{z})$.

condition of optimality, provided, of course, that the internal cone $IC(\hat{z},\Omega)$ is not the origin [recall from theorem (4) that the cone $IC(\hat{z},\Omega)$ is a conical approximation of the first kind to Ω at $\hat{z} \in \Omega$].

19 **Theorem.** If $\hat{z}$ is an optimal solution to the nonlinear programming problem (1.1) and assumption (16) is satisfied, then there exists a nonzero vector $\psi = (\psi^0, \psi^1, \ldots, \psi^m)$ in E^{m+1}, with $\psi^0 \leq 0$, such that

20 $$\left\langle \psi, \frac{\partial F(\hat{z})}{\partial z} \delta z \right\rangle \leq 0 \quad \text{for all } \delta z \in \overline{IC}(\hat{z},\Omega),$$

where, by theorem (4), $\overline{IC}(\hat{z},\Omega) = \{\delta z: \langle \nabla q^i(\hat{z}), \delta z \rangle \leq 0 \text{ for all } i \in I(\hat{z})\}$ and $F(z) = (f(z), r^1(z), r^2(z), \ldots, r^m(z))$. Alternatively, with (20) in expanded form, ψ satisfies

21 $$\left\langle \left(\psi^0 \nabla f(\hat{z}) + \sum_{i=1}^{m} \psi^i \nabla r^i(\hat{z}) \right), \delta z \right\rangle \leq 0 \quad \text{for all } \delta z$$
$$\in \overline{IC}(\hat{z},\Omega). \quad \square$$

Using this theorem and Farkas' lemma (A.5.34), we obtain necessary conditions for optimality in the more familiar form of a multiplier rule, as shown below.

22 **Theorem.** If $\hat{z}$ is an optimal solution to the nonlinear programming problem (1.1) and assumption (16) is satisfied, then there exist a nonzero vector $\psi = (\psi^0, \psi^1, \ldots, \psi^m)$ in E^{m+1}, with $\psi^0 \leq 0$, and a vector $\mu = (\mu^1, \mu^2, \ldots, \mu^k)$ in E^k, with $\mu^i \leq 0$ for $i = 1, 2, \ldots, k$, such that

23 $$\psi^0 \nabla f(\hat{z}) + \sum_{i=1}^{m} \psi^i \nabla r^i(\hat{z}) + \sum_{i=1}^{k} \mu^i \nabla q^i(\hat{z}) = 0$$

and

24 $$\mu^i q^i(\hat{z}) = 0 \quad i = 1, 2, \ldots, k.$$

Proof. From theorem (19), there exists a nonzero vector $\psi = (\psi^0, \psi^1, \ldots, \psi^m)$ in E^{m+1}, with $\psi^0 \leq 0$, such that

25 $$\left\langle \left(\psi^0 \nabla f(\hat{z}) \sum_{i=1}^{m} \psi^i \nabla r^i(\hat{z}) \right), \delta z \right\rangle \leq 0$$

for all δz such that $\langle \nabla q^i(\hat{z}), \delta z \rangle \leq 0$, with $i \in I(\hat{z})$. It therefore follows from Farkas' lemma that there exist scalars $\mu^i \leq 0$, with $i \in I(\hat{z})$, such that

26 $$\psi^0 \nabla f(\hat{z}) + \sum_{i=1}^{m} \psi^i \nabla r^i(\hat{z}) + \sum_{i \in I(\hat{z})} \nabla q^i(\hat{z}) = 0.$$

Let $\mu^i = 0$ for $i \in \bar{I}(\hat{z})$, the complement of $I(\hat{z})$ in $\{1,2,\ldots,k\}$; then the vectors $\psi = (\psi^0, \psi^1, \ldots, \psi^m)$ in E^{m+1} and $\mu = (\mu^1, \mu^2, \ldots, \mu^k)$ in E^k satisfy (23) and (24). □

Most of the other well-known necessary conditions for nonlinear programming problems can be obtained from theorem (22) by means of additional assumptions on the functions r and q. For example, the following corollaries are immediate consequences.

27 **Corollary.** If assumption (16) is satisfied and the vectors $\nabla r^i(\hat{z})$ for $i = 1, 2, \ldots, m$ are linearly independent, then any vectors $\psi \in E^{m+1}$ and $\mu \in E^k$ which satisfy the conditions of theorem (22) are such that $(\psi^0, \mu) \neq 0$.

Proof. Let $\psi = (\psi^0, \psi^1, \ldots, \psi^m)$ and $\mu = (\mu^1, \mu^2, \ldots, \mu^k)$ be vectors satisfying the conditions of theorem (22). Then, by assumption, $\psi \neq 0$. Hence if $\psi^0 = 0$, then $\sum_{i=1}^{m} \psi^i \nabla r^i(\hat{z}) \neq 0$, since not all $\psi^i = 0$ and the vectors $\nabla r^i(\hat{z})$ for $i = 1, 2, \ldots, m$ are linearly independent. It therefore follows that if $\psi^0 = 0$, then $\mu \neq 0$. Now suppose that $\mu = 0$; clearly, we cannot have $\psi^0 = 0$ and $\psi \neq 0$ and still have equation (23) satisfied. We therefore conclude that $(\psi^0, \mu) \neq 0$. □

28 **Corollary.** If $\nabla r^i(\hat{z})$ for $i = 1, 2, \ldots, m$, together with $\nabla q^i(\hat{z})$ for $i \in I(\hat{z})$, are linearly independent vectors, then any vector $\psi \in E^{m+1}$ satisfying the conditions of theorem (22) also satisfies $\psi^0 < 0$ [cf (3.17)].

Proof. Let $\psi = (\psi^0, \psi^1, \ldots, \psi^m)$ and $\mu = (\mu^1, \mu^2, \ldots, \mu^k)$ be vectors satisfying the conditions of theorem (22). Since $\psi \neq 0$, not all $\psi^i = 0$, with $i = 1, 2, \ldots, m$. Consequently, since the vectors $\nabla r^i(\hat{z})$ and $\nabla q^i(\hat{z})$ appearing in (23) (with nonzero multipliers) are linearly independent, it follows that if $\psi^0 = 0$, then

29
$$\psi^0 \nabla f(\hat{z}) + \sum_{i=1}^{m} \psi^i \nabla r^i(\hat{z}) + \sum_{i=1}^{k} \mu^i \nabla q^i(\hat{z}) \neq 0,$$

which contradicts (23). Hence $\psi^0 \neq 0$, and therefore $\psi^0 < 0$. □

30 **Remark.** It is clear that when $\psi^0 = 0$ in (23), the necessary condition expressed by (23) does not involve the cost function $f(\cdot)$ and hence cannot carry too much information. Indeed, if we find that $\psi^0 = 0$ for *all* vectors ψ satisfying theorem (22), then we are faced with a degeneracy in the problem (see Section 2.3). □

This remark points out the importance of the preceding corollary, as well as that of the following ones.

31 Corollary. If there exists a nonzero vector $v \in IC(\hat{z},\Omega)$ such that $\langle \nabla r^i(\hat{z}),v \rangle = 0$ for $i = 1, 2, \ldots, m$, and the vectors $\nabla r^1(\hat{z}), \nabla r^2(\hat{z}), \ldots, \nabla r^m(\hat{z})$ are linearly independent, then any vector $\psi \in E^{m+1}$ satisfying the conditions of theorem (22) also satisfies $\psi^0 < 0$ [cf (3.26)].

Proof. Let $\psi \in E^{m+1}$ and $\mu \in E^k$ be any vectors satisfying (23) and (24), and let $0 \neq v \in IC(\hat{z},\Omega)$ be any vector satisfying $\langle \nabla r^i(\hat{z}),v \rangle = 0$ for $i = 1, 2, \ldots, m$. Then we get from (23) that

$$\text{32} \qquad \psi^0 \langle \nabla f(\hat{z}),v \rangle + \sum_{i=1}^{m} \psi^i \langle \nabla r^i(\hat{z}),v \rangle + \sum_{i=1}^{k} \mu^i \langle \nabla q^i(\hat{z}),v \rangle = 0.$$

Eliminating in (32) the terms which are obviously zero, we get

$$\psi^0 \langle \nabla f(\hat{z}),v \rangle + \sum_{i=1}^{k} \mu^i \langle \nabla q^i(\hat{z}),v \rangle = 0.$$

It follows from corollary (27) that $(\psi^0,\mu) \neq 0$, since the vectors $\nabla r^i(\hat{z})$ for $i = 1, 2, \ldots, m$ are linearly independent. Therefore, if $\mu = 0$, then $\psi^0 < 0$. Alternatively, suppose that $\mu \neq 0$; then, since $0 \neq v \in IC(\hat{z},\Omega)$ and $\mu^i \leq 0$, it is clear that $\Sigma_{i=1}^{k} \mu^i \langle \nabla q^i(\hat{z}),v \rangle > 0$. This implies that $\psi^0 < 0$, and thus the lemma is proved. □

33 Corollary. If the functions $r^i(\cdot)$ for $i = 1, 2, \ldots, m$ and $q^i(\cdot)$ for $i = 1, 2, \ldots, k$ are affine, then there exists a vector $\psi \in E^{m+1}$ satisfying the conditions of theorem (22) which also satisfies $\psi^0 < 0$ [cf (3.19)].

Proof. If the functions $q^i(\cdot)$ for $i = 1, 2, \ldots, k$ are affine, then, by exercise (14), we may omit assumption (16) in theorem (19), since in place of the internal cone we may use the radial cone

$$RC(\hat{z},\Omega) = \{\delta z : \langle \nabla q^i(\hat{z}),\delta z \rangle \leq 0 \text{ for all } i \in I(\hat{z})]\}.$$

Since $RC(\hat{z},\Omega)$ is a closed convex polyhedron and $\partial F(\hat{z})/\partial z$ is a linear map, the set

$$K(\hat{z}) = \left\{ \frac{\partial F(\hat{z})}{\partial z} \delta z : \delta z \in RC(\hat{z},\Omega) \right\}$$

is closed (see, e.g., ref. [6]). Thus, by corollary (2.2.45), we may choose $\psi^0 < 0$. □

When the set $\Omega' = \{z: r(z) = 0,\ q(z) \leq 0\}$ has even more specialized structure than is implied by assumption (16), it is sometimes possible to recast theorem (19) as a "maximum principle." The reader may explore this possibility in the exercises below.

34 Definition. We say that a set Ω is *star shaped* at $\hat{z} \in \Omega$ if for every $z \in \Omega$ $(\lambda z + (1 - \lambda)\hat{z}) \in \Omega$, with $0 \leq \lambda \leq 1$. □

35 Exercise. Let $\hat{z}$ be a solution of the nonlinear programming problem (1.1), and suppose that assumption (16) is satisfied. Show that when the see $\Omega = \{z: q(z) \leq 0\}$ is *star shaped* at $\hat{z}$, then (21) is equivalent to the maximum principle

36 $$\left\langle \left(\psi^0 \nabla f(\hat{z}) + \sum_{i=1}^{m} \psi^i \nabla r^i(\hat{z})\right), \hat{z}\right\rangle \geq \left\langle \left(\psi^0 \nabla f(\hat{z}) + \sum_{i=1}^{m} \psi^i \nabla r^i(\hat{z})\right), z\right\rangle \quad \text{for all } z \in \Omega. \quad \Box$$

37 Exercise. Suppose that the set $\Omega' = \{z: r(z) = 0,\ q(z) \leq 0\}$ is *star shaped* at $\hat{z}$, a solution of the nonlinear programming problem (1.1). Show that

38 $$\langle -\nabla f(\hat{z}), \hat{z}\rangle \geq \langle -\nabla f(\hat{z}), z\rangle \qquad \text{for all } z \in \Omega'$$

independently of whether (16) is satisfied or not. [*Hint:* Use Theorem (2.2.12).] □

We now go on to nonlinear programming problems whose constraint sets are not assumed to satisfy a constraint qualification.

3.5 A FURTHER GENERALIZATION

In the preceding section we assumed that the internal cone $IC(\hat{z}, \Omega)$ [see (4.3)] was not the origin, and we then established the multiplier rule, theorem (4.22). That is, we showed that if $\hat{z}$ is an optimal solution to the nonlinear programming problem (1.1), then there exist Lagrange multipliers satisfying (4.23) and (4.24). The multipliers associated with the equality constraints were of indefinite sign, while the multipliers associated with the inequalities were nonpositive. We shall now show that even when assumption (4.16) is not satisfied, we can still get a condition of the same form as theorem (4.24), but it now carries considerably less information.

The necessary condition we are about to discuss was first proved by Mangasarian and Fromovitz [5], using the implicit-function theorem and a lemma by Motzkin. We begin by showing that whenever assumption (4.16) is not satisfied, it is possible to sum the vectors $\nabla q^i(\hat{z})$, with $i \in I(\hat{z})$ [see (4.1)], to zero with nonpositive scalars, not all of which are zero. This is established in the following lemma.

1   Lemma   Suppose that assumption (4.16) is not satisfied for the set $\Omega = \{z\colon q(z) \leqq 0\}$. Then there exists a nonzero vector $\mu \in E^k$, with $\mu^i \leqq 0$ for $i = 1, 2, \ldots, k$, such that

$$2 \qquad \sum_{i=1}^{k} \mu^i \, \nabla q^i(\hat{z}) = 0$$

and

$$3 \qquad \mu^i q^i(\hat{z}) = 0 \qquad i = 1, 2, \ldots, k.$$

Proof.   Suppose that $I(\hat{z}) = \{i_1, i_2, \ldots, i_\alpha\}$ and consider the linear subspace of E^α,

$$4 \qquad L = \{v\colon v = (\langle \nabla q^{i_1}(\hat{z}), h \rangle, \ldots, \langle \nabla q^{i_\alpha}(\hat{z}), h \rangle), \; h \in E^n\}.$$

By hypothesis, L has no rays in common with the convex cone

$$5 \qquad C = \{v = (v^1, v^2, \ldots, v^\alpha)\colon v^j < 0, j = 1, 2, \ldots, \alpha\}.$$

Hence L can be separated from $\bar{C}$; that is, there exists a nonzero vector $\beta \in E^\alpha$ such that

$$6 \qquad \langle \beta, v \rangle \geqq 0 \qquad \text{for all } v \in \bar{C}$$
$$7 \qquad \langle \beta, v \rangle \leqq 0 \qquad \text{for all } v \in L.$$

It is obvious from (6) that $\beta \leqq 0$. Since L is a linear subspace, $\langle \beta, v \rangle = 0$ for all $v \in L$, which implies that

$$8 \qquad \Big\langle \sum_{j=1}^{\alpha} \beta^j \, \nabla q^{i_j}(\hat{z}), h \Big\rangle = 0 \qquad \text{for all } h \in E^n,$$

and therefore

$$9 \qquad \sum_{j=1}^{\alpha} \beta^j \, \nabla q^{i_j}(\hat{z}) = 0.$$

If we now let $\mu^i = \beta^j$ when $i = i_j \in I(\hat{z})$ and $\mu^i = 0$ when $i \in \bar{I}(\hat{z})$, then $\mu = (\mu^1, \mu^2, \ldots, \mu^k)$ is the desired vector.   □

10   Corollary.   A sufficient condition for assumption (4.16) to be satisfied is that the vectors $\nabla q^i(\hat{z})$, with $i \in I(\hat{z})$, be linearly independent.   □

Lemma (1) may now be combined with theorem (4.19) to give a necessary condition for optimality which does not require that

(4.16) be satisfied. Thus for the most general case of the nonlinear programming problem (1.1) we obtain the following necessary condition for optimality.

11   **Theorem.** If $\hat{z}$ is an optimal solution to the nonlinear programming problem (1.1), then there exists a vector $\psi = (\psi^0, \psi^1, \ldots, \psi^m)$ in E^{m+1} and a vector $\mu = (\mu^1, \mu^2, \ldots, \mu^k)$ in E^k, with $\psi^0 \leq 0$ and $\mu^i \leq 0$ for $i = 1, 2, \ldots, k$, and with ψ and μ not both zero, such that

12
$$\psi^0 \nabla f(\hat{z}) + \sum_{i=1}^{m} \psi^i \nabla r^i(\hat{z}) + \sum_{i=1}^{k} \mu^i \nabla q^i(\hat{z}) = 0$$

13 $\mu^i q^i(\hat{z}) = 0 \quad i = 1, 2, \ldots, k.$

Proof. If (4.16) is satisfied, then theorem (11) follows directly from theorem (4.22) and is seen to be a slightly weaker statement of theorem (4.22). If (4.16) is not satisfied, let μ be the vector specified in lemma (1), and let $\psi = 0$. □

14   **Remark.** Theorem (11) is essentially trivial when assumption (4.16) is not satisfied, because, by lemma (1), when (4.16) is not satisfied, we can find a $\mu \neq 0$ satisfying (2) and (3) without depending in any way on the optimality of $\hat{z}$, and we can then satisfy (12) and (13) with this μ and $\psi = 0$. □

Under suitable assumptions equation (12) may be converted into a maximum principle, and it is left to the reader to develop this idea in the following exercises.

15   **Exercise.** Consider the nonlinear programming problem (1.1). Let $Q = \{q^{i_1}, q^{i_2}, \ldots, q^{i_j}\}$ be any subset of components of the inequality constraint function q with the property that the see $\Omega_1 = \{z: q^{i_\alpha}(z) \leq 0, \alpha = 1, 2, \ldots, j\}$ has a nonempty interior cone $IC(\hat{z}, \Omega_1)$ at $\hat{z}$, an optimal solution to problem (1.1). Renumber the functions $q^1, q^2, \ldots, q^k$ so that $q^i \notin Q$ for $i = 1, 2, \ldots, l$, with $l = k - j$, and $q^i \in Q$ for $i = l+1, l+2, \ldots, k$; that is, $\Omega_1 = \{z: q^i(z) \leq 0, i = l+1, l+2, \ldots, k\}$. Show that if $\hat{z}$ is an optimal solution to (1.1), then there exist a vector $\psi = (\psi^0, \psi^1, \ldots, \psi^m)$ in E^{m+1}, with $\psi^0 \leq 0$, and a vector $\mu = (\mu^1, \mu^2, \ldots, \mu^l)$ in E^l, with $\mu^i \leq 0$ for $i = 1, 2, \ldots, l$, such that $(\psi, \mu) \neq 0$ and

16
$$\left\langle \left(\psi^0 \nabla f(\hat{z}) + \sum_{i=1}^{m} \psi^i \nabla r^i(\hat{z}) + \sum_{i=1}^{l} \mu^i \nabla q^i(\hat{z}) \right), \delta z \right\rangle$$
$$\leq 0 \quad \text{for all } \delta z \in IC(\hat{z}, \Omega_1)$$

17 $\mu^i q^i(\hat{z}) = 0 \quad i = 1, 2, \ldots, l.$

Show that if, in addition, the set Ω_1 is star shaped at $\hat{z}$, [that is, for all $z \in \Omega_1$, $(\lambda z + (1 - \lambda)\hat{z}) \in \Omega_1$, with $0 \leq \lambda \leq 1$], then (16) becomes

18 $$\langle \psi^0 \nabla f(\hat{z}) + \sum_{i=1}^{m} \psi^i \nabla r^i(\hat{z}) + \sum_{i=1}^{l} \mu^i \nabla q^i(\hat{z}), z - \hat{z} \rangle \leq 0 \quad \text{for all } z \in \Omega_1,$$

which again is a condition in the form of a maximum principle. □

19 Remark. Finally, note that if we let $r \equiv 0$ in problem (1.1), then theorem (11) becomes the well-known *Fritz John necessary condition for optimality* [1]. □

We have thus seen that most of the known necessary conditions for nonlinear programming problems can be obtained simply by applying the Fundamental Theorem (2.3.12) and Farkas' lemma (A.5.34).

3.6 A SUFFICIENT CONDITION

Whenever a continuously differentiable real-valued function is convex or concave, we find that it assumes its unique extreme value, if it has one, at every point at which it is stationary. We are therefore led to believe that the necessary conditions expressed by theorem (3.5) may become sufficient whenever all the functions entering the description of the nonlinear programming problem (1.1) are convex or concave.

1 Theorem. Consider the nonlinear programming problem (1.1) [minimize $f(z)$ subject to $r(z) = 0$ and $q(z) \leq 0$], and suppose that $f: E^n \to E^1$ is convex, that $r: E^n \to E^m$ is affine, and that $q^i: E^n \to E^1$ for $i = 1, 2, \ldots, k$ (the components of the function $q: E^n \to E^k$) are convex. If $\hat{z} \in E^n$ is a feasible solution to (1.1) [that is, $r(\hat{z}) = 0$ and $q(\hat{z}) \leq 0$], and if there exist vectors $\lambda = (\lambda^1, \lambda^2, \ldots, \lambda^m)$ in E^m and $\mu = (\mu^1, \mu^2, \ldots, \mu^k)$ in E^k, with $\mu^i \leq 0$ for $i = 1, 2, \ldots, k$, such that

2 $$-\nabla f(\hat{z}) + \sum_{i=1}^{m} \lambda^i \nabla r^i(\hat{z}) + \sum_{i=1}^{k} \mu^i \nabla q^i(\hat{z}) = 0$$

and

3 $\mu^i q^i(\hat{z}) = 0 \quad i = 1, 2, \ldots, k,$

then $\hat{z}$ is an optimal solution.

SEC. 3.6 A SUFFICIENT CONDITION

Proof. Suppose $\hat{z}$ is a feasible solution for which there exist $\lambda \in E^m$ and $\mu \in E^k$ satisfying (2) and (3). Consider the function $L: E^n \to E^1$, defined by

(4) $$L(z) = -f(z) + \sum_{i=1}^{m} \lambda^i r^i(z) + \sum_{i=1}^{k} \mu^i q^i(z).$$

By assumption, the functions $f(\cdot)$ and $q^i(\cdot)$ for $i = 1, 2, \ldots, k$ are convex and the functions $r^i(\cdot)$ for $i = 1, 2, \ldots, m$ are affine. Furthermore, by assumption, $\mu^i \leq 0$ for $i = 1, 2, \ldots, k$. Consequently, the function $L(\cdot)$ is concave, and therefore [see theorem (A.6.37)] it assumes its maximum value at its stationary point $\hat{z}$; that is, for all $z \in E^n$

(5) $$-f(\hat{z}) + \sum_{i=1}^{m} \lambda^i r^i(\hat{z}) + \sum_{i=1}^{k} \mu^i q^i(\hat{z})$$
$$\geq -f(z) + \sum_{i=1}^{m} \lambda^i r^i(z) + \sum_{i=1}^{k} \mu^i q^i(z).$$

Since (5) is true for all $z \in E^n$, it must be true for all z satisfying $r(z) = 0$ and $q(z) \leq 0$. Since, by (3), $\mu^i q^i(\hat{z}) = 0$ for $i = 1, 2, \ldots, k$, it follows that for all z satisfying $r(z) = 0$ and $q(z) \leq 0$

(6) $$-f(\hat{z}) \geq -f(z) + \sum_{i=1}^{k} \mu^i q^i(z).$$

But $q^i(z) \leq 0$ and $\mu^i \leq 0$ for $i = 1, 2, \ldots, k$, and hence

(7) $$-f(\hat{z}) \geq -f(z) \quad \text{for all } z \in \{z: r(z) = 0, q(z) \leq 0\},$$

which proves that $\hat{z}$ is an optimal solution to the nonlinear programming problem (1.1). □

(8) **Corollary.** Suppose that the functions f, r, and q are as in theorem (1). If $\hat{z}$ is a feasible solution and the internal cone $IC(\hat{z},\Omega)$, with $\Omega = \{z: q(z) \leq 0\}$, is not the origin [see (4.3)], and there exists a vector $\lambda = (\lambda^1, \lambda^2, \ldots, \lambda^m)$ in E^m such that

(9) $$\left\langle \left(-\nabla f(\hat{z}) + \sum_{i=1}^{m} \lambda^i \nabla r^i(\hat{z})\right), \delta z \right\rangle \leq 0 \quad \text{for all } \delta z \in \overline{IC}(\hat{z}, \Omega)$$

then $\hat{z}$ is an optimal solution to the nonlinear programming problem (1.1).

Proof. By assumption, (9) is satisfied for all δz such that

(10) $$\langle \nabla q^i(\hat{z}), \delta z \rangle \leq 0 \quad i \in I(\hat{z})$$

[see (4.3) and (4.5)]. Hence, by Farkas' lemma (A.5.34), there exists a vector $\mu = (\mu^1, \mu^2, \ldots, \mu^k)$ in E^k, with $\mu^i \leq 0$ for $i = 1, 2, \ldots, k$, such that

11
$$-\nabla f(\hat{z}) + \sum_{i=1}^{m} \lambda^i \nabla r^i(\hat{z}) + \sum_{i=1}^{k} \mu^i \nabla q^i(\hat{z}) = 0$$

and

12
$$\mu^i q^i(\hat{z}) = 0 \qquad i = 1, 2, \ldots, k.$$

Hence (9) implies (2) and (3). □

13 **Exercise.** Give an independent proof for corollary (8). □

It is possible to relax somewhat the assumptions made in theorem (1). The reader might explore this possibility by proving theorem (20).

An Extension to Quasi-convex and pseudoconvex Functions

In the case the reader is not familiar with quasi-convex and pseudoconvex functions, let us begin with their definitions. Incidentally, the concepts of quasi- and pseudo-convexity suffer considerably from the fact that in general it is extremely difficult to establish whether a given function is quasi-convex or pseudoconvex.

14 **Definition.** A function $g(\cdot)$ mapping E^n into the reals is said to be *quasi-convex* if the set

15
$$S_\alpha = \{z : g(z) \leq \alpha\}$$

is convex for all real α. □

16 **Definition.** A continuously differentiable function $g(\cdot)$ mapping E^n into the reals is said to be *pseudoconvex* if

17
$$g(z_1) \geq g(z_2)$$

for every pair (z_1, z_2) satisfying

18
$$\langle \nabla g(z_2), z_1 - z_2 \rangle \geq 0. \quad \square$$

19 **Exercise.** Prove that if $\hat{z}$ is a feasible solution for the nonlinear programming problem (1.1), $f(\cdot)$ is pseudoconvex and $\nabla f(\hat{z}) = 0$, then $\hat{z}$ is optimal. □

20 **Exercise.** Prove the following theorem.

21 **Theorem.** Consider the nonlinear programming problem (1.1). Suppose $\hat{z}$ is a feasible solution, $\nabla f(\hat{z}) \neq 0$, $f(\cdot)$ is quasi-convex, $r(\cdot)$ is

affine, the set $\Omega = \{z: q(z) \leq 0\}$ is star shaped at $\hat{z}$, and that there exist multipliers $\psi^1, \psi^2, \ldots, \psi^m$, and $\mu^1, \mu^2, \ldots, \mu^k$ with $\mu^i \leq 0$ $i = 1, 2, \ldots, k$, such that

22
$$-\nabla f(\hat{z}) + \sum_{i=1}^{m} \psi^i \nabla r^i(\hat{z}) + \sum_{i=1}^{k} \mu^i \nabla q^i(\hat{z}) = 0$$

$$\mu^i q^i(\hat{z}) = 0 \quad i = 1, 2, \ldots, k$$

Then $\hat{z}$ is an optimal solution.

Hint: The proof should be constructed in two steps. First, show that $r(\cdot)$ affine and Ω star-shaped at $\hat{z}$, in conjunction with (23), gives the condition

23
$$\langle \nabla f(\hat{z}), z - \hat{z} \rangle \geq 0$$

for any feasible z. Next show that for any $f(\cdot)$, quasi-convex and differentiable,

24
$$\langle \nabla f(\hat{z}), z - \hat{z} \rangle \leq 0 \quad \text{for all } z \in \{z: f(z) \leq f(\hat{z})\},$$

and that, as a result, if $\nabla f(\hat{z}) \neq 0$,

25
$$f(z) > f(\hat{z}) \quad \text{for all } z \in \{z: \langle \nabla f(\hat{z}), z - \hat{z} \rangle > 0\}.$$

Finally, for any z feasible, use the sequence $z_i = z + \dfrac{1}{i} \nabla f(\hat{z}), i = 1, 2, \ldots$, together with (24) and (26) to show that $f(z) \geq f(\hat{z})$. □

This concludes our discussion of nonlinear programming problems, and we now proceed to optimal control problems, which, by their very nature, have considerably more structure, and hence gives rise to more complex conditions of optimality.

REFERENCES

1. F. John: Extremum Problems with Inequalities as Side Conditions, in K. O. Friedrichs, O. W. Neugebauer, and J. J. Stoker (eds.), "Studies and Essays: Courant Anniversary Volume," pp. 187–204, Interscience Publishers, Inc., New York, 1948.
2. H. W. Kuhn and A. W. Tucker: Nonlinear Programming, *Proc. Second Berkeley Symp. Math. Stat. Probability*, pp. 481–492, University of California Press, Berkeley, Calif., 1951.
3. K. J. Arrow, L. Hurwicz, and H. Uzawa, Constraint Qualifications in Maximization Problems, *Naval Res. Logistics Quart.*, 8:175–191 (1961).
4. K. J. Arrow, L. Hurwicz, and H. Uzawa (eds.): "Studies in Linear and Nonlinear Programming, "Stanford Mathematical Studies in the Social Sciences, II, Stanford University Press, Stanford, Calif., 1958.

5. O. L. Mangasarian and S. Fromovitz: The Fritz John Necessary Optimality Conditions in the Presence of Equality and Inequality Constraints, *Shell Development Company Paper* P1433, 1965.
6. A. J. Goldman: Resolution and Separation Theorems for Polyhedral Convex Sets, *Ann. Math. Studies* 38, pp. 41–51, Princeton University Press, Princeton, N.J., 1956.
7. W. F. Fleming: "Functions of Several Variables," McGraw-Hill Book Company, New York, 1965.

4
Discrete optimal control problems

4.1 THE GENERAL CASE

Discrete optimal control problems were first defined in Section 1.2, and in Section 1.5 it was shown that fixed-time discrete optimal problems can be transcribed into the form of the Basic Problem (1.4.1), for which necessary conditions are given by the Fundamental Theorem (2.3.12). Consequently, theorem (2.3.12) also gives necessary conditions for the optimality of a solution to any fixed-time discrete optimal control problem whose defining quantities satisfy the assumptions of this theorem. However, as in the case of nonlinear programming problems, the considerable structure of optimal control problems may lead to substantially more detailed necessary conditions than either the original statement of the Fundamental Theorem (2.3.12) or theorem (3.4.22).

To illustrate this fact, let us examine a very general optimal control problem whose formulation is governed by the following two considerations. First, we want to obtain necessary conditions in separated form, that is conditions stated in terms of the time index i (with $i = 0, 1, \ldots, k - 1$), the optimal state $\hat{x}_i$, and the optimal control $\hat{u}_i$, at time i only. This precludes trajectory constraints of the form $h(\mathfrak{X},\mathfrak{U}) \in D$ [see (1.2.4)] which cannot be separated into the form $x_i \in X_i$ for $i = 0, 1, \ldots, k$ and $u_i \in U_i$ for $i = 0, 1, \ldots, k-1$. Second, we wish to obtain easily specifiable transversality conditions, and therefore we can allow only differentiable equality and inequality constraints on the state vectors. These restrictions lead to the following particular case of the discrete optimal control problem (1.2.8).

1 Optimal control problem. Consider a dynamical system described by the difference equation

2
$$x_{i+1} - x_i = f_i(x_i, u_i) \quad \text{for } i = 0, 1, \ldots, k-1,$$

where $x_i \in E^n$ is the state of the system at time i, $u_i \in E^m$ is the input to the system at time i, and $f_i(\cdot,\cdot)$ is a function mapping $E^n \times E^m$ into E^n. Find a control sequence $\hat{\mathfrak{u}} = (\hat{u}_0, \hat{u}_1, \ldots, \hat{u}_{k-1})$ and a corresponding trajectory $\mathfrak{X} = (\hat{x}_0, \hat{x}_1, \ldots, \hat{x}_k)$, determined by (2), which minimize the sum

$$\sum_{i=0}^{k-1} f_i^0(x_i, u_i),$$

where the $f_i^0(\cdot,\cdot)$ map $E^n \times E^m$ into the reals.

This minimization is subject to the following constraints, which we write as the intersection of inequality and equality constraints whenever appropriate. The *control constraints* are

3
$$u_i \in U_i \subset E^m \quad i = 0, 1, \ldots, k-1.$$

the *initial boundary constraints* are

4
$$x_0 \in X_0 = X_0' \cap X_0'' \quad X_0' = \{x : q_0(x) \leqq 0\}, X_0''$$
$$= \{x : g_0(x) = 0\},$$

where $q_0(\cdot)$ maps E^n into E^{m_0} and $g_0(\cdot)$ maps E^n into E^{l_0}. The *terminal boundary constraints* are

5
$$x_k \in X_k = X_k' \cap X_k'' \quad X_k' = \{x : q_k(x) \leqq 0\}, X_k''$$
$$= \{x : g_k(x) = 0\},$$

where $q_k(\cdot)$ maps E^n into E^{m_k} and $g_k(\cdot)$ maps E^n into E^{l_k}. The *state-space constraints* are

6
$$x_i \in X_i = X_i' \cap E^n = \{x : q_i(x) \leqq 0\} \quad i = 1, 2, \ldots, k-1,$$

where $q_i(\cdot)$ maps E^n into E^{m_i}.

(Note that this formulation allows the initial and final sets, X_0 and X_k, to be manifolds, manifolds with boundaries, or "solid figures," i.e., sets with nonempty interiors. The state-space-constraint sets X_i, with $i = 1, 2, \ldots, k - 1$, usually have nonempty interiors.)

To complete the formulation of the problem we now introduce some assumptions on the various sets and functions above:

7   For $i = 0, 1, \ldots, k - 1$ the functions $f_i \colon E^n \times E^m \to E^n$ and $f_i^0 \colon E^n \times E^m \to E^1$ are continuously differentiable.

8   For every $u \in U_i$ and $i \in \{0, 1, \ldots, k - 1\}$ there exists a conical approximation $C(u, U_i)$ [see (2.3.1) and (2.3.3)] for the set U_i at u.

9   The functions $g_0 \colon E^n \to E^{l_0}$ and $g_k \colon E^n \to E^{l_k}$ are continuously differentiable, and *unless they are identically zero*, they have jacobian matrices $\partial g_0(x)/\partial x$ and $\partial g_k(x)/\partial x$ which are of maximum rank[1] for all x in X_0 and X_k, respectively.

10   For all $x_i \in X_i'$ and $i = 0, 1, \ldots, k$ the gradients of the active inequality constraints (that is, $\nabla q_i^j(x_i)$, with $j \in \{j \colon q_i^j(x_i) = 0, j \in \{1, 2, \ldots, l^i\}\}$) are linearly independent vectors.[2]   ☐

As we saw in Sec. 1.5, this problem may be recast in the form of the Basic Problem (1.4.1) [minimize $f(z)$ subject to $r(z) = 0$ and $z \in \Omega$] by making the following identifications. Let $z = (x_0, x_1, \ldots, x_k, u_0, u_1, \ldots, u_{k-1}) \in E^{(k+1)n + km}$ and let f, r, and Ω be defined by

11   $$f(z) = \sum_{i=0}^{k-1} f_i^0(x_i, u_i)$$

12   $$r(z) = \begin{bmatrix} x_1 - x_0 - f_0(x_0, u_0) \\ \cdots\cdots\cdots\cdots\cdots\cdots \\ x_k - x_{k-1} - f_{k-1}(x_{k-1}, u_{k-1}) \\ g_0(x_0) \\ g_k(x_k) \end{bmatrix}$$

13   $$\Omega = X_1' \times X_1' \times X_2' \times \cdots \times X_{k-1}' \times X_k' \times U_0 \times U_1 \times \cdots \times U_{k-1}.$$

Clearly, f and r have the required differentiability properties. Now, suppose that $\hat{z} \in \Omega$ is an optimal solution to the optimal control

---

[1] When this assumption is not satisfied it is possible for all the costates $p_0, p_1, \ldots, p_k$ and the scalar p^0 appearing in theorem (17) to be zero, in which case this theorem becomes trivial.

[2] This assumption guarantees the existence of conical approximations of the first kind for the sets X_i' for $i = 0, 1, \ldots, k$ [see corollary (3.5.10)].

problem under consideration. Then the cone

(14) $$C(\hat{z},\Omega) = IC(\hat{x}_0, X'_0) \times IC(\hat{x}_1, X'_1) \times \cdots \times IC(\hat{x}_{k-1}, X'_{k-1}) \\ \times IC(\hat{x}_k, X'_k) \times C(\hat{u}_0, U_0) \times \cdots \times C(\hat{u}_{k-1}, U_{k-1}),$$

which is the cartesian product of internal cones, first defined in (3.4.2), and of conical approximations, which exist by assumption (8), is obviously a conical approximation for the set Ω at $\hat{z}$. This follows from the fact that assumption (10) and corollary (3.5.10) guarantee that the cones $IC(\hat{x}_i, X'_i)$, with $i = 0, 1, \ldots, k$, are nonempty, while theorem (3.4.4) ensures that they are conical approximations of the first kind. We may therefore apply the Fundamental Theorem (2.3.12) to the optimal control problem under consideration and conclude that if $\hat{z}$ is an optimal solution, then there exists a nonzero vector $\psi = (p^0, \pi)$, with $p^0 \leq 0$ and $\pi = (-p_1, -p_2, \ldots, -p_k, \mu_0, \mu_k)$, where $p_i \in E^n$, $\mu_0 \in E^{l_0}$, and $\mu_k \in E^{l_k}$, such that, with $F = (f, r)$,

(15) $$\left\langle \psi, \frac{\partial F(\hat{z})}{\partial z} \delta z \right\rangle = p^0 \langle \nabla f(\hat{z}), \delta z \rangle + \left\langle \pi, \frac{\partial r(\hat{z})}{\partial z} \delta z \right\rangle \\ \leq 0 \qquad \text{for all } \delta z \in \check{C}(\hat{z}, \Omega).$$

Substituting for f and r in (15) from (11) and (12) and expanding, we get

(16) $$p^0 \left(\sum_{i=0}^{k-1} \left\langle \frac{\partial f_i^0(\hat{x}_i, \hat{u}_i)}{\partial x}, \delta x_i \right\rangle + \sum_{i=0}^{k-1} \left\langle \frac{\partial f_i^0(\hat{x}_i, \hat{u}_i)}{\partial u}, \delta u_i \right\rangle \right) \\ + \sum_{i=0}^{k-1} \left\langle -p_{i+1}, \left[\delta x_{i+1} - \delta x_i - \frac{\partial f_i(\hat{x}_i, \hat{u}_i)}{\partial x} \delta x_i - \frac{\partial f_i(\hat{x}_i, \hat{u}_i)}{\partial u} \delta u_i \right] \right\rangle \\ + \left\langle \mu_0, \frac{\partial g_0(\hat{x}_0)}{\partial x} \delta x_0 \right\rangle + \left\langle \mu_k, \frac{\partial g_k(\hat{x}_k)}{\partial x} \delta x_k \right\rangle \leq 0$$

for every $\delta z = (\delta x_0, \ldots, \delta x_k, \delta u_0, \ldots, \delta u_{k-1}) \in \bar{C}(\hat{z},\Omega)$.

This condition can be refined by considering special cases for the vector $\delta z \in \bar{C}(\hat{z},\Omega)$. The result is the theorem below, which converts condition (16) into a classical form involving conditions on a hamiltonian at each sampling instant $i = 0, 1, \ldots, k-1$, with the hamiltonian defined by means of *costate vectors* which satisfy an adjoint equation depending on the state-space constraints and transversality conditions.

(17) **Theorem.** If $\hat{z} = (\hat{x}_0, \hat{x}_1, \ldots, \hat{x}_k, \hat{u}_0, \hat{u}_1, \ldots, \hat{u}_{k-1})$ is an optimal solution to optimal control problem (1), then there exist *costate vectors* $p_0, p_1, \ldots, p_k$ in E^n, multiplier vectors $\lambda_0, \lambda_1, \ldots, \lambda_k$,

with $\lambda_i \in E^{m_i}$ satisfying $\lambda_i \leq 0$, multiplier vectors $\mu_0 \in E^{l_0}$ and $\mu_k \in E^{l_k}$, and a scalar $p^0 \leq 0$ with the following properties:

18. Not all the quantities $p^0, p_0, p_1, \ldots, p_k, \mu_0, \mu_k$ are zero.

19. The costate vectors p_i satisfy the "forced" adjoint equation

$$p_i - p_{i+1} = \left[\frac{\partial f_i(\hat{x}_i, \hat{u}_i)}{\partial x}\right]^T p_{i+1} + p^0 \left[\frac{\partial f_i^0(\hat{x}_i, \hat{u}_i)}{\partial x}\right]^T + \left[\frac{\partial q_i(\hat{x}_i)}{\partial x}\right]^T \lambda_i \quad i = 0, 1, \ldots, k-1.$$

20. At the initial set X_0, the transversality condition

$$p_0 = \left[\frac{\partial g_0(\hat{x}_0)}{\partial x}\right]^T \mu_0$$

is satisfied.

21. At the terminal set X_k, the transversality condition

$$p_k = \left[\frac{\partial g_k(\hat{x}_k)}{\partial x}\right]^T \mu_k + \left[\frac{\partial q_k(\hat{x}_k)}{\partial x}\right]^T \lambda_k$$

is satisfied.

22. For $j = 1, 2, \ldots, m_i$ and $i = 0, 1, \ldots, k$,

$$\lambda_i{}^j q_i{}^j(\hat{x}_i) = 0$$

[equivalently, $\langle \lambda_i, q_i(\hat{x}_i) \rangle = 0$ for $i = 0, 1, \ldots, k$].

23. For $i = 0, 1, \ldots, k-1$ the hamiltonian $H: E^n \times E^m \times E^n \times E^1 \times \{0, 1, \ldots, k-1\} \to E^1$, defined by

$$H(x, u, p, p^0, i) = p^0 f_i^0(x, u) + \langle p, f_i(x, u) \rangle$$

satisfies the condition

24. $$\left\langle \frac{\partial H(\hat{x}_i, \hat{u}_i, p_{i+1}, p^0, i)}{\partial u}, \delta u_i \right\rangle \leq 0 \quad \text{for all } \delta u_i \in \bar{C}(\hat{u}_i, U_i),$$

which becomes, in expanded form,

25. $$p^0 \left\langle \frac{\partial f_i^0(\hat{x}_i, \hat{u}_i)}{\partial u}, \delta u_i \right\rangle + \left\langle p_{i+1}, \frac{\partial f_i(\hat{x}_i, \hat{u}_i)}{\partial u} \delta u_i \right\rangle \leq 0 \quad \text{for all } \delta u_i \in \bar{C}(\hat{u}_i, U_i).$$

Proof. Condition (18) was established prior to the statement of the theorem. We therefore proceed with the remaining conditions.

First let us prove (25). Let

$$\delta z = (0,0, \ldots, 0, 0, \delta u_i, 0, 0, \ldots, 0) \qquad \delta u_i \in \bar{C}(\hat{u}_i, U_i).$$

Then, from (16), we immediately get (25). Now let $\delta z = (0, 0, \ldots, \delta x_k, 0, 0, \ldots, 0)$, with $\delta x_k \in IC(\hat{x}_k, X'_k)$. Then (16) yields

(26) $$\langle -p_k, \delta x_k \rangle + \left\langle \mu_k, \frac{\partial g_k(\hat{x}_k)}{\partial x} \delta x_k \right\rangle \leq 0 \qquad \text{for all } \delta x_k \in \overline{IC}(\hat{x}_k, X'_k).$$

But if $\delta x_k \in \overline{IC}(\hat{x}_k, X'_k)$, then $\langle [\partial q_k{}^j(\hat{x}_k)/\partial x], \delta x_k \rangle \leq 0$ for all $j \in \{j: q_k{}^j(\hat{x}) = 0 \text{ and } j \in \{1, 2, \ldots, m_k\}\}$. Applying Farkas' lemma (A.5.34), we find that there exists a vector $\lambda_k \in E^{m_k}$, with $\lambda_k \leq 0$, such that

(27) $$-p_k + \left[\frac{\partial g_k(\hat{x}_k)}{\partial x}\right]^T \mu_k + \left[\frac{\partial q_k(\hat{x}_k)}{\partial x}\right]^T \lambda_k = 0$$

and $\lambda_k{}^j q_k{}^j(\hat{x}_k) = 0$ for $j = 1, 2, \ldots, m_k$; that is, conditions (21) and (22) are satisfied for $j = k$. Similarly, for $i = 0, 1, \ldots, k - 1$, letting $\delta z = (0, 0, \ldots, \delta x_i, 0, \ldots, 0, 0, \ldots, 0)$, with $\delta x_i \in \overline{IC}(\hat{x}_i, X'_i)$, and again applying Farkas' lemma, we get (19) and (22). Condition (20) is now seen to be simply the result of an arbitrary but consistent definition. □

This theorem is still too general for most applications, and we shall therefore specialize it further by considering special cases in the following corollaries.

(28) **Corollary.** If the functions $q_i \equiv 0$ for $i = 1, 2, \ldots, k - 1$ and $\hat{z} = (\hat{x}_0, \hat{x}_1, \ldots, \hat{x}_k, \hat{u}_0, \hat{u}_1, \ldots, \hat{u}_{k-1})$ is an optimal solution to problem (1), then there exist costate vectors $p_0, p_1, \ldots, p_k$ in E^n, multiplier vectors $\lambda_0 \leq 0$ in E^{m_0} and $\lambda_k \leq 0$ in E^{m_k} and $\mu_0 \in E^{l_0}$ and $\mu_k \in E^{l_k}$, and a scalar $p^0 \leq 0$ with the following properties:

(29) Not all quantities $p^0, p_0, p_1, \ldots, p_k$, and μ_0 and μ_k are zero.

(30) $$p_i - p_{i+1} = \left[\frac{\partial f_i(\hat{x}_i, \hat{u}_i)}{\partial x}\right]^T p_{i+1} + p^0 \left[\frac{\partial f_i(\hat{x}_i, \hat{u}_i)}{\partial x}\right]^T \qquad \text{for } i = 0, 1, \ldots, k - 1;$$

(31) $$p_0 = -\left[\frac{\partial g_0(\hat{x}_0)}{\partial x}\right]^T \mu_0;$$

(32) $$p_k = \left[\frac{\partial g_k(\hat{x}_k)}{\partial x}\right]^T \mu_k + \left[\frac{\partial q_k(\hat{x}_k)}{\partial x}\right]^T \lambda_k;$$

(33) $\lambda_i{}^j q_i{}^j(\hat{x}_i) = 0$ for $j = 1, 2, \ldots, m_i$ and $i = 0, k$.

34
$$p^0 \left\langle \frac{\partial f_i^0(\hat{x}_i, \hat{u}_i)}{\partial u}, \delta u_i \right\rangle + \left\langle p_{i+1}, \frac{\partial f_i(\hat{x}_i, \hat{u}_i)}{\partial u} \delta u_i \right\rangle \leq 0 \quad \text{for all} \quad \delta u_i \in$$
$\bar{C}(\hat{u}_i, U_i)$ and $i = 0, 1, \ldots, k-1$. □

35 **Corollary.** If X_0 and X_k are singletons (points in the state space), then (31) and (32) carry no information, and corollary (28) becomes identical with theorem (2.4.28).

Proof. Let $X_0 = \{\hat{x}_0\}$ and $X_k = \{\hat{x}_k\}$. Then we have $g_0(x) = x - \hat{x}_0$, $q_0 \equiv 0$, $g_k(x) = x - \hat{x}_k$, and $q_k \equiv 0$. Hence $\partial g_0(\hat{x}_0)/\partial x = \partial g_k(\hat{x}_k)/\partial x = I$, the identity matrix, and (31) and (32) reduce to $p_0 = -\mu_0$ and $p_k = \mu_k$, which is hardly enlightening. □

36 **Corollary.** If $X_k = E^n$, or if $g_k(x) \equiv 0$ and $q_k(\hat{x}_k) < 0$, then corollary (28) holds, with $p^0 = -1$ and $p_k = 0$.

Proof. Suppose that $X_k = E^n$. This implies that both g_k and q_k are identically zero, that $g_k(x)$ does not appear in (12) (making $\mu_k = 0$), and that $IC(\hat{x}_k, X_k') = X_k' = E^n$. Hence, from (32), $p_k = 0$. Now suppose that $q_k(\hat{x}_k) < 0$ and $g_k \equiv 0$. Then again $\mu_k = 0$, from (33) $\lambda_k = 0$, and from (32) $p_k = 0$. Now suppose that $p^0 = 0$; then from (30) $p_0 = p_1 = p_2 = \cdots = p_{k-1} = p_k = 0$, and from (31) $\mu_0 = 0$. But this contradicts the fact that the vector $\psi = (p^0, -p_1, -p_2, \ldots, -p_k, \mu_0, \mu_k)$ is not zero in (15). □

Theorem (17) and its corollaries show how the Fundamental Theorem (2.3.12) can be interpreted to obtain necessary conditions for optimal control problems with various types of constraint. A specific characteristic of these theorems is the condition on the hamiltonian given by (25). At first glance, one might conclude that (25) is some sort of a maximum condition on the hamiltonian. That this is not necessarily so is demonstrated by the following example.

37 **Example.** Consider the system described by the scalar difference equation

38 $$x_{i+1} - x_i = u_i \quad i = 0, 1,$$

and suppose that we wish to minimize the cost function

39 $$f(\mathfrak{X}, \mathfrak{U}) = \sum_{i=0}^{1} \{x_i^2 - \tfrac{1}{2}(-2)^i[(u_i + 2)^2 - 1]\}$$

subject to the constraints that $x_0 = 1$ and $-3 \leq u_i \leq 1$ for $i = 0, 1$. This problem is simple enough to solve explicitly. From (38), $x_1 = (1 + u_0)$, and hence, with $x_0 = 1$, the cost function becomes

39 $$f(\mathfrak{X}, \mathfrak{U}) = \tfrac{1}{2}u_0^2 + (u_1 + 2)^2 - \tfrac{1}{2}.$$

Obviously, (39) is minimized by setting $\hat{u}_0 = 0$ and $\hat{u}_1 = -2$, both of these values satisfying the constraint $-3 \leq u_i \leq +1$ for $i = 0, 1$. The corresponding trajectory is seen to be $\hat{x}_0 = 1$, $\hat{x}_1 = 1$, $\hat{x}_2 = -1$.

Now let us go back to corollary (28). Since $g_2 \equiv 0$ and $q_2 \equiv 0$, we find that $p_2 = 0$. Equation (30) becomes

40
$$p_i - p_{i+1} = p^0 \cdot 2\hat{x}_i \qquad i = 0, 1.$$

Since $X_2 = E^1$ (the real line), we may set $p^0 = -1$, by corollary (36). Therefore, from (40), $p_1 = -2$. We can now evaluate the hamiltonian $H(\hat{x}_i, u_i, p_{i+1}, p^0, i)$, defined in condition (23), for $i = 0, 1$. Thus

41
$$H(\hat{x}_0, u_0, p_1, 0) = -\{1 - \tfrac{1}{2}[(u_0 + 2)^2 - 1]\} - 2u_0 = \tfrac{1}{2}u_0^2 + \tfrac{1}{2}$$

and

42
$$H(\hat{x}_1, u_1, p_2, 1) = -(u_1 + 2)^2.$$

It is clear by inspection that $H(\hat{x}_0, u_0, p_1, 0)$ is minimized by $u_0 = 0$ and $H(\hat{x}_1, u_1, p_2, 1)$ is maximized by $u_1 = -2$. Thus the optimal control sequence $(\hat{u}_0, \hat{u}_1)$ does not necessarily maximize the hamiltonian H. ☐

In the next section we shall consider a class of problems for which we do indeed obtain a maximum condition for the hamiltonian H.

Conditions for $p^0 < 0$

As we have already pointed out, it may be quite important to know whether or not $p^0 = 0$ in theorem (17). The following set of exercises is designed to shed some light on this matter.

43 **Exercise.** Suppose that for optimal control problem (1) we have $X_0 = \{x : g_0(x) = 0\}$, $x_k = \{x : g_k(x) = 0\}$, and $X_i = E^n$ for $i = 1, 2, \ldots, k-1$, and the sets U_i are convex for $i = 0, 1, \ldots, k-1$. Furthermore, suppose that $\hat{x}_0, \hat{x}_1, \ldots, \hat{x}_k, \hat{u}_0, \hat{u}_1, \ldots, \hat{u}_{k-1}$ is an optimal solution to this problem, and that the matrices A_i and B_i for $i = 0, 1, \ldots, k-1$ and $\tilde{Q}_c$ are defined by

44
$$A_i = I + \frac{\partial f_i}{\partial x}(\hat{x}_i, \hat{u}_i) \qquad i = 0, 1, \ldots, k-1$$

45
$$B_i = \frac{\partial f_i}{\partial u}(\hat{x}_i, \hat{u}_i) \qquad i = 0, 1, \ldots, k-1$$

46
$$\tilde{Q}_c = [B_{k-1} \mid A_{k-1}B_{k-2} \mid \cdots \mid A_{k-1}A_{k-2} \cdots A_1 B_0].$$

Assume (a) that the linear system

47 $\delta x_{i+1} = A_i \delta x_i + B_i \delta u_i$ $i = 0, 1, \ldots, k - 1$,

with boundary conditions $[\partial g_0(\hat{x}_0)/\partial x] \delta x_0 = 0$ and $[\partial g_k(\hat{x}_k)/\partial x] \delta x_k = 0$, has a solution $\delta x_0^*, \ldots, \delta x_k^*, \delta u_0^*, \ldots, \delta u_{k-1}^*$, with δu_i^* in the interior of $RC(\hat{u}_i, U_i)$, for $i = 0, 1, \ldots, k - 1$, (b) that $\partial g_0(\hat{x}_0)/\partial x$ and $\partial g_k(\hat{x}_k)/\partial x$ have maximum rank, and (c) that the controllability matrix $\bar{Q}_c$ has rank n. Show that if costate vectors $p_0, p_1, \ldots, p_k$, multiplier vectors μ_0 and μ_k, and a scalar $p^0 \leq 0$ satisfy the conditions of theorem (17), with $C(\hat{u}_0, U_i) = RC(\hat{u}_i, U_i)$ for $i = 1, 0, \ldots, k - 1$, then $p^0 < 0$. [*Hint:* Show that, under the above assumptions, with the optimal control problem transcribed into Basic Problem form, $\partial r(\hat{z})/\partial z$ has maximum rank and there is a vector δz^* in the interior of $C(\hat{z}, \Omega)$ such that $[\partial r(\hat{z})/\partial z] \delta z = 0$. Then use (2.3.50).] □

48 **Exercise.** Consider the two-point-boundary-value optimal control problem, $X_0 = \{\hat{x}_0\}$, $X_k = \{\hat{x}_k\}$, and $X_i = E^n$ for $i = 1, 2, \ldots, k - 1$. Suppose that the functions $f_i(\cdot, \cdot)$ are linear for $i = 0, 1, \ldots, k - 1$ [that is, $f_i(x, u) = A_i x + B_i u$] and that the sets U_i are convex for $i = 0, 1, \ldots, k - 1$. Show that there are a set of costates $p_0, p_1, \ldots, p_k$ and a scalar $p^0 < 0$ satisfying the conditions of theorem (17), whenever there exists a control sequence $u_0, u_1, \ldots, u_{k-1}$, with u_i in the relative interior of U_i, such that the corresponding trajectory $\hat{x}_0, x_1, \ldots, x_k$ satisfies the boundary conditions; that is, $x_0 = \hat{x}_0$ and $x_k = \hat{x}_k$. [*Hint:* Show that, under the above assumptions, with the optimal control problem transcribed into Basic Problem form, there exists a δz^* in the relative interior of $C(\hat{z}, \Omega)$ such that $[\partial r(\hat{z})/\partial z] \delta z^* = 0$. Show that this fact, with the convexity of U_i for $i = 0, 1, \ldots, k - 1$ and the linearity of $r(\cdot)$, can be used to verify that

$$TC_s(\hat{z}, \Omega') = TC_s(\hat{z}, \Omega) \cap \left\{ \delta z : \frac{\partial r(\hat{z})}{\partial z} \delta z = 0 \right\};$$

that is, use the results of exercise (2.3.47).] □

49 **Remark.** Note that if *all the constraint functions* appearing in the statement of the optimal control problem are linear, or affine, then, by corollary (3.4.33), the conditions of theorem (17) can be satisfied with $p^0 \ (= \psi^0) \neq 0$. □

This concludes our discussion of the general discrete optimal control problem. In the next section we shall consider a special class of optimal control problems for which the hamiltonian H does in fact assume a global maximum with respect to the controls, along an optimal trajectory.

4.2 A MAXIMUM PRINCIPLE

Convexity is the magic word in optimization theory. In the presence of convexity, local extrema become global extrema, sufficient conditions of optimality can be obtained, and conditions for the existence and uniqueness of an optimal solution come within our reach. Actually, as was shown by Holtzman [3] in his extension of Halkin's maximum principle [2] for discrete optimal control problems, we can often relax the requirement that a set be convex to the requirement that it be directionally convex.

1 Definition. Let e be any vector in E^s. A set $S \subset E^s$ is said to be *e-directionally convex* if for every vector z' in the convex hull of S there exists a vector z in S such that

2 $$z = z' + \beta e \qquad \beta \geqq 0. \qquad \square$$

A couple of examples of directionally convex sets are shown in Figure 1. We now establish the most relevant property of directionally convex sets as far as the following discussion is concerned.

3 Theorem. Let $e_1, e_2, \ldots, e_s$ be any basis for E^s. If the set $S \subset E^s$ is e_1-directionally convex, then its projection S_L onto the subspace L spanned by the vectors $e_2, e_3, \ldots, e_s$ is convex.[1]

Proof. Let z_1 and z_2 be any two vectors in S. We can decompose them uniquely as follows: $z_1 = z_1' + z_1''$ and $z_2 = z_2' + z_2''$, where $z_1' = \beta^1 e_1$, $z_2' = \beta^2 e_1$, and z_1'' and z_2'' are vectors in S_L. Let λ be any scalar satisfying $0 \leqq \lambda \leqq 1$; then, since S is e_1-directionally convex, we have, by (2), that

4 $$\{[\lambda z_1'' + (1 - \lambda)z_2''] + [\lambda \beta^1 + (1 - \lambda)\beta^2]e_1\} + \beta e_1 \in S \qquad \text{for some } \beta \geqq 0.$$

Hence, by inspection, $[\lambda z_1'' + (1 - \lambda)z_2''] \in S_L$ for $0 \leqq \lambda \leqq 1$, and therefore S_L is convex. $\square$

Before we can return to our discussion of optimal control problems, we need to extend the Fundamental Theorem (2.3.12) to

[1] Given a basis $e_1, e_2, \ldots, e_s$ for the space E^s and the subspace L spanned by $e_2, e_3, \ldots, e_s$, we say that S_L is the *projection* of S onto L if $S_L \subset L$ and if every vector $z \in S$ can be uniquely represented as $z = z' + z''$, with $z' = \beta e_1$ and $z'' \in S_L$.

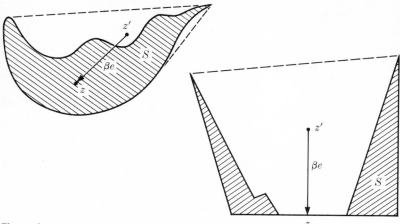

Figure 1

a more general form. First, after examining theorem (2.3.12) and its proof, we find that it can be restated in the following alternative form:

5 **Theorem.** Suppose that $\hat{z}$ is a feasible solution to the Basic Problem (1.4.1), that is, $\hat{z} \in \Omega$ and $r(\hat{z}) = 0$, and that $C(\hat{z},\Omega)$ is a conical approximation to Ω at $\hat{z}$ [see (2.3.3)]. If the ray $R = \{y: y = (-1,0,\ldots,0), \beta \geqq 0\}$ is an interior ray of the cone $[\partial F(\hat{z})/\partial z]C(\hat{z},\Omega)$, where $F = (f,r)$, then there exists a $z^* \in \Omega$ such that $r(z^*) = r(\hat{z}) = 0$ and $f(z^*) < f(\hat{z})$. □

Note that this version of theorem (2.3.12) does not require that $\hat{z}$ be optimal. It is merely a theorem which establishes a property of continuously differentiable functions. We can now prove an extension of the Fundamental Theorem (2.3.12).

6 **Theorem.** Let $\Omega' \subset E^n$ be any set with the property that for every $z' \in \Omega'$ there exists a $z \in \Omega$ satisfying $r(z) = r(z')$ and $f(z) \leqq f(z')$. If $\hat{z}$ is an optimal solution to the Basic Problem (1.4.1), if $\hat{z} \in \Omega'$, and if $C(\hat{z},\Omega')$ is a conical approximation to the set Ω' at $\hat{z}$, then there exists a nonzero vector $\psi = (\psi^0, \psi^1, \ldots, \psi^m) \in E^{m+1}$, with $\psi^0 \leqq 0$, such that for all $\delta z \in C(\hat{z},\Omega')$

7 $$\left\langle \psi, \frac{\partial F(\hat{z})}{\partial z} \delta z \right\rangle \leqq 0.$$

Proof. Suppose that the ray $R = \{y: y = \beta(-1,0,0,\ldots,0), \beta > 0\}$ is an interior ray of the cone $[\partial F(\hat{z})/\partial z]C(\hat{z},\Omega')$. Then it follows from theorem (5) that there exists a $z' \in \Omega'$ such that $r(z') = 0$ and $f(z') < f(\hat{z})$. But, by assumption, this implies that there exists a $z^* \in \Omega$ such that $r(z^*) = 0$ and $f(z^*) \leq f(z') \leq f(z)$, which contradicts the optimality of $\hat{z}$. Hence the ray R and the convex cone $[\partial F(\hat{z})/\partial z]C(\hat{z},\Omega')$ must be separated. The existence of a vector ψ satisfying (7) then follows from the definition of separation. $\square$

8 Directionally convex optimal control problem.

Given a dynamical system described by the difference equation

9
$$x_{i+1} - x_i = f_i(x_i, u_i) \qquad i = 0, 1, \ldots, k-1,$$

where $x_i \in E^n$ is the system state at time i, $u_i \in E^m$ is the system input at time i, and $f_i(\cdot,\cdot)$ is a function mapping $E^n \times E^m$ into E^n, find a control sequence $\hat{\mathfrak{U}} = (\hat{u}_0, \hat{u}_1, \ldots, \hat{u}_{k-1})$ and a corresponding trajectory $\hat{\mathfrak{X}} = (\hat{x}_0, \hat{x}_1, \ldots, \hat{x}_k)$, determined by (9), which minimize the sum

$$\sum_{i=0}^{k-1} f_i^0(x_i, u_i),$$

where the $f_i^0(\cdot,\cdot)$ map $E^n \times E^m$ into the reals, subject to constraints (1.3) to (1.6).

However, we now introduce a set of assumptions which are somewhat different from the ones made in Section 1.

10   For $i = 0, 1, \ldots, k-1$ and every $u \in U_i$ the functions $f_i(\cdot, u)$ are continuously differentiable on E^n. [Note that we no longer require the functions $f_i(\cdot,\cdot)$ to be differentiable in u; in fact, we do not even require the $f_i(\cdot,\cdot)$ to be continuous in u.]

11   Let $b_0 \in E^{n+1}$ be the vector $(-1,0,0,\ldots,0)$, and for $i = 0, 1, \ldots, k-1$ let $\mathbf{f}_i(\cdot,\cdot)$ be the function from $E^n \times E^m$ into E^{n+1}, defined by $\mathbf{f}_i(x,u) = (f_i^0(x,u), f_i(x,u))$. For $i = 0, 1, \ldots, k-1$ and every $x \in E^n$ the sets $\mathbf{f}_i(x, U_i)$ are b_0-directionally convex; that is, for every $\mathbf{f}_i(x,u')$ and $\mathbf{f}_i(x,u'')$ in $\mathbf{f}_i(x, U_i)$ (that is, $u', u'' \in U_i$) and $0 \leq \lambda \leq 1$ there exists a $u(\lambda) \in U_i$ such that $f_i(x, u(\lambda)) = \lambda f_i(x, u') + (1-\lambda) f_i(x, u'')$ and $f_i^0(x, u(\lambda)) \leq \lambda f_i^0(x, u') + (1-\lambda) f_i^0(x, u'')$.

12   The functions $g_0: E^n \to E^{l_0}$ and $g_k: E^n \to E^{l_k}$ are continuously differentiable and have jacobian matrices $\partial g_0(x)/\partial x$ and $\partial g_k(x)/\partial x$ which are of maximum rank for all x in X_0 and X_k, respectively.

13   For all $x \in X_i'$ and $i = 0, 1, \ldots, k$ the gradients of the active inequality constraints, $\nabla q_i^j(x)$, with $j \in \{j: q_i^j(x) = 0\}$, are linearly independent vectors. $\square$

To transcribe this optimal control problem into the form of the Basic Problem (1.4.1), we make the following substitutions, which are somewhat different from the ones we used in the preceding section.[1]

For $i = 0, 1, \ldots, k-1$ let $\mathbf{v}_i = (v_i^0, v_i) \in E^{n+1}$, where $v_i = (v_i^1, v_i^2, \ldots, v_i^n) \in E^n$. Equation (9) is now seen to be equivalent to

(14) $$x_{i+1} - x_i = v_i \qquad i = 0, 1, \ldots, k-1,$$

with $v_i \in f_i(x_i, U_i)$ for $i = 0, 1, \ldots, k-1$. Finally, let $z = (x_0, x_1, \ldots, x_k, \mathbf{v}_0, \mathbf{v}_1, \ldots, \mathbf{v}_{k-1}) \in E^{n(k+1)+k(n+1)}$. We can now define the functions f and r and the constraint set Ω as follows:

(15) $$f(z) = \sum_{i=0}^{k-1} v_i^0$$

(16) $$r(z) = \begin{bmatrix} x_1 - x_0 - v_0 \\ x_2 - x_1 - v_1 \\ \vdots \\ x_k - x_{k-1} - v_{k-1} \\ g_0(x_0) \\ g_k(x_k) \end{bmatrix}$$

(17) $$\Omega = \{z = (x_0, x_1, \ldots, x_k, x_{k-1}, \mathbf{v}_0, \mathbf{v}_1, \ldots, \mathbf{v}_{k-1}) : x_i \in X_i',$$
$$i = 0, 1, \ldots, k, \mathbf{v}_i \in \mathbf{f}_i(x_i, U_i), i = 0, 1, \ldots, k-1\}.$$

Thus in Ω the $\mathbf{v}_i$ are constrained not only by the sets U_i, but also by the values of the $x_i \in X_i'$. Note, in addition, that for $z \in \Omega$, $v_i^0 = f_i^0(x_i, u_i)$ for some $x_i \in X_i'$ and $u_i \in U_i$. Hence, from (15),

(18) $$f(z) = \sum_{i=0}^{k-1} v_i^0 = \sum_{i=0}^{k-1} f_i^0(x_i, u_i) \qquad z \in \Omega,$$

and (18) becomes identical in form with the sum in problem (8). This completes our transcription of the optimal control problem into the form of the Basic Problem (1.4.1).

Unfortunately, we do not know how to construct a conical approximation for the set Ω. Let us therefore introduce a more

---

[1] The first reason for using a different substitution in this section is that, by definition, in (1.4.1) $r(z)$ must be continuously differentiable. Hence the terms $f_i(x_i, u_i)$ which may not be differentiable in u_i cannot appear in the definition of $r(z)$. The second reason is that substitution (16) is particularly suitable for exploiting the convexity of the sets U_i.

tractable set, Ω', which we can use in conjunction with theorem (6). Let Ω' be the set defined by

19
$$\Omega' = \{z = (x_0, x_1, \ldots, x_k, \mathbf{v}_0, \mathbf{v}_1, \ldots, \mathbf{v}_{k-1}): x_i \in X_i',$$
$$i = 0, 1, \ldots, k, \mathbf{v}_i \in \text{co } \mathbf{f}_i(x_i, U_i), i = 0, 1, \ldots, k-1\}.$$

Now let $z^* = (x_0^*, x_1^*, \ldots, x_k^*, \mathbf{v}_0^*, \mathbf{v}_1^*, \ldots, \mathbf{v}_{k-1}^*)$ be any point in Ω'. Then, since the sets $\mathbf{f}_i(x_i^*, U_i)$ are b_0-directionally convex, there exist points $\tilde{\mathbf{v}}_i \in \mathbf{f}_i(x_i^*, U_i)$, with $i = 0, 1, \ldots, k-1$, such that $\tilde{v}_i = v_i^*$ and $v_i{}^0 \leq v_i^{*0}$. Hence for $\tilde{z} = (x_0^*, x_1^*, \ldots, x_k^*, \tilde{\mathbf{v}}_0, \tilde{\mathbf{v}}_1, \ldots, \tilde{\mathbf{v}}_{k-1})$ we have $r(\tilde{z}) = r(z^*)$ and $f(\tilde{z}) \leq f(z^*)$, where r and f are as defined by (16) and (15), respectively. We thus see that the set Ω' satisfies the hypotheses of theorem (6) with respect to the set Ω, defined in (17), and the functions f and r defined in (15) and (16), respectively.

Now suppose that $\hat{z} = (\hat{x}_0, \hat{x}_1, \ldots, \hat{x}_k, \hat{\mathbf{v}}_0, \hat{\mathbf{v}}_1, \ldots, \hat{\mathbf{v}}_{k-1})$ is an optimal solution to problem (8); that is, $(\hat{x}_0, \hat{x}_1, \ldots, \hat{x}_k)$ is an optimal trajectory and $\hat{\mathbf{v}}_i = \mathbf{f}_i(\hat{x}_i, \hat{u}_i)$ for $i = 0, 1, \ldots, k-1$, where $(\hat{u}_0, \hat{u}_1, \ldots, \hat{u}_{k-1})$ is an optimal control sequence. We shall now show that the set

20
$$C(\hat{z}, \Omega') = \{\delta z = (\delta x_0, \delta x_1, \ldots, \delta x_k, \delta \mathbf{v}_0, \delta \mathbf{v}_1, \ldots, \delta \mathbf{v}_{k-1}):$$
$$\delta x_i \in IC(\hat{x}_i, X_i'),$$
$$(\delta \mathbf{v}_i - [\partial \mathbf{f}_i(\hat{x}_i, \hat{u}_i)/\partial x] \delta x_i) \in RC(\hat{v}_i, \text{co } \mathbf{f}_i(\hat{x}_i, U_i))\}$$

is a conical approximation to the set Ω' at $\hat{z}$.

First we must prove that $C(\hat{z}, \Omega')$ is a convex cone. Let δz be any nonzero vector in $C(\hat{z}, \Omega')$, and let $\lambda \geq 0$ be an arbitrary scalar. Then

21
$$\lambda \, \delta x_i \in IC(\hat{x}_i, X_i')$$

and

$$\left[\lambda \, \delta \mathbf{v}_i - \frac{\partial \mathbf{f}_i(\hat{x}_i, \hat{u}_i)}{\partial x} \lambda \, \delta x_i\right] \in RC(\hat{\mathbf{v}}_i, \text{co } \mathbf{f}_i(\hat{x}_i, U_i)).$$

Hence $C(\hat{z}, \Omega')$ is a cone. Now let $\delta z'$ and $\delta z''$ be any two points in $C(\hat{z}, \Omega')$. It is simple to show that $\lambda \, \delta z' + (1 - \lambda) \, \delta z''$ is also a vector in $C(\hat{z}, \Omega')$ for $0 \leq \lambda \leq 1$. Hence $C(\hat{z}, \Omega')$ is convex.

Finally [see (2.3.3)], for every finite collection $\{\delta z_1, \delta z_2, \ldots, \delta z_p\}$ of linearly independent vectors in $C(\hat{z}, \Omega')$ we must exhibit the existence of an $\epsilon > 0$ and a continuous map ζ from co $\{\hat{z}, \hat{z} + \epsilon \, \delta \hat{z}_1, \ldots, \hat{z} + \epsilon \, \delta z_{\hat{p}}\}$ into Ω' such that $\zeta(\hat{z} + \delta z) = \hat{z} + \delta z + o(\delta z)$ where $\|o(\delta z)\|/\|\delta z\| \to 0$ as $\|\delta z\| \to 0$. Let $\delta z_1, \delta z_2, \ldots, \delta z_p$ be any finite collection of linearly independent vectors in $C(\hat{z}, \Omega')$, with $\delta z_j = (\delta x_{0j}, \delta x_{1j}, \ldots, \delta x_{kj}, \delta \mathbf{v}_{0j}, \delta \mathbf{v}_{1j}, \ldots, \delta \mathbf{v}_{k-1j})$. Hence, by definition (20),

(22) $\quad \delta x_{ij} \in IC(\hat{x}_i, X_i') \qquad i = 0, 1, \ldots, k, j = 1, 2, \ldots, p$

and

(23) $\quad \delta \mathbf{v}_{ij} = \dfrac{\partial \mathbf{f}_i(\hat{x}_i, \hat{u}_i)}{\partial x} \delta x_{ij} + (\mathbf{v}_{ij} - \hat{\mathbf{v}}_i) \qquad i = 0, 1, \ldots, k - 1,$

$$j = 1, 2, \ldots, p,$$

where $(\mathbf{v}_{ij} - \hat{\mathbf{v}}_i) \in RC(\hat{\mathbf{v}}_i, \text{co } \mathbf{f}_i(\hat{x}_i, U_i))$.

Since both internal cones and radial cones are conical approximations of the first kind [see (2.3.1)], there must exist an $\epsilon > 0$ such that for any scalars $\mu^1, \mu^2, \ldots, \mu^p$ which satisfy $\mu^j \geqq 0$ for $j = 1, 2, \ldots, p$, and $\Sigma_{j=1}^p \mu^j \leqq 1$, $(\hat{x}_i + \epsilon \Sigma_{j=1}^p \mu^j \delta x_{ij}) \in X_i'$ for $i = 0, 1, \ldots, k$ and $[\hat{\mathbf{v}}_i + \epsilon \Sigma_{j=1}^p \mu^j (\mathbf{v}_{ij} - \hat{\mathbf{v}}_i)] \in \text{co } \mathbf{f}_i(\hat{x}_i, U_i)$ for $i = 0, 1, \ldots, k$. We claim that the above-indicated $\epsilon > 0$ will serve our purpose and that we can therefore proceed to construct the map ζ (we shall justify this claim in the process of construction). We construct the map ζ in two steps. First, we obtain a representation for vectors $z = \hat{z} + \delta z$ in co $\{\hat{z}, \hat{z} + \epsilon \, \delta z_1, \ldots, \hat{z} + \epsilon \, \delta z_p\}$ in terms of vectors in X_i' and in $\mathbf{f}_i(\hat{x}_i, U_i)$. Then, since we know that the map ζ must have an expansion of the form $\zeta(\hat{z} + \delta z) = \hat{z} + \delta z + o(\delta z)$, we define ζ in such a way that its linear part has the representation obtained in the first step and, furthermore, $\zeta(\hat{z} + \delta z)$ is in Ω'. This last step is arrived at essentially by inspection.

Thus we let $C = \text{co } \{\hat{z}, \hat{z} + \epsilon \, \delta z_1, \hat{z} + \epsilon \, \delta z_2, \ldots, \hat{z} + \epsilon \, \delta z_p\}$. Then for any $z \in C$ we have

(24) $\quad \delta z = z - \hat{z} = \epsilon \sum_{j=1}^{p} \mu^j(z) \, \delta z_j \qquad \mu^j(z) \geqq 0, \; \sum_{j=1}^{p} \mu^j(z) \leqq 1.$

Furthermore, since the vectors δz_j are linearly independent, for any $z \in C$ the scalars $\mu^j(z)$ for $j = 1, 2, \ldots, p$ are uniquely determined by (24).

Now, since $[\hat{\mathbf{v}}_i + \epsilon(\mathbf{v}_{ij} - \hat{\mathbf{v}}_i)] \in \text{co } \mathbf{f}_i(\hat{x}_i, U_i)$ for $i = 0, 1, \ldots, k - 1$ and $j = 1, 2, \ldots, p$, there exist controls $u_{ij}{}^\alpha \in U_{ij}$ with $\alpha = 1, 2, \ldots, s_i$, such that

(25) $\quad \hat{\mathbf{v}}_i + \epsilon(\mathbf{v}_{ij} - \hat{\mathbf{v}}_i) = \sum_{\alpha=1}^{s_i} \lambda_i{}^\alpha \mathbf{f}_i(\hat{x}_i, u_{ij}{}^\alpha) \qquad i = 0, 1, \ldots, k - 1,$

$j = 1, 2, \ldots, p$, where $\lambda_i{}^\alpha \geqq 0$ and $\sum_{\alpha=1}^{s_i} \lambda_i{}^\alpha = 1$.

Consequently, for any $\delta z = (\delta x_0, \delta x_1, \ldots, \delta x_k, \delta \mathbf{v}_0, \delta \mathbf{v}_1, \ldots, \delta \mathbf{v}_{k-1}) = z - \hat{z}$, where $z \in C$, we have, from (24) and (25),

(26) $\quad \delta x_i = \epsilon \sum_{j=1}^{p} \mu^j(z) \, \delta x_{ij} \qquad i = 0, 1, \ldots, k - 1$

90 DISCRETE OPTIMAL CONTROL PROBLEMS

and

27
$$\delta\mathbf{v}_i = \frac{\partial \mathbf{f}_i(\hat{x}_i,\hat{u}_i)}{\partial x}\,\delta x_i + \epsilon \sum_{j=1}^{p} \mu^j(z)(\mathbf{v}_{ij} - \hat{\mathbf{v}}_i)$$
$$= \frac{\partial \mathbf{f}_i(\hat{x}_i,\hat{u}_i)}{\partial x}\,\delta x_i + \sum_{j=1}^{p} \mu^j(z)\bigg[\sum_{\alpha=1}^{s_i} \lambda_i{}^\alpha \mathbf{f}_i(\hat{x}_i,u_{ij}{}^\alpha) - \hat{\mathbf{v}}_i\bigg]$$
$$i = 0, 1, \ldots, k-1.$$

Expressions (26) and (27) give the desired representations for vectors in C in terms of vectors in X_i and in $\mathbf{f}_i(\hat{x}_i, U_i)$.

We now define the map ζ from the set C into Ω' as follows. Let $z = (x_0, x_1, \ldots, x_k, \mathbf{v}_0, \mathbf{v}_1, \ldots, \mathbf{v}_{k-1}) \in C$ be arbitrary, and let $\delta z = z - \hat{z}$; then

28
$$\zeta(z) = (y_0, y_1, \ldots, y_k, \mathbf{w}_0, \mathbf{w}_1, \ldots, \mathbf{w}_{k-1}),$$

where

29
$$y_i(z) = x_i \qquad i = 0, 1, \ldots, k$$

and

30
$$\mathbf{w}_i(z) = \mathbf{f}_i(x_i, \hat{u}_i) + \sum_{j=1}^{p} \mu^j(z)\bigg[\sum_{\alpha=1}^{s_i} \lambda_i{}^\alpha \mathbf{f}_i(x_i, u_{ij}{}^\alpha) - \mathbf{f}_i(x_i, \hat{u}_i)\bigg],$$

with the $\mu^j(z)$ determined uniquely by (24). First we observe that, by construction, for every $z = (x_0, x_1, \ldots, x_k, \mathbf{v}_0, \mathbf{v}_1, \ldots, \mathbf{v}_{k-1}) \in C$ we have $x_i \in X_i'$, with $i = 0, 1, \ldots, k$. Hence y_i is also in X_i'. Since the controls $u_{ij}{}^\alpha$ are in U_i and the scalars $\mu^j(z)$ satisfy $\mu^j(z) \geq 0$ and $\Sigma_{j=1}^{p} \mu^j(z) \leq 1$, it is clear that the $\mathbf{w}_i$, as determined by (30), are in co $\mathbf{f}_i(x_i, U_i)$.

Thus the map ζ does indeed map C into Ω'. We must still show that it can be written in the form $\zeta(\hat{z} + \delta z) = \hat{z} + \delta z + o(\delta z)$. We therefore expand (30) about $\hat{z}$ and get, with $z = \hat{z} + \delta z$,

31
$$\mathbf{w}_i(\hat{z} + \delta z) = \mathbf{f}_i(\hat{x}_i, \hat{u}_i) + \frac{\partial \mathbf{f}_i(\hat{x}_i, \hat{u}_i)}{\partial x}\,\delta x_i + \sum_{j=1}^{p} \mu^j(z)$$
$$\bigg[\sum_{\alpha=1}^{s_i} \lambda_i{}^\alpha \mathbf{f}_i(x_i, u_{ij}{}^\alpha) - \mathbf{f}_i(x_i, \hat{u}_i)\bigg] + o(\delta z)$$
$$i = 0, 1, \ldots, k-1,\dagger$$

† Let $\mu(z) = (\mu^1(z), \mu^2(z), \ldots, \mu^p(z))$; then we conclude from (24) that there exists a matrix Y with p rows such that $\mu(z) = Y(z - \hat{z})$. Let $Z_i(z)$ be a matrix with p columns whose jth column is $\Sigma_{\alpha=1}^{s_i} \lambda_i{}^\alpha \mathbf{f}_i(x_i, u_{ij}{}^\alpha) - \mathbf{f}_i(x_i, \hat{u}_i)$; then, from (30), $\mathbf{w}_i(\hat{z} + \delta z) = \mathbf{f}_i(\hat{x}_i + \delta x_i, \hat{u}_i) + z_i(\hat{z} + \delta z)Y\,\delta z$. Now, $\mathbf{f}_i(\hat{x}_i + \delta x_i, \hat{u}_i) = \mathbf{f}_i(\hat{x}_i, \hat{u}_i) + [\partial \mathbf{f}_i(\hat{x}_i, \hat{u}_i)/\partial x]\,\delta x_i + o(\delta x_i)$ and $Z_i(\hat{z} + \delta z)Y\,\delta z = Z_i(\hat{z})Y\,\delta z + o(\delta z)$, and hence (31) follows.

where $\|o(\delta z)\|/\|\delta z\| \to 0$ as $\|\delta z\| \to 0$. Comparing (31) with (27) and taking into account (29), we see that for any $z = \hat{z} + \delta z \in C$, $\zeta(\hat{z} + \delta z) = \hat{z} + \delta_z + o(\delta z)$, where $o(\delta z)$ is obviously continuous and satisfies $\|o(\delta z)\|/\|\delta z\| \to 0$ as $\|\delta z\| \to 0$. Hence $C(\hat{z}, \Omega')$ is indeed a conical approximation to the set Ω' at $\hat{z}$.

Now that we have established that the set Ω' defined by (19) satisfies the conditions of theorem (6) with respect to the set Ω defined by (17) and the functions $f(\cdot)$ and $r(\cdot)$ defined by (15) and (16), respectively, and that the set $C(\hat{z}, \Omega')$ defined by (20) is a conical approximation to the set Ω' at $\hat{z}$, we conclude from theorem (6) that there exists a nonzero vector $\psi = (p^0, \pi)$, with $p^0 \leq 0$ and $\pi = (-p_1, -p_2, \ldots, -p_k, \mu_0, \mu_k)$, where $p_i \in E^n$, $\mu_0 \in E^{l_0}$, and $\mu_k \in E^{l_k}$, such that

32
$$p^0 \langle \nabla f(\hat{z}), \delta z \rangle + \left\langle \pi, \frac{\partial r(\hat{z})}{\partial z} \delta z \right\rangle \leq 0 \quad \text{for all } \delta z \in C(\hat{z}, \Omega').$$

Substituting for f and r from (15) and (16), respectively, into (32), for all $\delta z \in C(\hat{z}, \Omega')$ we get

33
$$p^0 \sum_{i=0}^{k-1} \delta v_i^0 + \sum_{i=0}^{k-1} \langle -p_{i+1}, (\delta x_{i+1} - \delta x_i - \delta v_i) \rangle$$
$$+ \left\langle \mu_0, \frac{\partial g_0(x_0)}{\partial x} \delta x_0 \right\rangle + \left\langle \mu_k, \frac{\partial g_k(x_k)}{\partial x} \delta x_k \right\rangle \leq 0.$$

As in the preceding section, we now convert the result expressed by (33) into a classical form for necessary conditions, involving a costate, a hamiltonian, and transversality conditions.

34 **Theorem: the maximum principle.** If $(\hat{u}_0, \hat{u}_1, \ldots, \hat{u}_{k-1})$ is an optimal control sequence and $(\hat{x}_0, \hat{x}_1, \ldots, \hat{x}_k)$ is an optimal trajectory for control problem (8), then there exist costate vectors $p_0, p_1, \ldots, p_k$ in E^n, multiplier vectors $\lambda_0, \lambda_1, \ldots, \lambda_k$, with $\lambda_i \in E^{m_i}$ satisfying $\lambda_i \leq 0$, $\mu_0 \in E^{l_0}$, and $\mu_k \in E^{l_k}$, and a scalar $p^0 \leq 0$ such that:

35 Not all the quantities $p^0, p_0, p_1, \ldots, p_k$, and μ_0, μ_k are zero.

36 The costate vectors p_i satisfy the "forced" adjoint equation

$$p_i - p_{i+1} = \left[\frac{\partial f_i(\hat{x}_i, \hat{u}_i)}{\partial x} \right]^T p_{i+1} + p^0 \left[\frac{\partial f_i^0(\hat{x}_i, \hat{u}_i)}{\partial x} \right]^T$$
$$+ \left[\frac{\partial q_i(\hat{x}_i)}{\partial x} \right]^T \lambda_i \quad i = 0, 1, \ldots, k-1.$$

92 DISCRETE OPTIMAL CONTROL PROBLEMS

37   At the initial set X_0 the transversality condition

$$p_0 = -\left[\frac{\partial g_0(\hat{x}_0)}{\partial x}\right]^T \mu_0$$

is satisfied.

38   At the terminal set X_k the transversality condition

$$p_k = \left[\frac{\partial g_k(\hat{x}_k)}{\partial x}\right]^T \mu_k + \left[\frac{\partial q_k(\hat{x}_k)}{\partial x}\right]^T \lambda_k$$

is satisfied.

39 $\langle \lambda_i, q_i(\hat{x}_i) \rangle = 0 \qquad i = 0, 1, \ldots, k.$

40   For $i = 0, 1, \ldots, k - 1$ the hamiltonian $H: E^n \times E^m \times E^n \times E^1 \times \{0, 1, \ldots, k - 1\} \to E^1$, defined by

$$H(x,u,p,p^0,i) = p^0 f_i^0(x,u) + \langle p, f_i(x,u) \rangle,$$

satisfies the maximum condition

41 $H(\hat{x}_i, \hat{u}_i, p_{i+1}, p^0, i) \geqq H(\hat{x}_i, u_i, p_{i+1}, p^0, i) \qquad$ for all $u_i \in U_i.$

This last condition is known as a *maximum principle*.

Proof. Condition (35) was established prior to the statement of the theorem. Now let us turn to (41). Suppose $\delta z = (0,0,\ldots,0,\delta v_i,0,\ldots,0)$ is in $C(z,\Omega')$. Then, from (33), we have

42 $p^0 \, \delta v_i^0 + \langle p_{i+1}, \delta v_i \rangle \leqq 0$ for all $\delta v_i \in RC(\hat{v}_i, \text{co } f_i(\hat{x}_i, U_i)).$

But $f_i(\hat{x}_i, U_i) \subset \text{co } f_i(\hat{x}_i, U_i) \subset RC(\hat{v}_i, \text{co } f_i(\hat{x}_i, U_i))$ and therefore we obtain from (42) that

43 $p^0[f_i^0(\hat{x}_i, \hat{u}_i) - f_i^0(\hat{x}_i, u_i)] + \langle p_{i+1}, [f_i(\hat{x}_i, \hat{u}_i) - f_i(x_i, u_i)] \rangle \geqq 0,$

which is readily recognized as an expansion of (41). We now observe that the closure of set $RC(\hat{v}_i, \text{co } f_i(\hat{x}_i, U_i)) + \{\hat{v}_i\}$ contains the set co $f_i(\hat{x}_i, U_i)$, since the latter is convex, and hence (43) is true for all $u_i \in U_i$, which proves (41).

Now let $\delta z = (0,0,\ldots,0,\delta x_k,0,0,\ldots,0)$, with $\delta x_k \in \overline{IC}(x_k, X'_k)$. Then (33) yields

44 $\left\langle -p_k + \left[\frac{\partial g_k(\hat{x}_k)}{\partial x}\right]^T \mu_k, \delta x_k \right\rangle \leqq 0 \qquad$ for all $\delta x_k \in \overline{IC}(\hat{x}_k, X'_k),$

that is, for all δx_k satisfying $[\partial q_k^j(\hat{x}_k)/\partial x] \, \delta x_k \leqq 0$, with $q_k^j(\hat{x}_k) = 0$ and $j \in \{1, 2, \ldots, m_k\}$. Applying Farkas' lemma (A.5.34), we find that there exists a vector $\lambda_k \leq 0$ in E^{m_k} such that[1]

45
$$-p_k + \left[\frac{\partial g_k(\hat{x}_k)}{\partial x}\right]^T \mu_k = -\left[\frac{\partial q_k(\hat{x}_k)}{\partial x}\right]^T \lambda_k$$

and $\langle \lambda_k, g_k(\hat{x}_k) \rangle = 0$, that is, such that (38) and (39) are satisfied. Similarly, let $\delta z = (0, 0, \ldots, \delta x_i, 0, \ldots, 0, \delta v_i, 0, \ldots, 0)$ in $C(z, \Omega')$, $i \neq 0$, $i \neq k$, with $\delta v_i = [\partial \mathbf{f}_k(\hat{x}_i, \hat{u}_i)/\partial x]\, \delta x_i$. Then, from (33), we get

46
$$p^0 \left\langle \frac{\partial f_i^0(\hat{x}_i, \hat{u}_i)}{\partial x}, \delta x_i \right\rangle + \left\langle p_{i+1}, \frac{\partial f_i(\hat{x}_i, \hat{u}_i)}{\partial x} \delta x_i \right\rangle + \langle p_{i+1}, \delta x_i \rangle$$
$$- \langle p_i, \delta x_i \rangle \leq 0 \qquad \text{for all } \delta x_i \in \overline{IC}(\hat{x}_i, X_i'),$$

that is, for all δx_i satisfying $[\partial q_i{}^j(\hat{x}_i)/\partial x_i]\, \delta x_i \leq 0$, when $q_i{}^j(\hat{x}_i) = 0$, with $j \in \{1, 2, \ldots, m_i\}$. Again applying Farkas' lemma, we get (36) and (39). Finally, condition (37) is seen to be merely an arbitrary but consistent definition. ☐

It is clear at this stage that we can proceed to specialize theorem (34) even further, exactly as we have done in the previous section. However, this is left as an exercise for the reader.

47 **Exercise.** Show that for the problem considered in example (2.4.38) the maximum principle (34) gives the same control law as theorem (2.4.28). ☐

At first glance, the maximum principle (34) appears to be a considerably stronger result than theorem (1.17). In practice, however, it is difficult to find a problem for which it is more useful than theorem (1.17). However, as will be seen from the following example, the maximum principle (34) may be somewhat easier to apply (provided its assumptions are satisfied, of course).

48 **Example: a minimum-fuel problem.** Given the system

49
$$x_{i+1} - x_i = A x_i + b u_i \qquad i = 0, 1, \ldots, k-1,$$

where A is an $n \times n$ matrix and b is an $n \times 1$ matrix (that is, $x_i \in E^n$ for $i = 0, 1, \ldots, k$ and $u_i \in E^1$ for $i = 0, 1, \ldots, k-1$), minimize

50
$$\sum_{i=0}^{k-} |u_i|$$

---

[1] We can now see the effect of assumption (12), that $\nabla g_k{}^j(x)$ for $j = 1, 2, \ldots, l_k$ are linearly independent. When they are dependent, we may choose $\mu_k \neq 0$ such that $[\partial g_k(x_k)/\partial x]^T \mu_k = 0$ and let $p_0 = p_1 = \cdots = p_k = 0$, $p^0 = 0$, and $\mu_0 = 0$, satisfying the theorem trivially.

subject to the constraint on the input

51 $\quad |u_i| \leq 1 \quad i = 0, 1, \ldots, k - 1$

and to the boundary constraints

52 $\quad x_0 = \hat{x}_0 \quad$ and $\quad x_k = \hat{x}_k,$

where $\hat{x}_0$ and $\hat{x}_k$ are given vectors in E^n and x_k is the solution of system (49) at time k corresponding to the initial conditions $x_0 = \hat{x}_0$ and the input sequence $u_0, u_1, \ldots, u_{k-1}.$

First approach. Since $|u_i|$ is not a continuously differentiable function, we cannot apply theorem (1.17) to this problem directly. Hence let us make the following substitution:

53 $\quad u_i = v_i - w_i \quad i = 0, 1, \ldots, k - 1$
54 $\quad |u_i| = v_i + w_i \quad i = 0, 1, \ldots, k - 1$
55 $\quad 0 \leq v_i \leq 1, \quad 0 \leq w_i \leq 1 \quad i = 0, 1, \ldots, k - 1.$

Note that (54) is valid only if either v_i or w_i is zero for each i. We shall show that this is indeed the case if $\hat{v}_i$ and $\hat{w}_i,$ with $i = 0, 1, \ldots, k - 1,$ are an optimal control sequence for the problem:

56 $\quad \text{Minimize} \sum_{i=0}^{k-1} (v_i + w_i) \text{ subject to (55), subject to}$

57 $\quad x_{i+1} - x_i = Ax_i + b(v_i - w_i),$

and subject to (52).

Obviously, we can now apply theorem (1.17) to obtain, for $i = 0, 1, \ldots, k - 1,$

58 $\quad p^0(\delta v_i + \delta w_i) + \langle p_{i+1}, b \rangle (\delta v_i - \delta w_i) \leq 0$

for all $\delta v_i \in RC(\hat{v}_i, \Omega)$ and $\delta w_i \in RC(\hat{w}_i, \Omega),$ where $\Omega = \{\alpha : 0 \leq \alpha \leq 1\},$ p_i is the costate vector defined as in theorem (1.17), and $\hat{v}_i$ and $\hat{w}_i$ for $i = 0, 1, \ldots, k - 1$ are optimal control sequences.

By remark (1.49), $p^0 = -1$ in (58), and hence (58) becomes

59 $\quad (-1 + \langle p_{i+1}, b \rangle) \delta v_i - (1 + \langle p_{i+1}, b \rangle) \delta w_i \leq 0.$

Suppose that $\langle p_{i+1}, b \rangle > 1.$ Then for $\delta w_i = 0$ (59) can be satisfied for all $\delta v_i \in RC(\hat{v}_i, \Omega)$ only if

60 $\quad \delta v_i \leq 0 \quad \text{for all } \delta v_i \in RC(\hat{v}_i, \Omega),$

that is, only if

61 $\quad \hat{v}_i = +1.$

Similarly, if we set $\delta v_i = 0$, we find that (58) can be satisfied for all $\delta w_i \in RC(\hat{w}_i, \Omega)$ only if

$$\delta w_i \geqq 0 \qquad \text{for all } \delta w_i \in RC(\hat{w}_i, \Omega);$$

that is, we must have

62 $\qquad \hat{w}_i = 0.$

Thus when $\langle p_{i+1}, b \rangle > 1$, we must have $\hat{v}_i = +1$ and $\hat{w}_i = 0$. Similarly, when $\langle p_{i+1}, b \rangle < -1$ we find that

63 $\qquad \hat{v}_i = 0 \qquad \text{and} \qquad \hat{w}_i = +1.$

Now suppose that $|\langle p_{i+1}, b \rangle| < 1$. Then, putting, $\delta v_i = 0$ and $\delta w_i = 0$, consecutively, we find that all must have

64 $\qquad \hat{v}_i = \hat{w}_i = 0.$

Finally, suppose that $|\langle p_{i+1}, b \rangle| = 1$. If $\langle p_{i+1}, b \rangle = 1$, then for (59) to hold we must have

65 $\qquad \hat{w}_i = 0$

(since δw_i must be positive), but $\hat{v}_i$ is not determinable from (59). However, if $\langle p_{i+1}, b \rangle = -1$, then, from (59),

66 $\qquad \hat{v}_i = 0$

(since δv_i must be positive), but $\hat{w}_i$ is not determinable from (59).

The indeterminacy caused by $|\langle p_{i+1}, b \rangle| = 1$ must be resolved by recourse to boundary conditions. Thus (62) to (66) indicate the extent to which theorem (1.17) determines the optimal control sequence $\hat{u}_i = \hat{v}_i - \hat{w}_i$ for $i = 0, 1, \ldots, k-1$. Note that either $\hat{v}_i = 0$ or $\hat{w}_i = 0$ in (62) to (66), and hence our substitutions (53) and (54) are valid.

Second approach. Applying the maximum principle (34) to our minimum-fuel problem (48), we find that if $\hat{u}_i$ for $i = 0, 1, \ldots, k-1$ is an optimal control sequence, then it must maximize the hamiltonian

67 $\qquad p^0 |u_i| + \langle p_{i+1}, A x_i \rangle + \langle p_{i+1}, b \rangle u_i$

over $u_i \in [-1, +1]$, where, as before, we can show that $p^0 = -1$. Hence with $p^0 = -1$ we get

68 $\qquad \hat{u}_i = \begin{cases} +1 & \text{if } \langle p_{i+1}, b \rangle > 1 \\ -1 & \text{if } \langle p_{i+1}, b \rangle < -1 \\ 0 & \text{if } |\langle p_{i+1}, b \rangle| < 1, \end{cases}$

and again we find that (67) does not define $\hat{u}_i$ when $|\langle p_{i+1}, b \rangle| = 1$.

Both approaches yield exactly the same amount of information in this case, but the second approach is obviously more efficient. □

69 Exercise. Suppose that the cost functions $f_i{}^0(\cdot,\cdot)$ are of the form

70 $$f_i{}^0(x,u) = f_i{}^{0\prime}(x) + f_i{}^{0\prime\prime}(u) \qquad i = 0, 1, \ldots, k - 1,$$

where $f_i{}^{0\prime}$ and $f_i{}^{0\prime\prime}$ are convex functions, but not necessarily differentiable. Furthermore, suppose that the functions $f_i(\cdot,\cdot)$ are of the form

71 $$f_i(x,u) = A_i x + B_i u \qquad i = 0, 1, \ldots, k - 1,$$

where A_i is an $n \times n$ matrix and B_i is an $n \times m$ matrix. Assuming that the sets X_i for $i = 0, 1, \ldots, k$ and U_i for $i = 0, 1, \ldots, k - 1$ are convex, use result (2.3.59) to obtain a version of the maximum principle (34) directly for this case. □

72 Exercise. In addition to the assumptions made in exercise (69), suppose that the sets $X_i = E^n$ for $i = 1, \ldots, k - 1$, that $X_0 = \{\hat{x}_0\}$, that $X_k = \{\hat{x}_k\}$, and that there exists a control sequence $\mathfrak{U} = (u_0, u_1, \ldots, u_{k-1})$, with u_i in the relative interior of U_i, such that the resulting trajectory $\mathfrak{X} = (x_0, x_1, \ldots, x_k)$ satisfies $x_0 = \hat{x}_0$ and $\hat{x}_k = \hat{x}_k$. Show that there exist costate vectors $p_0, p_1, \ldots, p_k$ and a scalar $p^0 < 0$ satisfying the conditions of theorem (34). [*Hint:* Proceed as in exercise (1.48), but now with respect to the function $\mathbf{r}(\cdot)$ and the set Ω^* defined in exercise (2.3.55).] □

This concludes our discussion of necessary conditions of optimality for optimal control problems. The reader interested in sufficient conditions will find that he can easily adapt to optimal control problems all the sufficient conditions developed for nonlinear programming problems with convex functions in Section 3.6. The remainder of this book is devoted to computational methods.

REFERENCES

1. M. D. Canon, C. D. Cullum, and E. Polak: Constrained Minimization Problems in Finite Dimensional Spaces, *SIAM J. Control*, **4**:528–547 (1966).
2. H. Halkin: A Maximum Principle of the Pontryagin Type for Systems Described by Nonlinear Difference Equations, *SIAM J. Control*, **4**:90–111 (1966).
3. J. M. Holtzman: On the Maximum Principle for Nonlinear Discrete-time Systems, *IEEE Trans. Automatic Control*, **4**:528–547 (1966).

4. J. B. Rosen: Optimal Control and Convex Programming, *IBM Symp. Control Theory Applications*, pp. 223–237, Yorktown Heights, N.Y., October, 1964.
5. A. I. Propoi: The Maximum Principle for Discrete Control Systems, *Avtomatica i Telemechanica*, **7**:1177–1187 (1965).
6. B. W. Jordan and E. Polak: Theory of a Class of Discrete Optimal Control Systems, *J. Electron. Control*, **17**:697–713 (1964).

5
Optimal control and linear programming

5.1 INTRODUCTION

Beginning with this chapter, we shall examine in detail certain classes of discrete optimal control problems, together with methods for their solution. These problems are characterized by *convex* cost functions, *linear* dynamics, *linear* equality constraints and *convex* inequality constraints. The reason for this restriction is that, by and large, these are the only problems for which satisfactory methods of solution exist. We shall group problems into classes according to the properties of the corresponding Basic Problem (see Sections 1.4 and 1.5).

In this chapter we shall treat a class of discrete optimal control problems for which the corresponding Basic Problem is a linear programming problem of the following form.

1   The linear programming problem. Minimize $\langle d, z \rangle$ subject to $Rz = c$, $Pz \leqq \nu$, and $\alpha \leqq z \leqq \beta$, where $z, d, \alpha,$ and β are in E^n, R is an $m \times n$ matrix, P is an $l \times n$ matrix, and $c \in E^m$ and $\nu \in E^l$ (we shall always assume that $\alpha^i < \beta^i$ for $i = 1, 2, \ldots, n$ and that it is possible to have $\alpha^i = -\infty$ and $\beta^j = \infty$ for any i and j in $\{1, 2, \ldots, n\}$). □

5.2 LINEAR CONTROL PROBLEMS

The first class of problems we shall consider consists of problems in which the dynamics, the cost, and the constraints are all linear or affine.

1 The linear control problem. Consider a dynamical system described by the linear difference equation

2
$$x_{i+1} - x_i = Ax_i + Bu_i \qquad i = 0, 1, \ldots, k-1,$$

where $x_i \in E^n$ is the system state at time i, $u_i \in E^m$ is the system input at time i, A is an $n \times n$ matrix, and B is an $n \times m$ matrix. Find a control sequence $\hat{\mathfrak{u}} = (\hat{u}_0, \hat{u}_1, \ldots, \hat{u}_{k-1})$ and a corresponding trajectory $\hat{\mathfrak{X}} = (\hat{x}_0, \hat{x}_1, \ldots, \hat{x}_k)$, determined by (2), which minimize the linear cost function

3
$$\sum_{i=0}^{k} \langle d'_i, x_i \rangle + \sum_{i=0}^{k-1} \langle d''_i, u_i \rangle,$$

where $d'_i \in E^n$ for $i = 0, 1, \ldots, k$ and $d''_i \in E^m$ for $i = 0, 1, \ldots, k-1$. The minimization is subject to the constraints

4
5
$$u_i \in U_i = \{u : F_i u \leq w_i\} \qquad i = 0, 1, \ldots, k-1$$
$$x_i \in X'_i = \{x : H_i x \leq v_i\} \qquad i = 0, 1, \ldots, k,$$

and in addition, x_0 and x_k are required to satisfy

6
$$G_0 x_0 = c_0 \qquad G_k x_k = c_k,$$

where the F_i and w_i for $i = 0, 1, \ldots, k-1$, the H_i and v_i for $i = 0, 1, \ldots, k$, and G_0, G_k, c_0, and c_k are appropriate-dimensional matrices and vectors [of course, some of the matrices in (4) to (6) may be zero]. □

Obviously, problem (1) is a special case of the general optimal control problem (4.1.1). Note, however, that assumptions (4.1.8) to (4.1.10) are not needed for the linear case.

The most straightforward way to reduce the linear control problem (1) to the form of the linear programming problem (1.1) is to follow the method of Section 1.5 for obtaining an equivalent Basic

Problem. Thus, letting $z = (x_0, x_1, \ldots, x_k, u_0, u_1, \ldots, u_{k-1})$, we obtain from (1) and (7) that

7
$$r(z) = \begin{bmatrix} x_1 - x_0 - Ax_0 - Bu_0 \\ \vdots \\ x_k - x_{k-1} - Ax_{k-1} - Bu_{k-1} \\ G_0 x_0 - c_0 \\ G_k x_k - c_k \end{bmatrix} = Rz - c = 0$$

and from (4) and (5) that

8
$$z \in \Omega = \{z : Pz \leq \nu, \alpha \leq z \leq \beta\},$$

where in (7) R is a matrix with $(k+1)n + km$ columns and at least kn rows and in (8) P is a matrix defined by (4) and (5). The vectors α and β in (8) may or may not be finite.[1] Now, setting

9
$$f(z) = \langle d, z \rangle,$$

with

10
$$d = (d_0', d_1', \ldots, d_k', d_0'', d_1'', \ldots, d_{k-1}''),$$

and collecting expressions (3), (7), and (8), we see immediately that we have just rewritten problem (1) in the form of the linear programming problem (1.1).

Although the above method for transcribing problem (1) into the form of a linear programming problem is very direct, it results in a matrix R which usually has a very large number of rows. This makes problem (1) extremely difficult to solve by means of the simplex algorithm unless special codes are used which take advantage of the large number of zeros in R (usually such codes incorporate decomposition methods and are quite sophisticated).

It is possible to transcribe problem (1) into a linear programming problem with a matrix R which has considerably fewer rows than (9). However, this alternate transcription is obtained only at the expense of additional calculation. Thus, solving (2) for x_i, with $i = 1, 2, \ldots, k$, we obtain

11
$$x_i = (I - A)^i x_0 + \sum_{j=0}^{i-1} (I - A)^{i-j-1} B u_j.$$

---

[1] In converting the control problem to Basic Problem form it is advantageous not to include constraints for the form $u^i \leq \delta^i$ into the matrix P, but to collect them all together into the form $\alpha \leq z \leq \beta$.

SEC. 5.3 LINEAR CONTROL PROBLEMS WITH PIECEWISE-LINEAR COST

12 Exercise. Use (11), together with (3) to (6), to transcribe the linear control problem (1) into the form of problem (1.1), with $z = (x_0, u_0, u_1, \ldots, u_{k-1})$. (*Hint:* See Section 2.4 for guidance.) ☐

In two-point-boundary-value problems, as well as in other problems with few constraints, the transcription indicated in (12) offers great advantages. However, as the complexity of the constraints increases, this transcription may prove to be less attractive (e.g., see Section 4).

5.3 CONTROL PROBLEMS WITH LINEAR DYNAMICS, LINEAR CONSTRAINTS, AND PIECEWISE-LINEAR COST

The linear control problems by no means exhaust the family of control problems which can be reduced to linear programming problems. In fact, there are a number of important control problems with dynamics (2.2) and constraints (2.4) to (2.6), but with nonlinear cost functions, which also reduce to linear-programming-problem form. We shall consider only two such cases.

We begin by considering a class of optimal control problems with dynamics as in (2.2) and constraints as in (2.4) to (2.6), but with a piecewise-linear cost function, which can be transcribed by either of the techniques used in Section 2, into the form

1 $$\text{Minimize } f(z) = \sum_{i=1}^{\mu} |\langle d_i, z \rangle + \eta^i| \text{ subject to } Rz = c, Pz \leq \nu,$$

and $\alpha \leq z \leq \beta$.

When the first method discussed in Section 2 is used to transcribe the optimal control problem into the form (1), the quantities z, α, β, R and c in (1) are related to the control problem in exactly the same manner as in (2.7) and (2.8). The vectors $d_i \in E^{(k+1)n+km}$, the scalars η^i, and the integer μ are derived from the cost function originally given for the optimal control problem (2.1). Consider now the following situations where a piecewise-linear cost will arise.

2 Example. In a number of attitude-control systems for space vehicles the control force is obtained by ejecting fuel under pressure. Frequently, the magnitude of the force is proportional to the rate at which fuel is ejected. Under appropriate assumptions about the mode of operation of the system, the control system can be modeled as having dynamics of the form (2.2) and constraints of the form (2.4), (2.5), while the total amount of fuel expended is found to be proportional to the quantity

3 $$\sum_{i=0}^{k-1} \sum_{j=1}^{m} |u_{ij}|.$$

Thus an attempt to minimize the fuel expended in a certain maneuver leads to a cost of the form appearing in problem (1).

4 Example. In a number of practical control situations, we may be required to get the state of a dynamical system, such as (2.2), to agree "as closely as possible" with some desired value at a prespecified time k. If the criterion "as closely as possible" is a subjective one, it might be acceptable to take as the definition of magnitude the L_1 norm of the error $x_k - \xi_k$,

5
$$\sum_{i=1}^{n} |x_k{}^i - \xi_k{}^i|.$$

where ξ_k is the desired value of x_k, the state at time k. A comparison of the simplex algorithm to be presented in this chapter with the quadratic programming algorithm discussed in Chapter 6 indicates clearly that there are computational advantages in using (5) instead of the more usual euclidean norm.

The reduction of problem (1) to the form of the linear programming problem (1.1) is accomplished by showing that the optimal solutions of (1) can be obtained by solving the following problem.

6 Minimize $\sum_{i=1}^{\mu} (v^i + y^i)$ subject to $Rz = c$, $v^i - y^i = \langle d_i, z \rangle + \eta^i$ for $i = 1, 2, \ldots, \mu$, $Pz \leq \nu$, $\alpha \leq z \leq \beta$, and $v^i \geq 0$ and $y^i \geq 0$ for $i = 1, 2, \ldots, \mu$.

It should be clear that this is indeed a linear programming problem, readily identified with form (1.1).

7 Theorem. The vector $\hat{z}$ is an optimal solution to problem (1) if and only if $\hat{z}$, together with the vectors $\hat{v}$ and $\hat{y}$, defined by

8
$$\hat{v}^i = \begin{cases} \langle d_i, \hat{z} \rangle + \eta^i & \text{if } \langle d_i, \hat{z} \rangle + \eta^i > 0 \\ 0 & \text{if } \langle d_i, \hat{z} \rangle + \eta^i \leq 0 \end{cases}$$

and

9
$$\hat{y}^i = \begin{cases} 0 & \text{if } \langle d_i, \hat{z} \rangle + \eta^i \geq 0 \\ -(\langle d_i, \hat{z} \rangle + \eta^i) & \text{if } \langle d_i, \hat{z} \rangle + \eta^i < 0, \end{cases}$$

$i = 1, 2, \ldots, \mu$, constitute an optimal solution to problem (6).

Proof. Let z be any vector satisfying the constraints of problem (1), that is, $Rz = c$, $Pz \leq \nu$, and $\alpha \leq z \leq \beta$. If, for $i = 1, 2, \ldots, \mu$, we define v^i and y^i as in (8) and (9) above, with z taking the place of $\hat{z}$, then z, together with the v^i and y^i for $i = 1, 2, \ldots, \mu$, satisfies the constraints of problem (6). In addition, the cost function of (6) evaluated for this case becomes

$$\sum_{i=1}^{\mu} (v^i + y^i) = \sum_{i=1}^{\mu} |\langle d_i, z \rangle + \eta^i|,\qquad(10)$$

which is the same as the cost in (1).

Next we show that if $\hat{z}$, together with $\hat{v}^i$ and $\hat{y}^i$ for $i = 1, 2, \ldots, \mu$, is an optimal solution to problem (6), then these quantities must be related by (8) and (9). To show this it is sufficient to show that $\hat{v}^i$ and $\hat{y}^i$ cannot simultaneously be strictly positive for any $i \in \{1, 2, \ldots, \mu\}$. Then (8) and (9) follow from the constraint that

$$\hat{v}^i - \hat{y}^i = \langle d_i, \hat{z} \rangle + \eta^i.\qquad(11)$$

Thus suppose that for some i, $\hat{v}^i > 0$ and $\hat{y}^i > 0$. Let us first suppose that $\hat{v}^i \geqq \hat{y}^i$ and define the new variables $\dot{v}^i$ and $\dot{y}^i$ by

$$\dot{v}^i = \hat{v}^i - \hat{y}^i \geqq 0 \quad \text{and} \quad \dot{y}^i = \hat{y}^i - \hat{y}^i = 0.\qquad(12)$$

Then

$$\dot{v}^i - \dot{y}^i = \hat{v}^i - \hat{y}^i = \langle d_i, \hat{z} \rangle + \eta^i,\qquad(13)$$

and hence if we replace $\hat{v}^i$ and $\hat{y}^i$ by $\dot{v}^i$ and $\dot{y}^i$, respectively, in the optimal solution, we shall still satisfy the constraints. However,

$$\dot{v}^i + \dot{y}^i = \hat{v}^i - \hat{y}^i < \hat{v}^i + \hat{y}^i,\qquad(14)$$

since $\hat{y}^i > 0$. This contradicts the optimality of $\hat{v}^i$ and $\hat{y}^i$. The case where $\hat{y}^i \geqq \hat{v}^i$ can be disposed of similarly.

It follows immediately from (8), (9), and the above that the optimal cost in (6) is given by

$$\sum_{i=1}^{\mu} (\hat{v}^i + \hat{y}^i) = \sum_{i=1}^{\mu} |\langle d_i, \hat{z} \rangle + \eta^i|.\qquad(15)$$

The rest of the proof is obvious and we therefore omit it. □

Note that this reduction of problem (1) to a linear programming problem of form (1.1) is obtained only at the expense of increasing the dimensions of the problem. Specifically, we have added as many equality constraints as there are terms in the summation of the cost function, and twice this number of additional variables. This price is more than offset, however, by the fact that the problem can now be solved by means of the very efficient simplex algorithm.

Let us conclude our discussion of optimal control problems which are reducible to form (1.1) with the following, possibly less obvious, case. Consider an optimal control problem with dynamics as in (2.2) and constraints as in (2.4) to (2.6), which can be reduced, by the previously indicated techniques, to the form:

(16) Minimize $f(z) = \max\limits_{i \in \{1,2,\ldots,\mu\}} |\langle d_i, z \rangle + \eta^i|$ subject to $Rz = c$, $Pz \leq \nu$, and $\alpha \leq z \leq \beta$.

The criterion in (16) is yet another form of a piecewise-linear function.

(17) **Example.** Consider again the problem in example 4, where it was desired to minimize the distance between the actual terminal state x_k and some desired terminal state ξ_k. Instead of the L_1 norm (5), we could choose the L_∞ norm

(18) $$\max\limits_{i=1,\ldots,n} |x_k^i - \xi_k^i|$$

to indicate the deviation from the desired value. □

(19) **Example.** Consider a control problem where it is desired to approximate a given trajectory $(\xi_0, \xi_1, \ldots, \xi_k)$ as closely as possible. We could choose as the distance measure between the actual trajectory $(x_0, x_1, \ldots, x_k)$ and the desired trajectory the norm

(20) $$\max\limits_{\substack{j=0,\ldots,k \\ i=1,\ldots,n}} |x_j^i - \xi_j^i|.$$

The resulting optimal trajectory is the so-called *Tchebyshev approximation* to $(\xi_0, \xi_1, \ldots, \xi_k)$. □

Now consider the linear programming problem:

(21) Minimize y subject to $Rz = c$, $Pz \leq \nu$, $y + \langle d_i, z \rangle + \eta^i \geq 0$ and $y - (\langle d_i, z \rangle + \eta^i) \geq 0$ for $i = 1, 2, \ldots, \mu$, $\alpha \leq z \leq \beta$, and $y \geq 0$, where R, c, d_i, η^i, α, β, and μ are as defined in (16).

We now relate the solutions of problems (16) and (21).

(22) **Theorem.** The vector $\hat{z}$ is an optimal solution to problem (16) if and only if $\hat{z}$, together with

(23) $$\hat{y} \triangleq \max\limits_{i=1,\ldots,\mu} |\langle d_i, \hat{z} \rangle + \eta^i|,$$

is an optimal solution to problem (21).

Proof. Clearly, any vector z which is feasible for (16), taken together with a y defined as in (23), is feasible for (21), and vice versa. Suppose that $\hat{z}$ and $\hat{y}$ are an optimal solution for (21), and suppose that

$\hat{z}$ is not an optimal solution to (16). Let z be any optimal solution to (16). Then z, together with the corresponding y defined by (23), is feasible for (21), and it is necessarily true that $y < \hat{y}$, because (23) corresponds to the cost function in (16). This contradicts the optimality of $\hat{z}$ and $\hat{y}$ for (21).

The converse is proved in a similar manner. □

Note that we have considerably increased the dimensions of the problem in going from (16) to (21), which has 2μ inequalities more than (16). Since each of these inequalities must be converted to an equality eventually, we shall have to add still more variables. However, as before, we are now in a position to apply the powerful algorithms of linear programming to this problem.

24 **Exercise.** Perform the reduction to a linear programming problem for the case when $f(z) \triangleq \max_{i=1,\ldots,\mu} (\langle d_i, z \rangle + \eta_i)$ and the constraints are as in (16). □

5.4 THE CANONICAL LINEAR PROGRAMMING PROBLEM

The usual way to deal with a linear programming problem of form (1.1) is to transcribe it into the following form, which is then solved by the simplex algorithm.

1 **The canonical linear programming problem (CLP).** Given an $m \times n$ ($m \leq n$) full-rank matrix R and vectors $c \in E^m$ and $d \in E^n$, find a vector $\hat{z} \in E^n$ which minimizes $\langle z, d \rangle$ subject to the constraints

2 $\qquad Rz = c$
3 $\qquad z \geq 0.$ □

Obviously, this problem is a special case of problem (1.1). We shall now see that we can construct from (1.1) an equivalent problem of form (1) (but the resultant matrix R may not have full rank).

4 **Theorem.** Consider the linear programming problem (1.1):

5 $\qquad$ Minimize $\langle d, z \rangle$ subject to $Rz = c$, $Pz \leq v$, and $\alpha \leq z \leq \beta$. Let $\tilde{\alpha}$ and $\tilde{\beta}$ be vectors in E^n, defined for $i = 1, 2, \ldots, n$ by

6 $\qquad \tilde{\alpha}^i = \begin{cases} \alpha^i & \text{if } -\infty < \alpha^i \\ 0 & \text{if } -\infty = \alpha^i \end{cases}$

7 $\qquad \tilde{\beta}^i = \begin{cases} \beta^i & \text{if } \beta^i < \infty \\ 0 & \beta^i = \infty. \end{cases}$

Now consider the derived canonical linear programming problem:

8   Minimize $\langle d,w \rangle$ subject to $Rw = c - R\bar{\alpha}$, $Pw + y = \nu - P\bar{\alpha}$, $v + w = \tilde{\beta} - \bar{\alpha}$, and $v \geqq 0$, $w \geqq 0$, and $y \geqq 0$.

Then $\hat{z}$ is an optimal solution for (5) if and only if $\hat{w} \triangleq \hat{z} - \bar{\alpha}$, $\hat{v} \triangleq \tilde{\beta} - \hat{z}$, and $\hat{y} \triangleq \nu - P\hat{z}$ constitute an optimal solution for (8).

Proof. Suppose that $\hat{z}$ is an optimal solution to (5), but $\hat{w}$, $\hat{v}$, and $\hat{y}$, defined as above, are not an optimal solution to (8). Then there exists a feasible solution w^*, v^*, and y^* to (8) such that

9   $\langle d,w^* \rangle + \langle d,\bar{\alpha} \rangle < \langle d,\hat{w} \rangle + \langle d,\bar{\alpha} \rangle$.

Let $z^* = w^* + \bar{\alpha}$. Then from the relations in (8), $Rz^* = c$, $Pz^* \leqq \nu$, and $\alpha \leqq z^* \leqq \beta$; that is, z^* is feasible. Furthermore, by (9),

10   $\langle d,z^* \rangle < \langle d,\hat{z} \rangle$,

which contradicts the optimality of $\hat{z}$.

The converse is established in a similar manner. □

Now let $x = (v,w,y) \in E^{(2n+l)}$, $\tilde{c} = c - R\bar{\alpha}$, $\tilde{\nu} = \nu - P\bar{\alpha}$, $\delta = \tilde{\beta} - \bar{\alpha}$, and

11   $\tilde{d} = (0,d,0) \in E^{2n+l}$.

Then (8) becomes:

12   Minimize $\langle \tilde{d},x \rangle$ subject to $x \geqq 0$ and

$$\tilde{R}x = \begin{bmatrix} I & I & 0 \\ R & 0 & 0 \\ P & 0 & I \end{bmatrix} \begin{bmatrix} v \\ w \\ y \end{bmatrix} = \begin{bmatrix} \delta \\ \tilde{c} \\ \tilde{\nu} \end{bmatrix}.$$

The matrix $\tilde{R}$ is $(n + m + l) \times (2n + l)$ dimensional and will obviously have full rank if and only if the matrix R has full rank. Thus the full-rank property is preserved in the transcription; however, the number of equations and variables has increased considerably. Consequently, we shall examine, in Section 9, an algorithm which requires more logic operations than the simplex algorithm but which solves linear programming problems of the form: Minimize $\langle d,z \rangle$ subject to $Rz = c$ and $\alpha \leqq z \leqq \beta$ when both α and β are finite vectors. This form frequently arises in optimal control problems such as the minimum-fuel problem (4.2.48).

5.5 CHARACTERIZATION OF AN OPTIMAL SOLUTION TO THE CANONICAL LINEAR PROGRAMMING PROBLEM

We begin by obtaining necessary and sufficient conditions of optimality for the canonical linear programming problem (4.1). Note that we may appeal to theorems (3.3.5) and (3.6.1), because the cost function $\langle d,z \rangle$ is convex and the constraints $Rz = c$ and $z \geq 0$ are affine. In the context of CLP (4.1), theorems (3.3.5) and (3.6.1) combine to yield the following result.

Theorem. A feasible solution $z \in E^n$ to CLP (4.1) is an optimal solution if and only if there exists a vector $\psi = (\psi^1, \psi^2, \ldots, \psi^m) \in E^m$ such that for $i = 1, 2, \ldots, n$

$$\langle r_i, \psi \rangle - d^i = 0 \quad \text{if } z^i > 0$$
$$\langle r_i, \psi \rangle - d^i \leq 0 \quad \text{if } z^i = 0,$$

where r_i is the ith column of R and d^i is the ith component of d.

Proof. Applying theorems (3.3.5) and (3.6.1), we conclude that a feasible solution z to CLP (4.1) is optimal if and only if there exist vectors $\psi \in E^m$ and $\mu \in E^n$ such that for $i = 1, 2, \ldots, n$

$$-d^i + \langle r_i, \psi \rangle + \mu^i = 0 \qquad \mu^i \geq 0 \qquad \mu^i z^i = 0.$$

Obviously, (3) implies (2), and, by inspection, (2) implies the existence of a vector $\mu \in E^n$ satisfying (3). □

From a geometric point of view, it is clear that the constraint set $\Omega' = \{z \in E^n : Rz = c, z \geq 0\}$ is a convex polyhedron and that the equation $\langle d,z \rangle = $ constraint specifies a hyperplane. Consequently, assuming that Ω' is not empty, to solve CLP (4.1) we must shift this hyperplane, parallel to itself, as far as possible in the direction of the vector $-d$, while still maintaining contact with the set Ω'. Since the boundaries of Ω' are portions of linear manifolds (planes), it is clear that the optimal cost plane ($\langle d,z \rangle = \langle d,\hat{z} \rangle$, with $\hat{z}$ an optimal solution) may contact the set Ω' along an entire face or edge; i.e., there may be many solutions. It is also clear that when Ω' is unbounded the optimal cost $\langle d,z \rangle$ might possibly be $-\infty$, in which case we shall say that CLP (4.1) has no solution. The simplex algorithm which will be presented later has a feature to establish whether the set Ω' is empty and whether $\langle d,z \rangle$ is unbounded on Ω'. Consequently, there is no need for a separate test to establish the existence of optimal solutions.

It is clear from the above that if the canonical linear programming problem (4.1) has an optimal solution, then either this solution is an extreme point (see Appendix A) of the convex polyhedron Ω' or else there exists another optimal solution which is an extreme point of Ω'. We shall now obtain a characterization of extreme points of Ω' and establish formally the existence of extreme point optimal solutions.

4 Definition. Let z be a feasible solution to CLP (4.1); that is, $z \in \Omega' = \{z : Rz = c, z \geq 0\}$. Then the index sets $I(z)$ and $\bar{I}(z)$ are defined by

$$I(z) = \{i \in \{1,2,\ldots,n\} : z^i = 0\}$$
$$\bar{I}(z) = \{i \in \{1,2,\ldots,n\} : z^i > 0\};$$

that is, $I(z)$ is the *active-constraints index set* [cf. (3.4.1)]. □

5 Exercise. Show that a vector $\hat{z} \in \Omega' = \{z : Rz = c, z \geq 0\}$ is an extreme point of Ω' if and only if the set of vectors $\{r_i : i \in \bar{I}(z)\}$ are linearly independent, where r_i is the ith column of the matrix R. [*Hint:* Use the fact that for any $\hat{z} \in \Omega'$, $\hat{z} = \sum_{i \in \bar{I}(\hat{z})} \hat{z}^i e_i$, where for $i = 1, 2, \ldots, n$ the $e_i = (0,0,0,1,0,\ldots,0)$ are the usual unit basis vectors for E^n, and hence $R\hat{z} = \sum_{i=1}^{n} \hat{z}^i r_i = \sum_{i \in \bar{I}(\hat{z})} \hat{z}^i r_i$.] □

The existence of an extreme-point optimal solution for the canonical linear programming problem (4.1) is now concluded from the following result.

6 Theorem. If there exists an optimal solution to CLP (4.1), then there exists an optimal solution $\hat{z}$ such that the vectors in the set $\{r_i : i \in \bar{I}(\hat{z})\}$ are linearly independent.

Proof. Let z_0 be any optimal solution to CLP (4.1), and suppose that the vectors in the set $\{r_i : i \in \bar{I}(z_0)\}$ are linearly dependent. Then there exist scalars γ^i, with $i \in \bar{I}(z_0)$ and not all zero, such that

7
$$\sum_{i \in \bar{I}(z_0)} \gamma^i r_i = 0.$$

For any real number θ let $z(\theta) \in E^n$ be a vector whose components are

8
$$z^i(\theta) = \begin{cases} z_0^i = 0 & \text{if } i \in I(z_0) \\ z_0^i + \theta\gamma^i & \text{if } i \in \bar{I}(z_0). \end{cases}$$

It follows immediately that for any θ

9
$$Rz(\theta) = \sum_{i=1}^{n} z^i(\theta) r_i$$
$$= \sum_{i \in \bar{I}(z_0)} z_0^i r_i + \theta \sum_{i \in \bar{I}(z_0)} \gamma^i r_i = c.$$

Thus $z(\theta)$ satisfies the equality constraints of CLP (4.1) for any value of θ. Since $z(\theta)$ differs from z_0 only in those components of z_0 which originally satisfied $z_0^i > 0$, it is clear that $z(\theta)$ will satisfy

$z(\theta) \geq 0$ for all values of θ which are sufficiently small. Hence $z(\theta) \in \Omega'$ for all θ sufficiently small. By direct calculation,

10
$$\langle d,z(\theta) \rangle = \sum_{i=1}^{n} d^i z^i(\theta)$$
$$= \sum_{i \in I(z_0)} d^i z_0{}^i + \theta \sum_{i \in \bar{I}(z_0)} \gamma^i d_i = \langle d,z_0 \rangle + \theta \sum_{i \in \bar{I}(z_0)} \gamma^i d_i.$$

It now follows that $\sum_{i \in \bar{I}(z_0)} \gamma^i d^i = 0$, for otherwise we could choose a $\theta \neq 0$ such that $z(\theta)$ is feasible and $\langle d,z(\theta) \rangle < \langle d,z_0 \rangle$, which contradicts the optimality of z_0.

Now, there must be at least one $\gamma^i \neq 0$ in (7), and we may therefore assume that there is at least one $\gamma^i < 0$ for $i \in \bar{I}(z_0)$. Hence we may define

11
$$\theta_1 = \max \{\theta > 0 : z^i(\theta) = z_0{}^i + \theta \gamma^i \geq 0,\ i \in \bar{I}(z_0)\}.$$

Clearly, by (9) to (11), $z(\theta_1)$ is also an optimal solution to CLP (4.1). Furthermore, since there is at least one $\gamma^i < 0$ for $i \in \bar{I}(z_0)$, we must have, for $z_1 \triangleq z(\theta_1)$,

12
$$\bar{I}(z_1) \subset \bar{I}(z_0),$$

and the inclusion is proper, by (11). Hence the cardinality of $\bar{I}(z_1)$ is at least one less than the cardinality of $\bar{I}(z_0)$.

If z_1 does not satisfy the property that the vectors in the set $\{r_i : i \in I(z_1)\}$ are linearly independent, we may repeat the above process to obtain an optimal vector z_2, with the cardinality of $\bar{I}(z_2)$ again strictly smaller than the cardinality of $\bar{I}(z_1)$. Clearly, this process must terminate in a finite number of steps. □

Since the polyhedron Ω' must have a finite number of vertices, it is clear that an optimal solution (if it exists) can be found simply by enumeration, i.e., in a finite number of steps. The simplex algorithm we are about to examine is a systematic and very efficient procedure for finding an optimal extreme point.

5.6 SOME PRELIMINARY REMARKS ON THE SIMPLEX ALGORITHM

In Section 5 we observed that an optimal solution to the canonical linear programming problem (4.1) must be a point of the convex polyhedron $\Omega' = \{z : Rz = c,\ z \geq 0\}$ which is located as far as possible in the direction of the vector $-d$. In exercise (5.5) and theorem (5.6) we established that if CLP (4.1) has a solution, then it has a solution which is also an extreme point of the convex set Ω'. The simplex algorithm proceeds in the following way.

First we find an extreme point z_0 of Ω'. Then we calculate the direction cosines of all the edges of Ω' leading away from this point, in order to determine the edges of Ω' which make an acute angle with $-d$. By following such an edge from the extreme point z_0, either (a) we reach an adjacent extreme point z_1 of Ω', with $\langle d,z_1\rangle < \langle d,z_0\rangle$, or (b) along this edge $\langle d,z\rangle \to -\infty$. If (b) occurs, CLP (4.1) does not have an optimal solution (since we do not accept $-\infty$ as an optimal value). If (a) occurs, the above process is repeated until no edge emanating from the current extreme point makes an acute angle with $-d$, and CLP (4.1) is solved.

Conceptually, we see that this algorithm is rather elementary. The question, of course, is whether the calculations can be carried out in an efficient manner. We shall see that they can be.

The algorithm divides naturally into two subprocedures, each of which we shall consider separately in the following sections.[1] The first subprocedure solves the problem of finding an initial extreme point z_0 of Ω', which is required to initiate the simplex algorithm. The second subprocedure is used for finding extreme points z_i, adjacent[2] to a given extreme point z_0, with the property that $\langle d,z_i\rangle < \langle d,z_0\rangle$.

Before proceeding any further, let us introduce some new notation and terminology.

1 **Definition.** A vector $z \in E^n$ is said to be a *basic solution* to CLP (4.1) if z is a feasible solution and the vectors in the set $\{r_i : i \in \bar{I}(z)\}$ are linearly independent [see (5.4)]. □

2 **Definition.** A vector $z \in E^n$ is said to be a *nondegenerate basic solution* to CLP (4.1) if z is a basic solution and the vectors in the set $\{r_i : i \in \bar{I}(z)\}$ form a basis for E^m. □

3 **Remark.** It is clear from exercise (5.5) that a basic solution to CLP (4.1) is an extreme point of Ω', and hence we shall use the term *extreme point* and *basic solution* interchangeably. However, in many texts on linear programming the term *basic solution* is given preference. □

4 **Remark.** A basic solution z can have at most m positive components z^i, whereas a nondegenerate basic solution z has exactly m positive components z^i. □

[1] The division of the algorithm into subprocedures is convenient in exposition. For computational purposes the two subprocedures can be merged [2].

[2] Two extreme points of a closed convex polyhedron are called *adjacent* if the line connecting them is an edge of the polyhedron.

SEC. 5.7 DETERMINING AN INITIAL EXTREME POINT OF Ω'

Given a feasible solution z to CLP (4.1), we can always determine whether z is a basic solution (an extreme point of Ω') by checking the linear independence of the vectors $\{r_i : i \in \bar{I}(z)\}$. When z is a nondegenerate basic solution to CLP (4.1), these vectors form a basis for E^m, the significance of which will become apparent later. Let us note, however, that it is always possible to associate a set of m linearly independent columns of R with a basic solution z in the following way.

5 **Definition.** Let z be a basic solution to CLP (4.1). We shall mean by a *basis-indicator set* of z any index set $\bar{J}(z) \subset \{1, 2, \ldots, n\}$ with the following properties:

6     $\bar{I}(z) \subset \bar{J}(z)$.
7   The vectors $\{r_i : i \in \bar{J}(z)\}$ are a basis for E^m. □

(If z is a nondegenerate basic solution, then, of course, $\bar{J}(z) = \bar{I}(z)$.)

8 **Lemma.** Suppose z is a basic solution to CLP (4.1). Then there exists a basis-indicator set $\bar{J}(z) \subset \{1, 2, \ldots, n\}$ satisfying (6) and (7).

Proof. The vectors in the set $\{r_i : i \in \bar{I}(z)\}$ are linearly independent, since z is a basic solution. If the cardinality of $\bar{I}(z)$ is m, we are done. Since, by assumption, the rank of the matrix R is m (that is, there is at least one set of m linearly independent columns of R), we can always find a sufficient number of additional columns of R, say, $\{r_i : i \in K(z)\}$, with $K(z) \cap \bar{I}(z) = \phi$, such that the vectors in the set $\{r_i : i \in [\bar{I}(z) \cup K(z)] \triangleq \bar{J}(z)\}$ form a basis E^m. □

This concludes our preliminary remarks, and we are now ready to examine the problem of initializing the simplex algorithm.

5.7 PROCEDURE FOR DETERMINING AN INITIAL EXTREME POINT OF Ω'

As indicated in the preceding section, to apply the simplex algorithm to the canonical linear programming problem (4.1), we must first obtain an initial basic solution to constraints (4.2), (4.3), that is, a vector z_0 satisfying

1     $Rz = c$
2     $z \geqq 0$,

with the property that the columns r_i, with $i \in \bar{I}(z)$, of the matrix R are linearly independent. To compute such a z_0 we proceed as follows.

Step 1. Multiply every strictly negative component of c, together with the corresponding row of R in (1), by -1. At the end of this step we have a new system of equations of form (1), with the new $c \geq 0$. We therefore proceed on the assumption that $c \geq 0$ in (1).

Step 2. Use the simplex algorithm to solve the linear programming problem:

(3) $\qquad$ Minimize $\sum_{i=1}^{m} \xi^i$ subject to the constraints

(4) $\qquad Rz + \xi = c \qquad z \geq 0 \qquad \xi \geq 0.$

Note that the $m \times (n + m)$ matrix

(5) $\qquad [R \vdots I]$

has rank m, since the identity matrix I has rank m. Furthermore, note that $\bar{z} = 0$ and $\bar{\xi} = c$ constitute a basic solution to (4) which can be used as a starting point for the simplex algorithm in solving problem (3).

Note that $\sum_{i=1}^{m} \xi^i \geq 0$ for all (z,ξ) satisfying (4). Hence if there is a $z \geq 0$ such that $Rz = c$, and (z_0, ξ_0) is any optimal solution to problem (3), then $\xi_0 = 0$, and z_0 is a feasible solution to constraints (1) and (2). Furthermore, if (z_0, ξ_0) is obtained by means of the simplex algorithm, the columns of the matrix $[R \vdots I]$ associated with the strictly positive elements of (z_0, ξ_0) will be linearly independent; that is, z_0 will be a basic solution to (1) and (2). □

5.8 GENERATING IMPROVED ADJACENT EXTREME POINTS

Let z_0 be a basic solution to the canonical linear programming problem (4.1), and let $\bar{J}(z_0)$ be a basis-indicator set, defined as in (6.5); that is, $z_0^i \geq 0$ for $i \in \bar{J}(z)$ and the set of vectors $\{r_i : i \in \bar{J}(z_0)\}$ form a basis for E^m. Then we may write

(1) $\qquad z_0 = \sum_{i \in \bar{J}(z_0)} z_0^i e_i,$

where for $i = 1, 2, \ldots, n$, $e_i = (0,0, \ldots ,0,1,0, \ldots ,0) \in E^n$ are the usual unit basis vectors whose ith component is 1 and whose other components are zero.

Let $j \in J(z_0)$, the complement of $\bar{J}(z_0)$ in $\{1,2, \ldots ,n\}$. Then there exist scalars α_j^i for $i \in \bar{J}(z_0)$ such that

(2) $\qquad r_j = \sum_{i \in \bar{J}(z_0)} \alpha_j^i r_i.$

SEC. 5.8 GENERATING IMPROVED ADJACENT EXTREME POINTS

Since z_0 is a basic solution to CLP (4.1), we have from (1) that

3 $$Rz_0 = \sum_{i \in \bar{J}(z_0)} z_0{}^i r_i = c.$$

Hence if (2) is multiplied by any scalar $\theta \geqq 0$ and added to (3), we get, on rearranging terms,

4 $$\sum_{i \in \bar{J}(z)} (z_0{}^i - \theta \alpha_j{}^i) r_i + \theta r_j = c.$$

Suppose that $\alpha_j{}^i > 0$ for at least one $i \in \bar{J}(z_0)$.† Then let

5 $$\theta_0 = \min\left\{\frac{z_0{}^i}{\alpha_j{}^i} : i \in \bar{J}(z_0),\ \alpha_j{}^i > 0\right\} = \frac{z^k}{\alpha_j{}^k},$$

with the minimum taking place for $i = k$ (if ties occur, choose the smallest index k). It is easy to see that

6 $$z_0{}^k - \theta_0 \alpha_j{}^k = 0$$

and

7 $$z_0{}^i - \theta_0 \alpha_j{}^i \geqq 0 \qquad \text{for all } i \in \bar{J}(z_0).$$

Now let

8 $$z_1 = \sum_{i \in \bar{J}(z_0)} (z_0{}^i - \theta_0 \alpha_j{}^i) e_i + \theta_0 e_j.$$

From (6) and (7) we see that $z_1 \geqq 0$, and from (4) we see that $Rz_1 = c$; hence z_1 is a feasible solution to CLP (4.1).

We shall now show that z_1 is a basic solution, i.e., that the columns r_i of R, corresponding to the strictly positive components $z_1{}^i$, are linearly independent. To this end we define the index set $\bar{J}(z_1)$ by

9 $$\bar{J}(z_1) = \bar{J}(z_0) - \{k\} + \{j\},$$

where k is an index for which the minimum in (5) occurred and $j \in J(z_0)$ is as in (2) and (8). We shall show that $\bar{J}(z_1)$ conforms to definition (6.5), that is, that the vectors r_i for $i \in \bar{J}(z_1)$ form a basis for E^m. Note that if $i \in \bar{J}(z_1)$, then $z_1{}^i \geqq 0$, and if $i \in J(z_1)$, the complement of $\bar{J}(z_1)$ in $\{1, 2, \ldots, n\}$, then $z_1{}^i = 0$.

Suppose that the set of vectors $\{r_i : i \in \bar{J}(z_1)\}$ are linearly dependent. Then there exist scalars β^i for $i \in \bar{J}(z_1)$, not all zero, such that

$$\sum_{i \in \bar{J}(z_1)} \beta^i r_i = 0.$$

† When $\alpha_j{}^i \leqq 0$ for all the $i \in \bar{J}(z_0)$ and $j \in J(z_0)$ there are no extreme points adjacent to z_0 [see exercise (12)].

We claim that $\beta^j \neq 0$, for otherwise, by (9), we would have

$$\tag{10} \sum_{\substack{i \in \bar{J}(z_0) \\ i \neq k}} \beta^i r_i = 0,$$

which is impossible because the vectors $\{r_i: i \in \bar{J}(z_0)\}$ are linearly independent (and hence any subset of these vectors is linearly independent). We may therefore write

$$\tag{11} r_j = \sum_{\substack{i \in \bar{J}(z_0) \\ i \neq k}} \frac{\beta^i}{\beta^j} r_i.$$

Comparing (2) and (11) and again using the fact that the vectors in the set $\{r_i: i \in J(z_0)\}$ are linearly independent, we conclude that $\alpha_j{}^k = 0$ in (2). But this is a contradiction, since, by assumption, $\alpha_j{}^k > 0$. We must therefore conclude that the vectors in the set $\{r_i: i \in \bar{J}(z_1)\}$ are linearly independent, and hence that z_1 is a basic solution to CLP (4.1).

12 Exercise. Show that the extreme points z_0 and z_1 discussed above are adjacent extreme points. Also show that if all the $\alpha_j{}^i$ for $i \in \bar{J}(z_0)$ are negative or zero in (2) for all $j \in J(z_0)$, then z_0 is the only extreme point of $\Omega' = \{z: Rz = 0, z \geq 0\}$. □

We can summarize the above procedure as follows.

13 Procedure for computing adjacent extreme points. Let $z_0 = \sum_{i \in \bar{J}(z_0)} z_0^i e_i$ be a basic solution to CLP (4.1), and let $\bar{J}(z_0)$ be an associated basis indicator set.
  a. Find an index $j \in J(z_0)$ such that at least one $\alpha_j{}^i$, with $i \in \bar{J}(z_0)$, in (2) is positive.
  b. Compute θ_0 according to (5).
  c. Compute the adjacent extreme point according to (8).
  d. Compute the new basis-indicator set according to (9). □

14 Remark. Suppose that the extreme point z_1 was obtained from the extreme point z_0 by the above procedure. Referring to (9), we find that we can express this by saying "the extreme point z_1 was obtained from z_0 by exchanging the index $j \in J(z_0)$ for the index $k \in \bar{J}(z_0)$." □

It is important to note here that although z_0 and z_1, as computed above, are adjacent extreme points, it may very well be that

$z_0 = z_1$. To see why this is possible, let us return to expression (5),

$$\theta_0 = \min\left\{\frac{z_0{}^i}{\alpha_j{}^i} : i \in \bar{J}(z_0),\ \alpha_j{}^i > 0\right\}.$$

Now suppose that $z_0{}^k = 0$ for some $k \in \bar{J}(z_0)$ (that is, z_0 is a degenerate basic solution) and that $\alpha_j{}^k > 0$. Then, obviously, $\theta_0 = 0$, and according to (8), $z_1 = z_0$. If, however, $\theta_0 > 0$, then $z_1 \neq z_0$. It is common in mathematical programming literature to invoke a nondegeneracy assumption which precludes the occurrence of $\theta_0 = 0$ in (5). This assumption considerably simplifies the presentation of the simplex algorithm.

15 Nondegeneracy assumption. Every basic solution is a nondegenerate basic solution. □

16 Remark. When assumption (15) is satisfied, the vectors in the set $\{r_i : i \in \bar{I}(z_0)\}$ form a basis for E^m, and so $\bar{J}(z_0) = \bar{I}(z_0)$ in (2). Consequently, in (5), $z_0{}^i > 0$ for all $i \in \bar{J}(z_0)$, which means that $\theta > 0$, that is, that $z_1 \neq z_0$. □

After describing the simplex algorithm under the simplifying assumption (15), we shall examine a method for resolving degeneracies, i.e., for proceeding when some of the basic solutions are degenerate. In the meantime, to avoid confusion we shall state explicitly in theorems, remarks, etc., whether assumption (15) is required or not.

The following is one of the most important properties of procedure (13) when the nondegeneracy assumption applies. We shall make heavy use of this property in Section 6.11, where convergence of quadratic programming algorithms is considered.

17 Theorem. Suppose that all the basic solutions to the system $Rz = c$ and $z \geq 0$ are nondegenerate. If the extreme point z_1 of the set $\{z : Rz = c,\ z \geq 0\}$ was obtained from the extreme point z_0 by exchanging some index $\alpha \in I(z_0)$ for the index $\beta \in \bar{I}(z_0)$ [that is, $\bar{I}(z_1) = \bar{I}(z_0) + \{\alpha\} - \{\beta\}$], then z_0 is the unique extreme point which can be obtained from z_1 by exchanging the index $\beta \in I(z_1)$ for some index $\gamma \in \bar{I}(z_1)$ (that is, $\gamma = \alpha$). □

Since this is an easy theorem to establish, its proof is left as an exercise for the reader.

Now let us return to our initial basic solution introduced in (1). From (4) we define for $\theta \geq 0$

18
$$z(\theta) = \sum_{i \in \bar{J}(z_0)} (z_0{}^i - \theta\alpha_j{}^i)e_i + \theta e_j,$$

where $j \in J(z_0)$ and the $\alpha_j{}^i$ are as in (2). Hence the cost associated with this vector is

(19)
$$\langle d, z(\theta) \rangle = \sum_{i \in \bar{J}(z_0)} d^i(z_0{}^i - \theta \alpha_j{}^i) + \theta d^j$$
$$= \langle d, z_0 \rangle - \theta \Big(\sum_{i \in \bar{J}(z_0)} d^i \alpha_j{}^i - d^j \Big).$$

To interpret (19) we must bring in a little more detail.

(20) **Definition.** Given a basic solution z_0 to CLP (4.1) and its associated basis-indicator set $\bar{J}(z_0)$, we define $R_{\bar{J}(z_0)}$ to be the $m \times m$ nonsingular matrix whose columns are the linearly ordered vectors r_i for $i \in \bar{J}(z_0)$, $d_{\bar{J}(z_0)} \in E^m$ to be the vector whose components are the linearly ordered elements d^i for $i \in \bar{J}(z_0)$, and $\alpha_{\bar{J}(z_0)}$ for every $j \in J(z_0)$ to be the vector whose components are the linearly ordered coefficients $\alpha_j{}^i$ for $i \in \bar{J}(z_0)$, as determined by (2) (the ordering is on i). □

It follows immediately from (20) and (2) that

(21)
$$\alpha_{\bar{J}(z_0)} = R_{\bar{J}(z_0)}^{-1} r_j$$

and from (19) to (21) that

(22)
$$\sum_{i \in \bar{J}(z_0)} d^i \alpha_j{}^i = \langle d_{\bar{J}(z_0)}, \alpha_{\bar{J}(z_0)} \rangle = \langle (R_{\bar{J}(z_0)}^T)^{-1} d_{\bar{J}(z_0)}, r_j \rangle.$$

Now let

(23)
$$\psi_{\bar{J}(z_0)} \triangleq (R_{\bar{J}(z_0)}^T)^{-1} d_{\bar{J}(z_0)}.$$

Then, substituting from (23) into (22) and hence into (19), we obtain

(24)
$$\langle d, z(\theta) \rangle = \langle d, z_0 \rangle - \theta(\langle \psi_{\bar{J}(z_0)}, r_j \rangle - d^j).$$

We now come to the following conclusions.

Case 1. For a given $j \in J(z_0)$ suppose that there is at least one $\alpha_j{}^i > 0$, as defined by (2), and $\langle \psi_{\bar{J}(z_0)}, r_j \rangle - d^j > 0$. Then z_1, as given by (8), is an extreme point adjacent to z_0, with equal or lower associated cost

(25)
$$\langle d, z_1 \rangle = \langle d, z_0 \rangle - \theta_0(\langle \psi_{\bar{J}(z_0)}, r_j \rangle - d^j) \leq \langle d, z_0 \rangle.$$

If z_0 is a nondegenerate basic solution, then, by remark (16), $\theta_0 > 0$ in (25), and strict inequality holds; that is, $\langle d, z_1 \rangle < \langle d, z_0 \rangle$. When z_0 is degenerate θ_0 may well be zero.

Case 2. For a given $j \in J(z_0)$ suppose that $\alpha_j{}^i \leq 0$ for all $i \in \bar{J}(z_0)$ in (2) and that $\langle \psi_{J(z_0)}, r_j \rangle - d^j > 0$. Then every vector on the ray $\{z(\theta): \theta \geq 0\}$ defined by (19) is feasible [see (4) and (17)], and $\langle d, z(\theta) \rangle \to -\infty$ as $\theta \to \infty$, by (24). Hence CLP (4.1) has no solution (in the sense that we do not accept $-\infty$ as an optimal value).

Case 3. For every $j \in J(z_0)$ suppose that the conditions of case 2 do not arise and that

26 $$\langle \psi_{J(z_0)}, r_j \rangle - d^j \leq 0 \qquad \text{for all } j \in J(z_0).$$

Then z_0 is optimal, since (23), together with (26), is identical to (5.2).

27 **Remark.** The above result simply confirms the intuitively obvious fact that if the cost hyperplane $\{z: \langle d, z \rangle = \text{constant}\}$ touches the polyhedron $\Omega' = \{z: Rz = 0, z \geq 0\}$ at a vertex z_0, and all the vertices adjacent to and rays emanating from z_0 are on the higher-cost side of this hyperplane, then the entire polyhedron Ω' must be on the higher-cost side, and hence z_0 must be optimal. □

5.9 THE SIMPLEX ALGORITHM

We now combine the procedures described in the preceding sections into an algorithm for solving the canonical linear programming problem (4.1).

1 **Initialization.** Find a basic solution to the system of equations and inequalities

2 $$Rz = c \qquad z \geq 0.$$

If a basic solution is not readily available, use the simplex algorithm below to solve the linear programming problem:

3 $$\text{Minimize } \sum_{i=1}^{m} \xi^i \text{ subject to the constraints}$$
$$Rz + \xi = c \qquad z \geq 0 \qquad \xi \geq 0.$$

We may assume that $c \geq 0$ without loss of generality.[1] As a basic solution to (3) we may take $z = 0$ and $\xi = c$. If $(\hat{z}, \hat{\xi})$ is an optimal solution to (3) obtained by means of the simplex algorithm, and if $\hat{\xi} = 0$, then $\hat{z}$ is a basic solution to system (2). If $\hat{\xi} \neq 0$, then CLP (4.1) has no feasible solutions. □

---

[1] We can always ensure that $c \geq 0$ in (1) by multiplying the negative components of c, together with the corresponding rows of R, by -1.

4   The simplex algorithm. Let z_0 be a basic solution to CLP (4.1), and let $\bar{J}(z_0)$ be an associated basis-indicator set [see (6.8)].

Step 1. Compute $\psi_{\bar{J}(z_0)}$ according to

5
$$\psi_{\bar{J}(z_0)} = (R_{\bar{J}(z_0)}^T)^{-1} d_{\bar{J}(z_0)},$$

where $R_{\bar{J}(z_0)}^T$ is an $m \times m$ nonsingular matrix with rows r_{i_k} for $k = 1, 2, \ldots, m$, with $i_k \in \bar{J}(z_0)$, and $d_{\bar{J}(z_0)}$ is an m vector whose components are d^{i_k} for $k = 1, 2, \ldots, m$, with $i_k \in \bar{J}(z_0)$ [see (8.20)].

Step 2. If $\langle r_j, \psi_{\bar{J}(z_0)} \rangle - d^j \leq 0$ for all $j \in J(z_0)$, then stop, since z_0 is an optimal solution to CLP (4.1) [see (8.26)]. Otherwise let $j \in J(z_0)$ be the smallest (any) index such that $\langle r_j, \psi_{\bar{J}(z_0)} \rangle - d^j > 0$, and go to step 3.

Step 3. Determine the coefficients $\alpha_j{}^i$ for $i \in \bar{J}(z)$ satisfying

6
$$r_j = \sum_{i \in \bar{J}(z_0)} \alpha_j{}^i r_i,$$

where $j \in J(z_0)$ is the index obtained in step 2. Note that

7
$$\alpha_{\bar{J}(z_0)} = R_{\bar{J}(z_0)}^{-1} r_j,$$

where $\alpha_{\bar{J}(z_0)} = (\alpha_j{}^{i_1}, \alpha_j{}^{i_2}, \ldots, \alpha_j{}^{i_m})$ [see (8.21)]. If some $\alpha_j{}^i > 0$ for $i \in \bar{J}(z_0)$, go to step 4. Otherwise stop, since CLP (4.1) has no optimal solution.

Step 4. Determine θ_0 according to

8
$$\theta_0 = \min \left\{ \frac{z_0{}^i}{\alpha_j{}^i} : i \in \bar{J}(z_0), \, \alpha_j{}^i > 0 \right\} = \frac{z_0{}^k}{\alpha_j{}^k}.$$

If ties occur in (8), choose the smallest index k which satisfies (8). Set

9
$$z_1 = \sum_{i \in \bar{J}(z)} (z_0{}^i - \theta \alpha_j{}^i) e_i + \theta_0 e_j$$

and

10
$$\bar{J}(z_1) = \bar{J}(z_0) - \{k\} + \{j\}.$$

Then z_1 is a basic solution, and $\bar{J}(z_1)$ is an associated basis-indicator set. Set $z_0 = z_1$ and return to step 1. □

11 Theorem. Suppose that every basic solution to CLP (4.1) is a nondegenerate basic solution. Then the simplex algorithm reaches either the stop condition in step 2 (z_0 is an optimal solution) or the stop condition in step 3 (the problem has no solution) in a finite number of steps.

Proof. At each iteration, if z_0 is not an optimal solution, and if an edge $\{z(\theta): \theta \geqq 0\}$ of $\Omega' = \{z: Rz = 0, z \geqq 0\}$ is not encountered along which $\langle d, z(\theta) \rangle \to -\infty$ as $\theta \to \infty$ [that is, CLP (4.1) has no solution], then the simplex algorithm gives a procedure for generating a new extreme point z_1 of Ω', with $\langle d, z_1 \rangle < \langle d, z_0 \rangle$. It follows, therefore, that the algorithm is well defined and that it can never generate the same extreme point twice. Since there are only a finite number of extreme points of Ω', we conclude that the construction of new basic solutions must terminate in finitely many iterations either because the stop condition in step 2 is encountered, or because the stop condition in step 3 is encountered. □

The simplex algorithm (4) need not (but usually does) terminate in a finite number of steps when the nondegeneracy assumption is not satisfied. In the next section we shall describe a modified finite-step procedure which does not depend on the nondegeneracy assumption, but which is considerably more complex than the simplex algorithm and is rarely used.

5.10 RESOLUTION OF DEGENERACY

Let us suppose now that some of the basic solutions to the canonical linear programming problem (4.1) are in fact degenerate. Under these circumstances, a sequence of basic solutions $z_0, z_1, z_2, \ldots$ generated by the simplex algorithm (9.4) need not terminate after a finite number of vectors z_i. Indeed, since it is now possible to have $\langle d, z_i \rangle = \langle d, z_{i+1} \rangle$, the sequence $z_0, z_1, z_2, \ldots$ may cycle as follows: $z_0, z_1, \ldots, z_l, z_{l+1}, \ldots, z_p, z_l, z_{l+1}, \ldots, z_p, z_l, \ldots$ (for examples of such behavior see [3]). Although cycling is a phenomenon encountered only rarely in practice, it is nice to know that it can be avoided by means of a modification of the simplex algorithm (the procedure described below is due to Charnes [4]).

Consider again CLP (4.1):

1 Minimize $\langle d, z \rangle$ subject to the constraints

2 $$\sum_{i=1}^{n} z^i r_i = c, \quad z \geqq 0,$$

(that is, $Rz = c$), and suppose that we have an initial basic solution z_0 to (2) such that $\bar{J}(z_0) = \{1,2,\ldots,m\}$.† Now consider the perturbed linear programming problem:

3 $\qquad$ Minimize $\langle d,z \rangle$ subject to the constraints

$$\sum_{j=1}^{n} z^j r_j = c + \sum_{j=1}^{n} \epsilon^j r_j \triangleq c + \rho(\epsilon) \qquad z \geqq 0,$$

where $\epsilon > 0$ is sufficiently small to make what follows valid and ϵ^j denotes ϵ to the power j. The fact that such an $\epsilon > 0$ exists, as well as that we need not know its value, will become clear later on.

4 $\qquad$ **Properties of perturbed problem (3).** We shall show that:

a. For the perturbed problem (3) the simplex algorithm constructs a finite sequence of nondegenerate basic solutions $z_0(\epsilon), z_1(\epsilon), \ldots, z_N(\epsilon)$.
b. The original problem (1) has no solution if the perturbed problem (3) has no solution.
c. The vectors $z_i(\epsilon)$ for $i = 0, 1, \ldots$ are of the form $z_i(\epsilon) = z_i + \zeta_i(\epsilon)$, with z_i a basic but possibly degenerate solution to (2).
d. $\bar{I}(z_i(\epsilon)) = \bar{J}(z_i)$.
e. If $z_N(\epsilon) = z_N + \zeta_N(\epsilon)$ is an optimal solution to the perturbed problem (3), then z_N is an optimal solution to problem (1).

Furthermore, we shall see that we can calculate the successive index sets $\bar{I}(z_i(\epsilon)) = \bar{J}(z_i)$ without a knowledge of ϵ. Hence to obtain an optimal solution to the canonical linear programming problem (1) we need only establish the existence of a satisfactory $\epsilon > 0$.

Initialization. Recall that we already have a basic solution z_0 to (2), with $\bar{J}(z_0) = \{1,2,\ldots,m\}$. Hence for each $j \in \{1,2,\ldots,n\}$ there are scalars $\alpha_{0j}{}^i$ for $i \in \bar{J}(z_0)$ such that

5 $\qquad$ $r_j = \sum_{i \in \bar{J}(z_0)} \alpha_{0j}{}^i r_i.$

Hence, since $\sum_{i \in \bar{J}(z_0)} z_0{}^i r_i = c$, we have

6 $\qquad$ $\sum_{i \in \bar{J}(z_0)} z_0{}^i r_i + \sum_{j=1}^{n} \epsilon^j r_j = \sum_{i \in \bar{J}(z_0)} (z_0{}^i + \sum_{j=1}^{n} \alpha_{0j}{}^i \epsilon^j) r_i = c + \rho(\epsilon).$

† The requirement that $\bar{J}(z_0) = \{1,2,\ldots,m\}$ is only to facilitate exposition; in practice we simply put the indices in $\bar{J}(z_0)$ into one-to-one correspondence with $1, 2, \ldots, m$ and then proceed.

Now, for $j \in \bar{J}(z_0)$, $\alpha_{0j}{}^i = 1$ if $i = j$ and $\alpha_{0j}{}^i = 0$ if $i \neq j$. Hence for $i \in \bar{J}(z_0)$

(7) $$z_0{}^i(\epsilon) \triangleq z_0{}^i + \sum_{j=1}^{n} \alpha_{0j}{}^i \epsilon^j = z_0{}^i + \epsilon^i + \sum_{j \in J(z_0)} \alpha_{0j}{}^i \epsilon^j,$$

and therefore $z_0{}^i(\epsilon) \neq z_0{}^j(\epsilon)$ for all $i \neq j$ in $\bar{J}(z_0)$. Since every $j \in J(z_0)$ is larger than every $i \in \bar{J}(z_0)$, by construction, there is an $\epsilon'_0 > 0$ such that $z_0{}^i(\epsilon) > 0$ for all $\epsilon \in (0, \epsilon'_0]$ and $i \in \bar{J}(z_0)$. We now define the vector $z_0(\epsilon)$ by

(8) $$z_0{}^i(\epsilon) = \begin{cases} z_0{}^i + \sum_{j=1}^{n} \alpha_{0j}{}^i \epsilon^j & \text{if } i \in \bar{J}(z_0) \\ 0 & \text{if } i \in J(z_0). \end{cases}$$

Clearly, $z_0(\epsilon) \geq 0$ for all $\epsilon \in (0, \epsilon'_0]$, $\bar{I}(z_0(\epsilon)) = \bar{J}(z_0)$, and $\sum_{i \in \bar{I}(z_0(\epsilon))} z_0{}^i(\epsilon) r_i = c + \rho(\epsilon)$; that is, $z_0(\epsilon)$ is a nondegenerate basic solution to the perturbed problem (3) for all $\epsilon \in (0, \epsilon'_0]$. Setting $\zeta_0{}^i(\epsilon) = \sum_{j=1}^{n} \alpha_{0j}{}^i \epsilon^j$ for $i \in \bar{J}(z_0)$ and zero otherwise, we obtain

(9) $$z_0(\epsilon) = z_0 + \zeta_0(\epsilon)$$
(10) $$\langle d, z_0(\epsilon) \rangle = \langle d, z_0 \rangle + \langle d, \zeta_0(\epsilon) \rangle.$$

Computation of $z_1(\epsilon)$. We now proceed exactly as in Section 8. For $j \in J(z_0) = I(z_0(\epsilon))$ we multiply (5) by θ, add it to (6), and rearrange terms to obtain

(11) $$\sum_{i \in \bar{J}(z_0)} \left(z_0{}^i + \sum_{s=1}^{n} \alpha_{0s}{}^i \epsilon^s - \theta \alpha_{0j}{}^i \right) r_i + \theta r_j = c + \rho(\epsilon).$$

With (11) defining $z(\theta)$, we find that

(12) $$\langle d, z(\theta) \rangle = \sum_{i \in \bar{J}(z_0)} d^i \left(z_0{}^i + \sum_{s=1}^{n} \alpha_{0s}{}^i \epsilon^s - \theta \alpha_{0j}{}^i \right) + d^j \theta$$
$$= \langle d, z_0(\epsilon) \rangle - \theta(\langle r_j, \psi_{\bar{J}(z_0)} \rangle - d^j),$$

where, as in Section 5.8,

(13) $$\psi_{\bar{J}(z_0)} = (R_{\bar{J}(z_0)}^T)^{-1} d_{\bar{J}(z_0)}.$$

Again there are three possibilities to be considered.

Case 1. $\langle r_j, \psi_{\bar{J}(z_0)} \rangle - d^j \leq 0$ *for every* $j \in J(z_0)$. In this case $z_0(\epsilon)$ is an optimal solution to the perturbed problem (3), and since $\psi_{\bar{J}(z_0)}$ obviously does not depend on ϵ, z_0 is an optimal solution to problem (1), which proves e.

Case 2. $\langle r_j, \psi_{\bar{J}(z_0)} \rangle - d^j > 0$ *for some* $j \in \bar{J}(z_0)$ *and* $\alpha_{0j}{}^i \leqq 0$ *for every* $i \in \bar{J}(z_0)$. In this case neither the perturbed problem (3) nor problem (1) has an optimal solution, since $\langle d, z(\theta) \rangle \to -\infty$ as $\theta \to +\infty$ in (12) [cf. (8.27)], which proves b.

Case 3. *There is an* $l \in J(z_0)$ *such that for some* $i \in \bar{J}(z_0)$, $\alpha_{0l}{}^i > 0$ *and* $\langle r_l, \psi_{\bar{J}(z_0)} \rangle - d^l > 0$. Now [cf. (9.8)] let

14
$$\theta(\epsilon) = \min \left\{ \frac{z_0{}^i + \sum_{s=1}^{n} \alpha_{0s}{}^i \epsilon^s}{\alpha_{0l}{}^i} : i \in \bar{J}(z_0),\ \alpha_{0l}{}^i > 0 \right\}.$$

Since $z_0{}^i + \sum_{s=1}^{n} \alpha_{0s}{}^i \epsilon^s > 0$ for all $i \in \bar{J}(z_0)$, it is clear that $\theta(\epsilon) > 0$.

Now each of the polynomials appearing in (14) contains a power of ϵ which cannot be found in any of the other polynomials in (14). Consequently, there exists an $\epsilon_0'' > 0$ such that for all $\epsilon \in (0, \epsilon_0'']$ one of these polynomials is strictly smaller than all the others. Therefore, for all $\epsilon \in (0, \epsilon_0'']$, $\theta(\epsilon)$ is attained at a unique index i, say $i = k \in \bar{J}(z_0)$, and therefore the vector $z_1(\epsilon)$ defined by

15
$$z_1{}^i(\epsilon) = \begin{cases} z_0{}^i + \sum_{s=1}^{n} \alpha_{0s}{}^i \epsilon^s - \theta(\epsilon) \alpha_{0l}{}^i & \text{if } i \in \bar{J}(z_0),\ i \neq k \\ \theta(\epsilon) & \text{if } i = l \\ 0 & \text{otherwise,} \end{cases}$$

is an improved nondegenerate basic solution to the perturbed problem (3).

16 **Remark.** To determine which ϵ-polynomial in (14) is the smallest, it is not necessary to choose a specific value of $\epsilon \in (0, \epsilon_0'']$. Indeed, for all $\epsilon \in (0, \epsilon_0'']$ and $i, j \in \bar{J}(z_0)$, $\left(z_0{}^i + \sum_{s=1}^{n} \alpha_{0s}{}^i \epsilon^s \right) / \alpha_{0l}{}^i < \left(z_0{}^j + \sum_{s=1}^{n} \alpha_{0s}{}^j \epsilon^s \right) / \alpha_{0l}{}^j$ if and only if $z_0{}^i / \alpha_{0l}{}^i < z_0{}^j / \alpha_{0l}{}^j$, or $z_0{}^i / \alpha_{0l}{}^i = z_0{}^j / \alpha_{0l}{}^j$ and $\alpha_{01}{}^i / \alpha_{0l}{}^i < \alpha_{01}{}^j / \alpha_{0l}{}^j$, etc. That is, we need only compare the coefficients of ϵ^s in (14), starting with the coefficient of ϵ^0. If we now set $\epsilon_0 = \min [\epsilon_0', \epsilon_0'']$, then for all $\epsilon \in (0, \epsilon_0]$ both $z_0(\epsilon)$ and $z_1(\epsilon)$ will be nondegenerate basic solution to problem (3), which proves the first part of a.

We shall now show that $z_1(\epsilon)$ can be expressed as $z_1(\epsilon) = z_1 + \zeta_1(\epsilon)$, with z_1 a basic solution to (2), and that $\bar{J}(z_1) = \bar{I}(z_1(\epsilon))$. Obviously, since $z_1(\epsilon)$ is a nondegenerate basic solution to (3), for

SEC. 5.10 RESOLUTION OF DEGENERACY

each $j \in \{1, 2, \ldots, n\}$ there are coefficients $\alpha_{1j}{}^i$ for $i \in \bar{I}(z_1(\epsilon))$ such that

17
$$r_j = \sum_{i \in \bar{J}(1(\epsilon))} \alpha_{1j}{}^i r_i.$$

Also, from (5), with k and l defined as in (15),

18
$$r_k = \frac{1}{\alpha_{0l}{}^k} r_l - \sum_{\substack{i \in \bar{J}(z_0) \\ i \neq k}} \frac{\alpha_{0l}{}^i}{\alpha_{0l}{}^k} r_i.$$

Substituting for r_k in (5), we obtain

19
$$r_j = \sum_{\substack{i \in \bar{J}(z_0) \\ i \neq k}} \alpha_0{}^i r_i + \alpha_{0j}{}^k r_k$$

$$= \sum_{\substack{i \in \bar{J}(z_0) \\ i \neq k}} \left(\alpha_{0j}{}^i - \alpha_{0j}{}^k \frac{\alpha_{0l}{}^i}{\alpha_{0l}{}^k} \right) r_i + \frac{\alpha_{0j}{}^k}{\alpha_{0l}{}^k} r_l.$$

Since the coefficients in this representation must be unique, comparing (17) with (19) and recalling (16), we conclude that

20
$$\alpha_{1j}{}^i = \alpha_{0j}{}^i - \alpha_{0j}{}^k \frac{\alpha_{0l}{}^i}{\alpha_{0l}{}^k} \qquad i \in \bar{J}(z_0), i \neq k$$

$$\alpha_{1j}{}^l = \frac{\alpha_{0j}{}^k}{\alpha_{0l}{}^k}.$$

Now, since

$$\theta(\epsilon) = \frac{z_0{}^k}{\alpha_{0l}{}^k} + \frac{\sum_{s=1}^{n} \alpha_{0s}{}^k \epsilon^s}{\alpha_{0l}{}^k},$$

we obtain from (15) that

21
$$z_1{}^i(\epsilon) = \left(z_0{}^i - \alpha_{0l}{}^i \frac{z_0{}^k}{\alpha_{0l}{}^k} \right) + \sum_{s=1}^{n} \left(\alpha_{0s}{}^i - \alpha_{0l}{}^i \frac{\alpha_{0s}{}^k}{\alpha_{0l}{}^k} \right) \epsilon^s$$
$$\qquad\qquad\qquad\qquad\qquad\qquad i \in \bar{J}(z_0), i \neq k$$

$$z_1{}^l(\epsilon) = \frac{z_0{}^k}{\alpha_{0l}{}^k} + \sum_{s=1}^{n} \frac{\alpha_{0s}{}^k}{\alpha_{0l}{}^k} \epsilon^s.$$

Defining the vector z_1 by

22
$$z_1{}^i = \begin{cases} z_0{}^i - \alpha_{0l}{}^i \dfrac{z_0{}^k}{\alpha_{0l}{}^k} & \text{if } i \in \bar{J}(z_0), i \neq k \\ z_0{}^k / \alpha_{0l}{}^k & \text{if } i = l \\ 0 & \text{otherwise,} \end{cases}$$

we find, making use of (20) and (21), that

(23) $$z_1{}^i(\epsilon) = z_1{}^i + \sum_{s=1}^{n} \alpha_{1s}{}^i \epsilon^s \qquad i \in \bar{I}(z_1(\epsilon)).$$

Note that (17) and (23) are of the same form as (5), (7) [obviously, $z_1{}^i(\epsilon) = 0$ for $i \in I(z_1(\epsilon))$]. Now, since $z_1{}^i(\epsilon) > 0$ for all $i \in \bar{I}(z_1(\epsilon))$ and all $\epsilon \in (0, \epsilon_0]$, we must have $z_1{}^i \geqq 0$ for $i = 1, 2, \ldots, n$. Furthermore, it is easy to verify that

(24) $$\sum_{i \in I(z_1(\epsilon))} z_1{}^i r_i = c.$$

Thus z_1 is a basic solution to problem (1), and $\bar{J}(z_1)$ may be taken to be the set $\bar{I}(z_1(\epsilon))$, which proves c and d.

We may now repeat this construction with $0 < \epsilon \leqq \epsilon_1 \leqq \epsilon_0$, since each of the polynomials $z_1{}^i(\epsilon)$ for $i \in \bar{I}(z_1(\epsilon))$ contains a power of ϵ not appearing in the others, and, $z_1{}^i(\epsilon) > 0$ for all $i \in \bar{I}(z_1(\epsilon))$ and $\epsilon \in (0, \epsilon_0]$, since ties did not occur in computing $\theta(\epsilon)$ in (14). Because the cost of the perturbed problem (3) decreases a nonzero amount at each iteration, it is clear that no basis-indicator set can be repeated, and therefore after a finite number of iterations we shall either compute an optimal solution to the canonical linear programming problem or else we shall establish that it has no solution, which proves the second half of a. □

(25) **Remark.** Note that in the above process we also generate a sequence of strictly positive numbers $\epsilon_0 \geqq \epsilon_1 \geqq \epsilon_2 \geqq \cdots \geqq \epsilon_N$, and provided $\epsilon \leqq \epsilon_N$ throughout the entire calculation, our arguments remain valid. Furthermore, $\theta(\epsilon)$ in (14) need not be computed; it is necessary only to compute the index k which leaves the original basis-indicator set; i.e., we need only know which k determines $\theta(\epsilon)$. As already pointed out, for sufficiently small $\epsilon > 0$ we can determine the smallest element in the set (14) simply by comparing the coefficients of ϵ^s for $s = 0, 1, 2, \ldots, n$ in the polynomials, and hence we need never know the value of ϵ_N. □

When the original canonical linear programming problem has nondegenerate basic solutions only, it is clear that the simplex algorithm applied to the perturbed problem yields exactly the same sequence of basic solutions as when it is applied directly to the original problem. When some of the basic solutions of the original problem are degenerate, the more complex rule for computing the index k implied by (14) is operative only at the degenerate basic solutions. Thus we see that whether the canonical linear programming problem has nondegenerate basic solutions only or not, it is still possible to use the simplex algorithm, with a slight modification, to obtain an optimal solution.

5.11 BOUNDED-VARIABLE LINEAR PROGRAMMING

In Section 5.4 we chose to take as our canonical linear programming problem (4.1) a linear programming problem (1.1) in which all the components of z were constrained to be nonnegative. However, we have seen that many optimal control problems, such as (4.2.48), reduce to linear programming problems in which the variables are bounded from above and from below. We have also seen (Section 4) that bounded-variable linear programming problems can be reduced to the form of CLP (4.1) by suitable transformations, but only at the expense of considerably increasing the number of equations and variables. The purpose of this section is to modify the simplex algorithm so that it can be applied directly to bounded-variable linear programming problems. Most of the details are left as exercises for the reader.

1   **The bounded-variable linear programming problem (BLP).** Given an $m \times n$ $(m \leq n)$ full-rank matrix R and vectors $c \in E^m$ and $d \in E^n$, find a vector $\hat{z} \in E^n$ which minimizes $\langle z, d \rangle$ subject to the constraints

2   $$Rz = c \qquad 0 \leq z^i \leq 1 \qquad i = 1, 2, \ldots, n. \qquad \square$$

3   **Exercise.** Consider the linear programming problem: minimize $\langle x, d \rangle$ subject to the constraints

4   $$\tilde{R}x = \tilde{c}, \quad -\infty < \alpha^i \leq x^i \leq \beta^i < +\infty \qquad i = 1, 2, \ldots, n,$$

where $\tilde{R}$, $\tilde{d}$, and $\tilde{c}$ are $m \times n$ $(m \leq n)$, $n \times 1$, and $m \times 1$ matrices, respectively. Assuming that $\tilde{R}$ has rank m, transform this problem into the canonical form (1) (this can be done without increasing the number of equations or variables). $\square$

Note that z is a feasible solution to BLP (1) if $Rz = c$ and $0 \leq z^i \leq 1$ for $i = 1, 2, \ldots, n$ or, equivalently, if $z \in \Omega' = \{z : Rz = c, 0 \leq z^i \leq 1, i = 1, 2, \ldots, n\}$.

5   **Definition.** Let $z \in E^n$ be a vector such that $0 \leq z^i \leq 1$ for $i = 1, 2, \ldots, n$. Then the index sets $\bar{I}(z)$, $L(x)$, and $U(x)$ are defined as follows:

6   $$\bar{I}(z) = \{i \in \{1, 2, \ldots, n\} : 0 < z^i < 1\}$$
7   $$L(z) = \{i \in \{1, 2, \ldots, n\} : z^i = 0\}$$
8   $$U(z) = \{i \in \{1, 2, \ldots, n\} : z^i = 1\}.$$

Consequently, if, for $i = 1, 2, \ldots, n$, we let $e_i = (0,0,1,0,0, \ldots ,0)$ be the usual unit basis vectors in E^n (the ith component of e_i is 1 and all the other components are zero), then for any $z \in E^n$ which is a feasible solution to BLP (1) we have

9
$$z = \sum_{i=1}^{n} z^i e_i = \sum_{i \in \bar{I}(z)} z^i e_i + \sum_{i \in U(z)} z^i e_i$$

and

10
$$Rz = \sum_{i=1}^{n} z^i r_i = \sum_{i \in \bar{I}(z)} z^i r_i + \sum_{i \in U(z)} z^i r_i = c.$$

Note that $I(z) = L(z) \cup U(z)$ is, as usual, the active-inequality-constraints index set.

11 **Definition.** A vector z is said to be a *basic solution* to BLP (1) if z is a feasible solution and the vectors $\{r_i : i \in \bar{I}(z)\}$ are linearly independent. □

12 **Definition.** A vector z is said to be a *nondegenerate basic solution* to BLP (1) if z is a basic solution and the vectors $\{r_i : i \in \bar{I}(z)\}$ form a basis for E^m. □

13 **Exercise.** Show that a basic solution to BLP (1) is an extreme point of the set $\Omega' = \{z : Rz = c, 0 \leq z^i \leq 1, i = 1, 2, \ldots, n\}$. □

14 **Theorem.** If an optimal solution to BLP (1) exists, then there exists an optimal solution z which is an extreme point of Ω'. □

15 **Exercise.** Prove theorem (14). □

As in the case of the canonical linear programming problem (4.1), it is always possible to associate with any extreme point z of Ω' a set of m columns of R which form a basis for E^m.

16 **Definition.** Let z be a basic solution to BLP (1). We shall mean by a *basis-indicator set* any set $\bar{J}(z) \subset \{1,2, \ldots ,n\}$ with the following properties:

17
$\bar{I}(z) \subset \bar{J}(z)$, and vectors in the set $\{r_i : i \in \bar{J}(z)\}$ form a basis for E^m. □

18 **Lemma.** Suppose z is a basic solution of BLP (1). Then there exists a basis-indicator set $\bar{J}(z) \subset \{1,2, \ldots ,n\}$ satisfying (17). □

19 Exercise. Prove lemma (18). □

We can now describe an algorithm for solving the bounded-variable linear programming problem (1).

20 Initialization. Find a basic solution to the system of equations and inequalities

$$Rz = c \qquad 0 \leq z^i \leq 1 \qquad i = 1, 2, \ldots, n.$$

If a basic solution is not readily available, use the bounded-variable simplex algorithm below to solve the linear problem:

21 Minimize $\sum_{i=1}^{m} \xi^i$ subject to the constraints

22
$$Rz + \xi = c$$
$$0 \leq z^i \leq 1 \qquad i = 1, 2, \ldots, n$$
$$0 \leq \xi^i \leq 1 \qquad i = 1, 2, \ldots, m.$$

Assuming that we have multiplied all rows of R and all components of c by $1/\alpha$ or $-1/\alpha$, with $\alpha = \max |c^i|$, if necessary, we presume that $0 \leq c^i \leq 1$ for $i = 1, 2, \ldots, m$. Therefore, $z = 0$ and $\xi = c$ constitute a basic solution to (22). If $(\hat{z}, \hat{\xi})$ is a solution to (22) obtained by means of the bounded-variable simplex algorithm and $\hat{\xi} = 0$, then $\hat{z}$ is a basic solution to (2). If $\hat{\xi} \neq 0$, then BLP (1) has no feasible solutions. □

23 The bounded-variable simplex algorithm. Let z_0 be a basic solution, and let $\bar{J}(z_0)$ be an associated basis-indicator set.

Step 1. Compute $\psi_{\bar{J}(z_0)}$ according to

24
$$\psi_{\bar{J}(z_0)} = (R_{\bar{J}(z_0)}^T)^{-1} d_{\bar{J}(z_0)},$$

where $R_{\bar{J}(z_0)}^T$ is an $m \times m$ nonsingular matrix with whose kth row is r_{i_k} for $k = 1, 2, \ldots, m$ and $i_k \in \bar{J}(z_0)$ and $d_{\bar{J}(z_0)}$ is an m vector whose kth component is d^{i_k} for $k = 1, 2, \ldots, m$ and $i_k \in \bar{J}(z_0)$.

Step 2. If $\langle r_l, \psi_{\bar{J}(z_0)} \rangle - d^l \leq 0$ for all $l \in L(z_0)$ and if $\langle r_u, \psi_{\bar{J}(z_0)} \rangle - d^u \geq 0$ for all $u \in U(z_0)$, then stop, since z_0 is an optimal solution to BLP (1). Otherwise let $j \in L(z_0) \cup U(z_0)$ be the smallest (any) index such that

25 $\langle r_j, \psi_{\bar{J}(z_0)} \rangle - d^j > 0 \qquad$ if $j \in L(z_0)$
26 $\langle r_j, \psi_{\bar{J}(z_0)} \rangle - d^j < 0 \qquad$ if $j \in U(z_0)$,

and proceed to step 3.

128 OPTIMAL CONTROL AND LINEAR PROGRAMMING

Step 3. Compute the coefficients $\alpha_j{}^i$ for $i \in \bar{J}(z_0)$ satisfying

27 $$r_j = \sum_{i \in \bar{J}(z_0)} \alpha_j{}^i r_i,$$

where $j \in L(z_0) \cup U(z_0)$ is the index determined in step 2. Note that

28 $$\alpha_{\bar{J}(z_0)} = R_{\bar{J}}^{-1}(z_0) r_j,$$

where $\alpha_{\bar{J}(z_0)} = (\alpha_j{}^{i_1}, \alpha_j{}^{i_2}, \ldots, \alpha_j{}^{i_m})$.

Step 4. Compute θ_0 as follows. If $j \in L(z_0)$, then

29 $$\theta_0 = \min \left\{ \min_{\substack{i \in \bar{J}(z_0) \\ \alpha_j{}^i > 0}} \frac{z_0{}^i}{\alpha_j{}^i}, \min_{\substack{i \in \bar{J}(z_0) \\ \alpha_j{}^i < 0}} \frac{z_0{}^i}{\alpha_j{}^i}, 1 \right\} \geqq 0.$$

If $j \in U(z)$, then

30 $$\theta_0 = \min \left\{ \min_{\substack{i \in \bar{J}(z_0) \\ \alpha_j{}^i < 0}} \frac{z_0{}^i}{\alpha_j{}^i}, \min_{\substack{i \in \bar{J}(z_0) \\ \alpha_j{}^i > 0}} \frac{z_0{}^i}{\alpha_j{}^i}, -1 \right\} \leqq 0.$$

Compute the adjacent extreme point

31 $$z_1 = \sum_{i \in \bar{J}(z_0)} (z_0{}^i - \theta_0 \alpha_j{}^i) e_i + \theta_0 e_j + \sum_{i \in (U(z_0)/U(z_0) \cap \bar{J}(z_0))} z_0{}^i e_i,$$

where $U(z_0)/U(z_0) \cap \bar{J}(z_0) = \{i: i \in U(z_0), i \notin \bar{J}(z_0)\}$. Determine the basis-indicator set $\bar{J}(z_1)$ associated with z_1 as follows. If $|\theta_0| = 1$, then set

32 $$J(z_1) = J(z_0).$$

If $|\theta_0| < 1$, and $k \in \bar{J}(z_0)$ is an index for which the minimum in (29) or in (30) occurred (if ties occur, take the smallest index k), then set

33 $$\bar{J}(z_1) = \bar{J}(z_0) - \{k\} + \{j\}.$$

Set $z_0 = z_1$ and return to step 1. □

34 **Theorem.** Suppose that every basic solution of BLP (1) is a nondegenerate basic solution. Then the bounded-variable simplex algorithm (23) determines an optimal solution to BLP (2) in a finite number of steps. □

35 **Exercise.** Prove theorem (34). [*Hint:* Show that the algorithm never repeats a basic solution, so that it stops after a finite number of iterations. Establish a necessary-and-sufficient-condition theorem analogous to (5.1) and show that it coincides with the stop conditions in step 2.] □

36 Remark. It is possible to give a subprocedure for resolving degeneracies in the above case similar to the one given in Section 5.10 when the simplex algorithm was used to solve the canonical linear programming problem (4.1). The derivation of such a subprocedure is left as an exercise for the interested reader; such a subprocedure is usually not required in practice. □

37 Remark. When BLP (1) is converted to the form of CLP (4.1), and simplex algorithm (9.4) is used to solve the resulting problem, it is necessary to work with $(n + m) \times (n + m)$ matrices (i.e., to solve a system of $n + m$ simultaneous equations) of each step. By comparison, we need only work with $m \times m$ matrices to solve BLP (1) with the algorithm (23). This, quite obviously, is considerably to our advantage. □

REFERENCES

1. G. B. Dantzig: Maximization of a Linear Function of Variables Subject to Linear Inequalities, chap. XXI in T. C. Koopmans (ed.), "Activity Analysis of Production and Allocation," Monograph 13 of the Cowles Commission, New York, N.Y., 1951.
2. G. B. Dantzig: "Linear Programming and Extensions," Princeton University Press, Princeton, N.J., 1963.
3. A. J. Hoffman: Cycling in the Simplex Algorithm, *Natl. Bur. Std. Rept.* 2974, December 16, 1953.
4. A. Charnes: Optimality and Degeneracy in Linear Programming, *Econometrica*, **20**:160–170 (1952).
5. G. B. Dantzig, A. Orden, and P. Wolfe: The Generalized Simplex Method for Minimizing a Linear Form under Linear Inequality constraints, *Pac. J. Math.*, **5**:183–195 (1955).

6
Optimal control and quadratic programming

6.1 FORMULATION OF THE GENERAL CONTROL PROBLEM WITH QUADRATIC COST AND TRANSFORMATION TO A QUADRATIC PROGRAMMING PROBLEM

We shall now consider a class of discrete optimal control problems for which the corresponding Basic Problem [see (1.4.1)] is a quadratic programming problem of the form:

1 Minimize $\frac{1}{2}\langle z,Qz \rangle + \langle d,z \rangle$ subject to $Rz = c$, $Pz \leqq \nu$, and $\alpha \leqq z \leqq \beta$,

where d, α, and β are in E^n, Q is an $n \times n$ symmetric (usually positive-semidefinite) matrix, R is an $m \times n$ matrix, P is an $l \times n$ matrix, and $c \in E^m$ and $\nu \in E^l$ (we shall always assume that $\alpha^i < \beta^i$ for $i = 1, 2, \ldots, n$ and that it is possible to have $\alpha^i = -\infty$ and $\beta^j = \infty$ for any i and j in $\{1,2,\ldots,n\}$).

Our formulation of this class of problems is intended to be sufficiently general to include all the discrete optimal control problems with quadratic costs which are discussed in the literature. However, the possible variations on such problems are so extensive that it is virtually impossible to formulate a class general enough to

SEC. 6.1 FORMULATION OF THE QUADRATIC CONTROL PROBLEM

include all possible cases. Some examples of control problems outside our formulation which reduce to quadratic programming problems are included in the exercises; the reader can probably generate others. The following formulation is the simplest one that will include all the standard problems.

2 The quadratic control problem (QCP). Consider a dynamical system described by the linear difference equation

3 $$x_{i+1} - x_i = Ax_i + Bu_i \qquad i = 0, 1, \ldots, k - 1,$$

where $x_i \in E^n$ is the system state at time i, $u_i \in E^m$ is the system input at time i, A is an $n \times n$ matrix, and B is an $n \times m$ matrix. Find a control sequence $\hat{\mathcal{U}} = (\hat{u}_0, \hat{u}_1, \ldots, \hat{u}_{k-1})$ and a corresponding trajectory $\hat{\mathcal{X}} = (\hat{x}_0, \hat{x}_1, \ldots, \hat{x}_k)$, determined by (3), which minimize the quadratic cost function

4 $$\tfrac{1}{2}\langle z, Qz \rangle + \langle d, z \rangle,$$

where $z \triangleq (\mathcal{X}, \mathcal{U}) = (x_0, x_1, \ldots, x_k, u_0, u_1, \ldots, u_{k-1})$, Q is a $[(k+1)n + km] \times [(k+1)n + km]$ symmetric positive-semidefinite matrix, and $d \in E^{(k+1)n+km}$. The minimization is subject to the constraints

5 $$u_i \in U_i = \{u : F_i u_i \leq w_i\} \qquad i = 0, 1, \ldots, k - 1$$
6 $$x_i \in X_i' = \{x : H_i x \leq v_i\} \qquad i = 0, 1, \ldots, k,$$

and in addition, x_0 and x_k are required to satisfy

7 $$G_0 x_0 = c_0 \qquad G_k x_k = c_k,$$

where the F_i and w_i for $i = 0, 1, \ldots, k$, the H_i and v_i for $i = 0, 1, \ldots, k - 1$, and G_0, G_k, c_0, and c_k are appropriate matrices and vectors [some of the matrices in (4) to (7) may be zero]. $\square$

We now reduce the QCP (2) to a Basic Problem by making the usual identifications (see Section 1.5). Let z be defined as in (4) and let f and r defined by

8 $$f(z) = \tfrac{1}{2}\langle z, Qz \rangle + \langle d, z \rangle$$

9 $$r(z) = \begin{bmatrix} x_1 - x_0 - Ax_0 - Bu_0 \\ \vdots \\ x_k - x_{k-1} - Ax_{k-1} - Bu_{k-1} \\ G_0 x_0 - c_0 \\ G_k x_k - c_k \end{bmatrix} = Rz - c,$$

where c is of the form $c \triangleq (0, \ldots, 0, c_0, c_k)$. Le Ω be defined by

(10) $$\Omega = X_0' \times X_1' \times X_2' \times \cdots \times X_k' \times U_0 \times U_1 \\ \times \cdots \times U_{k-1};$$

hence Ω is of the form

(11) $$\Omega = \{z : Pz \leqq \nu, \alpha \leqq z \leqq \beta\},$$

where P, ν, α and β are derived from (5) and (6).

Obviously, with f, r, and Ω defined by (8), (9), and (11), respectively, we have reduced problem (2) to the form of problem (1).

(12) **Remark.** We saw in Section 5.2 that there is an alternative method for transcribing control problems into a Basic Problem which leads to a reduced number of unknown variables. This method entails solving the difference equation (3) for each of the states x_i, with $i = 1, \ldots, k$, in terms of the initial state x_0 and the control sequence $\mathfrak{u} = (u_0, u_1, \ldots, u_{k-1})$, and then eliminating these states from the problem by substituting for them in the cost function and the constraints. □

(13) **Exercise.** Transcribe QCP (2) into a quadratic programming problem by means of the method indicated in (12). □

Note that not only the number of variables, but also the number of equality constraints, is reduced when the alternative method for transcribing QCP into Basic Problem form is used. We shall make this transcription for a number of important special cases of the quadratic control problem when we discuss these problems in detail.

6.2 THE EXISTENCE OF AN OPTIMAL SOLUTION TO THE QUADRATIC PROGRAMMING PROBLEM

Let us now turn our attention to the question of when an optimal solution to the quadratic programming problem (1.1) exists. In this section we shall see that we need only a minimum of additional assumptions to guarantee the existence of an optimal solution. First, however, we introduce a form of the quadratic programming problem (1.1) which will facilitate our task, and some terminology:

(1) **The quadratic programming problem (QP).** Minimize $f(z) = \frac{1}{2}\langle z, Qz \rangle + \langle d, z \rangle$ subject to the constraints $Rz = c$ and $\alpha \leqq z \leqq \beta$, where Q is an $n \times n$ symmetric positive-semidefinite matrix, $d \in E^n$, R is an

$m \times n$ matrix, and α and β are vectors in E^n whose components may take on the values $\pm \infty$. □

2 **Exercise.** Transcribe the quadratic programming problem (1.1) into the form (1). By definition, (1) is already of the form (1.1). (*Hint:* See section (5.4) on canonical linear programming problems.) □

Again it will be recollected that a vector z which satisfies the constraints $Rz = c$ and $\alpha \leq z \leq \beta$ of problem (1) is a *feasible solution* to the problem.

We now state and prove a theorem concerning the existence of an optimal solution to QP (1).

3 **Theorem.** If $\langle d,z \rangle = 0$ for every $z \in E^n$ satisfying $Qz = 0$ and $Rz = 0$ and there exists a feasible solution to QP (1), then there exists an optimal solution to QP (1).

Proof. Let $\Omega' = \{z: Rz = c, \alpha \leq z \leq \beta\}$; that is, Ω' is the set of feasible solutions to QP (1). Ω' is nonempty, by assumption. We shall prove the theorem in two parts. First we shall show that $f(z)$ is bounded from below for $z \in \Omega'$, and hence that there exists a finite number γ which is the *infimum* of $f(z)$ for $z \in \Omega'$. Then we shall show that there exists a $z^* \in \Omega'$ such that $f(z^*) = \gamma$.

To begin, let us denote by $\mathfrak{N}(R)$ and $\mathfrak{N}(Q)$ the null spaces of R and Q, respectively. By assumption, $d \in [\mathfrak{N}(R) \cap \mathfrak{N}(Q)]^\perp$, the orthogonal complement of $\mathfrak{N}(R) \cap \mathfrak{N}(Q)$. This is equivalent to saying that

$$d \in \left[\mathfrak{N}\left(\begin{bmatrix} Q \\ \hdashline R \end{bmatrix} \right) \right]^\perp,$$

where $\begin{bmatrix} Q \\ \hdashline R \end{bmatrix}$ is the $(n + m) \times n$ matrix, with Q and R as submatrices, as indicated. Since d is orthogonal to the null space of $\begin{bmatrix} Q \\ \hdashline R \end{bmatrix}$, it must be contained in the range of $\begin{bmatrix} Q \\ \hdashline R \end{bmatrix}^T = [Q \vdots R^T]$. Therefore we must have

4 $\qquad d = Q\mu + R^T\nu,$

where μ and ν are some vectors in E^n and E^m, respectively. Substituting (4) into the expression for $f(z)$, we obtain

5 $\qquad \begin{aligned} f(z) &= \langle z,Qz \rangle + \langle Q\mu + R^T\nu, z \rangle \\ &= \langle z,Qz \rangle + \langle \mu,Qz \rangle + \langle \nu,Rz \rangle. \end{aligned}$

Now, we are interested only in those vectors $z \in E^n$ which satisfy $Rz = c$. It follows, therefore, that

6
$$f(z) = \langle z + \mu, Qz \rangle + \langle \nu, c \rangle \qquad \text{for all } z \in \Omega'.$$

Since Q is symmetric and positive semidefinite, we obtain for Q the spectral expansion

7
$$Q = \sum_{i=1}^{n} \lambda^i \xi_i \rangle \langle \xi_i,$$

where the $\lambda^i \geqq 0$ for $i = 1, 2, \ldots, n$ are the eigenvalues of Q, the $\xi_i = (\xi_i^1, \xi_i^2, \ldots, \xi_i^n)$ are a corresponding set of orthonormal eigenvectors of Q, and $\xi_i \rangle \langle \xi_i$ is the $n \times n$ (dyad) matrix whose lkth element is $\xi_i^l \xi_i^k$. Hence

8
$$\langle z + \mu, Qz \rangle = \sum_{i=1}^{n} \lambda^i (\langle \xi_i, z \rangle^2 + \langle \xi_i, \mu \rangle \langle \xi_i, z \rangle),$$

where the quantities in the parentheses,

$$\langle \xi_i, z \rangle^2 + \langle \xi_i, \mu \rangle \langle \xi_i, z \rangle \qquad i = 1, 2, \ldots, n,$$

are of the form $x^2 + ax$, which has a finite minimum for every finite value of a. Since $\lambda^i \geqq 0$ for $i = 1, 2, \ldots, n$, we can conclude that $\langle z + \mu, Qz \rangle$ must be bounded from below for all z. We have already seen that $f(z) = \langle z + \mu, Qz \rangle + \langle \nu, c \rangle$ for all $z \in \Omega'$. Hence $f(z)$ is bounded from below on Ω'. Let γ be the greatest lower bound of $f(z)$ for $z \in \Omega'$; that is,

9
$$\gamma = \inf_{z \in \Omega'} f(z) > -\infty.$$

This completes the first half of the proof.

Since γ is the infimum of f over Ω', it follows that there exists a sequence of points $\{z_i\}_{i=1}^{\infty}$ in Ω' such that $f(z_i) \to \gamma$. If $\{z_i\}_{i=1}^{\infty}$ is convergent, or even if it has a convergent subsequence, we are done, since the limit point of this sequence would be in Ω' and would give the required value for f. However, this sequence need not be bounded, since Ω' need not be bounded, and hence it need not have a convergent subsequence. What we shall do now is project Ω' and the sequence $\{z_i\}_{i=1}^{\infty}$ onto a subspace in such a way that we can obtain a bounded sequence with certain nice properties. Note that since Q is symmetric, the orthogonal complement of $\mathfrak{N}(Q)$, the null space of Q, is $\mathfrak{R}(Q)$, the range of Q. Let P be an $n \times n$ matrix such that $Px = 0$ for all $x \in \mathfrak{N}(Q)$ and $Px = x$ for all $x \in \mathfrak{R}(Q)$; that is, P is an orthogonal-projection matrix. Let $\{z_i\}_{i=1}^{\infty}$ be a sequence in Ω' such that $f(z_i) \to \gamma$, and let

10 $$y_i = Pz_i \qquad i = 1, 2, \ldots.$$

Also let

11 $$\hat{\Omega}' = \{y \in E^n : y = Qz, z \in \Omega'\}.$$

Clearly, $y_i \in \hat{\Omega}$ for every i. Furthermore, if we define $h: E^n \to E^1$ by

12 $$h(z) = \langle z + \mu, Qz \rangle,$$

then

13 $$h(y_i) = h(z_i) = f(z_i) - \langle \nu, c \rangle \qquad \text{for all } i,$$

since, by definition of y_i,

14 $$z_i = y_i + x_i,$$

where $x_i \in \mathfrak{N}(Q)$. It follows that

15 $$h(y_i) \to \gamma - \langle \nu, c \rangle$$

and therefore that the sequence $\{h(y_i)\}$ is bounded. Furthermore, Q restricted to $\mathfrak{R}(Q)$ is positive definite, and hence $\{h(y_i)\}_{i=1}^{\infty}$ is bounded if and only if $\{y_i\}_{i=1}^{\infty}$ is bounded. We therefore conclude from (15) that there exists at least one limit point y^* to which a subsequence of $\{y_i\}_{i=1}^{\infty}$ converges. If $\hat{\Omega}'$ is closed, we are done, since then y^* belongs to $\hat{\Omega}'$, and any corresponding vector $z^* \in \Omega'$ for which $y^* = Pz^*$ satisfies $f(z^*) = \gamma$.

Since it is not true in general that the image of a closed set under a linear map is closed, we must appeal to the special nature of the set Ω', namely, to the fact that it is a convex polyhedral set. The following representation theorem is stated here without proof, (for a proof see Goldman [4], theorem 1, p. 44).

16 **Theorem.** Any closed nonempty convex polyhedral set S in E^n of the form

17 $$S = \{x : Ax \leq b\}$$

can be expressed as the vector sum

18 $$C^\Delta + D^L = \{x : x = x_1 + x_2 : x_1 \in C^\Delta, x_2 \in D^L\}$$

of a closed and bounded convex polyhedron C^Δ of the form

19 $$C^\Delta = \{x : x = Cu, u \geq 0, \Sigma u_i = 1\}$$

and a closed convex polyhedral cone D^L of the form

20 $$D^L = \{x : x = Dw, w \geqq 0\}.$$

Conversely, any nonempty set of the form $C^\Delta + D^L$ is a closed convex polyhedral set.

If we represent S as $C^\Delta + D^L$, it is easy to see that the image of $C^\Delta + D^L$ under a linear transformation P is a set of the same form, that is, $P[C^\Delta + D^L] = \hat{C}^\Delta + \hat{D}^L$, where

21 $$\hat{C}^\Delta = \{y : y = (PC)u, u \geqq 0, \Sigma u_i = 1\}$$
22 $$\hat{D}^L = \{y : y = (PD)w, w \geqq 0\}.$$

From the converse of theorem (16) we now conclude that $\hat{C}^\Delta + \hat{D}^L$ is a closed convex polyhedral set.

Although our set Ω' does not have the specific form of S in the theorem, it may easily be verified that these forms are equivalent, and hence that $\hat{\Omega}'$ is a closed convex polyhedron. □

There are several special cases of this theorem, stated here in the form of corollaries.

23 **Corollary.** If $d = 0$ and there exists a feasible solution to QP (1), then there exists an optimal solution to QP (1). □

24 **Corollary.** If Q is positive definite and there exists a feasible solution to QP (1), then there exists an optimal solution to QP (1). □

25 **Corollary.** If R is nonsingular and there exists a feasible solution to QP (1), then there exists an optimal solution to QP (1). □

26 **Corollary.** If $\mathfrak{N}(R) \cap \mathfrak{N}(Q) = \{0\}$ and there exists a feasible solution to QP (1), then there exists an optimal solution to QP (1). □

This last corollary is of particular interest, since, as we shall see in the next section, the condition $\mathfrak{N}(R) \cap \mathfrak{N}(Q) = \{0\}$ is also a sufficient condition for the optimal solution to QP (1) to be unique.

As a final note, it should be emphasized that the conditions of theorem (3) are only sufficient conditions for the existence of an optimal solution. Optimal solutions may, in fact, exist even if these conditions are not satisfied. Indeed, referring to the proof of theorem (3), we see that the assumption

$$d \in \left[\mathfrak{N}\left(\begin{bmatrix} Q \\ \vdots \\ R \end{bmatrix}\right)\right]^\perp$$

SEC. 6.3 A SUFFICIENT CONDITION FOR A UNIQUE OPTIMAL SOLUTION

was needed only in order to show that $f(z)$ is bounded from below for $z \in \Omega'$. Consequently, we may substitute for theorem (2) the following more general result.

27 Theorem. If $f(z)$ is bounded from below for $z \in \Omega'$ and Ω' is not empty, then there exists an optimal solution to QP (1). □

This concludes our discussion of conditions ensuring the existence of solutions to the quadratic programming problem.

6.3 A SUFFICIENT CONDITION FOR A UNIQUE OPTIMAL SOLUTION TO THE QUADRATIC PROGRAMMING PROBLEM

In this section we shall always assume that an optimal solution to the quadratic programming problem [see (1.1) or (2.1)] exists, and we shall seek to establish conditions which will guarantee that there is a unique optimal solution. As we shall see later in the chapter, the quadratic programming problem is considerably easier to solve when there is a unique optimal solution.

There does not appear to be any set of conditions which are both necessary and sufficient for the quadratic programming problem to have a unique solution. Of the many possible sets of sufficient conditions, we shall examine one set which not only has the advantage of being relatively straightforward to check, but also immediately implies the nonsingularity of certain matrices which appear in algorithms for the solution of the quadratic programming problem. We now state these conditions in the form of a theorem. The terminology is the same as in Section 2.

1 Theorem. If $\mathfrak{N}(Q) \cap \mathfrak{N}(R) = \{0\}$ and the QP (2.1) has a feasible solution, then it has a unique optimal solution.

Proof. Note first that if a feasible solution to QP (2.1) exists, then an optimal solution exists by corollary (2.26). Therefore we need only show that there cannot be two distinct optimal solutions.

We prove this by contradiction. Suppose that z_1 and z_2, where $z_1 \neq z_2$, are both optimal solutions to QP (2.1). Then $z_2 = z_1 + z_0$ for some $z_0 \neq 0$, and $z_0 \in \mathfrak{N}(R)$. Again denoting the set of all feasible solutions by Ω', we find that

$$z(\lambda) = \lambda z_1 + (1 - \lambda)z_2 \in \Omega' \quad \text{for every } \lambda \in [0,1]$$

because Ω' is a convex set. Thus $z(\lambda)$ is a feasible solution to QP (2.1) for every $\lambda \in [0,1]$. We shall show that for all $\lambda \in (0,1)$, $f(z(\lambda)) <$

$f(z_1) = f(z_2)$, which is a contradiction of the assumption that z_1 and z_2 are both optimal. First note that

(2) $$f(z_1) = f(z_2) = f(z_1 + z_0) = f(z_1) + f(z_0) + \langle z_1, Q z_0 \rangle,$$

and hence that

(3) $$f(z_0) = -\langle z_1, Q z_0 \rangle.$$

After some tedious algebraic manipulation it can be shown that

(4) $$f(z(\lambda)) = f(z_1) + (1 - \lambda)[\langle z_1, Q z_0 \rangle + \tfrac{1}{2}(1 - \lambda)\langle z_0, Q z_0 \rangle + \langle z_0, d \rangle].$$

Using (3), we reduce this to

(5) $$f(z(\lambda)) = f(z_1) - \frac{\lambda(1 - \lambda)}{2} \langle z_0, Q z_0 \rangle.$$

Since $\mathfrak{N}(R) \cap \mathfrak{N}(Q) = \{0\}$ and $z_0 \neq 0$ belongs to $\mathfrak{N}(R)$, we conclude that $z_0 \notin \mathfrak{N}(Q)$, that is, that $\langle z_0, Q z_0 \rangle > 0$ (this follows upon writing Q in spectral-expansion form). Therefore $f(z(\lambda)) < f(z_1)$ for all $\lambda \in (0,1)$. □

The condition that the null spaces of Q and R have only the zero vector in common is relatively easy to check, since it is equivalent to verifying that the $(n + m) \times n$ matrix $\begin{bmatrix} Q \\ R \end{bmatrix}$ has rank n. If we make the not unreasonable restriction that R have rank m, that is, that the equality constraints form a linearly independent set of equations, we get an even stronger result, which is stated below in the form of a theorem.

(6) **Theorem.** Suppose that the matrix R has rank m; then $\mathfrak{N}(Q) \cap \mathfrak{N}(R) = \{0\}$ if and only if the $(n + m) \times (n + m)$ matrix

(7) $$D = \begin{bmatrix} -Q & R^T \\ R & 0 \end{bmatrix}$$

is nonsingular.

Proof. Every vector $u \in E^n \times E^m$ can be partitioned into two components, $u = (x, v)$, where $x \in E^n$ and $v \in E^m$. If D is nonsingular, then for every vector u of the form $u = (x, 0)$, with $x \neq 0$, we have

(8) $$Du = \begin{bmatrix} -Qx \\ Rx \end{bmatrix} \neq 0,$$

which implies that $\mathfrak{N}(Q) \cap \mathfrak{N}(R) = \{0\}$.

SEC. 6.4 NECESSARY AND SUFFICIENT CONDITIONS OF OPTIMALITY

Conversely, suppose that $\mathfrak{N}(Q) \cap \mathfrak{N}(R) = \{0\}$ and that $u^* = (x^*,v^*)$ satisfies $Du^* = 0$. Then, from (7),

9 $\qquad R^T v^* = Qx^*$
10 $\qquad Rx^* = 0.$

Using equations (9) and (10), we obtain

11 $\qquad \langle x^*, Qx^* \rangle = \langle x^*, R^T v^* \rangle = \langle Rx^*, v^* \rangle = 0,$

and hence $x^* \in \mathfrak{N}(Q)$. This, together with (10), implies that $x^* = 0$, since $\mathfrak{N}(Q) \cap \mathfrak{N}(R) = \{0\}$. We therefore conclude from (9) that

12 $\qquad R^T v^* = 0.$

Since R has rank m, $R^T v^* = 0$ implies that $v^* = 0$. Therefore the only solution to $Du^* = 0$ is $u^* = 0$, and hence D is nonsingular. □

On the surface, this result may not appear to be any more significant than the previously noted fact that $\begin{bmatrix} Q \\ R \end{bmatrix}$ must have rank n. However, the matrix D in (7) plays a prominent role when we apply necessary and sufficient conditions for optimality to the quadratic programming problem, and the fact that it is nonsingular under the assumptions stated above is highly useful.

6.4 NECESSARY AND SUFFICIENT CONDITIONS OF OPTIMALITY FOR THE QUADRATIC PROGRAMMING PROBLEM

For the purpose of deriving algorithms it is convenient to restrict ourself to one of the following two canonical forms of the quadratic programming problem (1.1):

1 $\qquad$ Minimize $\frac{1}{2}\langle z, Qz \rangle + \langle d, z \rangle$ subject to $Rz = c$ and $\alpha \leq z \leq \beta$.
2 $\qquad$ Minimize $\frac{1}{2}\langle z, Qz \rangle + \langle d, z \rangle$ subject to $Rz = c$ and $z \geq 0$.

3 **Exercise.** Show that by adding variables we can transcribe the form (1) into the form (2) [(2) is a special case of (1)]. □

When applied to problem (1), the necessary-and-sufficient conditions theorems (3.3.5) and (3.6.1) combine to give the following result.

4 **Theorem.** If $\hat{z}$ is a feasible solution to problem (1), then $\hat{z}$ is an optimal solution if and only if there exists a vector $\psi = (\psi^1, \ldots, \psi^m) \in E^m$ such that for $i = 1, 2, \ldots, n$

5 $\qquad \begin{array}{ll} \langle r_i, \psi \rangle - \langle q_i, \hat{z} \rangle - d^i = 0 & \text{if } \alpha^i < \hat{z}^i < \beta^i \\ \langle r_i, \psi \rangle - \langle q_i, \hat{z} \rangle - d^i \leq 0 & \text{if } \hat{z}^i = \alpha^i \\ \langle r_i, \psi \rangle - \langle q_i, \hat{z} \rangle - d^i \geq 0 & \text{if } \hat{z}^i = \beta^i, \end{array}$

where for $i = 1, 2, \ldots, n$, r_i is the ith column of R, q_i is the ith column of Q, and d^i is the ith component of d. □

6 Exercise. Prove theorem (4). □

6.5 APPLICATIONS TO UNBOUNDED CONTROL PROBLEMS

We shall now consider several special cases of the quadratic control problem (1.2) which have no inequality constraints. The corresponding quadratic programming problems also have no inequality constraints. These problems have been chosen not because of their great practical significance, but because they are among the few which have closed-form solutions. This permits us to compare and evaluate various methods of attack. The observations we shall make in this section will also be pertinent to more general forms of the quadratic control problem.

As a first example, we shall consider the following special case of the quadratic control problem.

1 The minimum-energy problem: two-point-boundary form. Consider a dynamical system described by the linear difference equation

2 $$x_{i+1} - x_i = Ax_i + Bu_i \qquad i = 0, 1, \ldots, k-1,$$

where $x_i \in E^n$ is the system state at time i, $u_i \in E^m$ is the system input at time i, A is an $n \times n$ matrix, and B is an $n \times m$ matrix. Find a control sequence $\hat{\mathfrak{U}} = (\hat{u}_0, \hat{u}_1, \ldots, \hat{u}_{k-1})$ which transfers system (2) from the initial state $\hat{x}_0 = c_0$ to the terminal state $\hat{x}_k = c_k$ and minimizes

3 $$f(\mathfrak{U}) = \sum_{i=0}^{k-1} \langle u_i, u_i \rangle = \sum_{i=0}^{k-1} \|u_i\|^2 \qquad □$$

4 Assumption. $k \geqq n$. □

5 Assumption. System (2) is completely controllable; i.e., the $(n \times nm)$-dimensional matrix $[B \mid (I + A)B \mid \cdots \mid (I + A)^{n-1}B]$ has rank n. □

Solution of (1) by Direct Transcription

Obviously, problem (1) is a special case of the quadratic control problem, and we may therefore reduce it to a quadratic programming problem of the following form:

SEC. 6.5 APPLICATIONS TO UNBOUNDED CONTROL PROBLEMS

6   Minimize $\tfrac{1}{2}\langle z, Qz \rangle$ subject to the constraint $Rz = c$, where $z = (x_0, x_1, \ldots, x_k, u_0, u_1, \ldots, u_{k-1})$,

7
$$Q = \begin{bmatrix} 0 & 0 \\ 0 & 2I \end{bmatrix}$$

is a $[(k+1)n + km] \times [(k+1)n + km]$ in which I is the $km \times km$ identity matrix,

8
$$R = \begin{bmatrix} -(I+A) & I & 0 & \cdots & 0 & -B & 0 & \cdots & \cdots & 0 \\ 0 & -(I+A) & I & 0 & 0 & 0 & -B & 0 & \cdots & 0 \\ \cdots & \cdots & \cdots & \cdots & \cdots & \cdots & \cdots & \cdots & \cdots & \cdots \\ 0 & \cdots & 0 & -(I+A) & I & 0 & 0 & \cdots & 0 & -B \\ I & 0 & \cdots & \cdots & \cdots & 0 & 0 & \cdots & \cdots & 0 \\ 0 & \cdots & \cdots & \cdots & \cdots & I & 0 & \cdots & \cdots & 0 \end{bmatrix}$$

is a $(k+2)n \times [(k+1)n + km]$ matrix and

9 $c = (0, 0, \ldots, c_0, c_k) \in E^{(k+2)n}$. □

Note that there is no linear term in the cost function of problem (6), and hence an optimal solution to (6) exists whenever a feasible solution exists [corollary (2.23)]. Since there are no constraints in (6) of the form $\alpha^i \leq z^i \leq \beta^i$, a feasible solution exists if R has rank $(k+2)n$.

10   **Exercise.** Show that assumptions (4) and (5) imply that R has rank $(k+2)n$. □

The only question we might wish to investigate before attempting to solve problem (6) is whether it has a unique optimal solution.

11   **Lemma.** Problem (6) [and hence problem (1)] has a unique optimal solution.

Proof. According to theorem (3.1), we need only show that $\mathfrak{N}(Q) \cap \mathfrak{N}(R) = \{0\}$. Clearly, the null space of Q is the set of vectors $\tilde{z}$ of the form $\tilde{z} = (x_0, x_1, \ldots, x_k, 0, \ldots, 0)$. Thus we must examine the equation

$$R\tilde{z} = \begin{bmatrix} -(I+A) & I & 0 & \cdots & & & 0 \\ 0 & -(I+A) & I & 0 & & \cdots & 0 \\ \cdots & \cdots & \cdots & \cdots & \cdots & \cdots & \cdots \\ 0 & \cdots & & \cdots & -(I+A) & I & \\ I & 0 & \cdots & & & & 0 \\ 0 & \cdots & & \cdots & & 0 & I \end{bmatrix} \begin{bmatrix} x_0 \\ x_1 \\ \cdot \\ \cdot \\ \cdot \\ x_k \end{bmatrix} = 0.$$

Note that the next-to-the-bottom block of rows in (12) yield

13 $x_0 = 0.$

We may now start with the top block of rows and successively determine that

14
$$x_1 = (I + A)x_0 = 0$$
$$x_2 = (I + A)x_1 = 0$$
$$\dots\dots\dots\dots$$
$$x_k = (I + A)x_{k-1} = 0.$$

Consequently, $\mathfrak{N}(Q) \cap \mathfrak{N}(R) = \{0\}$, which is sufficient to guarantee that (6) has a unique optimal solution. □

We now apply theorem (4.4) to obtain a set of equations from which the optimal solution $\hat{z}$ to problem (6) can be computed. Since for this case $\alpha^i = -\infty$ and $\beta^i = +\infty$ for $i = 1, 2, \dots,$ $(k+1)n + km$, we obtain from (4.4) that a vector $\hat{z}$ satisfying

15 $$R\hat{z} = c$$

is optimal if and only if there exists a vector $\psi \in E^{(k+1)n+km}$ such that

16 $$\langle r_i, \psi \rangle - \langle q_i, \hat{z} \rangle = 0 \qquad i = 1, 2, \dots, (k+1)n + km.$$

We may write (16) in matrix form as

17 $$R^T\psi - Q\hat{z} = 0.$$

Combining (15) and (17), we see that a solution $(\hat{z}, \psi)$ to the matrix equation

18
$$\begin{bmatrix} -Q & R^T \\ R & 0 \end{bmatrix} \begin{bmatrix} \hat{z} \\ \psi \end{bmatrix} = \begin{bmatrix} 0 \\ c \end{bmatrix}$$

will yield the desired optimal solution $\hat{z}$. It should now be clear why we stopped to establish the rank of R and the uniqueness of the optimal solution $\hat{z}$ to problem (6), for it now follows immediately from theorem (3.6) that the matrix on the left-hand side of (18) is nonsingular. Therefore we obtain

19
$$\begin{bmatrix} \hat{z} \\ \psi \end{bmatrix} = \begin{bmatrix} -Q & R^T \\ R & 0 \end{bmatrix}^{-1} \begin{bmatrix} 0 \\ c \end{bmatrix}.$$

The actual computation of $\hat{z}$ requires the inversion of a matrix whose dimensions can be quite large even for a relatively simple problem. However, this matrix contains many zeros, which helps in the numerical inversion. The following example should give some idea of the degree of difficulty involved in solving the problem by this method.

SEC. 6.5 APPLICATIONS TO UNBOUNDED CONTROL PROBLEMS

Example. Let $n = 2$, $m = 1$, and $k = 4$, and let system (2) be of the specific form

$$x_{i+1} - x_i = \begin{bmatrix} -1 & 1 \\ -2 & -2 \end{bmatrix} x_i + \begin{bmatrix} 0 \\ 1 \end{bmatrix} u_i.$$

The initial state is taken as $c_0 = (2,2)$, and the terminal state is taken as $c_4 = (0,0)$. It is easily verified that (21) is a completely controllable system.

In this case R is the (12×14)-dimensional matrix

$$R = \begin{bmatrix} 0 & -1 & 1 & 0 & 0 & 0 & 0 & 0 & 0 & 0 & 0 & 0 & 0 & 0 \\ 2 & 1 & 0 & 1 & 0 & 0 & 0 & 0 & 0 & 0 & -1 & 0 & 0 & 0 \\ 0 & 0 & 0 & -1 & 1 & 0 & 0 & 0 & 0 & 0 & 0 & 0 & 0 & 0 \\ 0 & 0 & 2 & 1 & 0 & 1 & 0 & 0 & 0 & 0 & 0 & -1 & 0 & 0 \\ 0 & 0 & 0 & 0 & 0 & -1 & 1 & 0 & 0 & 0 & 0 & 0 & 0 & 0 \\ 0 & 0 & 0 & 0 & 2 & 1 & 0 & 1 & 0 & 0 & 0 & 0 & -1 & 0 \\ 0 & 0 & 0 & 0 & 0 & 0 & -1 & 1 & 0 & 0 & 0 & 0 & 0 & 0 \\ 0 & 0 & 0 & 0 & 0 & 2 & 1 & 0 & 1 & 0 & 0 & 0 & 0 & -1 \\ 1 & 0 & 0 & 0 & 0 & 0 & 0 & 0 & 0 & 0 & 0 & 0 & 0 & 0 \\ 0 & 1 & 0 & 0 & 0 & 0 & 0 & 0 & 0 & 0 & 0 & 0 & 0 & 0 \\ 0 & 0 & 0 & 0 & 0 & 0 & 0 & 1 & 0 & 0 & 0 & 0 & 0 & 0 \\ 0 & 0 & 0 & 0 & 0 & 0 & 0 & 0 & 1 & 0 & 0 & 0 & 0 & 0 \end{bmatrix}.$$

The matrix

$$\begin{bmatrix} -Q & R^T \\ R & 0 \end{bmatrix}$$

is 26×26 and would require such a large space to write down that it is omitted here. The interested reader may pursue this to verify that the optimal control sequence is

$$\hat{u} = (-14\tfrac{2}{23}, 10\tfrac{5}{23}, 22\tfrac{2}{23}, 28\tfrac{5}{23}). \quad \square$$

It is clear from this example that even simple problems become fairly cumbersome with this approach. One reason is that no account has been taken of the particular nature of the equality constraints. We have already pointed out in (1.12) that it may sometimes be more convenient to use an alternative approach for converting a control problem into Basic Problem form. We shall now see what simplifications can be achieved by taking into account the structure of the system equation.

Solution of (1) by Alternative Transcription

Solving problem (2) for x_k, with $x_0 = c_0$, we obtain

$$25 \quad x_k = (I + A)^k c_0 + \sum_{i=0}^{k-1} (I + A)^{k-i-1} B u_i,$$

and problem (1) is now seen to be equivalent to the problem:

26   Minimize $\frac{1}{2}\langle \mathcal{U}, Q\mathcal{U}\rangle$ subject to the constraint $R\mathcal{U} = c$, where $Q = 2I$, I being the $km \times km$ identity matrix

$$27 \quad R = [(I + A)^{k-1}B \mid (I + A)^{k-2}B \mid \cdots \mid (I + A)B \mid B],$$

and

$$28 \quad c = c_k - (I + A)^k c_0. \quad \square$$

Problem (26) is clearly of the form of QP (2.1).

Note the several significant differences in this reformulation. First (since $k \geq n$), the matrix appearing in the assumption of complete controllability (5) is a submatrix of R, as given by (27). It follows immediately that R has rank n. Thus a feasible solution to this problem always exists. Second, Q is a positive-definite matrix, and so the optimal solution exists and is unique. Obviously, these facts are now much easier to establish than they were previously.

Now consider the solution of this problem. It can easily be verified from (4.4) that a necessary and sufficient condition for a control sequence $\hat{\mathcal{U}}$ to be optimal is that there exist a vector ψ such that

$$29 \quad \begin{bmatrix} -Q & R^T \\ R & 0 \end{bmatrix} \begin{bmatrix} \hat{\mathcal{U}} \\ \psi \end{bmatrix} = \begin{bmatrix} 0 \\ c \end{bmatrix}$$

is satisfied. As in (18), the matrix in (29) is nonsingular, and hence it may be inverted to find $\hat{\mathcal{U}}$. Note, however, the dimension of the matrix in (29) is $(km + n) \times (km + n)$, which is significantly lower than $[km + (2k + 3)n] \times [km + (2k + 3)n]$, the dimension of the matrix in (18). For example (20) this works out to be 6×6 in comparison with 26×26 in (22).

However, we still have not taken full advantage of the structure of the problem. Since $Q = 2I$, we may solve the first set of equations in (29) for $\hat{\mathcal{U}}$, which yields

$$30 \quad \hat{\mathcal{U}} = \tfrac{1}{2} R^T \psi.$$

This, in turn, may be substituted into the second set of equations in (29) to obtain

$$31 \quad R\hat{\mathcal{U}} = \tfrac{1}{2} R R^T \psi = c.$$

SEC. 6.5 APPLICATIONS TO UNBOUNDED CONTROL PROBLEMS

Since R has full row rank, the matrix RR^T is positive definite[1] and hence is nonsingular. Therefore

(32) $\quad \psi = 2(RR^T)^{-1}c,$

and substitution into (30) yields

(33) $\quad \hat{u} = R^T(RR^T)^{-1}c.$

Thus by taking full advantage of the structure of this problem we need only invert the $n \times n$ matrix RR^T to obtain the optimal solution $\hat{u}$. This is indeed a significant improvement over the previous approach, since the dimension RR^T is independent of k. To illustrate the difference let us again solve the problem in example (20).

(34) **Example.** Let $n = 2$, $m = 1$, and $k = 4$, and let system (2) be of the specific form

$$x_{i+1} - x_i = \begin{bmatrix} -1 & 1 \\ -2 & -2 \end{bmatrix} x_i + \begin{bmatrix} 0 \\ 1 \end{bmatrix} u_i,$$

with $c_0 = (2,2)$ and $c_4 = (0,0)$.

Evaluating R and c, we obtain

(35) $\quad R = \begin{bmatrix} 7 & 3 & 1 & 0 \\ 15 & 7 & 3 & 1 \end{bmatrix} \quad c = (-2,-2).$

Consequently,

(36) $\quad RR^T = \begin{bmatrix} 59 & 129 \\ 129 & 284 \end{bmatrix}$

and

(37) $\quad \hat{u} = (-14\tfrac{2}{23}, 10\tfrac{2}{23}, 22\tfrac{2}{23}, 28\tfrac{2}{23}).\quad \square$

Thus, while the alternative transcription of an optimal control problem into Basic Problem form may not always be practical for nonlinear problems, it does offer significant advantages for many problems with linear dynamics.

Before we leave the miminum-energy problem (1), however, we must consider one more approach to its solution.

The Control Approach to the Minimum-energy Problem

By the "control approach" to the mimimum-energy problem (1) we shall mean the examination of this problem by means of the neces-

---

[1] For any $x \neq 0$ the quadratic form $\langle x, RR^T x \rangle = \|R^T x\|^2 > 0$, since $R^T x \neq 0$, by the rank assumption.

sary conditions for optimality developed in Sections 4.1 and 4.2 [which can also be shown to be sufficient for problem (1)]. We may use either theorem (4.1.17) or theorem (4.2.34), for, as we shall see shortly, the hypotheses of both theorems are satisfied. For no reason other than to demonstrate its application, we shall use theorem (4.2.34), the discrete *maximum principle*.

Recall that in order to apply theorem (4.2.34) we must first examine the sets $\mathbf{f}_i(x_i, U_i)$ for $i = 0, 1, \ldots, k - 1$, which for problem (1) are given by

38 $$\mathbf{f}_i(x, U_i) = \{\mathbf{v} = (v^0, v): v^0 = \|u_i\|^2,$$
$$v = Ax + Bu \text{ for some } u \in U_i\}.$$

It may be verified that for every $x \in E^n$ these sets are b_0-directionally convex [see (4.2.1)], where $b_0 = (-1, 0, \ldots, 0) \in E^{n+1}$, as required in assumption (4.2.11). All the other assumptions made in Section 4.1 are obviously satisfied. Therefore we may apply the necessary-conditions theorem (4.2.34) to problem (1).

39 **Exercise.** Show that in applying theorem (4.2.34) to problem (1) we must take $p^0 \neq 0$, and hence we may choose $p^0 = -1$. [*Hint:* See (4.1.45). What role does controllability play here?] □

40 **Exercise.** Show that if a feasible control sequence for problem (1) satisfies the conditions of theorem (4.2.34) with $p_i{}^0 = -1$ for $i = 0, 1, \ldots, k$, then it must be optimal; i.e., the maximum principle (4.2.34) is also a sufficient condition of optimality for this problem [*Hint:* Adapt corollary (3.6.9)]. □

Applying theorem (4.2.34), together with the results in (39) and (40), to problem (1), we find that a control sequence $\hat{\mathfrak{u}} = (\hat{u}_0, \hat{u}_1, \ldots, \hat{u}_{k-1})$ and the corresponding trajectory $\hat{\mathfrak{X}} = (\hat{x}_0, \hat{x}_1, \ldots, \hat{x}_k)$ are optimal if and only if $\hat{x}_0 = c_0$, $\hat{x}_k = c_k$, and there exist n-dimensional costate vectors $p_0, p_1, \ldots, p_k$ satisfying

41 $$p_i - p_{i+1} = A^T p_{i+1} \qquad i = 0, 1, \ldots, k - 1$$

such that

42 $$H(\hat{x}_i, \hat{u}_i, p_{i+1}, p^0, i) = \max_{u_i \in U_i} H(\hat{x}_i, u_i, p_{i+1}, p^0, i)$$
$$i = 0, 1, \ldots, k - 1,$$

where for this case the hamiltonian $H(\hat{x}_i, u_i, p_{i+1}, p^0, i) = -\|u_i\|^2 + \langle p_{i+1}, A\hat{x}_i + Bu_i \rangle$, since $p^0 = -1$.

SEC. 6.5 APPLICATIONS TO UNBOUNDED CONTROL PROBLEMS

Since the hamiltonian $H(\hat{x}_i, u_i, p_{i+1}, -1, i)$ consists of a negative-definite quadratic form in u_i and of a linear term in u_i, it is concave in u_i. Thus, since u_i is unconstrainted, $\hat{u}_i$ must satisfy

(43) $$\frac{\partial}{\partial u} H(\hat{x}_i, \hat{u}_i, p_{i+1}, -1, i) = -2\hat{u}_i + B^T p_{i+1} = 0$$

$$i = 0, 1, \ldots, k - 1.$$

Consequently,

(44) $$\hat{u}_i = \tfrac{1}{2} B^T p_{i+1} \quad i = 0, 1, \ldots, k - 1.$$

To obtain the optimal control sequence $\hat{\mathbf{u}} = (\hat{u}_0, \hat{u}_1, \ldots, \hat{u}_{k-1})$ we now solve equations (2), (41), and (44), with the boundary conditions $x_0 = c_0$ and $x_k = c_k$.

Substituting (41) in (2), we obtain

(45) $$\hat{x}_{i+1} - \hat{x}_i = A\hat{x}_i + \tfrac{1}{2} BB^T p_{i+1} \quad i = 0, 1, \ldots, k - 1.$$

Solving (41) for p_{i+1} in terms of p_k, we have

(46) $$p_{i+1} = [(I + A)^T]^{k-i-1} p_k \quad i = 0, 1, \ldots, k - 1.$$

We may now use (46) to eliminate p_{i+1} in (45), obtaining

(47) $$\hat{x}_{i+1} - \hat{x}_i = A\hat{x}_i + \tfrac{1}{2} BB^T [(I + A)^T]^{k-i-1} p_k$$

Next we solve (47) for $\hat{x}_k$ in terms of $\hat{x}_0$ and $\hat{p}_k$ and obtain

(48) $$\hat{x}_k = (I + A)^k \hat{x}_0 + \tfrac{1}{2} \sum_{i=0}^{k-1} (I + A)^{k-i-1} BB^T [(I + A)^{k-i-1}]^T p_k.$$

If we let

(49) $$M = \tfrac{1}{2} \sum_{i=0}^{k-1} (I + A)^{k-i-1} BB^T [(I + A)^{k-i-1}]^T$$

and substitute for $\hat{x}_k$ and $\hat{x}_0$ in (48), we arrive at the following equation for p_k:

(50) $$Mp_k = c_k - (I + A)^k c_0.$$

It is left as an exercise to verify that equation (50) is equivalent to equation (31) (with $\psi = p_k$); that is to say,

$$M = \tfrac{1}{2} RR^T,$$

where R is as defined in (27). Therefore M is invertible, and

$$p_k = M^{-1} c,$$

where $c = c_k - (I + A)^k c_0$. From (46) and (44) we now determine that

(51) $$\hat{u}_i = \tfrac{1}{2} B^T [(I + A)^T]^{k-i-1} M^{-1} c \quad i = 0, 1, \ldots, k - 1.$$

148 OPTIMAL CONTROL AND QUADRATIC PROGRAMMING

52 Exercise. Consider the dynamical system

* $\quad x_{i+1} - x_i = f_i(x_i, u_i), \; i = 0, 1, \ldots, k-1,$

with $x_i \in E^n$, $u_i \in E^m$, and the $f_i(\cdot,\cdot)$ continuously differentiable in both arguments. Suppose that $\hat{x}_0, \hat{x}_1, \ldots, \hat{x}_k$ is the nominal trajectory which the system is required to follow and that this trajectory can be achieved by applying the control sequence $\hat{u}_0, \hat{u}_1, \ldots, \hat{u}_{k-1}$ to (*), with $x_0 = \hat{x}_0$. Now suppose that there are some difficulties in setting the initial state of (*) and that we must accept $x_0 = \hat{x}_0 + \delta\hat{x}_0$, where $\delta\hat{x}_0$ is measured precisely and is known to be small. To account for this discrepancy in the initial state, the entire control sequence must be trimmed to a new set of values $\hat{u}_0 + \delta\hat{u}_0, \hat{u}_1 + \delta\hat{u}_1, \ldots, \hat{u}_{k-1} + \delta\hat{u}_{k-1}$. A set of control corrections $\delta\hat{u}_0, \ldots, \delta\hat{u}_{k-1}$, can be computed to first order accuracy by solving the following problem:

** $\quad$ Minimize $\frac{1}{2} \sum_{i=0}^{k-1} \langle \delta u_i, Q_i \delta u_i \rangle + \frac{1}{2} \sum_{i=1}^{k} \langle \delta x_i, R_i \delta x_i \rangle$, subject to

$$\delta x_{i+1} - \delta x_i = A_i \delta x_i + B_i \delta u_i \quad i = 0, \ldots, k-1,$$
$$\delta x_0 = \delta \hat{x}_0,$$

where $A_i = \partial f_i(\hat{x}_i, \hat{u}_i)/\partial x_i$, $B_i = \partial f_i(\hat{x}_i, \hat{u}_i)/\partial u_i$, and Q_i, R_i are positive definite, symmetric matrices which can be used as a design parameter.

Show that the optimal control sequence $\delta\hat{u}_0, \delta\hat{u}_1, \ldots, \delta\hat{u}_{k-1}$ for (**) is unique, and if $\delta\hat{x}_0, \delta\hat{x}_1, \ldots, \delta\hat{x}_k$ is the corresponding optimal trajectory, then, provided all the inverses used exist,

$$\delta\hat{u}_i = Q_i^{-1} B_i^T K_{i+1} (I - B_i Q_i^{-1} B_i^T K_{i+1})^{-1} (A_i + I) \delta\hat{x}_i,$$
$$i = 0, 1, \ldots, k-1,$$

where for $i = 0, 1, \ldots, k-1$, K_{i+1} is an $n \times n$ matrix defined by the Riccati type equation

$$K_i = -R_i + (A_i + I)^T K_{i+1} (I - B_i Q_i^{-1} B_i^T K_{i+1})^{-1} (A_i + I)$$
$$K_k = R_k.$$

Thus, for problem (**), the control approach yields the optimal control sequence in linear feedback form. □

It is clear from the preceding that for the minimum energy problem, the control approach works as well as the better of the two approaches discussed before it. This is also true for the particular minimum-quadratic-cost problem which was presented in exercise (52). The control approach depends entirely on our ability to express the optimal controls $\hat{u}_i$ in terms of the optimal states $\hat{x}_i$ and of the costates p_i. When the elimination of the controls $\hat{u}_i$ from the necessary conditions equations cannot be performed readily, as in problem (53), the control approach cannot be applied.

53 **The minimum-quadratic-cost problem.** Consider the dynamical system governed by the difference equation

54
$$x_{i+1} - x_i = Ax_i + Bu_i \qquad i = 0, 1, \ldots, k-1,$$

where the notation is the same as before. The initial state x_0 is required to belong to a linear manifold described by

55
$$G_0 x_0 = c_0,$$

where G_0 is an $l_0 \times n$ matrix of rank l_0 and $c_0 \in E^{l_0}$ is a fixed vector, and the terminal state x_k must belong to the linear manifold

56
$$G_k x_k = c_k,$$

where G_k is an $l_k \times n$ matrix of rank l_k and $c_k \in E^{l_k}$ is a fixed vector. Find a control sequence $\hat{\mathfrak{U}} = (\hat{u}_0, \hat{u}_1, \ldots, \hat{u}_k)$ and a corresponding trajectory $\hat{\mathfrak{X}} = (\hat{x}_0, \hat{x}_1, \ldots, \hat{x}_k)$ which minimize

57
$$\tfrac{1}{2}\langle \mathfrak{X}, Q_x \mathfrak{X}\rangle + \tfrac{1}{2}\langle \mathfrak{U}, Q_u \mathfrak{U}\rangle,$$

subject to the constraints (54) to (56), where Q_x and Q_u are symmetric positive-semidefinite matrices. □

As in the minimum-energy problem (1), we shall make the following assumptions:

58 **Assumption.** $k \geq n$. □

59 **Assumption.** System (54) is completely controllable. □

Obviously, problem (53) is a special case of the quadratic control problem (1.2). The matrix Q in (1.4) assumes for this case the form

60
$$Q = \begin{bmatrix} Q_x & 0 \\ 0 & Q_u \end{bmatrix}.$$

The conversion of problem (53) into a quadratic programming problem should by now be routine. Thus the cost function becomes

61
$$f(z) = \tfrac{1}{2}\langle z, Qz \rangle,$$

where $z = (x_0, x_1, \ldots, x_k, u_0, u_1, \ldots, u_{k-1})$ and Q is as given in (60). The equality constraints become

62
$$Rz = c,$$

where R is the matrix

$$
R = \begin{bmatrix}
-(I+A) & I & 0 & \cdots & & \cdots & 0 & -B & 0 & \cdots & \cdots & 0 \\
0 & -(I+A) & I & 0 & & \cdots & 0 & 0 & -B & 0 & \cdots & 0 \\
0 & \cdots & & \cdots & -(I+A) & I & 0 & \cdots & & & 0 & -B \\
G_0 & 0 & & \cdots & & \cdots & 0 & 0 & \cdots & & & 0 \\
0 & \cdots & & \cdots & 0 & G_k & 0 & \cdots & & & & 0
\end{bmatrix}
$$

and $c = (0,0,\ldots,0,c_0,c_k)$.

Since there is no linear term in the cost function, it follows immediately from corollary (2.23) that an optimal solution to this problem exists whenever a feasible solution exists. Moreover, assumptions (58) and (59), together with the fact that there are no inequality constraints present, guarantee that a feasible solution exists. Hence an optimal solution to this problem always exists.

There is very little that can be said about uniqueness of the optimal solutions to problem (53) other than that we can check the conditions of theorem (3.1); that is, we must show that $\mathfrak{N}(Q) \cap \mathfrak{N}(R) = \{0\}$. This may be done by examining the matrix

$$
\begin{bmatrix} Q \\ \hdashline R \end{bmatrix}
$$

to see if it has full rank. Alternatively, since R has full rank [because of assumptions (58) and (59), together with the full-rank assumption on G_0 and G_k], we may, as a consequence of theorem (2.6), determine if the matrix

$$
\begin{bmatrix} Q & R^T \\ R & 0 \end{bmatrix}
$$

is nonsingular.

Because of the special form of Q [see (60)], the rank of (64) becomes easier to check when either Q_x or Q_u is positive definite. For instance, if Q_u is positive definite, then the rank of the matrix in (64) is the same as the rank of the matrix

$$
\begin{bmatrix}
\multicolumn{5}{c}{Q_x} \\ \hdashline
-(I+A) & I & 0 & \cdots & 0 \\
\vdots & & & & \vdots \\
0 & \cdots & \cdots & -(I+A) & I \\
G_0 & \cdots & \cdots & \cdots & 0 \\
0 & \cdots & \cdots & \cdots & G_k
\end{bmatrix}
$$

Solution of (53) by Direct Transcription

The solution of the quadratic programming problem formulated above is so similar in procedure to the corresponding solution of the

SEC. 6.5 APPLICATIONS TO UNBOUNDED CONTROL PROBLEMS

minimum-energy problem (1) that we shall not repeat it. Suffice it to say that we eventually arrive at the point where we must solve the equation

66
$$\begin{bmatrix} -Q & R^T \\ R & 0 \end{bmatrix} \begin{bmatrix} \hat{z} \\ \psi \end{bmatrix} = \begin{bmatrix} 0 \\ c \end{bmatrix}.$$

Consequently, whenever the optimal solution to problem (53) is unique, we can solve (66) by inverting the matrix on the left-hand side.

It is appropriate to note that (66) may be solved for an optimal $\hat{z}$ even when the matrix appearing on the left-hand side is singular. The existence of an optimal solution to problem (53), together with the fact that (66) expresses a necessary and sufficient condition for optimality, guarantees that (66) has at least one solution. Furthermore, every solution of (66) is also an optimal solution to (53).

Solution of (53) by Alternative Transcription

In the alternate transcription of problem (53) into a quadratic programming problem, we eliminate all the x_i vectors other than x_0 by solving for them by means of (54). In the minimum-energy problem (1) this was very easy to do because only the terminal constraint involved any of these x_i. Now, however, the x_i also appear in the cost function, which makes the conversion to a quadratic programming problem much more complicated.

As before, we solve (54) for the x_i vectors, obtaining

67
$$x_i = (I + A)^i x_0 + \sum_{j=0}^{i-1} (I + A)^{i-j-1} B u_j \qquad i = 1, 2, \ldots, k.$$

If we define the vector z in E^{km+n} to be

68
$$z = (x_0, u_0, u_1, \ldots, u_{k-1}),$$

then expression (67) becomes

69
$$x_i = M_i z \qquad i = 1, 2, \ldots, k,$$

where M_i is the $n \times (km + n)$ matrix

70
$$M_i = [(I + A)^i \,|\, (I + A)^{i-1} B \,|\, \cdots \,|\, (I + A) B \,|\, B \,|\, 0 \,|\, \cdots \,|\, 0]$$
$$\text{for } i = 1, 2, \ldots, k.$$

Hence the trajectory $\mathcal{X} = (x_0, x_1, \ldots, x_k)$ is given by

71
$$\mathcal{X} = Mz.$$

where M is the $(k+1)n \times (km+n)$ matrix

(72) $$M = \begin{bmatrix} I & 0 & \cdots & 0 \\ & M_1 & & \\ & M_2 & & \\ \vdots & & & \\ & M_k & & \end{bmatrix} = \begin{bmatrix} I & 0 & \cdots & \cdots & \cdots & 0 \\ I+A & B & 0 & \cdots & \cdots & 0 \\ (I+A)^2 & (I+A)B & B & 0 & \cdots & 0 \\ \vdots & & & & & \\ (I+A)^k & (I+A)^{k-1}B & & \cdots & \cdots & B \end{bmatrix}.$$

The cost function (57) becomes

(73) $$\langle \mathfrak{X}, Q_x \mathfrak{X} \rangle + \langle \mathfrak{U}, Q_u \mathfrak{U} \rangle = \langle z, Pz \rangle,$$

where

(74) $$P = M^T Q_x M + N$$

and
$$N = \begin{bmatrix} 0 & 0 \\ 0 & Q_u \end{bmatrix}.$$

It is easily verified that P is a symmetric positive-semidefinite matrix.

The boundary constraints may now be reformulated in terms of z. The initial constraint manifold (55) becomes

(75) $$G_0 M_0 z = c_0,$$

where M_0 is the $n \times (km+n)$ matrix

(76) $$M_0 = [I \vdots 0 \vdots \cdots \vdots 0].$$

The terminal constraint manifold (56), together with (69), yields

(77) $$G_k M_k z = c_k,$$

where M_k is as defined in (70).

The matrix M_0 is clearly of rank n, and assumptions (58) and (59) guarantee that M_k has rank n. It may be verified that $G_0 M_0$ and $G_k M_k$ have rank l_0 and l_k, respectively, as a consequence.

From this point we proceed to minimize (73) subject to the equality constraints (75) and (77). It should be clear that we shall once again arrive at the expression

(78) $$\begin{bmatrix} -P & R^T \\ R & 0 \end{bmatrix} \begin{bmatrix} \hat{z} \\ \psi \end{bmatrix} = \begin{bmatrix} 0 \\ c \end{bmatrix},$$

where $\hat{z}$ is the optimal solution, ψ is the corresponding multiplier, R is the $(l_0 + l_k) \times (km + n)$ matrix given by

(79) $$R = \begin{bmatrix} G_0 M_0 \\ \hdashline G_k M_k \end{bmatrix},$$

and $c = (c_0, c_k)$.

SEC. 6.5 APPLICATIONS TO UNBOUNDED CONTROL PROBLEMS

Note that it is not immediately obvious that a unique solution to optimization problem (53) implies the nonsingularity of the matrix in (78), since R may not have full rank.

80 Exercise. Show that the uniqueness of the optimal solution to problem (53) implies that R in (79) has rank $(l_0 + l_k)$. [*Hint:* Show that $\mathfrak{N}(G_0) \cap \mathfrak{N}(G_k(I + A)^k) = \{0\}$ follows from the uniqueness of the optimal solution and is sufficient for full rank of R.] □

There will not, in general, be any way to further decompose (78), as we did in the minimum-energy problem. Hence the solution will require the inversion of a $(km + n + l_0 + l_k) \times (km + n + l_0 + l_k)$-dimensional matrix. This is still quite a reduction in dimension in comparison with the previous approach.

We conclude this section with a summary of some of our observations concerning the relative advantages and disadvantages of various approaches to solving unconstrained quadratic control problems.

The approach which is based on a direct transcription into an equivalent quadratic programming problem has the disadvantage that the dimensions of the matrices involved may get quite large, even for a relatively simple problem. However, these matrices have many zeros, and there are currently being developed methods of handling sparse matrices which may at least partially negate this disadvantage. This method has the advantage that the quadratic programming problem is set up and solved in a very straightforward manner which does not vary with the particular form of the control problem. It can also be applied to the general case of QCP (1.2).

The alternative method for setting up the quadratic programming problem entails first eliminating the state variables x_i for $i = 1, 2, \ldots, k$ by solving for them from the difference equation. In the two classes of problems considered in this section, this approach substantially reduced the dimensions of the resulting problem. As was demonstrated in the second class of problems, in some cases the initial setting up of the quadratic programming problem may be somewhat complicated. However, if computation speed is a critical factor, this disadvantage is greatly outweighed by the reduction in the dimensions of the problem. This approach can also be applied to the general case of QCP (1.2).

The control approach takes advantage of the specialized form of the necessary conditions which were developed for control problems. While the control approach is very useful for solving simple, unconstrained problems, it is not a satisfactory tool for dealing with problems that include state space constraints, or problems with control constraints and cost functions depending on the states, etc. Since this text is primarily concerned with constrained minimization problems, we leave to the reader the extension of the results indicated in (52) to more general situations.

6.6 QUADRATIC CONTROL PROBLEMS WITH INEQUALITY CONSTRAINTS

In Section 6.1 it was shown that the most general form of the quadratic control problem can be reduced to a quadratic programming problem. Section 6.5 contained some examples of how we would proceed if there were no inequality constraints. The remainder of this chapter will be devoted to the presentation of an algorithm for solving the quadratic programming problem (under suitable assumptions, which will be stated later) in the case when inequality constraints are present. However, before we proceed, let us consider the nature of the quadratic programming problems resulting from different special cases of the quadratic control problem. For our purposes, the two classes of problems considered in Section 6.5, with some additional inequality constraints, will serve nicely.

1 **The constrained minimum-energy problem.** (Compare with problem (5.1).) Consider a dynamical system described by

2 $$x_{i+1} - x_i = Ax_i + Bu_i \qquad i = 0, 1, \ldots, k-1,$$

with the initial and terminal constraints

3 $$x_0 = c_0 \quad \text{and} \quad x_k = c_k,$$

and with the control constraints

4 $$|u_i{}^j| \leq 1 \qquad j = 1, 2, \ldots, m, \, i = 0, 1, \ldots, k-1,$$

where $u_i = (u_i{}^1, u_i{}^2, \ldots, u_i{}^m)$ for $i = 0, 1, \ldots, k-1$. (The bound on the components of the control variables has been set at unity, but this is not really important, since we can always normalize a finite bound to unity.) Find a control sequence $\hat{\mathfrak{u}} = (\hat{u}_0, \hat{u}_1, \ldots, \hat{u}_{k-1})$ which minimizes

5 $$\sum_{i=0}^{k-1} \langle u_i, u_i \rangle,$$

subject to (2), (3) and (4). □

6 **Assumption.** $k \geq n$. □

7 **Assumption.** System (2) is completely controllable. □

SEC. 6.6 QUADRATIC CONTROL PROBLEMS WITH INEQUALITY CONSTRAINTS

The transformation of problem (1) into a quadratic programming problem may again be accomplished by first eliminating the x_i for $i = 1, 2, \ldots, k$ by means of the difference equation, as was worked out in detail in Section 5. The resulting equivalent quadratic programming problem has the form

(8) Minimize $\tfrac{1}{2}\langle \mathfrak{u}, Q\mathfrak{u}\rangle$ subject to the constraints
(9) $\quad R\mathfrak{u} = c$
(10) $\quad -1 \leq u^j \leq 1 \qquad j = 1, 2, \ldots, km,$

where $Q = 2I$,

(11) $$R = [(I + A)^{k-1}B \mid (I + A)^{k-2}B \mid \cdots \mid (I + A)B \mid B],$$

and

(12) $$c = c_k - (I + A)^k c_0. \quad \square$$

Note that (8) differs from (5.6) only by the inequality constraints (10).

There are two points about this problem which are significant with respect to the algorithms for solving quadratic programming problems. First, the variables in this problem are all bounded [see (10)]. Second, the matrix R appearing in (9) has rank n, as is easily deduced from (7). This is an assumption which is required by the algorithms discussed later.

Now let us consider what happens to the minimum-quadratic-cost problem (5.53) when we add bounds for the control variables.

(13) **The constrained minimum-quadratic-cost problem.** Consider a dynamical system governed by the difference equation

(14) $$x_{i+1} - x_i = Ax_i + Bu_i \qquad i = 0, 1, \ldots, k - 1,$$

with initial and terminal constraints

(15) $\quad G_0 x_0 = c_0 \qquad G_k x_k = c_k,$

where G_0 is an $l_0 \times n$ matrix of rank l_0 and G_k is an $l_k \times n$ matrix of rank l_k; with control constraints

(16) $\quad |u_i^j| \leq 1 \qquad j = 1, 2, \ldots, m, \; i = 0, 1, \ldots, k - 1,$

where $u_i = (u_i^1, u_i^2, \ldots, u_i^m)$, and with cost function

(17) $\quad \langle \mathfrak{X}, Q_x \mathfrak{X}\rangle + \langle \mathfrak{u}, Q_u \mathfrak{u}\rangle,$

where Q_x and Q_u are symmetric positive-semidefinite matrices, $\mathfrak{X} = (x_0, x_1, \ldots, x_k)$ and $\mathfrak{U} = (u_0, u_1, \ldots, u_{k-1})$. Find a control sequence $\mathfrak{U}$ which minimizes (17) subject to (14), (15), and (16). □

18 Assumption. $k \geqq n$. □

19 Assumption. System (14) is completely controllable. □

We convert this problem to a quadratic programming problem by the alternative transcription method described in Section 5, beginning with equation (67). It can easily be verified that if we define

20 $\qquad z \triangleq (x_0, u_1, \ldots, u_{k-1})$,

then the resulting quadratic programming problem is:

21 $\qquad$ Minimize $\langle z, Pz \rangle$ subject to the constraints
22 $\qquad Rz = c$
23 $\qquad |z^i| \leqq 1 \qquad i = n+1, n+2, \ldots, n+km,$

where P, R, and c are as defined in (5.74) and (5.79). □

Note that it is no longer true that all the variables are bounded, even though the controls are bounded. Specifically, the first n components of z corresponding to x_0 are unbounded. In general, then, not all quadratic programming problems which arise from a control problem have all variables bounded.

In the following sections we shall derive two algorithms for solving quadratic programming problems. Like the simplex algorithm, these algorithms are combinational in nature, and they either determine an optimal solution or determine that an optimal solution does not exist.

6.7 OPTIMALITY CONDITIONS FOR THE CANONICAL QUADRATIC PROGRAMMING PROBLEM

The algorithms we are about to discuss apply to quadratic programming problems of the following form.

1 The canonical quadratic programming problem (CQP). Given an $m \times n (m \leqq n)$ full-rank matrix R, an $n \times n$ symmetric positive-semidefinite matrix Q, and vectors $c \in E^m$ and $d \in E^n$, find a $\hat{z} \in E^n$ which minimizes

2 $\qquad \frac{1}{2} \langle z, Qz \rangle + \langle z, d \rangle$

subject to the constraints

3 $\qquad Rz = c \qquad z \geqq 0.$ □

4 **Remark.** It follows from exercises (2.6) and (4.2) that the quadratic programming problem (1.1) can be reduced to the above canonical form. □

For the purpose of describing the algorithms for the solution of (1), the necessary and sufficient conditions for optimality as stated in theorems (3.3.5) and (3.6.1) are more convenient to work with than theorem (4.1). Thus, combining (3.3.5) and (3.6.1), we obtain the following result.

5 **Theorem.** A vector $z \in E^n$ is an optimal solution to CQP (1) if and only if there exist multipliers $\psi = (\psi^1, \psi^2, \ldots, \psi^m)$ and $\xi = (\xi^1, \xi^2, \ldots, \xi^n)$ such that

6
$$Qz + R^T\psi - \xi + d = 0 \qquad Rz - c = 0$$
$$z \geqq 0 \qquad \xi \geqq 0 \qquad \langle z, \xi \rangle = 0. \quad \Box$$

The above system of equations and inequalities can be written in a more symmetric form by decomposing the vector ψ into a positive part, which we denote by ψ_+, and a negative part, which we denote by $-\psi_-$. Thus let

7
$$\psi = \psi_+ - \psi_-,$$

with the restriction that

8
$$\psi_+ \geqq 0 \qquad \psi_- \geqq 0,$$
9
$$\psi_+^i \psi_-^i = 0 \qquad i = 1, 2, \ldots, m.$$

Note that (9) permits either ψ_+^i or ψ_-^i to be strictly positive, but not both at the same time. Because of (8), (9) can be written more compactly as

10
$$\langle \psi_+, \psi_- \rangle = 0.$$

Substituting from (7), (8), and (10) into (6) leads to the following system of equations and inequalities:

11
$$Qz - R^T\psi_- - \xi + R^T\psi_+ = -d \qquad Rz = c$$
$$z \geqq 0 \qquad \psi_- \geqq 0 \qquad \xi \geqq 0 \qquad \psi_+ \geqq 0$$
$$\langle z, \xi \rangle = \langle \psi_+, \psi_- \rangle = 0.$$

6.8 THE DERIVED MINIMIZATION PROBLEM

The effect of theorem (7.5) was to reduce the problem of solving the canonical quadratic programming problem (7.1) to that of solving system (7.11). In turn, a solution to this system can be obtained by solving the derived minimization problem stated below. It might be noted at this point that it is common practice in mathematical programming to find solutions to systems of equations and inequalities by solving a related optimization problem for which there exist efficient algorithms. As a case in point, we shall see later that a slight modification of the simplex algorithm can be used to solve the problem we are about to define, and hence to obtain a solution to CQP (7.1).

1 The derived problem. Minimize the linear form $\langle l,y \rangle$ subject to the constraints

2 $\quad Ax = g \quad x \geqq 0$
3 $\quad \langle v,w \rangle = 0,$

where

$$x = (z,\psi_-,\xi,\psi_+,y) = (v,w,y) \in E^{2(n+m)+n}$$
$$v = (z,\psi_-) \in E^{n+m}, \quad w = (\xi,\psi_+) \in E^{n+m}$$

4 $\quad A = \begin{bmatrix} Q & -R^T & -I & R^T & K \\ R & 0 & 0 & 0 & 0 \end{bmatrix} \quad (n+m) \times [2(n+m)+n]$

$$g = (-d,c) \in E^{n+m} \quad l = (1,1,\ldots,1) \in E^n$$

The matrices Q, R, d, and b are as defined in (7.1); the $n \times n$ matrix K will be specified in (9.2). □

In the remainder of this chapter we shall be using the terms *basic solution, nondegenerate basic solution,* etc., which are all defined and discussed in Section 5.6. The reader is strongly urged to review this section as well as Section 5.8, in which the procedure for generating adjacent extreme points is described.

5 Definition. A vector x is said to be a *usable basic solution* to the derived problem (1) if x is a basic solution to (2) and, in addition, satisfies (3). A usable basic solution which is a nondegenerate basic solution to (2) will be called a *usable nondegenerate basic solution.* □

6 Remark. Clearly, every usable basic solution x to problem (1) is a feasible solution; that is,

$$x \in \Omega' = \Omega \cap \{x \colon \langle v,w \rangle = 0\},$$

where

7 $\quad \Omega = \{x \colon Ax = g, x \geqq 0\}.$ □

8 Lemma. Suppose that the system of equations and inequalities (7.11) has a solution. Then $\hat{x} = (\hat{v},\hat{w},\hat{y})$ is an optimal solution for the derived problem (1) if and only if $(\hat{v},\hat{w})$ satisfies system (7.11).

Proof. For every $x \in \Omega'$, $\langle l,y \rangle \geq 0$. Furthermore, $\langle l,y \rangle = 0$ if and only if $y = 0$. The claimed result then follows. □

9 Exercise. Show that there is an extreme point of Ω, defined as in (7), at which the objective function $\langle l,y \rangle$ of the derived problem (1) attains its minimum. [*Hint:* The proof is almost identical to the proof of theorem (5.5.6).] □

The importance of exercise (9) lies in the fact that to find a solution to the derived problem we need only search over the extreme points of Ω. Since the simplex algorithm can be used to explore extreme points, the possibility of using a modification of the simplex algorithm readily comes to mind.

6.9 THE SIMPLEX ALGORITHM FOR QUADRATIC PROGRAMMING

We shall now consider an algorithm due to Wolfe [1] for solving the canonical quadratic programming problem (7.1). The popularity of this algorithm is unquestionably due to the ease with which it lends itself to computer coding. It is somewhat restricted in that it converges (demonstrably) only if the matrices appearing in the statement of CQP (7.1) satisfy the following additional condition:

1 Assumption. $\langle d,\eta \rangle = 0$ for every $\eta \in E^n$ satisfying $Q\eta = 0$ and $R\eta = 0$. □

Several sufficient conditions for the satisfaction of assumption (1) were given in Sections 6.2 and 6.3. Recall that, by theorem (2.3), this is a sufficient condition for CQP (7.1) to have an optimal solution.

We are now ready to consider the algorithm.

2 Initialization. First we must obtain a usable basic solution x_0 to the derived problem (8.1). This is done in two steps. We use the simplex algorithm (5.9.4) to obtain a basic feasible solution to the system

$$Rz = c \quad z \geqq 0.$$

Let z_0 be the basic solution thus obtained, and let $\bar{J}(z_0) \subset \{1,2, \ldots ,n\}$ be the associated basis-indicator set [see (5.6.5)]. We can now define the remaining components of x_0, as well as the $n \times n$ diagonal matrix $K = [k_{ii}]$ appearing in (8.4), as follows:

For $i = 1, 2, \ldots, n$, $\xi_0^i = 0$.
For $i = 1, 2, \ldots, m$, $\psi_{-0}^i = \psi_{+0}^i = 0$.
For $i = 1, 2, \ldots, n$

$$k_{ii} = \begin{cases} \delta^i \triangleq (-d - Qz_0 + R^T\psi_{-0} + \xi_0 - R^T\psi_{+0})^i & \text{if } \delta^i \neq 0 \\ 1 & \text{if } \delta^i = 0. \end{cases}$$

$$y_0^i = \begin{cases} 1 & \text{if } \delta^i \neq 0 \\ 0 & \text{if } \delta^i = 0. \end{cases}$$

160   OPTIMAL CONTROL AND QUADRATIC PROGRAMMING

3   **Exercise.** Let x_0 and K be as defined above. Show that x_0 is a usable basic solution to the derived problem (8.1) and that $\bar{J}(x_0) = \bar{J}(z_0) \cup \{2(n+m)+1, \ldots, 2(n+m)+n\}$ is an associated basis-indicator set. □

4   **The simplex algorithm for quadratic programming.** Suppose that x_N is a usable basic solution to the derived problem (8.1). Let $\bar{J}(x_N) \subset \{1, 2, \ldots, 2(n+m)+n\}$ be the basis-indicator set associated with x_N.

   Step 1. Use the basic solution x_N in the simplex algorithm (5.9.4)† to compute an improved basic solution x_{N+1} to the problem:

5   Minimize $\langle l, y \rangle$, subject to the constraints $Ax = g$ $x \geq 0$, with the following side condition:
6   If $j \in \bar{J}(x_N) \cap \{1, 2, \ldots, n+m\}$ do not admit $j + n + m$ into the basis-indicator set $\bar{J}(x_{N+1})$ for x_{N+1} and if $j \in \bar{J}(x_N) \cap \{n+m+1, \ldots, 2(n+m)\}$ do not admit $j - (n+m)$ into the basis-indicator set $\bar{J}(x_{N+1})$.

   If a basic feasible solution x_{N+1} cannot be constructed in accordance with (6), then stop; otherwise proceed to step 2. Note that x_{N+1} is a usable basic solution.

   Step 2. If $\langle l, y_{N+1} \rangle = 0$, then stop. If $\langle l, y_{N+1} \rangle > 0$, set $x_N = x_{N+1}$ and return to step 1. □

   To summarize, algorithm (4) can reach a stop command in one of two ways.

7   **Stop 1.** The algorithm, under condition (6), does not yield an improved basic solution to problem (5). □

8   **Remark.** We shall see later that whenever assumption (1) is satisfied, stop condition (7) coincides with the following stop condition. □

9   **Stop 2.** At some iteration $\langle l, y_N \rangle = 0$. In this case x_N is an optimal solution to the derived problem (8.1). The first n components of x_N, that is, $z_N{}^i$ for $i = 1, 2, \ldots, n$, are an optimal solution to the canonical quadratic programming problem. □

10 **Exercise.** Show that algorithm (4) generates a finite sequence of usable basic solutions to the derived problem (8.1); that is, one of the

   † Use the procedure in Section 5.10 whenever cycling occurs because of degeneracies.

stop commands must be executed in a finite number of steps. [*Hint:* First eliminate the possibility of obtaining an unbounded ray, $\{x(\theta): \theta \in [0, \infty)\}$, of the polyhedron Ω (8.7).] □

By exercise (10), algorithm (4) generates a finite number of usable basic solutions to the derived problem. The generation of new usable basic solutions stops when either condition (7) or condition (9) is reached. If condition (9) occurs, we are done. We shall now see that whenever condition (7) occurs, condition (9) occurs at the same time.

11   **Theorem.** Suppose that assumption (1) is satisfied. Let $\hat{x} = (\hat{z}, \hat{\psi}_-, \hat{\xi}, \hat{\psi}_+, \hat{y})$ be the last usable basic solution generated by algorithm (4) at which condition (7) has occurred. Then $\hat{y} = 0$, and $\hat{z}$ is an optimal solution to CQP (7.1).

Proof. Consider the linear programming problem:

12   Minimize $\langle l, y \rangle$ subject to the constraints
$$Ax = g \quad \langle \hat{z}, \xi \rangle = \langle \hat{\xi}, z \rangle = \langle \hat{\psi}_+, \psi_- \rangle = \langle \hat{\psi}_-, \psi_+ \rangle = 0$$
$$x \geqq 0,$$

where A, g, and x are as defined in (8.4). Expanding (12), we get:

13   Minimize $\langle l, y \rangle$ subject to the constraints

14   $$\begin{bmatrix} Q & -R^T & -I & R^T & K \\ R & 0 & 0 & 0 & 0 \\ \hat{\xi} & \hat{\psi}_+ & \hat{z} & \hat{\psi}_- & 0 \end{bmatrix} \begin{bmatrix} z \\ \psi_- \\ \xi \\ \psi_+ \\ y \end{bmatrix} = \begin{bmatrix} -d \\ c \\ 0 \end{bmatrix}$$
$$z \geqq 0 \quad \psi_- \geqq 0 \quad \xi \geqq 0 \quad \psi_+ \geqq 0 \quad y \geqq 0,$$

where $\hat{\xi}$, $\hat{\psi}_+$, $\hat{z}$, and $\hat{\psi}_-$ are considered to be row matrices.

Since algorithm (4) has examined all usable basic solutions to (5) which are adjacent to $\hat{x}$ and satisfy (6), and has found none of lower cost, we conclude that $\hat{x}$ must be an optimal solution to (14). We shall now see that this implies that $\hat{y} = 0$.

Because $\hat{x}$ is optimal for (14), there must exist, by theorem (5.5.1), multiplier vectors λ_1, λ_2, λ_3, and μ_1, μ_2, μ_3, μ_4, μ_5, of appropriate dimensions, such that

15   a. $Q\lambda_1 + R^T\lambda_2 + \hat{\xi}\lambda_3 - \mu_1 = 0$
b. $-R\lambda_1 + \hat{\psi}_+\lambda_3 - \mu_2 = 0$
c. $-\lambda_1 + \hat{z}\lambda_3 - \mu_3 = 0$
d. $R\lambda_1 + \hat{\psi}_-\lambda_3 - \mu_4 = 0$
e. $l + K^T\lambda_1 - \mu_5 = 0$

and

16 a. $\mu_1 \geq 0$ $\langle\mu_1,\hat{z}\rangle = 0$
 b. $\mu_2 \geq 0$ $\langle\mu_2,\hat{\psi}_-\rangle = 0$
 c. $\mu_3 \geq 0$ $\langle\mu_3,\hat{\xi}\rangle = 0$
 d. $\mu_4 \geq 0$ $\langle\mu_4,\hat{\psi}_+\rangle = 0$
 e. $\mu_5 \geq 0$ $\langle\mu_5,\hat{y}\rangle = 0.$

Note that $\hat{y} = 0$ if and only if $\langle l,\hat{y}\rangle = 0$. We shall now see that $\langle l,\hat{y}\rangle = 0$ whenever the following identities hold:

17 $Q\lambda_1 = 0$
18 $R\lambda_1 = 0$
19 $\langle\lambda_1,\hat{\xi}\rangle = 0.$

Note that (17) and (18), with assumption (1), imply that

20 $\langle d,\lambda_1\rangle = 0.$

Indeed, suppose that equations (17) to (19) hold. Then, taking the scalar product of both sides of (15e) with the vector $\hat{y}$, we obtain [with (16e)]

21 $\langle\hat{y},l\rangle = -\langle K\hat{y},\lambda_1\rangle.$

Since $\hat{x}$ is a feasible solution, $A\hat{x} = g$; that is,

22 $Q\hat{z} - R^T\hat{\psi}_- - \hat{\xi} + R^T\hat{\psi}_+ + K\hat{y} = -d.$

Substituting $K\hat{y}$ in (21) from (22) and making use of (17) to (20), we obtain

23 $\langle\hat{y},l\rangle = 0.$

Therefore we shall now prove (17) to (19). Taking the scalar product of both sides of (15b) and (15d) with $\hat{\psi}_-$ and $\hat{\psi}_+$, respectively, and recalling that $\langle\hat{\psi}_+,\hat{\psi}_-\rangle = 0$, we obtain [with the use of (16b) and (16d)]

24 $\langle\hat{\psi}_+,R\lambda_1\rangle = \langle\hat{\psi}_-,R\lambda_1\rangle = 0.$

Now, taking the scalar product of the left-hand side of (15b) with (15d) and making use of (16) and (24), we obtain

25 $-\langle R\lambda_1,R\lambda_1\rangle + \langle R\lambda_1, \mu_4 - \mu_2\rangle + \langle\mu_2,\mu_4\rangle = 0.$

Subtracting (15b) from (15d), we obtain

26 $2R\lambda_1 + (\hat{\psi}_- - \hat{\psi}_+)\lambda_3 - \mu_4 + \mu_2 = 0.$

Now, substituting for $\mu_4 - \mu_2$ from (26) into (25) and using (24), we obtain

(27) $\quad \langle R\lambda_1, R\lambda_1 \rangle + \langle \mu_2, \mu_4 \rangle = 0.$

Since $\mu_2 \geqq 0$ and $\mu_4 \geqq 0$, it follows that $\langle R\lambda_1, R\lambda_1 \rangle \leqq 0$, and hence that $R\lambda_1 = 0$, which proves (18).

Taking the scalar product of both sides of (15c) with $\hat{\xi}$ and making use of (16c) and the fact that $\langle \hat{z}, \hat{\xi} \rangle = 0$, we obtain (19).

Finally, taking the scalar product of both sides of (15a) with λ_1 and making use of (18) and (19), we obtain

(28) $\quad \langle \lambda_1, Q\lambda_1 \rangle = \langle \mu_1, \lambda_1 \rangle.$

Now, from (15c) and (16a), we obtain

(29) $\quad \langle \mu_1, \lambda_1 \rangle + \langle \mu_1, \mu_3 \rangle = 0,$

and since $\mu_1 \geqq 0$ and $\mu_3 \geqq 0$, we have $\langle \mu_1, \lambda_1 \rangle \leqq 0$, and therefore $\langle \lambda_1, Q\lambda_1 \rangle \leqq 0$. But Q is symmetric and positive semidefinite, which implies that $Q\lambda_1 = 0$, that is, that (17) holds. □

6.10 A FURTHER GENERALIZATION

The purpose of this section is to indicate how algorithm (9.4) can be generalized so that it applies to the canonical quadratic programming problem (7.1) independently of whether or not assumption (9.1) is satisfied. Furthermore, an initial feasible solution will not be required. The reader who is already familiar with the quadratic programming algorithms of Dantzig and Cottle [2] and Lemke [3] will readily see the similarity between their algorithms and the one we are about to discuss here.

The first step toward our goal is to make a deceptively simple observation. Suppose that instead of introducing n variables y^i for $i = 1, 2, \ldots, n$, as we did for the derived problem (8.1), we introduce only one variable, which we again denote by y. Thus, instead of original derived problem, consider the new derived problem:

(1) $\qquad$ Minimize y subject to the constraints

(2) $\qquad Ax = g \qquad x \geqq 0 \qquad \langle v, w \rangle = 0,$

where now

(3) $\qquad A = \begin{bmatrix} Q & -R^T & -I & R^T & k_1 \\ R & 0 & 0 & 0 & k_2 \end{bmatrix} \qquad (n+m) \times [2(n+m)+1]$

and, as before,

4
$$x = (z, \psi_-, \xi, \psi_+, y) = (v, w, y) \in E^{2(n+m)+1}$$
$$v = (z, \psi_-) \quad w = (\xi, \psi_+) \quad g = (-d, c). \quad \square$$

In the above formulation k_1 is an $n \times 1$ matrix and k_2 is an $m \times 1$ matrix. If, for the moment, we assume that k_2 is identically zero, then the only difference in formulation between the derived problem (8.1) and problem (1) is that we have collapsed the $n \times n$ diagonal matrix K in (8.4) into an $n \times 1$ matrix k_1. Eventually we shall assign specific values to k_1 and k_2 to facilitate the initialization of the algorithm.[1]

Since the restriction $\langle v, w \rangle = 0$ in (2) requires us to examine the components of v and w in pairs, we find that the discussion is considerably simplified by the following notation.

5   **Definition.**   Given a usable basic solution $x = (v, w, y)$ to (2), with an associated basis-indicator set $\bar{J}(x)$, we shall say that v^i for $i \in \{1, 2, \ldots, n + m\}$ is a *basic variable* if $i \in \bar{J}(x)$ and that w^i for $i \in \{1, 2, \ldots, n + m\}$ is a *basic variable* if $i + n + m \in \bar{J}(x)$. $\square$

Clearly, algorithm (9.4) can be applied to the new derived problem (1) without modification. However, an important simplification occurs in the procedure, which, together with a new set of stop rules, results in an important generalization, or extension of algorithm (9.4).

Thus suppose that x_N is a usable basic solution to (2), with $y_N > 0$, such that

6
v_N^i and w_N^i for $i = 1, 2, \ldots, n + m$ are not both basic variables. $\square$

That is, if v_N^i is basic, then w_N^i is not, and vice versa. Note that because of condition (9.6), algorithm (9.4) generates a sequence of usable basic solutions for the new derived problem (1) which satisfy condition (6). Since, by assumption, $y_N > 0$ and x_N is a usable basic solution, there are exactly $n + m - 1$ basic variables among the $2(n + m)$ variables of v_N and w_N. Furthermore, in view of (6), there is only one index $j \in \{1, 2, \ldots, n + m\}$ with the property that neither v_N^j nor w_N^j is a basic variable. Now, to construct an improved usable basic solution x_{N+1} in accordance with step 1 of algorithm (9.4), we use the simplex algorithm (5.9.4) with side condition (9.6):

7
For $i = 1, 2, \ldots, n + m$, if v_N^i is a basic variable, then w_{N+1}^i may not be a basic variable, and if w_N^i is a basic variable, then v_{N+1}^i may not be a basic variable. $\square$

---

[1] It will be seen that by using the vector k_2 we avoid the need for finding a basic solution to $Rz = c$ and $z \geq 0$, which was required to initialize the algorithm in the preceding section.

SEC. 6.10 A FURTHER GENERALIZATION

Consequently, the construction of x_{N+1} requires us to make either v_{N+1}^j or w_{N+1}^j a basic variable [that is, j or $j + n + m$ are the *only* elements in $J(x_N)$ which can be transferred into $\bar{J}(x_{N+1})$; there are no other possibilities]. We could determine which of these variables should be made basic by routinely following steps 1 and 2 of the simplex algorithm (5.9.4). However, a little reflection shows that these calculations are not necessary, because in the sequence of usable basic solutions $x_0, x_1, \ldots, x_{N-1}, x_N$, which we have implicitly assumed to have been constructed by algorithm (9.4), one of the following two situations must have occurred:

a. v_{N-1}^j was a basic variable.
b. w_{N-1}^j was a basic variable.

The conclusion should now be obvious. If v_{N-1}^j (w_{N-1}^j) was a basic variable, then make w_{N+1}^j (v_{N+1}^j) a basic variable.

Thus we see that in applying algorithm (9.4) to problem (1) we can eliminate the calculations involved in steps 1 and 2 of the simplex algorithm (5.9.4). Note, however, that the stop conditions of algorithm (9.4) become inoperative in this process. Thus, while algorithm (9.4) was originally conceived with the idea of reducing the value of a linear form at each step, the above-indicated modification results in a purely combinatorial procedure which makes no use at all of the values of this linear form. Our description of the new algorithm is still incomplete, since we have not yet stated the stop conditions which are to replace those of (9.4). As we shall see, it is the change in the stop conditions that increases the range of applicability of the new algorithm. Before we state our new algorithm with its stop conditions, let us consider how to initialize it. One method which readily comes to mind is to place an upper bound on the variable y whose optimal value should be zero. If we take this bound to be 1, problem (1) transforms into the following one:

8   Minimize y subject to the constraints

9   $$Ax = g \qquad x \geqq 0 \qquad \langle v,w \rangle = 0,$$

where x has been augmented by one component over the x in (2) and A and g have also been suitably modified. Thus

10
$$x = (z,\psi_-,\xi,\psi_+,y,s) = (v,w,y,s) \in E^{2(n+m)+2}$$
$$v = (z,\psi_-) \qquad w = (\xi,\psi_+)$$
$$g = (-d,b,1) \in E^{n+m+1}$$
$$A = \begin{bmatrix} Q & -R^T & -I & R^T & k_1 & 0 \\ R & 0 & 0 & 0 & k_2 & 0 \\ 0 & 0 & 0 & 0 & 1 & 1 \end{bmatrix} \quad \begin{array}{l} (n + m + 1) \\ \times [2(n+m)+2] \end{array} \quad \square$$

Note that, by (9) and (10), $y + s = 1$, $s \geq 0$, and $y \geq 0$, and we have $0 \leq y \leq 1$.

For reasons which will become clear later [see theorem (11.33)], we shall henceforth assume that *no column of the matrix*

$$\begin{bmatrix} Q \\ R \\ 0 \end{bmatrix}$$

is zero. This is not a restriction, for if, say, the ith column were zero, then $r_i = 0$ and $q_i = 0$, and consequently, without loss in generality, we may assume that $d^i = 0$ and restate CQP (7.1) with one less variable [note that if $d^i < 0$, then CQP (7.1) has no solution, since $d^i z^i \to -\infty$ as $z^i \to +\infty$].

11 Initialization. We must obtain a usable basic solution to system (9). Since R has rank m, by assumption, we may assume that the columns of R are arranged so that its first m columns are linearly independent. Set

12 $\qquad x_0 = (v_0, w_0, y_0, s_0),$

with

$$z_0{}^i = \begin{cases} 1 & i = 1, 2, \ldots, m \\ 0 & i = m+1, \ldots, n \end{cases}$$

$$\xi_0{}^i = \begin{cases} 0 & i = 1, 2, \ldots, m \\ 1 & i = m+1, \ldots, n \end{cases}$$

$$\psi_{-0}{}^i = 0 \qquad i = 1, 2, \ldots, m$$
$$\psi_{+0}{}^i = 1 \qquad i = 1, 2, \ldots, m$$
$$y_0 = 1 \qquad s_0 = 0,$$

and let the vectors $k_1 \in E^n$ and $k_2 \in E^m$ be defined by

13 $\qquad k_1 = -Qz_0 + R^T\psi_{0-} + \xi_0 - R^T\psi_{0+} - d \qquad k_2 = -Rz_0 + c.$

14 Exercise. Let k_1 and k_2 be defined as in (13). Show that x_0, defined as in (12), is a usable nondegenerate basic solution to (9).

As is our convention, let $\bar{I}(x_0) = \{i : x_0{}^i > 0\}$. Since x_0 is a nondegenerate basic solution to (9), $\bar{I}(x_0)$ is a basis-indicator set [see (5.6.5)]. We now use procedure (5.8.13) for computing adjacent extreme points to construct a new usable basic solution x_1 by exchanging the index $2(n + m) + 2 \in I(x_0)$ for some index in $\bar{I}(x_0)$ which will result in s_1 being a basic variable.[1] If $y_1 = 0$, then the

---

[1] We find it convenient at this point to extend definition (5) as follows: given a usable basic solution x to (9), with basis-indicator set $\bar{J}(x)$, we shall say that a component x^i of x is *basic variable* if $i \in \bar{J}(x)$.

This step is in keeping with the spirit of algorithm (9.4), since the linear form y will necessarily decrease due to the constraint $y + s = 1$.

SEC. 6.10 A FURTHER GENERALIZATION

first n components of x_1, that is, the vector z_1, is an optimal solution to the canonical quadratic programming problem (7.1). Otherwise x_1 is a usable basic solution satisfying the following conditions:

15 $\qquad 0 < y_1 < 1.$

16 $\qquad$ For $i = 1, 2, \ldots, n$, if v_1^i is a basic variable, then w_1^i is not, and if w_1^i is a basic variable, then v_1^i is not.

This concludes the initialization. $\square$

17 **Algorithm.** Suppose that $x_0, x_1, \ldots, x_N$, with x_0 and x_1 as constructed in (11), are usable basic solutions generated by the algorithm such that for $j = 1, 2, \ldots, N$

18 $\qquad 0 < y_j < 1;$

19 $\qquad$ For $i = 1, 2, \ldots, n$, if v_j^i is a basic variable, then w_j^i is not, and if w_j^i is a basic variable, then v_j^i is not.

Step 1. Note that there is a unique index $h \in \{1, 2, \ldots, n + m\}$ such that neither w_N^h nor v_N^h is a basic variable. Since a similar situation must also hold for x_{N-1} for with $N = 2, 3, \ldots$, we are left with two alternatives:

a. If v_{N-1}^h was a basic variable, use procedure (5.8.13) for computing adjacent extreme points to make w_{N+1}^h basic; i.e., attempt to exchange the index $h + n + m \in J(x_N)$ for some index in the basis indicator set $\bar{J}(x_N)$. Go to step 2.

b. If w_{N-1}^h was a basic variable, use procedure (5.8.13) to make v_{N+1}^h basic; i.e., attempt to exchange the index $h \in J(x_N)$ for some index in $\bar{J}(x_N)$. Go to step 2.

Step 2. There are two possible outcomes for step 1:

a. A new usable basic solution x_{N+1} was obtained in step 1. If either $y_{N+1} = 0$ or $y_{N+1} = 1$, then stop. Otherwise $0 < y_{N+1} < 1$; set $N = N + 1$ and return to step 1.

b. A new usable basic solution was not obtained in step 1; that is, an infinite ray $\{x(\theta): \theta \in [0, \infty)\}$ of the polyhedron $\Omega' = \{x: Ax = g, x \geq 0\}$ was encountered.[1] In this case stop. $\square$

Termination. Algorithm (17) incorporates three stop conditions.

20 **Stop 1.** $x_{N+1}^{2(n+m)+1} = y_{N+1} = 0$. In this case the first n components of x_{N+1}, that is, the vector z_{N+1}, are an optimal solution to CQP (7.1).

[1] This corresponds to all the coefficients α_j^i being nonpositive in (5.8.5). See also (5.8.27).

21   **Stop 2.** At some step N an infinite ray of the polyhedron Ω was obtained. We shall see in the next section that in this case CQP (7.1) does not have a solution, either because it has no feasible solution or because the constrained infimum of the cost function is $-\infty$.

22   **Stop 3.** At some step $N \geq 1$, $x_{N+1}{}^{2(n+m)+1} = 1$, that is, $y_{N+1} = 1$. We shall see in the next section that as long as the cost function of CQP (7.1) is a positive-semidefinite quadratic form, this stop condition can never occur. Thus $0 \leq y_N < 1$ for every $N \geq 1$. This stop command is included to eliminate the need for establishing whether the matrix Q is positive semidefinite or not (usually a difficult task), and the stop on $y_N = 1$ is designed to ensure that calculations will stop after a finite number of iterations. Note that when Q is not positive semidefinite, an optimal solution to the new derived problem (1) yields only a feasible solution to the quadratic programming problem, satisfying the necessary conditions for optimality. □

6.11 CONVERGENCE

In this section we shall prove that algorithm (10.17) must reach one of the stop conditions (10.20) to (10.22) and must therefore terminate after a finite number of iterations. In doing so, we shall prove that stop condition (10.22) never occurs in solving the canonical quadratic programming problem (7.1); that is, for each $N \geq 1$ $0 \leq y_N < 1$. Finally, we shall prove that if algorithm (10.17) terminates on stop condition (10.21), then CQP (7.1) does not have a solution.

It is interesting to note that in the convergence proofs of the algorithms we have seen so far we have relied heavily on the fact that the cost function is reduced at every iteration. This device is not available to us now. Indeed, it is not hard to generate examples for which the scalar y_N, formally the cost function for algorithm (9.4), increase at some iterations and decreases at others, due to the fact that the stop conditions of algorithm (10.17) are different from those of (9.4). For this reason the convergence proof becomes intricate. However, our task can be simplified somewhat by invoking a nondegeneracy assumption.

1   **Nondegeneracy assumption.** Every usable basic solution to the system

2   $$Ax = g \qquad x \geq 0 \qquad \langle v,w \rangle = 0$$

is a usable nondegenerate basic solution. □

Recall that a similar assumption (5.8.16) was made when the simplex algorithm was described in Chapter 5, and later a rule was given for resolving degeneracy, thus putting the algorithm on a firm theoretical foundation. It will subsequently become clear that the degeneracies (or ties) which arise in algorithm (10.17) can be resolved in the same manner as for the simplex algorithm (see Section 5.10).

As a consequence of this assumption, recall from definition (5.6.5) that if x is a basic solution to (2), then $\bar{I}(x) = \{i: x^i > 0\}$ is the basis-indicator set for x.

3 Theorem. Suppose that assumption (1) is satisfied by problem (10.1). Let $x_0, x_1, x_2, \ldots$ be a sequence of usable basic solutions to (10.9) generated by algorithm (10.17). Then $x_i \neq x_j$ for $i \neq j$, with $i,j = 0, 1, 2, \ldots$.

Proof. Suppose the theorem is false. Then some x_i must be repeated in the sequence $x_0, x_1, x_2, \ldots$. Let x_k be the first element in this sequence, with the property that there exists an x_l, with $l > k + 1$,† such that $x_l = x_k$ and $x_i \neq x_j$ for $i, j \in \{k + 1, k + 2, \ldots, l - 1\}$. Obviously, such an element x_k must exist. We shall consider two cases.

Case 1. $k = 0$. Recall that, by construction of x_0 in (10.11), the index $2(n + m) + 2$ is not in $\bar{I}(x_0)$; that is, s_0 is not a basic component. Also, none of the stop conditions for algorithm (10.17) were encountered at $x_1, x_2, \ldots, x_{l-1}$, for otherwise the sequence $x_0, x_1, \ldots, x_l$ could not have been constructed. Consequently, for $i = 1, 2, \ldots, l - 1$ we must have $0 < y_i < 1$ and $0 < s_i < 1$; that is, indices $2(n + m) + 1$ and $2(n + m) + 2$ must both belong to $\bar{I}(x_i)$ for $i = 1, 2, \ldots, l - 1$. Now, since $x_l = x_0$ and the index $2(n + m) + 2$ is in $\bar{I}(x_{l-1})$ but not in $\bar{I}(x_l)$, we must have [see (5.8.13) and (5.8.9)]

$$\bar{I}(x_0) = \bar{I}(x_l) = \bar{I}(x_{l-1}) + \{\alpha_{l-1}\} - \{2(n + m) + 2\},$$

where α_{l-1} is some index in $I(x_{l-1})$. By theorem (5.8.13), since all the x_i for $i = 0, 1, 2, \ldots$ are supposed to be usable nondegenerate basic solutions, x_{l-1} must be the unique adjacent extreme point resulting from procedure (5.8.13) [see (5.8.14)] when the index $2(n + m) + 2$ is exchanged for some index in $\bar{I}(x_0)$. But this is exactly how we constructed x_1; hence $x_1 = x_{l-1}$, which contradicts the assumption that $x_i \neq x_j$ for $i, j \in \{1, 2, \ldots, l - 1\}$.

† It is clear from the algorithm, with the nondegeneracy assumption, that $x_k = x_{k+1}$ cannot hold.

Case 2. $k > 0$. Since no stop conditions were reached at x_k, we must have $0 < y_k < 1$ and $0 < s_k < 1$; that is, both are basic. Hence there exists a unique index $j_v \in \{1, 2, \ldots, n + m\}$ such that neither j_v nor $j_w = j_v + (n + m)$ is in $\bar{I}(x_k)$. Furthermore, it is clear from step 1 of algorithm (10.17) that either j_v or j_w must belong to $\bar{I}(x_{k-1})$. So, without loss of generality, let us suppose that $j_v \in \bar{I}(x_{k-1})$; that is,

5
$$\bar{I}(x_k) = \bar{I}(x_{k-1}) + \{\alpha_{k-1}\} - \{j_v\},$$

where α_{k-1} is some index in $I(x_{k-1})$. To construct x_{k+1} according to (10.17), since $j_v \in \bar{I}(x_{k-1})$, we must exchange the index j_v for some index β_k in $\bar{I}(x_k)$; that is,

6
$$\bar{I}(x_{k+1}) = \bar{I}(x_k) + \{j_w\} - \{\beta_k\}.$$

Now, since $x_k = x_l$, by the same argument as above, either the index j_v or the index j_w must also belong to $\bar{I}(x_{l-1})$. If $j_v \in \bar{I}(x_{l-1})$ and we exchange the index $j_v \in I(x_l)$ for some index in $\bar{I}(x_l)$ according to procedure (5.8.13), then, by theorem (5.8.18), the resulting usable solution is unique and must be x_{l-1}. But, by (5), this operation should yield the point x_{k-1}, and hence if $j_v \in \bar{I}(x_{l-1})$, we must have $x_{k-1} = x_{l-1}$. Obviously, this contradicts the assumption that x_k was the first repeated point with the properties stated.

Consequently, let us suppose that $j_w \in \bar{I}(x_{l-1})$. Then we must be able to reconstruct x_{l-1} from x_k by exchanging the index $j_w \in I(x_l)$ for some index β'_k in $\bar{I}(x_l)$. Since, by (5.8.13), this process yields a unique result, we must conclude from (6) that $\beta_k = \beta'_k$ and $x_{k+1} = x_{l-1}$, which again contradicts the assumptions on x_k, in particular that $x_i \neq x_j$ for $i, j \in \{k + 1, k + 2, \ldots, l - 1\}$. □

7 **Corollary.** Algorithm (10.17) terminates in a finite number of steps; i.e., one of stop conditions (10.20) to (10.22) is encountered after a finite number of steps.

Proof. With every usable basic solution to (10.2) there are associated $n + m + 1$ linearly independent columns of the matrix A. Since A has a finite number of columns, it follows that there are a finite number of usable basic solutions. Hence, since the algorithm never repeats a usable basic solution and will construct a new one unless a stop condition is reached, it must reach one of stop conditions (10.20) to (10.22) after a finite number of steps. □

8 **Theorem.** Let k_1 and k_2 be as defined in (10.13), with x_0 as given by (10.12). If we set $s = 0$, then x_0 is the unique solution to the system of equations and inequalities (10.2).

Proof. Clearly, with s fixed at zero, $y_0 = 1$ is the only value of y which satisfies

(9) $$y_0 + s = 1 \qquad y_0 \geq 0.$$

Consequently, we need only show that $(z_0, \psi_{-_0}, \xi_0, \psi_{+_0})$ is the only solution to the system

(10) $$\begin{aligned} Qz - R^T\psi_- - \xi + R^T\psi_+ &= -k_1 - d \qquad Rz = -k_2 + c \\ z \geq 0 \qquad \psi_- \geq 0 \qquad \xi &\geq 0 \qquad \psi_+ \geq 0 \\ \langle z, \xi \rangle = \langle \psi_+, \psi_- \rangle &= 0 \end{aligned}$$

Suppose that $(\tilde{z}, \tilde{\psi}_-, \tilde{\xi}, \tilde{\psi}_+)$ also satisfies (10). Then, by subtraction, we obtain

(11) $$Q(z_0 - \tilde{z}) - R^T(\psi_{-_0} - \tilde{\psi}_-) - (\xi_0 - \tilde{\xi}) + R^T(\psi_{+_0} - \tilde{\psi}_+) = 0$$
(12) $$R(z_0 - \tilde{z}) = 0.$$

Taking the scalar product of both sides of (11) with the vector $z_0 - \tilde{z}$, we obtain

(13) $$\langle (z_0 - \tilde{z}), Q(z_0 - \tilde{z}) \rangle + \langle (z_0 - \tilde{z}), R^T(\psi_{+_0} - \psi_{-_0} - \tilde{\psi}_+ + \tilde{\psi}_-) \rangle \\ - \langle (z_0 - \tilde{z}), (\xi_0 - \tilde{\xi}) \rangle = 0.$$

Since the second term in (13) is zero, by (12), this reduces to

(14) $$\langle (z_0 - \tilde{z}), Q(z_0 - \tilde{z}) \rangle + \langle z_0, \tilde{\xi} \rangle + \langle \xi_0, \tilde{z} \rangle - \langle z_0, \xi_0 \rangle - \langle \tilde{z}, \tilde{\xi} \rangle = 0.$$

Now, the first term in (14) is nonnegative, since Q is a positive-semidefinite matrix, and the last two terms are zero, by (10). It follows that

(15) $$\langle z_0, \tilde{\xi} \rangle + \langle \xi_0, \tilde{z} \rangle \leq 0.$$

But each term in (15) is nonnegative, so that

(16) $$\langle z_0, \tilde{\xi} \rangle = 0 \qquad \langle \xi_0, \tilde{z} \rangle = 0.$$

Keeping in mind the definitions of z_0 and ξ_0 as given in (10.11), we conclude that

(17) $$\tilde{\xi}^i = 0 \qquad i = 1, 2, \cdots, m$$
(18) $$\tilde{z}^i = 0 \qquad i = m+1, \cdots, n$$

Now, the first m components of z_0 are positive, and the remaining components of z_0 are zero. Hence, by (10) and (18),

(19) $$\sum_{i=1}^{m} (z_0{}^i - \tilde{z}^i) r_i = 0,$$

where r_i is the ith column of R. Since the first m columns of R were assumed to be linearly independent, it follows that $z_0{}^i = \tilde{z}^i$ for $i = 1, 2, \ldots, m$, and hence that $z_0 = \tilde{z}$. Equation (11) now becomes

20 $$R^T([(\psi_{+_0} - \psi_{-_0}) - (\tilde{\psi}_+ - \tilde{\psi}_-)]) = (\xi_0 - \tilde{\xi}).$$

The first m components of ξ_0 are zero; consequently, by (17) the first m components of $\tilde{\xi}$ are zero. Hence

21 $$\langle r_i, [(\psi_{+_0} - \psi_{-_0}) - (\tilde{\psi}_+ - \tilde{\psi}_-)]\rangle = 0 \qquad i = 1, 2, \ldots, m.$$

Again, since the vectors r_i for $i = 1, 2, \ldots, m$ are a basis for E^m, we conclude that

22 $$\psi_{+_0} - \psi_{-_0} = \tilde{\psi}_+ - \tilde{\psi}_-.$$

But $\psi_{-_0} = 0$ and $\psi_{+_0}{}^i = 1$ for $i = 1, 2, \ldots, m$, so that

23 $$\tilde{\psi}_+{}^i - \tilde{\psi}_-{}^i = 1 \qquad i = 1, 2, \ldots, m.$$

Since $\tilde{\psi}_+{}^i$ and $\tilde{\psi}_-{}^i$ cannot both be positive, by (10), we have

24 $$\tilde{\psi}_-{}^i = 0 \qquad i = 1, 2, \ldots, m$$
25 $$\tilde{\psi}_+{}^i = 1 \qquad i = 1, 2, \ldots, m.$$

Therefore $\tilde{\psi}_+ = \psi_{+_0}$ and $\tilde{\psi}_- = \psi_{-_0}$, and hence, from (20), $\xi_0 = \tilde{\xi}$. □

26 **Corollary.** Let x_N for $N = 0, 1, 2, \ldots$ be the sequence of usable basic solutions generated by algorithm (10.17). Then for $N \geq 1$, $x_N{}^{2(n+m)+1} = y_N < 1$; that is, stop condition (10.22) is never executed.

Proof. Suppose the corollary is false. Then for some $N \geq 1$, $x_N{}^{2(n+m)+1} = y_N = 1$. Hence

27 $$x_N = (v_N, w_N, y_N, s_N) \qquad s_N = 0$$
28 $$x_0 = (v_0, w_0, y_0, s_0) \qquad s_0 = 0.$$

It now follows from the above theorem that $x_N = x_0$, which is a contradiction of theorem (3). □

We are now left with showing that the canonical quadratic programming problem (7.1) does not have a solution if stop condition (10.21) occurs. Toward this goal, the following lemmas will prove useful.

29 **Lemma.** If the system $Rz = c$ and $z \geq 0$ has a solution, then every vector $\eta \in E^m$ satisfying the system of inequalities $R^T \eta \geq 0$ also satisfies $\langle \eta, c \rangle \geq 0$. □

30 **Exercise.** Prove lemma (29). □

31 **Lemma.** Suppose that CQP (7.1) has a feasible solution, i.e., that the system $Rz = c$ and $z \geqq 0$ has a solution. If there is a vector $\eta \in E^n$ such that

$$R\eta = 0 \qquad Q\eta = 0 \qquad \eta \geqq 0 \qquad \langle d,\eta \rangle < 0,$$

then the constrained infimum of the cost function in CQP (7.1) is $-\infty$. □

32 **Exercise.** Prove lemma (31). □

33 **Theorem.** Suppose that algorithm (10.17) terminates on stop command (10.21). Then either CQP (7.1) does not have a feasible solution, or else the constrained infimum of its cost function is $-\infty$.

Proof. The proof is by contradiction. Suppose that CQP (7.1) has a feasible solution, that the constrained infimum of its cost function is finite, and that stop condition (10.21) is encountered at some iteration, say, N. Referring to procedure (5.8.13) for computing adjacent extreme points, and in particular to equation (5.8.2), we conclude that there exist scalars $\alpha^i_{P_{N-1}}$ for $i \in \bar{I}(x_N)$, *not all zero*,[1] and a column $a_{P_{N-1}}$ of the matrix A in (10.10), with $P_{N-1} \in I(x_N)$, such that

34
$$a_{P_{N-1}} = \sum_{i \in \bar{I}(x_N)} \alpha^i_{P_{N-1}} a_i$$
$$\alpha^i_{P_{N-1}} \leqq 0 \qquad \text{for all } i \in \bar{I}(x_N).$$

Using the definition of the matrix A in (10.9), we conclude that there is a non zero vector $\delta x = (\delta z, \delta \psi_-, \delta \xi, \delta \psi_+, \delta y, \delta s)$, made up of the $\alpha^i_{P_{N-1}}$, such that

35
$$Q\,\delta z - R^T\,\delta\psi_- - \delta\xi + R^T\,\delta\psi_+ + k_1\,\delta y = 0$$
$$R\,\delta z + k_2\,\delta y = 0 \qquad \delta y + \delta s = 0 \qquad \delta x \geqq 0.$$

In addition, we have

36
$$\langle \delta z, \delta \xi \rangle = \langle \delta\psi_-, \delta\psi_+ \rangle = \langle \delta\xi, z_N \rangle = \langle \delta z, \xi_N \rangle = 0,$$

---

[1] If $\alpha^i_{P_{N-1}} = 0$ for all $i \in \bar{I}(x_N)$, then the column $a_{P_{N-1}}$ of A is identically zero. An examination of the matrix A in (10.10) shows that this column is either a column of $\begin{bmatrix} R^T \\ 0 \\ 0 \end{bmatrix}$, which is impossible, since R has rank m, or a column of $\begin{bmatrix} Q \\ R \\ 0 \end{bmatrix}$, which is not allowed [see the discussion in (10.11)].

where $x_N = (z_N, \psi_{-N}, \xi_N, \psi_{+N}, \lambda_N, \sigma_N)$ is the usable basic solution determined at iteration N. Using (35) and (36), we can easily deduce the following, in the order given:

37 $\quad\quad\quad \delta y = \delta s = 0$
38 $\quad\quad\quad R\,\delta z = 0 \quad\quad \delta z \geq 0$
39 $\quad\quad\quad Q\,\delta z = 0$
40 $\quad\quad\quad R^T \delta \psi = \delta \xi \geq 0 \quad\quad \delta \psi \triangleq (\delta \psi_+ - \delta \psi_-).$

In obtaining (39) we first deduced from (37), (38), and the first equation in (35) that $\langle \delta z, Q\,\delta z \rangle = 0$.

Since x_N is a usable basic solution,

41 $\quad\quad\quad \begin{aligned} Qz_N - R^T\psi_{-N} - \xi_N + R^T\psi_{+N} + k_1 y_N &= -d \\ Rz_N + k_2 y_N &= c \\ y_N + s_N &= 1 \\ x_N &\geq 0 \end{aligned}$

and

42 $\quad\quad\quad \langle z_N, \xi_N \rangle = \langle \psi_{-N}, \psi_{+N} \rangle = 0$

(note that $0 < y_N < 1$).

Taking the scalar product of the first equation in (41) with the vector δz and using (36), (38), and (39), we obtain

43 $\quad\quad\quad \langle \delta z, k_1 \rangle y_N = -\langle d, \delta z \rangle.$

From the definition of k_1 in (10.13) this reduces to

44 $\quad\quad\quad y_N \langle \delta z, \xi_0 \rangle = (y_N - 1)\langle \delta z, d \rangle,$

where, it is recalled, $x_0 = (z_0, \psi_{-0}, \xi_0, \psi_{+0}, y_0, s_0)$ is the initial usable basic solution used in algorithm (10.17). Since $\delta z \geq 0$, $\xi_0 \geq 0$, $y_N > 0$, and $(y_N - 1) < 0$, it follows that $\langle \delta z, d \rangle \leq 0$. But by (38) and (39), $R\delta z = 0$, $Q\delta z = 0$, and $\delta z \geq 0$, and so if the constrained infimum of the cost function of CQP (7.1) is finite, then $\langle \delta z, d \rangle \geq 0$, by lemma (31). Consequently, we must have $\langle \delta z, d \rangle = 0$, and so

45 $\quad\quad\quad \langle \delta z, \xi_0 \rangle = 0.$

Next, taking the scalar product of the second equation in (41) with the vector $\delta \psi = (\delta \psi_+ - \delta \psi_-)$ and using (36), (40), and the definition of k_2 in (10.13), we obtain

46 $\quad\quad\quad y_N \langle z_0, \delta \xi \rangle = (y_N - 1)\langle \delta \psi, c \rangle.$

Since $y_N > 0$, $\delta \xi \geq 0$, $z_0 \geq 0$, and $(y_N - 1) < 0$, it follows that $\langle \delta \psi, b \rangle \leq 0$. But, by (40), $R^T \delta \psi \geq 0$, and so if CQP (7.1) has a feasible solution, then we must have $\langle \delta\psi, b\rangle \geq 0$, by Lemma (29). Hence $\langle \delta\psi, b\rangle = 0$, and so

47 $\quad\langle \delta\xi, z_0\rangle = 0.$

Now, the first m components of z_0 are positive [see (10.12)], so that, by (47), the first m components of $\delta\xi$ are zero. Thus, by (40), we have

$$\langle r_i, \delta\psi_+ - \delta\psi_-\rangle = 0 \quad i = 1, 2, \ldots, m,$$

where r_i is the ith column of R. Again recall that the first m columns of R were so arranged that they were linearly independent vectors; hence

48 $\quad\delta\psi_+ = \delta\psi_-,$

which implies that

49 $\quad\delta\psi_+ = \delta\psi_- = 0,$

since $\delta\psi_+ \geq 0$, $\delta\psi_- \geq 0$, and $\langle \delta\psi_+, \delta\psi_-\rangle = 0$.

The proof is now almost complete, for with (45), (47), and (49) we shall deduce that there are two solutions to the system of equations and inequalities (2) when $s = 0$, in contradiction to theorem (8). By construction, $x_0 = (z_0, \psi_{0-}, \xi_0, \psi_{0+}, y_0, s_0)$, with $y_0 = 1$, is a solution to (2) [see (10.11)]. We shall show that $x_0 + \theta \, \delta x$ is also a solution to this system when $y_0 = 1$, for any value of $\theta > 0$.

First, note that $x_0 + \theta \, \delta x \geq 0$ for all $\theta > 0$, since $\delta x \geq 0$. Next, we see from (36), (45), and (47) that

50 $\quad\langle z_0 + \theta \, \delta z, \xi_0 + \theta \, \delta\xi\rangle = \langle z_0, \xi_0\rangle + \theta(\langle \delta z, \xi_0\rangle + \langle \delta\xi, z_0\rangle)$
$\quad\quad\quad\quad\quad\quad\quad\quad\quad\quad\quad\quad\quad\quad + \theta^2 \langle \delta\xi, \delta z\rangle = 0$

and from (49) that

51 $\quad\langle \psi_{+_0} + \theta \, \delta\psi_+, \psi_{-_0} + \theta \, \delta\psi_-\rangle = \langle \psi_{+_0}, \psi_{-_0}\rangle = 0.$

Thus it only remains to show that

52 $\quad\begin{aligned}Q(z_0 + \theta \, \delta z) &- R^T(\psi_{-_0} + \theta \, \delta\psi_-) - (\xi_0 + \theta \, \delta\xi) \\ &+ R^T(\psi_{+_0} + \theta \, \delta\psi_+) + (y_0 + \theta \, \delta y)k_1 = -d \\ R(z_0 + \theta \, \delta z) &+ (y_0 + \theta \, \delta y)k_2 = c \\ (y_0 + \theta \, \delta y) &+ (s_0 + \theta \, \delta s) = 1\end{aligned}$

and that

53 $\quad y_0 + \theta \, \delta y = 1.$

But (52) follows immediately from (35) and the fact that x_0 is a usable basic solution, and (53) follows from the fact that $\delta y = 0$ and $y_0 = 1$. □

54 Corollary. If CQP (7.1) has a feasible solution and the constrained infimum of its cost function is finite, then it has an optimal solution [cf. theorem (2.27)].

Proof. Under the hypothesis of this corollary, stop command (10.21) is never executed. By corollary (26), stop command (10.22) is never executed. Hence, by corollary (7), stop command (10.20) is executed in a finite number of steps; that is, CQP (7.1) has an optimal solution. □

This completes the proof of convergence for algorithm (10.17), as well as our discussion of quadratic programming problems.

REFERENCES

1. P. Wolfe: The Simplex Method for Quadratic Programming, *Econometrica*, **27**:382–397 (1959).
2. G. B. Dantzig and R. W. Cottle: Positive (Semi-) Definite Programming, in "Nonlinear Programming," J. Abadie (ed), North-Holland Publishing Company, Amsterdam, 1968.
3. C. E. Lemke: Bimatrix Equilibrium Points and Mathematical Programming, *Management Science*, **11**:681–689 (1965).
4. A. J. Goldman: Resolution and Separation Theorems for Polyhedral Convex Sets, *Ann. Math. Studies* 38, pp. 41–51, Princeton University Press, Princeton, N.J., 1956.
5. E. M. L. Beale: On Quadratic Programming, *Naval Res. Logistics Quart.*, **6**:227–243 (1959).

7
Convex programming algorithms

7.1 INTRODUCTION

This chapter will be devoted primarily to nonlinear programming problems of the following form.

1 The convex programming problem. Minimize $f(z)$ subject to $r(z) = 0$ and $q(z) \leqq 0$, where $f(\cdot)$ is a continuously differentiable, real valued, convex function defined on E^n, $r\colon E^n \to E^m$ is affine, and $q\colon E^n \to E^k$ is a continuously differentiable function whose components, $q^i(\cdot)$, are convex. $\square$

We shall consider a few representative algorithms for dealing with problem (1), and we shall discuss how, and to what extent, these algorithms can be applied to the general nonlinear programming problem (3.1.1). In discussing the algorithms it will occasionally be convenient to depart from the notation we have used thus far. This departure from our standard notation will be clearly indicated whenever it occurs.

A number of the algorithms we are about to examine require the knowledge of an initial feasible solution to problem (1). We shall now show that such a solution can be obtained by solving another convex programming problem for which an initial solution is trivially constructed. Note that since the function $r(\cdot)$ in problem (1) is affine, we must have $r(z) = Rz - c$, where $c \in E^m$ and R is an $m \times n$ matrix. Thus let us consider the set

2
$$\Omega' = \{z: q(z) \leq 0, Rz = c\}.$$

If the functions $q^i(\cdot)$ for $i = 1, 2, \ldots, k$ are affine, then the simplex algorithm can be used to determine (in a finite number of steps) either a point $z_0 \in \Omega'$ or whether Ω' is empty. Let us assume, therefore, that at least one of the functions $q^i(\cdot)$ is nonlinear.

Consider the following nonlinear programming problem:

3 Minimize σ subject to the constraints

$$q^i(z) - \sigma l^i \leq 0 \qquad i = 1, 2, \ldots, k$$
4
$$\langle r^j, z \rangle - c^j - \sigma e^j = 0 \qquad j = 1, 2, \ldots, m$$
$$\sigma \geq 0,$$

where l^i for $i = 1, 2, \ldots, k$ and e^j for $j = 1, 2, \ldots, m$ are scalars and r^j for $j = 1, 2, \ldots, m$ denotes the jth row of R. □

Suppose that the $m \times n$ matrix R has rank m. Assuming that the first m columns of the matrix R are linearly independent, and denoting this submatrix by $\tilde{R}$, we may define

5
$$z_0^i = (\tilde{R}^{-1}c)^i \qquad i = 1, 2, \ldots, m$$
$$z_0^i = 0 \qquad i = m+1, \ldots, n.$$

Thus $Rz_0 = c$. Now let

$$e^j = 0 \qquad j = 1, 2, \ldots, m$$
6
$$l^i = \begin{cases} 0 & \text{if } q^i(z_0) \leq 0 \\ 1 & \text{if } q^i(z_0) > 0 \end{cases}$$
$$\sigma_0 = \max_i q^i(z_0).$$

Note that if $\sigma_0 \leq 0$, then $z_0 \in \Omega'$. Otherwise, (z_0, σ_0) is a feasible solution to system (4), with e^j and l^i as defined above.

Thus problem (3) is a convex programming problem with an initial feasible solution (σ_0, z_0), as in (5) and (6). If the set Ω' is not empty, then for any $z' \in \Omega'$ system (4) is satisfied with $\sigma = 0$. Since $\sigma \geq 0$, problem (3) must have an optimal solution $\hat{\sigma} = 0$ and $\hat{z} \in \Omega'$; that is, we can find a point in Ω' by solving problem (3).

7 **Remark.** Suppose that there is a point z in Ω' such that $q^i(z) < 0$ for $i = 1, 2, \ldots, k$. Then, if we remove the restriction in (4) that $\sigma \geq 0$, we obtain a new problem whose optimal solution is $\hat{\sigma} < 0$. Since the procedures we shall consider start with the initial value $\sigma_0 > 0$ and produce a sequence σ_i for $i = 0, 1, 2, \ldots$ converging to

the value $\hat{\sigma} < 0$, there will be an integer p such that $\sigma_i < 0$ for all $i \geqq p$; that is, the corresponding vectors z_i for $i \geqq p$ will be in Ω'. Thus it will be possible to compute an initial feasible solution $\tilde{z} \in \Omega'$ in a finite number of steps. □

In the event that R does not have maximum rank, we can avoid the calculation defined by (5) as follows. Take any point z_0 in E^n, and for $i = 1, 2, \ldots, k$ set

8
$$l^i = \begin{cases} 0 & \text{if } q^i(z_0) \leqq 0 \\ 1 & \text{if } q^i(z_0) > 0 \end{cases}$$

in (4). Now set

9
$$\sigma_0 = \begin{cases} 1 & \text{if } q^i(z_0) \leqq 0 \text{ for every } i = 1, 2, \ldots, k \\ \max_i q^i(z_0) & \text{otherwise} \end{cases}$$

and

10
$$e^j = \frac{\langle r^j, z_0 \rangle - c^j}{\sigma_0}.$$

Then (σ_0, z_0) is an initial feasible solution for problem (3), with the e^j and l^i defined as above (of course, now z_0 does not necessarily satisfy the equality constraint $Rz_0 = c$). Again, if there is at least one vector $z \in \Omega'$, problem (3) will have an optimal solution $(\hat{\sigma}, \hat{z})$, with $\hat{\sigma} = 0$ and $\hat{z} \in \Omega'$.

We now proceed to a few of the better-known convex programming algorithms.

7.2 METHODS OF FEASIBLE DIRECTIONS

Let us consider an algorithm, or, to be more precise, a family of algorithms introduced by Zoutendijk [1] for solving convex programming problems. These algorithms can also be applied to the nonlinear programming problem (3.1.1) [minimize $f(z)$ subject to the constraints $r(z) = 0$ and $q(z) \leqq 0$], when the equality constraint function $r(\cdot)$ is affine, to compute points $\hat{z}$ satisfying necessary conditions of optimality. Obviously, in this case there is no guarantee that the algorithms will determine a global minimum. However, under some mild assumptions, it can be shown that they will compute a local minimum. Zoutendijk called these algorithms *methods of feasible directions*. Because they can be brought to bear on such a wide class of problems, we shall not assume in what follows that the cost function $f(\cdot)$ or the inequality-constraint functions $q^i(\cdot)$ are convex. Let us begin by defining the class of problems which we shall consider in this section.

1 **Problem.** Given $k+1$ real-valued continuously differentiable functions $q^i(\cdot)$ for $i = 0, 1, \ldots, k$ defined on E^n and an affine function $r(z) = Rz - c$, where R is an $m \times n (m \leqq n)$ full-rank matrix and $c \in E^n$, find a vector $\hat{z} \in E^n$ satisfying

2 $\qquad q^i(\hat{z}) \leqq 0 \qquad i = 1, 2, \ldots, k$
3 $\qquad r(\hat{z}) = 0$

such that for all vectors $z \in E^n$ satisfying (2) and (3)

4 $\qquad q^0(\hat{z}) \leqq q^0(z). \quad \square$

5 **Remark.** Note that the cost function in (4) is denoted by $q^0(\cdot)$ rather than by $f(\cdot)$, as has been our practice until now. We shall see that this device enables us to simplify notation. $\quad \square$

Concerning notation, we shall again denote by Ω' the set of feasible solutions to the nonlinear programming problem (1), that is,

6 $\qquad \Omega' = \{z\colon q^i(z) \leqq 0, i = 1, 2, \ldots, k, r(z) = 0\},$

and for any $z \in \Omega'$ the functions $q^i(\cdot)$ for $i \in I(z)$ are the active constraints at z; that is,

7 $\qquad I(z) = \{i\colon q^i(z) = 0, i \in \{1, 2, \ldots, k\}\}.$

Finally, we shall assume that the functions defining Ω' satisfy the following constraint qualification:†

8 **Assumption.** For every $z \in \Omega'$ the *internal cone* to Ω' at z, defined by

9 $\qquad IC(z, \Omega') = \{h\colon Rh = 0, \langle \nabla q^i(z), h \rangle < 0 \text{ for all } i \in I(z)\} \cup \{0\},$

contains some point other than the origin. $\quad \square$

10 **Remark.** By corollary (3.4.31), under assumption (8), theorem (3.4.22) provides a necessary condition for optimality, with $\psi^0 < 0$, for the nonlinear programming problem (1). $\quad \square$

11 **Exercise.** Referring to problem (1), suppose that the functions $q^i(\cdot)$ for $i = 1, 2, \ldots, k$ are convex, and that for every $i \in \{1, 2, \ldots, k\}$ there exists a point $z \in \Omega'$, possibly depending on i, such that $q^i(z) < 0$. Show that assumption (8) is satisfied. [*Hint:* First show that there exists a $z \in \Omega'$ such that $q^i(z) < 0$ for every $i = 1, 2, \ldots, k$, and then examine the proof of theorem (2.3.30).] $\quad \square$

† When some of the functions q^i for $i \in I(z)$ are affine, it suffices to require that there exist a vector $h \in E^n$ such that $\langle \nabla q^i(z), h \rangle \leqq 0$ for these functions and $\langle \nabla q^i(z), h \rangle < 0$ for the remaining functions q^i, with $i \in I(z)$, and that $Rh = 0$.

As might be suspected, there are no known algorithms for finding optimal solutions to the quite general programming problem (1). About all that can be said concerning the algorithms which can be brought to bear on problem (1) is that they compute a point where the necessary conditions for optimality are satisfied. As a practical matter, therefore, we shall consider instead the following problem.

12 **Problem: determination of a point where the necessary conditions are satisfied.** Given $k + 1$ continuously differentiable real-valued functions $q^i(\cdot)$ for $i = 0, 1, \ldots, k$ defined on E^n and an affine function $r(z) = Rz - c$, where R is an $m \times n (m \leq n)$ full-rank matrix, find a vector $\hat{z} \in E^n$ and multiplier vectors $\psi = (\psi^1, \psi^2, \ldots, \psi^m) \in E^m$ and $\mu = (\mu^1, \mu^2, \ldots, \mu^k) \in E^k$ such that

13 $\quad\hat{z} \in \Omega'$

14 $\quad -\nabla q^0(\hat{z}) + \sum_{i=1}^{k} \mu^i \nabla q^i(\hat{z}) + R^T\psi = 0$

15 $\quad \mu \leq 0, \quad \mu^i q^i(\hat{z}) = 0 \quad i = 1, 2, \ldots, k.$ $\square$

16 **Remark.** Relations (13) to (15) are a restatement of the necessary conditions for optimality [theorem (3.4.22), with $\psi^0 < 0$]. By remark (10), if $\hat{z}$ is a solution to the nonlinear programming problem (1) and assumption (8) is satisfied, then there exist multiplier vectors ψ and μ satisfying (13) to (15). Conversely, if $(\hat{z}, \psi, \mu)$ satisfy (13) to (15), then $\hat{z}$ may be a global minimum for problem (1) [see theorems (3.6.1) and (3.6.22)], or if $\nabla q^0(\hat{z}) \neq 0$, then $\hat{z}$ may be a local minimum or a local saddle point, or if $\nabla q^0(\hat{z}) = 0$, then $\hat{z}$ may be a local minimum, a local maximum, or a local saddle point. $\square$

For the purpose of describing the methods of feasible directions and proving their convergence, it is convenient to restate problem (12) in an alternative form. First, however, let us introduce some new notation.

17 **Definition.** Let $z \in \Omega'$ and $\epsilon \geq 0$ be given. We shall define the index set $I_\epsilon(z) \subset \{0, 1, \ldots, k\}$ by

18 $\quad I_\epsilon(z) = \{0\} \cup \{i: q^i(z) + \epsilon \geq 0, i \in \{1, 2, \ldots, k\}\}.$ $\square$

For $\epsilon = 0$ the index set $I_0(z)$ may be expressed as

19 $\quad I_0(z) = \{0\} \cup I(z),$

where $I(z)$ is as defined in (7).

20 **Definition.** We shall denote by S any subset of E^n of the form

21 $\quad S = S' \cap S'',$

where S' is a *compact convex subset of E^n containing the origin in its interior* and S'' is the *linear subspace* of E^n defined by

$$S'' = \{h: Rh = 0\}. \quad \square$$

22 **Remark.** S is a compact convex subset of E^n containing the origin, and it is the choice of S that distinguishes one algorithm from another in the family of methods of feasible directions. $\quad \square$

23 **Theorem.** Suppose that assumption (8) is satisfied. Then a vector $\hat{z} \in \Omega'$, together with multiplier vectors $\psi \in E^m$ and $\mu \in E^k$, solves problem (12) if and only if

24 $$\min_{h \in S} \max_{i \in I_0(\hat{z})} \langle \nabla q^i(\hat{z}), h \rangle = 0$$

for every subset S defined as in (20).

Proof. Suppose that $(\hat{z}, \psi, \mu)$ satisfies the necessary conditions of optimality (13) to (15), and suppose that (24) does not hold at $\hat{z}$ for some subset S defined as in (20). Now, $0 \in S$, from which it follows that if (24) does not hold, then

$$\min_{h \in S} \max_{i \in I_0(\hat{z})} \langle \nabla q^i(\hat{z}), h \rangle < 0.$$

Hence there exists a vector $h^* \in S$ such that

25 $$\langle \nabla q^i(\hat{z}), h^* \rangle < 0 \text{ for all } i \in I_0(\hat{z}) = \{0\} \cup I(\hat{z}).$$

From the definition of the set S in (20), and from (15), the expression obtained after taking the scalar product of both sides of (14) with the vector h^* simplifies to

26 $$-\langle \nabla q^0(\hat{z}), h^* \rangle + \sum_{i \in I(\hat{z})} \mu^i \langle \nabla q^i(\hat{z}), h^* \rangle = 0.$$

Since $\mu \geq 0$, this is clearly in contradiction to equation (25).

Conversely, suppose that condition (24) holds for some vector $\hat{z} \in \Omega'$ and any set S as in (20). Then for all $h \in S$

$$\max_{i \in I_0(\hat{z})} \langle \nabla q^i(\hat{z}), h \rangle \geq 0.$$

In particular, since, by assumption (8), $IC(\hat{z}, \Omega) \neq \{0\}$, we must have

27 $$\langle \nabla q^0(\hat{z}), h \rangle \geq 0 \quad \text{for all } h \in \overline{IC(\hat{z}, \Omega')}$$
$$= \{h: \langle \nabla q^i(\hat{z}), h \rangle \leq 0, i \in I(\hat{z}), Rh = 0\}.$$

We now conclude from Farkas' lemma (A.5.34) that there exist multiplier vectors $\psi \in E^m$ and $\mu \in E^k$ such that $(\hat{z},\psi,\mu)$ satisfies (13) to (15). □

Thus problem (12) is equivalent to the following problem:

28 **Problem.** Given $k+1$ continuously differentiable functions $q^i(\cdot)$ for $i = 0, 1, \ldots, k$ defined on E^n and an affine function $r(z) = Rz - c$ mapping E^n into E^m, with R having rank m, find a vector $\hat{z} \in \Omega'$ such that

29
$$\min_{h \in S} \max_{i \in I_0(\hat{z})} \langle \nabla q^i(\hat{z}), h \rangle = 0,$$

where S and $I_0(\hat{z})$ are as defined in (20) and (19), respectively. □

It is problem (28) to which we address ourselves from now on. Accordingly, *we shall not impose assumption* (8) unless we wish to use theorem (23) to relate the solutions of (28) to the solutions of (12). The reader may well ask at this point how to interpret condition (29) in the absence of assumption (8). This is the content of the following lemma and exercises.

30 **Lemma.** A vector $\hat{z} \in \Omega'$ solves problem (28) if and only if there exist multiplier vectors $\psi \in E^m$ and $\mu \in E^k$ and a scalar $\psi^0 \leq 0$, not all zero, such that

31
$$\psi^0 \nabla q^0(\hat{z}) + \sum_{i=1}^{k} \mu^i \nabla q^i(\hat{z}) + R^T \psi = 0$$

$$\mu \geq 0 \qquad \mu^i q^i(\hat{z}) = 0 \qquad i = 1, 2, \ldots, k. \quad \square$$

32 **Exercise.** Prove lemma (30). [*Hint:* If assumption (8) is satisfied, then lemma (30) is a slightly weaker statement than theorem (23). Show that (31) holds trivially (with $\psi^0 = 0$) when assumption (8) is not satisfied.] □

33 **Exercise.** Let z be any point in Ω' such that $IC(z,\Omega') = \{0\}$. Show that

$$\min_{h \in S} \max_{i \in I_0(z)} \langle \nabla q^i(z), h \rangle = 0. \quad \square$$

34 **Remark.** In view of lemma (30), problem (28) is equivalent to finding a point where the generalized necessary conditions for optimality [theorem (3.5.11)] are satisfied. Recall that these conditions are very weak, and that even for convex programming problems their satisfaction at a point z may have little or no bearing on the optimality of that point. This was made abundantly clear in exercise (33). □

Let us now see what has been gained by transcribing problem (12) into form (28). Suppose we are given a point $z_0 \in \Omega'$ and we wish to determine whether or not z_0 solves problem (12). By theorem (23), this is equivalent to determining whether or not

35 $$\min_{h \in S} \max_{i \in I_0(z_0)} \langle \nabla q^i(z_0), h \rangle = 0.$$

Consider the following nonlinear programming subproblem:

36 **Subproblem.** Given a $z_0 \in \Omega'$, find a real σ_0 and a vector $h_0 \in E^n$ which minimize the linear form σ subject to the constraints

$$\langle \nabla q^i(z_0), h \rangle \leq \sigma \quad \text{for all } i \in I_0(z_0)$$
$$h \in S = S' \cap \{h : Rh = 0\},$$

where S' is *any* compact convex subset of E^n containing the origin in its interior. □

37 **Exercise.** Show that if (σ_0, h_0) is an optimal solution to (36) for some set S', then

38 $$\min_{h \in S} \max_{i \in I_0(z_0)} \langle \nabla q^i(z_0), h \rangle = \sigma_0.$$

Conversely, show that if (38) holds for the above S', then there exists a vector h_0 such that (σ_0, h_0) solves (36). □

39 **Exercise.** Show that subproblem (36) always has an optimal solution, as we have already tacitly assumed. Hence a (σ_0, h_0) satisfying (38) always exists. □

Thus to determine whether or not (35) holds we must solve subproblem (36). In turn, since the set S' in (20) is quite arbitrary, subproblem (36) is made tractable by a proper selection of S'. If we let

40 $$S' = \{h. = (h^1, h^2, \ldots, h^n) : |h^i| \leq 1, i = 1, 2, \ldots, n\},$$

i.e., if S' is the unit hypercube in E^n, then S' satisfies all the properties stipulated in definition (20), and subproblem (36) becomes a linear programming problem. When transcribed into canonical form by appropriate substitutions, it can be solved by either the simplex algorithm (5.9.4) or the bounded-variable simplex algorithm (5.10.23). The reader can no doubt generate other examples of sets S' for which subproblem (36) reduces to a linear programming problem. Theoretically, of course, we have a great deal of freedom in choosing S'. Other simple choices of S' which lead to a tractable form of (36) are

$$S' = \{h: \langle h,h \rangle \leq 1\} \qquad S' = \Big\{h: \sum_{i=1}^{n} |h^i| \leq 1\Big\}.$$

41 **Exercise.** Transform the linear programming problem: minimize $\{\sigma: \langle \nabla q^i(z_0), h \rangle \leq \sigma, i \in I_0(z_0), Rh = 0, |h^i| \leq 1, i = 1, 2, \ldots, n\}$ into a bounded-variable linear program of the form: minimize $\{\langle d,w \rangle: Lw = b, -\infty < \alpha^i \leq w^i \leq \beta^i < +\infty\}$. (*Hint:* Use "slack" variables and determine artificial, but valid, bounds α^i and β^i where necessary.) □

42 **Exercise.** Transform the linear programming problem defined in exercise (41) into the form: minimize $\{\langle d,w \rangle: Lw = b, -\infty < \alpha^i \leq w^i\}$. □

In view of the above exercises, we may suppose that for a given $z_0 \in \Omega'$, (σ_0, h_0) is an optimal solution to subproblem (36) (with a simple S) obtained, for example, by using the bounded-variable simplex algorithm. Now, if $\sigma_0 = 0$, then, by exercise (37) z_0 solves problem (28). Suppose, therefore, that $\sigma_0 < 0$, that is, that

$$\min_{h \in S} \max_{i \in I_0(z_0)} \langle \nabla q^i(z_0), h \rangle < 0.$$

It follows, then, that we have determined a vector $h_0 \in S$ such that $\langle \nabla q^i(z_0), h_0 \rangle < 0$ for all $i \in I_0(z_0)$, which in expanded form becomes

43 $\quad \langle \nabla q^0(z_0), h_0 \rangle < 0 \qquad Rh_0 = 0$
$\quad \langle \nabla q^i(z_0), h_0 \rangle < 0 \qquad$ for all $i \in I(z_0)$.

44 **Exercise.** Consider the parametrized vector $z(\lambda) = z_0 + \lambda h_0$, where $z_0 \in \Omega'$ and h_0 satisfies (43). Show that there exists a scalar $\lambda_0 > 0$ such that $z(\lambda) \in \Omega'$ and $q^0(z(\lambda)) < q^0(z_0)$ for all $0 \leq \lambda \leq \lambda_0$. □

Thus for all values of λ sufficiently small $z_0 + \lambda h_0 \in \Omega'$ and $q^0(z_0 + \lambda h_0) < q^0(z_0)$. Put another way, moving in the *feasible direction* h_0, we have reduced the value of the cost function $q^0(\cdot)$. Having gone to all the trouble of calculating this feasible direction h_0, we shall make the most of it. Let

45 $\quad \gamma_0 = \max \{\gamma: q^i(z_0 + \beta h_0) \leq 0$ for all $\beta \leq \gamma$,
$$i = 1, 2, \ldots, k\};$$

that is, γ_0 (which may be $+\infty$) is the largest value of γ such that $z_0 + \beta h_0 \in \Omega'$ for all $0 \leq \beta \leq \gamma$. Next, let $\lambda_0 \in [0, \gamma_0]$ be selected such that

46
$$\lambda_0 = \max \{\lambda: q^0(z_0 + \lambda h_0) \leqq q^0(z_0 + \beta h_0)$$
$$\text{for all } 0 \leqq \beta \leqq \lambda, 0 \leqq \lambda \leqq \gamma_0\}.$$

47 **Remark.** By exercise (44), $\lambda_0 > 0$ and $q^0(z + \lambda_0 h_0) < q^0(z_0)$. It may turn out, however, that $\lambda_0 = +\infty$, or that $q^0(z_0 + \lambda h_0) \to -\infty$ as $\lambda \to \lambda_0$, or both may occur. In the latter case the nonlinear programming problem (1) does not have a solution. □

Computationally, the determination of γ_0 in (45) is carried out by means of a one-dimensional search over the parameter $\gamma \geqq 0$ for every $i \in \{1,2, \ldots ,k\}$. The same approach is usually used to determine λ_0 in (46). If it turns out that $\lambda_0 < +\infty$, we may set $z_1 = z_0 + \lambda_0 h_0$ and repeat the above procedure, with z_1 replacing z_0.

It would appear, therefore, that a sequence $z_0, z_1, z_2, \ldots$ of vectors in Ω', constructed as shown below, might converge to a solution of (28). Thus, starting at $z_0 \in \Omega'$, with some set S defined as in (20), let

48
$$z_{j+1} = z_j + \lambda_j h_j \qquad j = 0, 1, 2, \ldots ,$$

where $h_j \in S$ satisfies

49
$$\max_{i \in I_0(z_j)} \langle \nabla q^i(z_j), h_j \rangle = \min_{h \in S} \max_{i \in I_0(z_j)} \langle \nabla q^i(z_j), h \rangle,$$

$\lambda_j \in [0,\gamma_j]$ satisfies

50
$$\lambda_j = \max \{\lambda: q^0(z_0 + \lambda h_0) \leqq q^0(z_0 + \beta h_0)$$
$$\text{for all } 0 \leqq \beta \leqq \lambda, 0 \leqq \lambda \leqq \gamma_j\}$$

and finally, $\gamma_j = \min \{\gamma_j', \gamma_j''\}$, where

51
$$\gamma_j' = \max \{\gamma: q^i(z_j + \beta h_j) \leqq 0 \text{ for all } 0 \leqq \beta \leqq \gamma, i \in I(z_j)\}$$
$$\gamma_j'' = \max \{\gamma: q^i(z_j + \beta h_j) \leqq 0 \text{ for all } 0 \leqq \beta \leqq \gamma, i \in \bar{I}(z_j)\}.$$

Unfortunately, unless we take certain precautions, a sequence of feasible points $\{z_j\}$ generated as above may not converge to a solution of problem (28), and in addition, none of its subsequences need converge to a solution of (28) either. We shall now show heuristically why this is so.

From exercise (44) we know that if z_j is not a solution of (28), then $q^0(z_j + \lambda_j h_j) = q^0(z_{j+1}) < q^0(z_j)$. Now, the actual numerical difference between these two quantities depends on several factors, and it turns out that the most critical one is the scalar γ_j determined in (51). Clearly, if γ_j is small, then the interval $[0,\gamma_j]$ over which the minimization in (50) is to be performed is small, and hence $q^0(z_{j+1})$ differs only slightly from $q^0(z_j)$. The numerical value of γ_j is, in turn, a function of the two independent scalars γ_j' and γ_j'', as is evident in (51). The first quantity, γ_j', is not particularly critical, since, by (49), $\langle \nabla q^i(z_j), h_j \rangle < 0$ for all $i \in I(z_j)$; that is, to first-order effects h_j is indeed a feasible direction for the $q^i(\cdot)$, with $i \in I(z_j)$. It is the second term in (51), γ_j'', which is the source of difficulty. While, by definition, $q^i(z_j) < 0$ for $i \in \bar{I}(z_j)$, this number can be arbitrarily small, and should the circumstances be such that $\langle \nabla q^i(z_j), h_j \rangle > 0$ for an $i \in \bar{I}(z_j)$, it is intuitively clear that γ_j'' could be arbitrarily small. Hence γ_j, and, a fortiori, $|q^0(z_{j+1}) - q^0(z_j)|$, may be arbitrarily small (certainly smaller than first-order quantities). Examples have been constructed (see [2]) where precisely this set of circumstances occurs for infinitely many vectors z_j for $j \in \{0,1,2,\ldots\}$, the result being that the sequence $\{z_j\}_{j=1}^{\infty}$ converges to a vector $\hat{z}$, but $\hat{z}$ does not solve problem (28). Behavior of this type is called *zigzagging* or *jamming* in the mathematical programming literature.

One way to take into account the fact that some of the inactive constraints $q^i(\cdot)$ for $i \in \bar{I}(z_j)$ are "almost active" is to modify relation (49), which determines h_j, in the following manner. Suppose we agree that for $q^i(\cdot)$, with $i \in \bar{I}(z_j)$, to be "almost active" at z_j means that

52 $\qquad q^i(z_j) + \epsilon \geq 0,$

where $\epsilon > 0$ is preassigned and small. Relation (49), which determines h_j, now becomes

53 $\qquad \min_{h \in S} \max_{i \in I_\epsilon(z_j)} \langle \nabla q^i(z_j), h \rangle \triangleq \sigma_j,$

where, by (17),

$$I_\epsilon(z_j) = \{i: q^i(z_j) + \epsilon \geq 0, i \in \{1,2,\ldots,k\}\} \cup \{0\}.$$

Clearly, $I_\epsilon(z_j) \supset I(z_j)$ and in addition, $I_\epsilon(z_j)$ contains the indices $i \in \bar{I}(z_j)$ for which $q^i(z_j)$ is within ϵ of being zero. It should be rather obvious that all the remarks pertaining to the solution of relation (49) are applicable to the modified subproblem defined by (53). Suppose, therefore, that (σ_j, h_j) is a solution to (53). If $\sigma_j < 0$, it now follows that h_j has the property that

54 $\qquad \langle \nabla q^i(z_j), h_j \rangle < 0 \qquad \text{for all } i \in I_0(z_j)$

and, more important, that

55 $\qquad \langle \nabla q^i(z_j), h_j \rangle < 0 \qquad \text{for all } i \in \bar{I}(z_j) \text{ such that } q^i(z_j) + \epsilon \geq 0.$

Thus by taking into account inequality constraints which are almost active we have eliminated the cause of jamming.

There is one important matter left to clear up. Suppose that $\sigma_j = 0$ in (53). If $I_\epsilon(z_j) = I_0(z_j)$ (within computational tolerances), we have indeed solved problem (28). Otherwise all that can be said is that there does not exist a vector $h \in S$ with $\langle \nabla q^i(z_j), h \rangle < 0$ for all $i \in I_\epsilon(z_j)$; but this does not imply that the necessary conditions of optimality are satisfied. In this case we can replace ϵ by $\epsilon/2$ and resolve (53), repeating the halving of ϵ either until we obtain $\sigma_j < 0$ in (53) or until $\sigma_j = 0$ and $I_\epsilon(z_j) = I_0(z_j)$ for some value of $\epsilon = \epsilon/m$, with $m \in \{2, 2^2, 2^3, \ldots, \}$, whichever occurs first.† One of these two alternatives must occur after a finite number of steps.

Now that we are acquainted with a few of the subtleties involved, let us proceed with a rigorous treatment of the methods of feasible directions. Incidentally, at this point it should be quite clear that the reason for saying *methods* rather than *method* is that each different choice of the set S defines a different algorithm in this family.

56 Algorithm: methods of feasible directions. We are given two sets, Ω' and $S = S' \cap S''$, where

57 $\quad\quad\Omega' = \{z \colon Rz = c,\, q^i(z) \leq 0,\, i = 1, 2, \ldots, k\}$,
58 $\quad\quad S'' = \{h \colon Rh = 0\}$,

R, c, and q^i for $i = 1, 2, \ldots, k$ are as in problem (28), and S' is a (any) compact convex set containing the origin in its interior. For any $z \in \Omega'$ and for any $\alpha \geq 0$ let $\phi_\alpha \colon E^n \to E^1$ be defined by

59 $$\phi_\alpha(z) = \min_{h \in S} \max_{i \in I_\alpha(z)} \langle \nabla q^i(z), h \rangle.$$

Let $\epsilon > 0$ and $z_0 \in \Omega'$ be given, and suppose that $z_0, z_1, \ldots, z_j$ have been computed and are in Ω'.

Step 1. Using the given $\epsilon > 0$, set $\epsilon_j = \epsilon$.

Step 2. Let h_{ϵ_j} be such that

60 $$\phi_{\epsilon_j}(z_j) = \langle \nabla q^l(z_j), h_{\epsilon_j} \rangle \quad \text{for some } l \in I_{\epsilon_j}(z_j).$$

If $\phi_{\epsilon_j}(z_j) \leq -\epsilon_j < 0$, then set $h_j = h_{\epsilon_j}$ and go to step 4. If $\phi_{\epsilon_j}(z_j) > -\epsilon_j$, then go to step 3.

---

† The repeated halving of ϵ is a device which simplifies our exposition. We could equally well use any other factor $\beta \in (0,1)$ for multiplying ϵ. Computationally, of course, we would choose the largest number $\bar{\epsilon} = \beta^l \epsilon$, for some integer l, such that $I_{\bar\epsilon}(z_j)$ is a proper subset of $I_\epsilon(z_j)$, for resolving (53), rather than blindly resolve (53) for $\beta\epsilon, \beta^2\epsilon, \ldots, \beta^{l-1}\epsilon$, which all give the same σ_j as ϵ.

SEC. 7.2 METHODS OF FEASIBLE DIRECTIONS 189

Step 3. If $\phi_{\epsilon_j}(z_j) < 0$, or $I_{\epsilon_j}(z_j) \neq I_0(z_j)$, then set $\epsilon_j = \epsilon_j/2$ and return to step 2. Otherwise $\phi_{\epsilon_j}(z_j) = 0$ and $I_{\epsilon_j}(z_j) = I_0(z_j)$, in which case stop, since z_j solves problem (28).

Step 4. Select $\gamma_j > 0$ such that

(61) $$\gamma_j = \max \{\gamma: q^i(z_j + \beta h_j) \leq 0 \text{ for all } 0 \leq \beta \leq \gamma, \\ i = 1, 2, \ldots, k\}$$

Step 5. Select $\lambda_j > 0$ to satisfy

(62) $$\lambda_j = \max \{\lambda: q^0(z_j + \lambda h_j) \leq q^0(z_j + \beta h_j) \text{ for all } 0 \leq \beta \leq \lambda, \\ 0 \leq \lambda \leq \gamma_j\}.$$

If $\lambda_j = +\infty$, then stop. If $q^0(z_j + \lambda h_j) \to -\infty$ as $\lambda \to \lambda_j$, then stop, since in this case inf $\{q^0(z): z \in \Omega'\} = -\infty$. Otherwise let

(63) $$z_{j+1} = z_j + \lambda_j h_j \in \Omega',$$

set $j = j + 1$, and return to step 1. □

(64) **Remark.** In order to initialize algorithm (56) it is necessary to know a $z_0 \in \Omega$. Techniques for finding such a point are described in Section 1. □

(65) **Remark.** By exercise (44), if z_j does not solve problem (28), then the parameter λ_j determined in step 5 is such that $\lambda_j > 0$, $z_j + \lambda_j h_j \in \Omega'$, and $q^0(z_j + \lambda_j h_j) < q^0(z_j)$. The last property implies that the infinite sequence $\{q^0(z_j)\}_{j=0}^{\infty}$ is strictly monotonically decreasing. □

(66) **Exercise.** Show that if z_j does not solve problem (28), then either the halving procedure indicated in step 3 will, after a finite number of halvings, allow the algorithm to proceed to step 4 or, if z_j is a solution of (28), the stop command in step 3 will be reached in a finite number of steps. Thus the algorithm is well defined. □

(67) **Theorem.** Let $z_0, z_1, z_2, \ldots$ be any infinite sequence generated by algorithm (56). If $\{z_l\}$ for $l \in L$ is any subsequence converging to a point $\hat{z}$, then $\hat{z} \in \Omega'$ and

$$\phi_0(\hat{z}) = \min_{h \in S} \max_{i \in I_0(\hat{z})} \langle \nabla q^i(\hat{z}), h \rangle = 0;$$

that is, $\hat{z}$ solves problem (28).

Proof. Let $\{z_l\}$ for $l \in L$ be a convergent subsequence generated by (56), with limit point $\hat{z}$. The set $\Omega' = \{z: Rz = c, q^i(z) \leq 0, i = 1, 2,$

..., k} is closed because the functions $q^i(\cdot)$ for $i = 1, 2, \ldots, k$ and the function $r(z) = Rz - c$ is continuous. Consequently, since $z_l \in \Omega'$ and $z_l \to \hat{z}$ for $l \in L$, it follows that $\hat{z} \in \Omega'$.

We next prove that $\phi_0(\hat{z}) = 0$. Suppose that $\phi_0(\hat{z}) < 0$. To arrive at the desired contradiction it is sufficient to show that there is an integer $\bar{l} \in L$ and an $\bar{\epsilon} > 0$ such that for all $l \in L$ and $l \geq \bar{l}$

68
$$q^0(z_{l+1}) - q^0(z_l) \leq -\bar{\epsilon}.$$

Indeed, suppose that (68) holds. Then for any two successive points z_l and z_{l+j} of the subsequence, with $l, l+j \in L$ and $l \geq \bar{l}$, we have

69
$$q^0(z_{l+j}) - q^0(z_l) = [q^0(z_{l+j}) - q^0(z_{l+j-1})] \\ + \cdots + [q^0(z_{l+1}) - q^0(z_l)] < -\bar{\epsilon}.$$

But since $z_l \to \hat{z}$ for $l \in L$, $q^0(z_l) \to \alpha > -\infty$, which contradicts (69). We shall now see that if $\phi_0(\hat{z}) < 0$, then (68) is satisfied.

Since $z_l \to \hat{z}$ for $l \in L$ and the functions $q^i(\cdot)$ are continuous, it is easily verified that corresponding to every $\hat{\epsilon} > 0$ and sufficiently small there is an integer $l' \in L$ such that for $l \geq l'$ and $l \in L$

70
$$I_{\hat{\epsilon}}(z_l) \subset I_0(\hat{z}).$$

Since $\phi_0(\hat{z}) < 0$, we may assume that $\hat{\epsilon}$ is chosen such that (70) holds, and

71
$$\phi_0(\hat{z}) < -\hat{\epsilon},$$

and $\hat{\epsilon} = \epsilon/m$ for some $m \in \{2, 2^2, \ldots, 2^p\}$, where $\epsilon > 0$ is as specified at the beginning of algorithm (56). Let $M: E^n \to E^1$ be defined by

72
$$M(z) = \min_{h \in S} \max_{i \in I_0(\hat{z})} \langle \nabla q^i(z), h \rangle.$$

Then M is continuous, and there is an $l'' \in L$ such that for all $l \geq l''$ and $l \in L$

73
$$|M(z_l) - \phi_0(\hat{z})| \leq \frac{\hat{\epsilon}}{2}.$$

Let $\bar{l} = \max(l', l'')$; then, because of (70) to (72), for all $l \geq \bar{l}$ and $l \in L$

74
$$\phi_{\hat{\epsilon}}(z_l) \leq M(z_l) \leq -\frac{\hat{\epsilon}}{2}.$$

But $I_{\hat{\epsilon}/2}(z_l) \subset I_{\hat{\epsilon}}(z_l)$, and hence for all $l \geq \bar{l}$ and $l \in L$ we have

75
$$\phi_{\hat{\epsilon}/2}(z_l) \leq \phi_{\hat{\epsilon}}(z_l) \leq -\frac{\hat{\epsilon}}{2}.$$

We therefore conclude that for all $l \geq \hat{l}$ and $l \in L$ the algorithm will use a value of $\epsilon_l \geq \hat{\epsilon}/2$ in computing h_l in step 2;† that is, for all $l \geq \hat{l}$, $l \in L$, and for all $i \in I_{\epsilon_l}(z_l)$, $\langle \nabla q^i(z_l), h_l \rangle \leq -\hat{\epsilon}/2$.

Let $\rho > 0$ be such that $z_l + \xi h_l \in B(\hat{z}, \rho)$, a closed ball of radius ρ about $\hat{z}$, for all $l \geq \hat{l}$, $l \in L$, and $0 \leq \xi \leq 1$. Now, for every $l \in L$ and $i = 0, 1, \ldots, k$ we have, by the mean-value theorem, that

$$q^i(z_l + \beta h_l) = q^i(z_l) + \beta \langle \nabla q^i(z_l + \xi h_l), h_l \rangle, \qquad 76$$

where $\xi \in [0, \beta]$. Since the functions $\langle \nabla q^i(\cdot), \cdot \rangle$ are uniformly continuous on $B(\hat{z}, \rho) \times S$, there exist $0 < \beta_i \leq 1$ such that for $l \geq \hat{l}$

$$|\langle \nabla q^i(z_l + \xi h_l), h_l \rangle - \langle \nabla q^i(z_l), h_l \rangle| \leq \frac{\hat{\epsilon}}{4} \qquad \text{for all } \xi \in [0, \beta_i].$$

Similarly, since the functions $q^i(\cdot)$ are uniformly continuous on $B(\hat{z}, \rho)$, there exist $0 < \bar{\beta}_i \leq 1$ such that for $l \geq \hat{l}$, and $l \in L$,

$$|q^i(z_l + \xi h_l) - q^i(z_l)| \leq \frac{\hat{\epsilon}}{2} \qquad \text{for all } \xi \in [0, \bar{\beta}_i]. \qquad 77$$

Now, for all $l \geq \hat{l}$ and $l \in L$ and for each $i \in I_{\epsilon_l}(z_l)$, $\langle \nabla q^i(z_l), h_l \rangle \leq -\hat{\epsilon}/2$, and for each $i \in \bar{I}_{\epsilon_l}(z_l)$, $q^i(z_l) \leq -\hat{\epsilon}/2$. Hence, setting

$$\bar{\mu} = \min \{\beta_0, \beta_1, \ldots, \beta_k, \bar{\beta}_0, \bar{\beta}_1, \ldots, \bar{\beta}_k\}, \qquad 78$$

we have for all $l \geq \hat{l}$, $l \in L$, and $\beta \in [0, \bar{\mu}]$,

$$q^i(z_l + \beta h_l) - q^i(z_l) \leq -\beta \frac{\hat{\epsilon}}{4} \qquad \text{for all } i \in I_{\epsilon_l}(z_l) \qquad 79$$

$$q^i(z_l + \beta h_l) \leq 0 \qquad \text{for all } i \in \bar{I}_{\epsilon_l}(z_l).$$

It follows from (79) that $\lambda_l \geq \bar{\mu}$ (see step 5), so that we are led to the conclusion that for all $l \geq \hat{l}$ and $l \in L$

$$q^0(z_{l+1}) - q^0(z_l) \leq -\bar{\mu} \frac{\hat{\epsilon}}{4} \triangleq -\bar{\epsilon};$$

that is, (68) holds. □

80 **Exercise.** Suppose that for the nonlinear programming problem (1) there are no inequality constraint functions, that is, $q^i \equiv 0$ for $i = 1, 2, \ldots, k$, and hence subproblem (59) reduces to

$$\phi(z) = \min_{h \in S} \langle \nabla q^0(z), h \rangle.$$

Show that the antijamming precaution of algorithm (56) can be

† In (56) j is a dummy subscript. Substitute l for j to compute ϵ_l, h_l, and λ_l.

eliminated without changing the result of theorem (67). Thus steps 1 to 3 of algorithm (56) are replaced by the following one:

Step 1'. Let h_j be such that $\phi_0(z_j) = \langle \nabla q^0(z_j), h_j \rangle$. If $\phi_0(z_j) = 0$, then stop, since z_j solves problem (28). Otherwise proceed to step 4. □

The reader is cautioned not to read more into theorem (67) than is actually stated. In particular, it has not been shown that algorithm (56) necessarily generates an infinite sequence (at some point $\lambda_j = +\infty$ may occur), nor is there any guarantee that a convergent subsequence can be extracted. The following exercises are directed to this problem.

81 Exercise. Suppose that for every $y \in \Omega'$ the set $\{z: q^0(z) \leq q^0(y), z \in \Omega'\}$ is bounded. Show that algorithm (56) either determines a solution to problem (28) in a finite number of steps or else generates an infinite sequence $\{z_j\}_{j=0}^{\infty}$ which has a convergent subsequence. □

82 Exercise. Suppose that the functions $q^i(\cdot)$ for $i = 0, 1, \ldots, k$ are convex, that $m \triangleq \min \{q^0(z): z \in \Omega'\}$ exists, and that the set $\{z: q^0(z) = m, z \in \Omega'\}$ is bounded. Show that for every $y \in \Omega'$ the set $\{z: q^0(z) \leq q^0(y), z \in \Omega'\}$ is bounded. □

83 Exercise. In addition to the hypotheses in exercise (82), suppose that the function $q^0(z)$ is strictly convex and that assumption (8) is satisfied. Show that the infinite sequence $\{z_j\}_{j=0}^{\infty}$ generated by algorithm (56) converges to a point $\hat{z}$ satisfying $q^0(\hat{z}) = \min \{q^0(z): z \in \Omega'\}$. □

We shall now show how the algorithm (56) can be used for solving discrete optimal control problems. Thus, suppose we wish to solve the following problem:

84 Minimize

$$\sum_{i=0}^{k-1} f_i^0(x_i, u_i) \qquad x_i \in E^n, u_i \in E^1$$

subject to

a. $x_{i+1} - x_i = f_i(x_i, u_i) \qquad i = 0, 1, \ldots, k-1$,
b. $x_0 = c_0, \bar{q}^j(x_k) \leq 0 \qquad j = 1, 2, \ldots, m$,
c. $|u_i| \leq 1 \qquad i = 0, 1, \ldots, k-1$,

where the $f_i^0(\cdot, \cdot)$, $i = 0, 1, \ldots, k-1$, and $\bar{q}^j(\cdot)$, $j = 1, 2, \ldots, m$, are continuously differentiable in all their arguments. □

Setting $z = (u_0, u_1, \ldots, u_{k-1})$, problem (84) assumes the form:

85   Minimize $q^0(z)$ subject to $q^j(z) \leq 0$, $j = 1, 2, \ldots, m$, $|u_i| \leq 1$, $i = 0, 1, 2, \ldots, k - 1$, where

86
$$q^0(z) = \sum_{i=0}^{k-1} f_i^0(x_i(z), u_i)$$

and

87
$$q^j(z) = \bar{q}^j(x_k(z)) \qquad j = 1, 2, \ldots, m. \quad \square$$

In (86) and (87), the $x_i(z)$ are obtained by solving (84) with $x_0 = c_0$ and the controls $(u_0, u_1, \ldots, u_{k-1}) = z$. In order to apply the algorithm (56) to problem (85), we need to calculate frequently $\nabla q^0(z)$ and $\nabla q^i(z)$, $i \in \{1, 2, \ldots, m\}$. The dynamical structure (84) of this problem can be utilized to simplify this calculation, or to be more precise, to transform it into a sequence of relatively simple calculations. We begin with $\nabla q^0(z)$.

Now, we can break up $\nabla q^0(z)$ into block form as follows:

88
$$(\nabla q^0(z))^T = \frac{\partial q^0(z)}{\partial z} = \left(\frac{\partial q^0(z)}{\partial u_0} \,\bigg|\, \frac{\partial q^0(z)}{\partial u_1} \,\bigg|\, \cdots \,\bigg|\, \frac{\partial q^0(z)}{\partial u_{k-1}} \right)$$

and, by inspection of (84),

89
$$\frac{\partial q^0(z)}{\partial u_i} = \frac{\partial f_i^0(x_i(z), u_i)}{\partial u_i} + \sum_{j=i+1}^{k-1} \frac{\partial f_j^0(x_j(z), u_j)}{\partial x_j} \frac{\partial x_j(z)}{\partial u_i}$$

Referring to Section 2.4, we find that $\partial x_j(z) / \partial u_i$, $j > i$, can be expressed as follows. For $i = 0, 1, 2, \ldots, k$ and $k \geq j \geq i$, let $G_{j,i}$ be an $n \times n$ matrix, such that $G_{i,i} = I$, the identity matrix, and

90
$$G_{j+1,i} - G_{j,i} = \frac{\partial f_i(x_i(z), u_i)}{\partial x_i} G_{j,i}, \, j = 1, \, i+1, \, \ldots, \, k-1,$$

where $(u_0, u_1, \ldots, u_{k-1}) = z$. Then,

91
$$\frac{\partial x_j(z)}{\partial u_i} = \begin{cases} G_{j,i+1} \dfrac{\partial f_i(x_i(z), u_i)}{\partial u_i} & \text{for } j = i+1, i+2, \ldots, k \\ 0 & \text{for } j = 1, 2, \ldots, i, \end{cases}$$

and hence,

92
$$\frac{\partial q^0(z)}{\partial u_i} = \frac{\partial f_i^0(x_i(z), u_i)}{\partial u_i} + \left(\sum_{j=i+1}^{k-1} \frac{\partial f_j^0(x_j(z), u_j)}{\partial x_j} G_{j,i+1} \right) \frac{\partial f_i(x_i(z), u_i)}{\partial u_i}$$

Now, for $i = 1, 2, \ldots, k$, let p_i be the solution of

(93)
$$p_i - p_{i+1} = \left(\frac{\partial f_i(x_i(z), u_i)}{\partial x}\right)^T p_{i+1} + \left(\frac{\partial f_i{}^0(x_i(z), u_i)}{\partial x}\right)^T$$
$$i = 0, 1, \ldots, k-1$$

with $p_k = 0$. Then (see Section 2.4),

(94)
$$p_i = \sum_{j=i}^{k-1} G_{j,i}^T \left(\frac{\partial f_j{}^0(x_j(z), u_j)}{\partial x_j}\right)^T \quad i = 1, 2, \ldots, k.$$

Comparing (92) and (94), we find that

(95)
$$\frac{\partial q^0(z)}{\partial u_i} = \frac{\partial f_i{}^0(x_i(z))}{\partial u_i} + \left\langle p_{i+1}, \frac{\partial f_i(x_i(z), u_i)}{\partial u_i} \right\rangle.$$

Thus, to calculate $\partial q^0(z)/\partial u_i$, we solve (93) (or even more simply, its transpose) for the p_i, $i = 1, 2, \ldots, k$ and then calculate the scalars $\partial q^0(z)/\partial u_i$ by means of (95).

Next, to calculate $\nabla q^i(z)$, $i \in \{1, 2, \ldots, m\}$, we observe that

(96)
$$\nabla q^i(z)^T = \frac{\partial q^i(z)}{\partial z} = \frac{\partial \bar{q}^i(x_k(z))}{\partial x_k} \frac{\partial x_k(z)}{\partial z},$$

that is,

(97)
$$\frac{\partial q^i(z)}{\partial u_j} = \left\langle \frac{\partial x_k(z)}{\partial u_j}, \frac{\partial \bar{q}^i(x_k(z))}{\partial x_k} \right\rangle.$$

Substituting for $\partial x_k(z)/\partial u_j$ from (91), we obtain

(98)
$$\frac{\partial q^i(z)}{\partial u_j} = \left\langle G_{k,j+1} \frac{\partial f_j(x_j(z), u_j)}{\partial u_j}, \frac{\partial \bar{q}^i(x_k(z))}{\partial x_k} \right\rangle$$
$$= \left\langle \frac{\partial f_j(x_j(z), u_j)}{\partial u_j}, G_{k,j+1}^T \nabla \bar{q}^i(x_k(z)) \right\rangle$$

Now, let $p_{1,i}, p_{2,i}, \ldots, p_{k,i}, i \in \{1, 2, \ldots, m\}$ be the solution of

(99)
$$p_{j,i} - p_{j+1,i} = \left(\frac{\partial f_j(x_j, u_j)}{\partial x_j}\right)^T p_{j+1,i} \quad j = i, i+1, \ldots, k-1,$$

with $p_{k,i} = \nabla \bar{q}^i(x_k(z))$. Then

(100)
$$p_{j+1,i} = G_{k,j+1}^T \nabla \bar{q}^i(x_k(z))$$

and hence

(101)
$$\frac{\partial q^i(z)}{\partial u_j} = \left\langle \frac{\partial f_j(x_j(z), u_j)}{\partial u_j}, p_{j+1,i} \right\rangle,$$

which again indicates a method for computing $\nabla q^i(z)$.

Suppose set $S = \{h: |h^i| \leq 1$ in (56). Then $\phi_\epsilon(z)$ is the solution of

102 Minimize σ subject to

a. $\langle \nabla q^i(z), h \rangle - \sigma \leq 0$ for $i = 0$ and for all $i\epsilon \{1, \ldots, m\}$ such that $q^i(z) + \epsilon \geq 0$
b. $h^{i+1} - \sigma \leq 0$ for all i such that $u_i - 1 + \epsilon \geq 0$, $i\epsilon \{0, \ldots, k - 1\}$
c. $-h^{i+1} - \sigma \leq 0$ for all i such that $-u_i - 1 + \epsilon \geq 0$
d. $|h^i| \leq 1$, $i = 1, 2, \ldots, k$.

A solution to this problem could conveniently be obtained either by using the bounded-variable simplex algorithm (5.11.23), or preferably, by generalized upper bounding techniques which require the inversion of matrices whose dimension is governed only by the inequalities a in (102) [rather than by b, c, and d; see (12)].

103 **Exercise.** Consider again problem (1) and suppose that some of the inequalities in (2) are affine. For every $z \in \Omega'$ [defined in (6)] and every $\epsilon \geq 0$, let $I_\epsilon^A(z) \subset I_\epsilon(z)$ and $I_\epsilon^N(z) \subset I_\epsilon(z)$ [defined in (17)] be such that $I_\epsilon^A(z) \cap I_\epsilon^N(z) = \phi$, $I_\epsilon^A(z) \cup I_\epsilon^N(z) = I_\epsilon(z)$, $0\epsilon I_\epsilon^N(z)$, and, in addition, for every $i \in I_\epsilon^A(z)$, $q^i(\cdot)$ is affine. For $\epsilon \geq 0$, let $\eta_\epsilon: \Omega' \to E^1$ be defined by

104 $\eta_\epsilon(z) = \min \{\sigma: \sigma - \langle \nabla q^i(z), h \rangle \geq 0$ for $i \in I_\epsilon^N(z)$,
$- \langle \nabla q^i(z), h \rangle \geq 0$ for $i \in I_\epsilon^A(z)$, $Rh = 0$, $|h^i| \leq 1\}$.

Show that (a) for every $z \in \Omega'$ and $\epsilon \geq 0$, $\eta_\epsilon(z) \leq \varphi_\epsilon(z)$ (provided $S' = \{h: |h^i| \leq 1\}$); (b) for every $\hat{z}$ which is optimal for (1) $\eta_0(\hat{z}) = 0$; and (c) if $\eta_\epsilon(\cdot)$ is substituted for $\varphi_\epsilon(\cdot)$ in the algorithm (56), the convergence properties of this algorithm remain unaltered. In addition, show that when algorithm (56), with $\eta_\epsilon(\cdot)$ taking the place of $\varphi_\epsilon(\cdot)$, is applied to the previously discussed optimal control problem, the feasible direction h at z is now determined by solving

105 Minimize σ subject to

a. $\langle \nabla q^i(z), h \rangle - \sigma \leq 0$ for $i = 0$ and for all $i\epsilon \{1, \ldots, m\}$ such that $q^i(z) + \epsilon \geq 0$
b. $-h^{i+1} \leq 0$ for all i such that $u_i - 1 + \epsilon \geq 0$, $i\epsilon \{0, \ldots, k-1\}$
c. $h^{i+1} \leq 0$ for all i such that $-u_i - 1 + \epsilon \geq 0$, $i\epsilon \{0, \ldots, k-1\}$
d. $|h^i| \leq 1$, $i = 1, 2, \ldots, k$.

Note that (105) is a simpler problem than (102). □

The reader should now have no difficulty in obtaining similar derivations for more complex cases of the discrete optimal control problem. When reading the gradient projection methods in the next section, he should have no difficulty in utilizing the above developments for calculating the required gradients in control applications.

7.3 STEEPEST DESCENT AND GRADIENT PROJECTION

The methods of feasible directions [algorithm (2.56)] presented in the preceding section were characterized by the fact that at each iteration we had to solve two subsidiary optimization problems. The first one was solved to find a feasible direction and the second one to obtain the step size. Let us now examine a few algorithms which entail only one subsidiary optimization at each iteration, but which are really effective only when all the constraint functions are affine. We shall see that in these algorithms certain fairly simple calculations take the place of solving a subsidiary optimization problem. In view of the above remarks, we shall restrict ourselves to problems of the following form:

1 **Problem.** Given a real-valued continuously differentiable function $f(\cdot)$ defined on E^n and affine functions $r(z) = Rz - c$ and $q(z) = Qz - b$, where R, Q, c, and b are, respectively, $m \times n$, $k \times n$, $m \times 1$, and $k \times 1$ matrices, find a $\hat{z} \in E^n$ such that

2 $\quad R\hat{z} = c$
3 $\quad Q\hat{z} \leqq b$

and such that for all vectors $z \in E^n$ satisfying (2) and (3)

4 $\quad f(\hat{z}) \leqq f(z)$. □

Obviously, this is a special case of the nonlinear programming problem (2.1).

Throughout this section we shall assume that the following regularity condition holds:

5 **Assumption.** Let r^i for $i = 1, 2, \ldots, m$, q^i for $i = 1, 2, \ldots, k$, and b^i for $i = 1, 2, \ldots, m$ denote the rows of R, Q, and b, respectively. If $z \in E^n$ is any vector satisfying (2) and (3), then the vectors r^i for $i = 1, 2, \ldots, m$, taken together with the vectors q^i, with $i \in I(z) = \{i : \langle q^i, z \rangle - b^i = 0\}$, are linearly independent. □

6 **Remark.** We anticipate the need to invert certain matrices which will be nonsingular because of assumption (5). □

As we noted in Section 2, algorithm (2.56), defining the methods of feasible directions, is in fact a family of algorithms, where each different choice of the compact set S' [see (2.20)] defines a different algorithm. We shall now consider two well-known algorithms which are obtainable from (2.56) by considering specific forms of constraints (2) and (3), as well as of the set S', and which involve only one optimization at each iteration. We begin with the simplest possible case of problem (1), the unconstrained minimization problem.

Unconstrained Minimization Problems

Let us consider the case when all the matrices and vectors appearing in (2) and (3) are identically zero. Thus we wish to find a $\hat{z} \in E^n$ such that $f(\hat{z}) \leq f(z)$ for all $z \in E^n$, where $f(\cdot)$ is a real-valued continuously differentiable function defined on E^n.

The algorithm we are about to discuss is commonly known as the method of steepest descent [4], and under rather liberal assumptions, it can be used to find a zero of the gradient function $\nabla f(\cdot)$. The condition $\nabla f(\hat{z}) = 0$ is, of course, the first-order necessary condition for $\hat{z}$ to minimize $f(\cdot)$. If $f(\cdot)$ is a convex function, then we may assert that $f(\hat{z}) \leq f(z)$ for all $z \in E^n$ whenever $\nabla f(\hat{z}) = 0$, that is, that $\hat{z}$ is an optimal solution.

We shall apply algorithm (2.56) to our problem. Referring to equations (2.57) and (2.58), we see that for the unconstrained optimization problem

(7) $\quad \Omega' = \{z: Rz = c, Qz \leq b\} = E^n \qquad S'' = \{h: Rh = 0\} = E^n$
$\quad I_\alpha(z) = \{0\} \qquad \text{for all } \alpha \geq 0.$

Therefore subproblem (2.59) becomes:[1]

(8) $\quad$ Minimize $\langle \nabla f(z), h \rangle$, with $h \in S'$, where S' is any compact convex subset of E^n containing the origin as an interior point. $\square$

(9) $\quad$ If we let $S' = \{h: \|h\| \leq 1\}$,

then a solution to (8) is given by

(10) $\quad \hat{h} = \begin{cases} 0 & \text{if } \nabla f(z) = 0 \\ \dfrac{-\nabla f(z)}{\|\nabla f(z)\|} & \text{otherwise}. \end{cases}$

Thus we see that algorithm (2.56) simplifies considerably when it is applied to unconstrained optimization problems. For future reference the resulting algorithm is summarized below.

(11) **Algorithm: steepest descent.** Let $z_0 \in E^n$ be arbitrary. Suppose that $z_0, z_1, \ldots, z_j$ have been computed by the algorithm and that $\nabla f(z_j) \neq 0$.

---

[1] Observe that we now return to our usual convention of denoting the cost function by $f(\cdot)$ rather than by $q^0(\cdot)$, as in Section 2.

Step 1. Set $h_j = -\nabla f(z_j)$.†

Step 2. Select $\lambda_j > 0$ to satisfy

12 $\qquad \lambda_j = \max \{\lambda : f(z_j + \lambda h_j) \leqq f(z_j + \beta h_j) \text{ for all } 0 \leqq \beta \leqq \lambda\}$

[that is, if $f(\cdot)$ is convex, λ_j is the smallest positive real root of the equation

13 $\qquad \dfrac{d}{d\lambda} f(z_j + \lambda h_j) = \langle \nabla f(z_j + \lambda h_j), h_j \rangle = 0].$

If $\lambda_j = +\infty$, then stop. If $f(z_j + \lambda h_j) \to -\infty$ as $\lambda \to \lambda_j$, then stop. Otherwise let

14 $\qquad z_{j+1} = z_j + \lambda_j h_j.$

If $\nabla f(z_{j+1}) = 0$, then stop. Otherwise set $j = j + 1$ and go to step 1. □

15 **Theorem.** Let $z_0, z_1, z_2, \ldots$ be any infinite sequence generated by algorithm (11). If $\{z_l\}$ for $l \in L$ is any subsequence converging to a point $\hat{z}$, then $\nabla f(\hat{z}) = 0$.

Proof. This theorem is an immediate consequence of exercise (2.80) and theorem (2.67). □

It should be noted that a different selection of the set S' in (9) would lead to a different "steepest-descent" algorithm.

16 **Exercise.** For each of the following sets S' find an explicit solution to (8) and modify algorithm (11) accordingly:

$$S' = \{h : |h^i| \leqq 1, i = 1, 2, \ldots, n\}$$
$$S' = \Big\{h : \sum_{i=1}^{n} |h^i| \leqq 1\Big\}.$$

Would there be any computational advantages in using the second set? □

Concerning the convergence rate of algorithm (11), consider the following exercise.

17 **Exercise.** Consider the problem of minimizing the quadratic form $f(z) = \tfrac{1}{2}\langle z, Qz\rangle$, where Q is a symmetric positive-definite $n \times n$

† Actually, to be consistent with (10), we should set $\bar{h}_i = -\nabla f(x_i)/\|\nabla f(x_i)\|$ and compute $\bar{\lambda}_j$ in (12) with $\bar{h}_j$ replacing h_j. But since $1/\|\nabla f(x_i)\|$ is a positive constant, it can be absorbed into λ_j as in (12).

matrix. Let $z_0, z_1, z_2, \ldots$ be any infinite sequence generated when algorithm (11) is applied to $f(\cdot)$. Show that if $\gamma_{\min}$ and $\gamma_{\max}$ are the smallest and largest eigenvalues of the matrix Q, then

$$f(z_{l+1}) \leq f(z_l)\left(1 - \frac{\gamma_{\min}}{\gamma_{\max}}\right).$$

[*Hint:* Find an explicit expression for λ_j, express $f(z_{j+1})$ in terms of $f(z_j)$, and use the fact that Q is positive definite.] ☐

18 **Exercise.** Find an explicit solution for (8) when $S' = \{h: \langle h, Qh \rangle \leq 1\}$, where Q is a symmetric positive-definite $n \times n$ matrix, and modify algorithm (11) accordingly. [Note that whenever $\partial^2 f(z)/\partial z^2$ is positive definite for all $z \in E^n$ and is used in place of Q above, then algorithm (11) becomes *Newton's method* with a variable step size.] ☐

Gradient Projection: A Special Case

Next we shall consider the problem of minimizing $f(z)$ subject to the constraint $Rz = c$, where $f, R,$ and c are as defined in (1). We shall again apply algorithm (2.56). Referring to equations (2.57) and (2.58), we see that in this particular case

$$\Omega' = \{z: Rz = c\} \qquad S'' = \{h: Rh = 0\}$$
$$I_\alpha(z) = \{0\} \qquad \text{for all } \alpha \geq 0,$$

so that subproblem (2.59) becomes:

19   Minimize $\langle \nabla f(z), h \rangle$ subject to the constraints
20   $Rh = 0 \qquad h \in S'$. ☐

Again we shall take

$$S' = \{h: \langle h,h \rangle \leq 1\},$$

and hence (19) becomes a convex programming problem. By theorem (3.6.1), a sufficient condition for $\hat{h}$ to be an optimal solution to problem (19) is that there exist multipliers $\lambda \in E^m$ and $\mu \leq 0$, real, such that

21   $$-\nabla f(z) + R^T \lambda + 2\mu\hat{h} = 0 \qquad R\hat{h} = 0$$
$$\langle \hat{h}, \hat{h} \rangle - 1 \leq 0 \qquad \mu(\langle \hat{h}, \hat{h} \rangle - 1) = 0$$

From (21) we have

22   $$2\mu\hat{h} = \nabla f(z) - R^T \lambda,$$

and $R\hat{h} = 0$ implies that

(23) $$RR^T\lambda = R\,\nabla f(z).$$

Since R has full row rank, by assumption (5), the $m \times m$ matrix RR^T is nonsingular. Consequently, we obtain

(24) $$\lambda = (RR^T)^{-1}R\,\nabla f(z),$$

which, when substituted into (22), yields

(25) $$2\mu\hat{h} = [I - R^T(RR^T)^{-1}R]\,\nabla f(z) \triangleq w.$$

To determine the scalar $\mu \leq 0$ we proceed as follows. If $w = 0$, let $\hat{h} = 0$ and $\mu = 0$. If $w \neq 0$, let $\mu = -\tfrac{1}{2}\|w\|$; then, by (25), $\hat{h} = -w/\|w\|$, that is, $\|\hat{h}\|^2 = 1$, and again $(\|\hat{h}\|^2 - 1)\mu = 0$. Thus, an optimal solution to problem (19) is given by

(26) $$\hat{h} = \begin{cases} 0 & \text{if } [I - R^T(RR^T)^{-1}R]\,\nabla f(z) = 0 \\ \dfrac{-[I - R^T(RR^T)^{-1}R]\,\nabla f(z)}{\|[I - R^T(RR^T)^{-1}R]\,\nabla f(z)\|} & \text{otherwise.} \end{cases}$$

(27) **Exercise.** Consider the linear subspace $\mathcal{L}$ of E^n defined by $\mathcal{L} = \{h: Rh = 0\}$. Show that the orthogonal projection of $\nabla f(z) \in E^n$ onto $\mathcal{L}$ is given by

$$h = [I - R^T(RR^T)^{-1}R]\,\nabla f(z). \quad \square$$

Note that if $\hat{h} = 0$ in (26), then, by theorem (2.23), z satisfies the first-order necessary conditions of optimality for the problem under consideration. Indeed, if $\hat{h} = 0$, then we have

(28) $$\nabla f(z) = R^T(RR^T)^{-1}R\,\nabla f(z),$$

and so if we define $\lambda = (RR^T)^{-1}R\,\nabla f(z)$, we obtain

(29) $$-\nabla f(z) + R^T\lambda = 0;$$

that is, z satisfies the Lagrange multiplier rule of theorem (3.2.6). Let us now summarize the algorithm sketched out above.

(30) **Algorithm.** Set $z_0 = R^T(RR^T)^{-1}c$. Then $Rz_0 = c$; that is, z_0 is a feasible solution. Suppose that the points $z_0, z_1, \ldots, z_j$ have been generated by the algorithm, starting with z_0 defined as above, that $Rz_i = c$ for $i = 0, 1, \ldots, j$, and that $[I - R^T(RR^T)^{-1}R]\,\nabla f(z_j) \neq 0$.

Step 1. Set $h_j = -[I - R^T(RR^T)^{-1}R]\,\nabla f(z_j).$† Note that $Rh_j = 0$.

† Here again the constant $1/\|[I - R^T(RR^T)^{-1}R]\,\nabla f(z)\|$ has been absorbed into λ_j.

Step 2. Select $\lambda_j > 0$ to satisfy

31
$$\lambda_j = \max \{\lambda : f(z_j + \lambda h_j) \leqq f(z_j + \beta h_j) \text{ for all } 0 \leqq \beta \leqq \lambda\}.$$

If $\lambda_j = +\infty$, then stop. If $f(z_j + \lambda h_j) \to -\infty$ as $\lambda \to \lambda_j$, then stop. Otherwise set $z_{j+1} = z_j + \lambda_j h_j$ and go to step 3.

Step 3. If $[I - R^T(RR^T)^{-1}R] \nabla f(z_{j+1}) = 0$, then stop. Otherwise set $j = j + 1$ and go to step 1. □

32 **Remark.** It follows from exercise (2.80) and theorem (2.67) that if $\{z_l\}$ for $l \in L$ is any convergent subsequence generated by algorithm (30) and converging to a point $\hat{z}$, then $R\hat{z} = c$ and there is a multiplier $\lambda \in E^m$ such that $-\nabla f(\hat{z}) + R^T \lambda = 0$. Obviously, when $f(\cdot)$ is convex this implies that $-\hat{z}$ is an optimal solution. □

Gradient Projection: The General Case

We now turn our attention to the most general case of problem (1), the minimization of $f(z)$ subject to the constraints

33
$$z \in \Omega' \triangleq \{z : Rz = c, Qz \leqq b\},$$

defined as in (1). The rows of R, Q, and b will be denoted by r^i, q^i, and b^i, respectively, and as usual, $I(z)$ will be the index set identifying the active inequality constraints, that is, for $z \in \Omega'$,

34
$$I(z) = \{i : \langle q^i, z \rangle - b^i = 0\}.$$

The algorithm described below is due to Rosen [2] and is called the *gradient-projection algorithm*. It is very much akin to algorithm (30) in that at every iteration a feasible direction h is computed by projecting the gradient vector $-\nabla f(z)$ onto a subspace defined by the equations $Rh = 0$ and $\langle q^i, h \rangle = 0$ for $i \in K(z)$, where $K(z)$ is a subset of $I(z)$, which differs from $I(z)$ by at most one index.

To ensure convergence, we must incorporate in the gradient-projection algorithm certain very cumbersome features which prevent jamming, or zigzagging, a phenomenon discussed in some detail in the previous section. Fortunately, it has been found that in practice jamming occurs rather rarely and that a partial antijamming precaution is usually adequate. We shall therefore consider here a version of the gradient-projection algorithm which is reasonably simple and performs very well in practice, but which cannot be proved to converge, since it uses only a partial antijamming procedure.

We begin by developing the necessary projection operators.

35 Definition. Let $z \in \Omega'$, and let $L(z) = \{i_1, i_2, \ldots, i_\alpha\}$ be a subset of $I(z)$. We define the $(m + \alpha) \times n$ matrix $S_{L(z)}$ to be the matrix whose first m rows are the rows of R (r^i for $i = 1, 2, \ldots, m$) and whose remaining rows are the q^{i_j} for $j = 1, 2, \ldots, \alpha$, with $i_j \in L(z)$. □

36 Remark. It follows from assumption (5) that $S_L(z)$ has full row rank for every $z \in \Omega'$ and for every $L(z) \subset I(z)$.† Hence the $(m + \alpha) \times (m + \alpha)$ matrix $S_{L(z)} S_{L(z)}^T$ is nonsingular. □

37 Definition. Let $z \in \Omega'$ and let $L(z) \subset I(z)$. We define the $n \times n$ projection matrix $P_{L(z)}$ to be

$$P_{L(z)} = I - S_{L(z)}^T (S_{L(z)} S_{L(z)}^T)^{-1} S_{L(z)}. \quad \square$$

38 Remark. The matrix $P_{L(z)}$ has the following properties: (a) It is symmetric and $P_{L(z)} S_{L(z)}^T = S_{L(z)} P_{L(z)} = 0$. (b) If $l \in L(z)$, then $P_{L(z)/\{l\}} P_{L(z)} = P_{L(z)} P_{L(z)/\{l\}} = P_{L(z)}$, where $L(z)/\{l\} = \{i \in L(z): i \neq l\}$. □

Now, let $z_0 \in \Omega'$ be arbitrary and let

$$h_0 = -P_{I(z)} \nabla f(z_0).$$

Then, by exercise (27), h_0 is the projection of $-\nabla f(z_0)$ onto the linear subspace $\{h: S_{I(z_0)} h = 0\}$. Since $S_{I(z_0)} h_0 = 0$, it follows that for all real $\gamma \geq 0$

39
$$R(z_0 + \gamma h_0) = c$$
$$\langle q^i, z_0 + \gamma h_0 \rangle - b^i = 0 \quad \text{for all } i \in I(z_0).$$

Now, if $i \in \bar{I}(z_0)$, the complement of $I(z_0)$ in $\{1, 2, \ldots, k\}$, then $q^i(z_0) - b^i < 0$, so that for all $\gamma > 0$ sufficiently small

40
$$\langle q^i, z_0 + \gamma h_0 \rangle - b^i \leq 0 \quad \text{for all } i \in \bar{I}(z_0),$$

and therefore h_0 is a feasible direction. Furthermore, since h_0 is the projection of $-\nabla f(z_0)$ onto a linear subspace, we have

41
$$\langle \nabla f(z_0), h_0 \rangle \leq 0,$$

with equality holding if and only if $h_0 = 0$. Thus when $h_0 \neq 0$ a displacement in the feasible direction h_0 will reduce the value of $f(\cdot)$.

Now let us consider (in anticipation that we shall need it in a partial antijamming procedure) what happens when we increase the dimension of the subspace on which $-\nabla f(z_0)$ is being projected.

† To avoid having to continually consider special cases, we shall assume that $m > 0$, that is, that R is not identically zero.

Resolving $\nabla f(z_0)$ into orthogonal components, we obtain

(42) $$\nabla f(z_0) = P_{I(z_0)} \nabla f(z_0) + S_{I(z_0)}^T \pi$$

for some vector π. Making use of the definition of $S_{I(z_0)}$, we obtain

(43) $$\nabla f(z_0) = P_{I(z_0)} \nabla f(z_0) + \sum_{i=1}^{m} \psi^i r^i + \sum_{i \in I(z_0)} \mu^i q^i,$$

where the q^i are now used as column vectors, the components of the vector π are denoted by ψ^i and μ^i, and the μ^i are assumed to be numbered with indices in $I(z_0)$. Let $l \in I(z_0)$, and let $L(z_0) = I(z_0)/\{l\}$. Then from (43) and (38) we obtain

(44) $$\begin{aligned} P_{L(z_0)} \nabla f(z_0) &= P_{L(z_0)} P_{I(z_0)} \nabla f(z_0) + \mu^l P_{L(z_0)} q^l \\ &= P_{I(z_0)} \nabla f(z_0) + \mu^l P_{L(z_0)} q^l. \end{aligned}$$

Note that

(45) $$\langle P_{I(z_0)} \nabla f(z_0), P_{L(z_0)} q^l \rangle = \langle P_{I(z_0)} \nabla f(z_0), q^l \rangle = 0,$$

and hence that

(46) $$\begin{aligned} \|P_{L(z_0)} \nabla f(z_0)\|^2 &= \|P_{I(z_0)} \nabla f(z_0)\|^2 + (\mu^l)^2 \|P_{L(z_0)} q^l\|^2 \\ &> \|P_{I(z_0)} \nabla f(z_0)\|^2 \geq 0. \end{aligned}$$

The strict inequality in (47) is due to the fact that

(47) $$P_{L(z_0)} q^l \neq 0$$

[otherwise q^l would be a linear combination of the q^i for $i \neq l$, with $i \in I(z_0)$, and r^i for $i = 1, 2, \ldots, m$, contradicting assumption (5)].

Now suppose that $\mu^l > 0$; then, from (45) to (47),

(48) $$\langle P_{L(z_0)} \nabla f(z_0), q^l \rangle > 0$$
(49) $$\langle P_{L(z_0)} \nabla f(z_0), q^i \rangle = 0 \qquad \text{for all } i \in L(z_0).$$

Thus $-P_{L(z_0)} \nabla f(z_0)$ is also a feasible direction, and the value of $f(\cdot)$, to first-order terms, is reduced along it.

Hence when $P_{I(z_0)} \nabla f(z_0) = 0$, we can find a new feasible direction, $-P_{L(z_0)} \nabla f(z_0)$, along which the value of $f(\cdot)$ will again be reduced, provided, of course, that there is a $\mu^l > 0$ in (43).

(50) **Remark.** Suppose that all the μ^i in (43) are nonpositive; then the point z_0, which is feasible, satisfies the necessary conditions of optimality (3.4.22). If $f(\cdot)$ is convex, then the sufficiency conditions (3.6.1) are also satisfied, and z_0 is an optimal solution to the (convex) programming problem (1). □

51 Remark. A procedure which chooses as its feasible direction, at $z_i \in \Omega'$, the vector $-P_{I(z_i)} \nabla f(z_i)$, as long as this vector is not zero, can often be shown to take an infinite number of iterations to converge to a point z^* such that $-P_{I(z^*)} \nabla f(z^*) = 0$, but z^* does not satisfy the necessary conditions of optimality. Thus this procedure jams. A *partial* remedy to jamming, which works very well in practice, is to choose an $\epsilon > 0$ and then proceed as follows. Given $z_i \in \Omega'$, we compute the vector π_i, from (42), to be

52
$$\pi_i = (S_{I(z_i)} S_{I(z_i)}^T)^{-1} S_{I(z_i)} \nabla f(z_0)$$

and label its components as ψ^j for $j = 1, 2, \ldots, m$ and μ^j for $j \in I(z_i)$, as was done in (43). If

53
$$\|P_{I(z_i)} \nabla f(z_i)\| \geq \{\max \epsilon \mu^j \|P_{I(z_i)/\{j\}} q^j\| : \mu^j > 0, j \in I(z_i)\},$$

we choose the feasible direction $h_i = P_{I(z_i)} \nabla f(z_i)$. Otherwise we let $l \in I(z_i)$ be the index for which the minimum in (53) takes place and let $L(z_i) = I(z_i)/\{l\}$. Then we choose the feasible direction $h_i = -P_{L(z_i)} \nabla f(z_i)$. The effect of this is to increase the dimension of the subspace on which $\nabla f(z_i)$ is projected *before* its projection becomes arbitrarily small. □

It must be emphasized once again that although the precaution indicated above has been found to be quite adequate in practice, it is still not sufficient to guarantee convergence in all possible cases.

We now summarize the preceding discussion.

54 Algorithm: gradient projection. We are given an $\epsilon > 0$. Let $z_0 \in \Omega'$ be given, and suppose that $z_1, z_2, \ldots, z_j$ in Ω' have been computed by the algorithm.

Step 1. Compute the vectors

55 $\quad P_{I(z_j)} \nabla f(z_j)$
56 $\quad \pi_j = (S_{I(z_j)} S_{I(z_j)}^T)^{-1} S_{I(z_j)} \nabla f(z_j).$

Then

57
$$\nabla f(z_j) = P_{I(z_j)} \nabla f(z_j) + S_{I(z_j)}^T \pi_j.$$

Step 2. If $P_{I(z_j)} \nabla f(z_j) = 0$ and all the components μ_j^i associated with the columns q^i of $S_{I(z_j)}^T$, with $i \in I(z_j)$, are nonpositive, then stop, since z_j satisfies the necessary conditions of optimality. If there is at least one $\mu_j^i > 0$, $i \in I(z_j)$, and

58
$$\|P_{I(z_j)} \nabla f(z_j)\| \geq \max \{\epsilon \mu_j^i \|P_{I(z_j)/\{i\}} q^i\| : \mu_j^i > 0, i \in I(z_j)\},$$

or if $P_{I(z_j)} \nabla f(z_j) \neq 0$ and $\mu_j^i \leq 0$ for all $i \in I(z_j)$, then set

59 $$h_j = -P_{I(z_j)} \nabla f(z_j)$$

and go to step 3. If there is at least one $\mu_j^i > 0$ and (58) does not hold, then set

60 $$h_j = -P_{L(z_j)} \nabla f(z_j)$$

and go to step 3. In (60), $L(z_j) = I(z_j)/\{l\}$, where l is any index for which the maximum in (58) is achieved.

Step 3. Choose $\gamma_j > 0$ such that

61 $$\gamma_j = \max \{\gamma : Q(z_j + \beta h_j) \leq b \text{ for all } 0 \leq \beta \leq \gamma\},$$

and choose λ_j to satisfy

62 $$\lambda_j = \max \{\lambda : f(z_j + \lambda h_j) \leq f(z_j + \beta h_j) \text{ for all } 0 \leq \beta \leq \lambda \leq \gamma_j\}.$$

If $\lambda_j = +\infty$, then stop. If $f(z_j + \lambda h_j) \to -\infty$ as $\lambda \to \lambda_j$, then stop. Otherwise set

63 $$z_{j+1} = z_j + \lambda_j h_j;$$

then $z_{j+1} \in \Omega'$ and $f(z_{j+1}) < f(z_j)$. Set $j = j + 1$ and return to step 1. □

64 **Remark.** One of the reasons that we do not encounter difficulties with this algorithm in practice is that we never establish with infinite precision whether a quantity is zero. This tends to make us consider as active at z_j constraints which are only "almost" active, and as we recall from the preceding section, the mechanism to avoid jamming consists to a great extent in considering as active all constraints which satisfy a condition of the form $\langle q^i, z_j \rangle - b^i + \epsilon_j \geq 0$ for initially chosen values of $\epsilon_j > 0$. □

This concludes our discussion of gradient methods. In the next section we shall discuss two procedures for combining the cost function with the constraint functions in order to produce a new family of unconstrained minimization problems whose solutions converge to a solution of the original problem.

7.4 PENALTY FUNCTIONS

Some years ago Courant [6] proposed a method for solving constrained minimization problems which has recently been revived and considerably generalized. The gist of this method consists of solving, instead of the given constrained optimization problem, a sequence of derived unconstrained optimization problems.

For an intuitive understanding of the method, suppose that we wish to minimize $f(z)$ subject to $r(z) = 0$, with $f: E^n \to E^1$ and $r: E^n \to E^m$, both continuously differentiable. If we form the sequence of auxiliary cost functions

$$f_i(z) = f(z) + \lambda_i \|r(z)\|^2 \qquad i = 1, 2, \ldots, \tag{1}$$

with $\lambda_i > 0$, $\lambda_{i+1} > \lambda_i$, and $\lambda_i \to \infty$ as $i \to \infty$, then when we minimize $f_i(z)$ at some point z_i the term $\lambda_i \|r(z_i)\|^2 > 0$ penalizes us for choosing a z_i not in the set $\{z : r(z) = 0\}$. Intuition now leads us to believe that since the penalty for z_i, the point minimizing $f_i(z)$ and not belonging to $\{z : r(z) = 0\}$, goes to infinity with i, the z_i should converge to a z^* satisfying $r(z^*) = 0$, which is an optimal solution to the original problem.

An important aspect of the method (or rather, methods) we are about to discuss is that their success depends greatly on the particular unconstrained optimization algorithm eventually chosen for solving the sequence of problems that are generated. Unfortunately, the most successful unconstrained optimization algorithms are rather complex and are mostly based on heuristic considerations. Unconstrained minimization is rather a large subject and lies somewhat outside of the scope of this book. The interested reader is encouraged to consult references [9,10,11] at the end of the chapter.

We shall now examine two representative methods for choosing penalty functions.

Exterior Penalty Functions

Suppose that we wish to solve the following problem:

Problem. Let $f: E^n \to E^1$ be a continuous function, and let Ω' be a nonempty closed subset of E^n. Find a vector $\hat{z} \in \Omega'$ such that $f(\hat{z}) \leq f(z)$ for all $z \in \Omega'$. □ (2)

Definition. A sequence $\{p_i(\cdot)\}_{i=1}^{\infty}$ of real-valued functions defined on E^n is called a *sequence of penalty functions for the set* Ω' if for every $i = 1, 2, \ldots$ the following properties hold: (3)

$p_i(\cdot)$ is continuous on E^n. (4)
$p_i(z) = 0$ if and only if $z \in \Omega'$. (5)
$p_i(z) > 0$ for every $z \notin \Omega'$. (6)
$p_{i+1}(z) > p_i(z)$ for every $z \notin \Omega'$. (7)
$p_i(z) \to +\infty$ as $i \to +\infty$, for every fixed $z \notin \Omega'$. □ (8)

We now construct our derived problem:

9 **Derived problem.** Let $i \in \{1, 2, \ldots\}$ be fixed. Find a vector $z_i \in E^n$ such that $f(z_i) + p_i(z_i) \leq f(z) + p_i(z)$ for all $z \in E^n$. □

Now let

10 $$b = \inf \{f(z) : z \in \Omega'\}$$
11 $$b_i = \inf \{f(z) + p_i(z) : z \in E^n\} \qquad i = 1, 2, \ldots.$$

We shall assume that b_1 is finite.

12 **Lemma.** The sequence $\{b_i\}_{i=1}^{\infty}$ satisfies the ordering relation
$$b_1 \leq b_2 \leq \cdots \leq b_i \leq \cdots \leq b.$$

Proof. By definition of b_i in (11),

13 $$b_i \leq f(z) + p_i(z) \qquad \text{for all } z \in E^n.$$

Using (7), we obtain
$$b_i \leq f(z) + p_i(z) \leq f(z) + p_{i+1}(z) \qquad \text{for all } z \in E^n,$$
and hence
$$b_i \leq \inf \{f(z) + p_{i+1}(z) : z \in E^n\} = b_{i+1}.$$

Now, using (5) and (13), we get
$$b_i \leq f(z) + p_i(z) = f(z) \qquad \text{for all } z \in \Omega',$$
that is,
$$b_i \leq \inf \{f(z) : z \in \Omega'\} = b,$$
and the proof is complete. □

The next lemma is the key to the main theorem.

14 **Lemma.** Let $\{p_i(\cdot)\}_{i=1}^{\infty}$ be a sequence of penalty functions for the constraint set Ω', and let $\{z_i\}_{i=1}^{\infty}$ be a sequence in E^n. If

15 $\{z_i\}_{i=1}^{\infty}$ converges to a point z^*,
16 $z_i \notin \Omega'$ for $i = 1, 2, \ldots$,
17 the sequence $\{p_i(z_i)\}_{i=1}^{\infty}$ is bounded,

then $z^* \in \Omega'$.

Proof. We shall prove this lemma by contradiction. Suppose that z^* is not in Ω', and let $M > 0$ be the bound on $p_i(z_i)$; that is, let $0 < p_i(z_i) \leq M$ for $i = 1, 2, \ldots$. Since $z^* \notin \Omega'$, and, by (8), $p_i(z^*) \to$

$+\infty$, there exists an integer N such that $p_N(z^*) > 2M$. Now, since $p_N(\cdot)$ is continuous, there exists a ball B with center z^* such that for all $z \in B$

(18) $$p_N(z) \geq \frac{3M}{2}.$$

Note that $p_N(z) = 0$ for all $z \in \Omega'$, so that $B \cap \Omega' = \phi$, the empty set. Now, $z_i \to z^*$, and hence there is an integer N' such that $z_i \in B$ for all $i \geq N'$. Let $N'' = \max\{N, N'\}$; then for all $i \geq N''$, $z_i \in B$ and, by (7) and (18),

(19) $$p_i(z_i) \geq p_N(z_i) \geq \frac{3M}{2},$$

which is a contradiction, since $p_i(z_i) \leq M$. Hence $z^* \in \Omega'$. □

(20) **Theorem.** Suppose that the derived problem (9) has an optimal solution z_i for every $i = 1, 2, \ldots$. Then any cluster point of the sequence $\{z_i\}_{i=1}^{\infty}$ is an optimal solution to problem (2).

Proof. For notational simplicity, and with no loss in generality, assume that $z_i \to z^*$. First, if for any integer i_0, $p_{i_0}(z_{i_0}) = 0$, then $z_{i_0} \in \Omega'$, and z_{i_0} also solves problem (2) [since $f(z) \equiv f(z) + p_{i_0}(z)$ for $z \in \Omega'$]. Consequently, by lemma (12), $b_i = b$ for every $i \geq i_0$, and hence $p_i(z_i) = 0$ for all $i \geq i_0$, since $p_i(z) > 0$ for all $z \notin \Omega'$. Hence $z_i \in \Omega'$ for all $i \geq i_0$, and z_i is also an optimal solution to problem (2) for every $i \geq i_0$. Since Ω' is closed, $z^* \in \Omega'$, and since $f(\cdot)$ is continuous, $f(z^*) = b$; that is, z^* is an optimal solution to problem (2).

Now suppose that $z_i \notin \Omega'$ for $i = 1, 2, \ldots$. Then, by lemma (12),

(21) $$f(z_i) + p_i(z_i) \leq b = \inf\{f(z): z \in \Omega'\}.$$

Since $z_i \notin \Omega'$, $p_i(z_i) > 0$, and hence

$$f(z_i) < b \quad i = 1, 2, \ldots.$$

Since $f(\cdot)$ is continuous and $z_i \to z^*$,

(22) $$f(z^*) \leq b.$$

Now, $f(z_i) - f(z^*) \to 0$, so that there exists a positive integer L such that

(23) $$f(z^*) - f(z_i) < 1 \quad \text{for every } i \geq L.$$

Combining (21) and (23), we obtain

(24) $$p_i(z_i) \leq b - f(z_i) \leq b + 1 - f(z^*) < +\infty$$

for every $i \geq L$.

Therefore the sequence $\{p_i(z_i)\}_{i=1}^{\infty}$ is bounded. By lemma (14), $z^* \in \Omega'$, and so, by the definition of b,

(25) $$b \leq f(z^*).$$

But $b \geq f(z^*)$, by (22), and therefore $b = f(z^*)$. Since $z^* \in \Omega'$ and $f(z^*) = b$, z^* is an optimal solution to problem (2). □

We shall now consider some examples of penalty functions which satisfy the properties stipulated in definition (3).

(26) **Lemma.** Let $q^i: E^n \to E^1$ for $i = 1, 2, \ldots, k$ be continuous functions, and let

(27) $$\Omega' \triangleq \{z : q^i(z) \leq 0, i = 1, 2, \ldots, k\}.$$

For each $i = 1, 2, \ldots$ let the map $p_i: E^n \to E^1$ be defined as

(28) $$p_i(z) = \lambda_i \sum_{i=1}^{k} [\max\{q^i(z), 0\}]^\alpha,$$

where λ_i and α are scalars satisfying $\lambda_i > 0$ and $\alpha \geq 1$. If $\lambda_{i+1} > \lambda_i$ for $i = 1, 2, \ldots$ and $\lambda_i \to +\infty$, then $\{p_i(\cdot)\}_{i=1}^{\infty}$ is a sequence of penalty functions for the set Ω'.

Proof. First note that Ω' is closed, since the functions $q^i(\cdot)$ for $i = 1, 2, \ldots, k$ are continuous. We next show that $p_i(\cdot)$ is continuous on E^n for every $i = 1, 2, \ldots$. Since the finite sum of continuous functions is itself a continuous function and the function

(29) $$[\max\{q^i(z), 0\}]^\alpha \triangleq h^i(z)$$

is continuous, $p_i(\cdot)$ is continuous.

By (29), $h^i(z) = 0$ if and only if $q^i(z) \leq 0$ and $h^i(z) > 0$ if $q^i(z) > 0$. Hence, since $\lambda_i > 0$, $p_i(z) = 0$ if and only if $z \in \Omega'$ and $p_i(z) > 0$ for all $z \notin \Omega'$. Since $\lambda_{i+1} > \lambda_i$ for $i = 1, 2, \ldots$, $p_{i+1}(z) > p_i(z)$ for every $z \notin \Omega'$, and since $\lambda_i \to +\infty$, $p_i(z) \to +\infty$ for every $z \notin \Omega'$. Therefore, by definition (3), $\{p_i(\cdot)\}_{i=1}^{\infty}$ is a sequence of penalty functions for Ω'. □

(30) **Exercise.** Show that for every $i = 1, 2, \ldots$ the function $p_i(\cdot)$ defined in (28) is continuously differentiable on E^n if the functions $q^i(\cdot)$ for $i = 1, 2, \ldots, k$ are continuously differentiable on E^n and $\alpha \geq 2$. □

31 Exercise. Let $r: E^n \to E^m$ be a continuous function on E^n, and let
$$\Omega' = \{z: r(z) = 0\}.$$
For each $i = 1, 2, \ldots$ let the map $p_i: E^n \to E^1$ be defined as

32 $\qquad p_i(z) = \lambda_i \|r(z)\|^\alpha,$

where λ_i and α are scalars, with $\lambda_i > 0$ and $\alpha \geqq 1$. Show that if $\lambda_{i+1} > \lambda_i$ for $i = 1, 2, \ldots$ and $\lambda_i \to +\infty$, then $\{p_i(\cdot)\}_{i=1}^\infty$ is a sequence of penalty functions for Ω'. Show that $p_i(\cdot)$ is continuously differentiable on E^n if $r(\cdot)$ is continuously differentiable on E^n and $\alpha \geqq 2$. □

33 Exercise. Suppose that the functions $q^i(\cdot)$ introduced in lemma (26) are convex and that the function $r(\cdot)$ defined in exercise (31) is affine. Show that for $i = 1, 2, \ldots$ the function $p_i(\cdot)$ defined in (28) and the function $p_i(\cdot)$ defined in (32) are convex. □

34 Exercise. Show that if $\{p_i^1(\cdot)\}_{i=1}^\infty$ is a sequence of penalty functions for the set Ω_1 and $\{p_i^2(\cdot)\}_{i=1}^\infty$ is a sequence of penalty functions for the set Ω_2, then $\{p_i^1(\cdot) + p_i^2(\cdot)\}_{i=1}^\infty$ is a sequence of penalty functions for $\Omega_1 \cap \Omega_2$. Also show that $\{\min\{p_i^1(\cdot), p_i^2(\cdot)\}\}_{i=1}^\infty$ is a sequence of penalty functions for $\Omega_1 \cup \Omega_2$. □

35 Remark. Penalty functions can be used to transform a constrained optimization problem not only into a sequence of unconstrained minimization problems, but also into a sequence of constrained minimization problems more amenable to solution. For example, suppose we wish to minimize $f(z)$ subject to $r(z) = 0$ and $q(z) \leqq 0$, and the the function $r(\cdot)$ is not affine. Then we cannot use any of the methods of feasible directions. If $\{p_i(\cdot)\}_{i=1}^\infty$ is a sequence of penalty functions for the set $\{z: r(z) = 0\}$, then, under suitable assumptions, we can obtain a solution of the original problem by using a feasible-directions method to solve the sequence of problems: minimize $f(z) + p_i(z)$ subject to $q(z) \leqq 0$, with $i = 1, 2, \ldots$. □

36 Exercise. Consider the problem: minimize $f(z)$ subject to $r(z) = 0$ and $q(z) \leqq 0$, where $f: E^n \to E^1$, $r: E^n \to E^m$, and $q: E^n \to E^k$ are continuously differentiable. Let $\{p_i(\cdot)\}_{i=1}^\infty$ be any sequence of penalty functions for the set $\{z: r(z) = 0\}$ satisfying definition (3), and let z_i be an optimal solution to the problem: minimize $f(z) + p_i(z)$ subject to $q(z) \leqq 0$. Show that any cluster point z^* of $\{z_i\}_{i=1}^\infty$ is an optimal solution of the original problem. □

Interior Penalty Functions

We shall now consider a different type of penalty function which is particularly useful when all the functions in question are convex. However, rather than consider a problem as general as (2), let us restrict ourselves to the following special case.

37 Problem. Let $f(\cdot)$ and $q^i(\cdot)$ for $i = 1, 2, \ldots, k$ be strictly convex functions from E^n into E^1. Minimize $f(z)$ subject to $q^i(z) \leq 0$ for $i = 1, 2, \ldots, k$. [Note that the $q^i(\cdot)$ and $f(\cdot)$ are continuous.] □

We shall make the following assumptions.

38 Assumption. The interior of the set $\Omega' = \{z: q^i(z) \leq 0, i = 1, 2, \ldots, k\}$ is not empty. □

39 Assumption. For every real α $\{z: f(z) \leq \alpha\}$ is a bounded set. □

We now define a sequence of penalty functions for Ω'.

40 Definition. Let $\{\lambda_i\}_{i=0}^{\infty}$ be a strictly monotonically decreasing sequence of positive numbers which converges to zero. We define the penalty functions $\tilde{p}_i(\cdot)$ for $i = 0, 1, 2, \ldots$ from $\mathring{\Omega}'$, the interior of $\mathring{\Omega}'$, into E^1 by

41
$$\tilde{p}_i(z) = -\lambda_i \sum_{j=1}^{k} \frac{1}{q^j(z)}.$$

[Note that for every $z \in \mathring{\Omega}'$, $\tilde{p}_i(z) \to 0$ as $i \to \infty$.]

We can now formulate the derived problem:

42 Derived problem. Find a $z_i \in \mathring{\Omega}'$ such that
$$f(z_i) + \tilde{p}_i(z_i) \leq f(z) + \tilde{p}_i(z) \quad \text{for all } z \in \mathring{\Omega}'. \quad \Box$$

43 Exercise. Show that there exists an optimal solution to problem (37), and to problem (42) for every i. [*Hint:* Let z_0 be any point in $\mathring{\Omega}'$. Then the set $\{z \in \mathring{\Omega}': f(z) + p_i(z) \leq f(z_0) + p_0(z_0)\}$ is compact.] □

44 Exercise. Show that problem (37) has a unique optimal solution. □

45 Theorem. Let z_i for $i = 0, 1, 2, \ldots$ be a solution to problem (42). Then the sequence $\{z_i\}_{i=0}^{\infty}$ converges to the optimal solution $\hat{z}$ of problem (37).

Proof. First we note that for all $z \in \dot{\Omega}'$, the interior of Ω',

(46) $$\tilde{p}_i(z) > \tilde{p}_{i+1}(z) > 0.$$

Therefore, if we let

(47) $$b_i = \min_{z \in \dot{\Omega}'} f(z) + \tilde{p}_i(z) \qquad i = 0, 1, 2, \ldots,$$

then $b_i \geqq b_{i+1}$ for $i = 0, 1, 2, \ldots$. Now we let

(48) $$b = \min_{z \in \Omega'} f(z),$$

which gives us

(49) $$b_0 \geqq b_1 \geqq \cdots \geqq b_i \geqq b_{i+1} \cdots \geqq b.$$

Since the b_i form a bounded, monotonically decreasing sequence, they must converge; that is, $b_i \to b^*$. Suppose that $b^* \neq b$; then, from (49), $b^* > b$. Since $f(\cdot)$ is continuous, there exists a ball B with center $\hat{z}$, the optimal solution to problem (37), such that for all $z \in B \cap \dot{\Omega}'$

(50) $$f(z) < b^* - \tfrac{1}{2}(b^* - b).$$

Now take any $z' \in B \cap \dot{\Omega}'$; then for i sufficiently large, since $\tilde{p}_i(z') \to 0$ as $i \to \infty$,

(51) $$\tilde{p}_i(z') < \tfrac{1}{4}(b^* - b),$$

and hence for all i sufficiently large

(52) $$f(z') + \tilde{p}_i(z') < b^* - \tfrac{1}{4}(b^* - b) < b^*,$$

which is a contradiction. Thus

(53) $$b^* = b.$$

Since $b_i = f(z_i) + \tilde{p}_i(z_i) \leqq f(z_0) + \tilde{p}_0(z_0)$, we must have that

(54) $$f(z_i) \leqq f(z_0) + \tilde{p}(z_0) \qquad \text{for all } i = 1, 2, \ldots,$$

and hence, by assumptions (38) and (39), $\{z_i\}$ is a bounded sequence. Let $\{z_j\}$, with $j \in L \subset \{0,1,2, \ldots\}$, be any convergent subsequence of $\{z_i\}_{i=0}^{\infty}$, with limit point $z^* \in \Omega'$, and suppose that $z^* \neq \hat{z}$, the optimal solution to (37). Then $f(z^*) > f(\hat{z})$, since $\hat{z}$ is the unique optimal solution of (37), and the sequence

(55) $$\{[f(z_j) - f(\hat{z})] + \tilde{p}_j(z_j)\} \qquad j \in L$$

cannot converge to zero, which contradicts the fact that $b_i - b$ converges to zero. Hence we must have $z^* = \hat{z}$. Thus all convergent subsequences of $\{z_i\}$ converge to $\hat{z}$, and hence z_i converges to $\hat{z}$. [Note also that $\tilde{p}_j(z_j) \to 0$, since $f(z_j) - f(\hat{z}) + \tilde{p}_j(z_j) \to 0$ and $z_j \to \hat{z}$.] □

To utilize penalty functions of the above type we must have an initial feasible solution in the interior of Ω' as a starting point for the unconstrained optimization algorithm used for the solution of (42). Since $f_i(z) + p_i(z) \to +\infty$ as z approaches the boundary of Ω', the unconstrained optimization algorithm will then generate a sequence of points z_{ij} for $j = 1, 2, \ldots$ which will all be in the interior of Ω'. For a detailed and comprehensive description of this method and recommended unconstrained-optimization procedures see reference [7].

This concludes our introductory discussion of algorithms for nonlinear programming problems. Bear in mind that the literature of nonlinear programming is rather vast; for a study in depth the reader is referred to any of the number of texts which deal exclusively with this subject.

REFERENCES

1. G. Zoutendijk: "Methods of Feasible Directions," Elsevier Publishing Company, Amsterdam, 1960.
2. P. Wolfe: On the Convergence of Gradient Methods Under Constraints, *IBM Res. Rept.* RC 1752, Yorktown Heights, N.Y., Jan. 24, 1967.
3. E. Polak: On the Convergence of Optimization Algorithms, *Revue Française d'Informatique et de Recherche Opérationnelle Série Rouge*, **16** (1969).
4. H. B. Curry: The Method of Steepest Descent for Non-linear Minimization Problems, *Quart. Appl. Math.*, **2**:258–261 (1944).
5. J. B. Rosen: The Gradient Projection Method for Nonlinear Programming, part I. Linear Constraints, *SIAM J. Appl. Math.*, **8**:181–217 (1960).
6. R. Courant: "Calculus of Variations and Supplementary Notes and Exercises" (mimeographed notes), Supplementary Notes by Martin Kruskal and Hanah Rubin, rev. and amended by J. Moser, New York University, New York, 1956–1957.
7. A. V. Fiacco and G. P. McCormic: The Sequential Unconstrained Minimization Technique for Nonlinear Programming: A Primal Dual Method, *Management Science*, **10**:360–364 (1964).
8. W. Zangwill: Nonlinear Programming Via Penalty Functions, *Management Science*, **13**:344–358 (1967).
9. W. C. Davidon: Variable Metric Method for Minimization, *Argonne Natl. Lab.* ANL-5990 (rev.), November, 1959.
10. R. Fletcher and M. J. D. Powell: A Rapidly Convergent Descent Method for Minimization, *Computer J.*, **6**:163–168 (1963).
11. R. Fletcher and C. M. Reeves: Function Minimization by Conjugate Gradients, *Computer J.*, **7**:149–154 (1964).
12. R. M. Van Slyke: Generalized Upper Bounding Techniques, *J. of Computer and Systems Science*, **1** (3):213–226 (1967).

8
Free-end-time optimal control problems

8.1 DESCRIPTION OF THE FREE-END-TIME PROBLEM

So far we have dealt with optimal control problems in which the initial and final times were specified. However, it is clear that there are optimal control problems for which the final time cannot be fixed in advance. The best-known class of such problems is the class of minimum-time problems, in which it is required to go from some initial state or manifold to some terminal state or manifold in a minimum time. Of course, there are many other types of problems in which the final time cannot be fixed beforehand, and we shall classify all such problems as *free-time problems*.

The free-time problem has already been stated [see (1.2.14)], but it is repeated here for the sake of completeness.

1 The free-time optimal control problem. Given a dynamical system described by the difference equation

2 $$x_{i+1} - x_i = f_i(x_i, u_i) \qquad i = 0, 1, 2, \ldots,$$

together with subsets $X_i \subset E^n$ for $i = 0, 1, 2, \ldots$, subsets $U_i \subset$

E^m for $i = 0, 1, 2, \ldots$, subsets $D_k \subset E^s$ for $k = 0, 1, 2, \ldots$, a sequence of constraint functions $h_{(k)}(\cdot, \cdot)$ mapping $E^{n(k+1)} \times E^{mk}$ into E^s for $k = 0, 1, 2, \ldots$, and a sequence of real-valued cost functions $f_{(k)}(\cdot, \cdot)$ defined on $E^{n(k+1)} \times E^{mk}$ for $k = 0, 1, 2, \ldots$, find an integer $\hat{k}$, a control sequence $\hat{\mathfrak{U}}_k = (\hat{u}_0, \hat{u}_1, \ldots, \hat{u}_{k-1})$, and a corresponding trajectory $\hat{\mathfrak{X}}_k = (\hat{x}_0, \hat{x}_1, \ldots, \hat{x}_k)$ satisfying (2), with

3      $\hat{u}_i \in U_i$ $i = 0, 1, \ldots, \hat{k} - 1$
4      $\hat{x}_i \in X_i$ $i = 0, 1, \ldots, \hat{k}$
5      $h_{(\hat{k})}(\hat{\mathfrak{X}}_{\hat{k}}, \hat{\mathfrak{U}}_{\hat{k}}) \in D_{(\hat{k})}$,

such that for every $k = 0, 1, 2, \ldots$, every control sequence $\mathfrak{U}_k = (u_0, u_1, \ldots, u_{k-1})$, and every corresponding trajectory $\mathfrak{X}_k = (x_0, x_1, \ldots, x_k)$ satisfying (2) to (5), with k taking the place of $\hat{k}$,

6      $f_{(\hat{k})}(\hat{\mathfrak{X}}_{\hat{k}}, \hat{\mathfrak{U}}_{\hat{k}}) \leq f_{(k)}(\mathfrak{X}_k, \mathfrak{U}_k)$.   □

7   **Remark.** Note that the statement of problem (1) implies that the optimal time must be finite. Optimal control problems in which the optimal terminal time is infinite are optimization problems in infinite-dimensional spaces and hence are beyond the scope of this book.   □

8.2 THE FREE-END-TIME PROBLEM AS A SEQUENCE OF FIXED-TIME PROBLEMS

Since the number of variables in the free-end-time problem is itself a variable, we recognize these problems as being of the mixed-integer programming type. Although there does not appear to be any direct way for handling such problems, they can often be solved by the simple expedient of solving a sequence of fixed-time problems. Of course, in principle every free-time problem can be solved in this manner, since, by assumption, the optimal solution corresponds to some fixed final time. The difficulty, of course, is in knowing when to stop. Generally, there is no way of establishing the optimal final time. There are, however, a wide variety of problems for which we can establish an upper bound on the optimal time. It is to this type of problem that we now direct our attention.

1   **The free-time problem with a penalty on time.** Given the free-time problem (1.1), let the cost function $f_{(k)}(\mathfrak{X}_k, \mathfrak{U}_k)$ take the form

2      $f_{(k)}(\mathfrak{X}_k, \mathfrak{U}_k) = \phi_{(k)}(\mathfrak{X}_k, \mathfrak{U}_k) + \alpha k$ $k = 0, 1, 2, \ldots,$

where

3      $\phi_{(k)}(\mathfrak{X}_k, \mathfrak{U}_k) \geq 0$ for all $\mathfrak{X}_k, \mathfrak{U}_k$

and $\alpha > 0$ is a constant which may be thought of as a penalty on time. □

Consider solving a problem of this type as a sequence of fixed-time problems. Let k_1 be the first value of k for which problem (1) has a feasible solution. Let us assume, for simplicity, that an optimal solution exists for all the fixed-time problems defined by (1) with final time $k \geqq k_1$. For every $k \geqq k_1$ let $c(k)$ be the optimal cost for the fixed-time version of this problem, with final time given by k, that is, $c(k) = f_k(\hat{\mathfrak{X}}_k, \hat{\mathfrak{U}}_{d,k})$, where $\hat{\mathfrak{X}}_k$ and $\hat{\mathfrak{U}}_k$ are the optimal trajectory and control sequence for the corresponding fixed-time problem. In particular, $c(k_1)$ is the optimal cost corresponding to the fixed end time k_1, the first time at which there is a feasible solution.

From (2) and (3) it is clear that for every k

4 $$c(k) \geqq \alpha k.$$

Hence if

5 $$k > \frac{c(k_1)}{\alpha},$$

then, from (4),

6 $$c(k) > c(k_1).$$

It follows that the right-hand side of (5) is an upper bound on any optimal final time $\hat{k}$; that is,

7 $$\hat{k} < \frac{c(k_1)}{\alpha}.$$

8 **Remark.** We must say *any* optimal final time here, since it is quite possible that the minimum cost may be attained for more than one final time. However, the strict inequality of (7) guarantees that *all* optimal final times satisfy (7). □

The implications of inequality (7) are clear in terms of our stated objective of solving the free-time problem by solving a finite sequence of fixed-time problems. We begin by finding the first final time, k_1, for which there is a feasible solution. We then solve the optimal control problem with the final time fixed at k_1 (assuming that an optimal solution exists for this time). Finally, we obtain from (7) an upper bound on the optimal final time $\hat{k}$. Consequently, we need only solve a finite number of fixed-time problems, corresponding to all final times from k_1 to the bound in (7),† in order to obtain an optimal solution by finding the least cost in this range.

† The reader should recognize the fact that it is possible to update the bound in (7) (with $k_1 + 1$, $k_1 + 2$, . . . taking the place of k_1) as successive fixed-time problems are solved.

There are many simple examples of cost functions of the type specified in (2).

9 Example. $f_{(k)}(\mathfrak{X}_k, \mathfrak{U}_k) = \sum_{i=0}^{k-1} |u_i| + \alpha k$. This is the so-called *fuel-plus-time cost function*. In the case of linear constraints, we may solve a finite set of linear programming problems to obtain an optimal solution (see Chapter 5). □

10 Example. $f_{(k)}(\mathfrak{X}_k, \mathfrak{U}_k) = \sum_{i=0}^{k-1} u_i^2 + \alpha k$. This is the *energy-plus-time cost function*. In the case of linear constraints, we may solve a finite set of quadratic programming problems to obtain an optimal solution (see Chapter 6). □

11 Remark. In general, the existence of a feasible solution for any final time k will not guarantee the existence of an optimal solution for that final time, even if there is an optimal solution for the free-time problem. Hence the definition of k_1 given above should be modified to make k_1 the first time for which there is an optimal solution. Similarly, inequality (7) should be interpreted as holding for all times $k > c_0(k_1)/\alpha$ for which optimal solutions exist. □

12 Remark. The essence of form (2) of the cost function is that it can be bounded from below by $c(k_1)$ for all k; hence the optimal cost for any fixed k must also be so bounded. Clearly, the property which is really essential for establishing a bound on $c(k)$ is that the cost be bounded from below by some strictly monotone-increasing function of k. Thus we can also consider nonlinear penalties on time, provided they have the required growth properties. □

The discussion of problems with penalties on time brings us to the following group of optimal control problems, which, while not fitting the definition of the free-time problem (1), can nevertheless be handled in the same way.

13 The bounded-final-time optimal control problem. Given the free-time optimal control problem (1.1), make the additional assumption that we are given a positive integer k_2 such that the optimal time $\hat{k}$ must satisfy $\hat{k} \leq k_2$. □

Clearly, this problem can be solved by solving a finite set of fixed-time problems for all final times less than or equal to k_2, and hence it may be grouped with the free-time problems with a penalty on time for the purpose of discussing methods of solution.

8.3 THE FIRST TIME FOR WHICH A FREE-END-TIME PROBLEM HAS A FEASIBLE SOLUTION: THE MINIMUM-TIME PROBLEM

It is clear from the discussion in Section 2 that in solving free-time problems it may be necessary to find the first time k_1 for which there is a feasible solution, that is, a control-sequence trajectory pair $(\mathfrak{U}_{k_1}, \mathfrak{X}_{k_1})$ satisfying all the constraints of the problem. This subproblem is known as the *minimum-time problem*. It may also arise independently of any other problem, and there has probably been as much written on this one problem as on all other discrete optimal control problems combined. Before proceeding with a discussion, let us formally introduce the problem.

1 The minimum-time problem. Given the free-time optimal control problem (1.1), let the cost function take the form

2   $$f_{(k)}(\mathfrak{X}_k, \mathfrak{U}_k) = k. \quad \square$$

As has been already noted, minimizing (2) is equivalent to finding the first k for which problem (1.1) has a feasible solution.

Let us now concentrate our attention on the relevant fixed-time problem. In keeping with the philosophy of this book, we first convert this fixed-time problem into the Basic Problem form (1.4.1). The details of this task are left to the reader. The resulting problem has the following form:

3   With k a given positive integer, determine a vector $\hat{z} \in E^{n_k}$ which minimizes $f_{(k)}(z) = k$ subject to the constraints $r_{(k)}(z) = 0$ and $z \in \Omega_k$, where $r_{(k)}$ is a continuously differentiable function from E^{n_k} into E^{m_k} and Ω_k is a subset of E^{n_k}.† $\quad \square$

Note that if we are to obtain a solution to the minimum-time problem (1) by solving a sequence of fixed-time problems of form (3), we must be able to solve (3), for each $k = 0, 1, 2, \ldots$, in a finite number of steps. Hence we must consider the following problem:

4   Given the constraints $r_{(k)}(z) = 0$ and $z \in \Omega_k$, defined as in problem (3), determine *in a finite number of steps* either (a) a point $\hat{z}$ satisfying $r_{(k)}(\hat{z}) = 0$ and $\hat{z} \in \Omega_k$ or (b) that no such $\hat{z}$ exists. $\quad \square$

Note that part (b) of this problem is as important as part (a), since for all k smaller than the optimal time $\hat{k}$ no feasible solution to problem (3) will exist.

Problem (4) is far too difficult to yield to any general computational technique. In order to give specific procedures it is neces-

† Clearly, any feasible solution to problem (3) is an optimal solution.

sary to make some simplifying assumptions on $r_{(k)}$ and Ω_k. The more restrictive the assumptions we make, the more effective are the procedures we can bring to bear. Let us consider a few specific cases.

Case 1. $\Omega_k = \{z: q_k(z) \leq 0\}$. Here $q_k(z)$ is assumed to be a *continuously differentiable* function, and hence problem (4) becomes a special case of the nonlinear programming problem. The techniques of Chapter 7 may now be used to attempt a solution to the equations $r_{(k)}(z) = 0$ and the inequalities $q_k(z) \leq 0$. However, since no assumptions have been made about the linearity or convexity of r and q, convergence to a feasible solution can be expected only if a starting point sufficiently close to a feasible solution is chosen. Similarly, it is not possible to establish with certainty that a feasible solution does not exist. Hence for case 1 problem (4) must still be classified as essentially unsolvable. □

Case 2. $\Omega_k = \{z: q_k(z) \leq 0\}$, $q_k: E^{n_k} \to E^{l_k}$ is continuously differentiable and convex, and $r_k(z) = R_k z - c_k$, where R_k is an $m_k \times n_k$ matrix and c_k is an m-dimensional vector. The constraints here are the same as those of the convex programming problem (7.2.1), and hence in this case problem (4) can be solved by the methods of feasible directions, described in Section 7.2. □

Case 3. $\Omega_k = \{z: Q_k z - d_k \leq 0\}$, where Q_k is an $l_k \times n_k$ matrix, d_k is an l_k vector, and $r_{(k)}(z) = R_k z - c_k$, where R_k is an $m_k \times n_k$ matrix and c_k is an m_k vector. It is clear from inspection that this case can be solved by the linear-programming techniques for finding an initial feasible solution described in Section 5.7 (see this section for details). □

5 **Remark.** The problem of finding a vector z satisfying $R_k z = c_k$ and $Q_k z \leq d_k$ in case 3 does not, in general, have a unique solution, and this presents us with the opportunity of finding a vector z which not only satisfies the above constraints, but also minimizes some auxiliary cost criterion.

For such a scheme to make sense, the auxiliary cost criterion selected must have certain features. First, the auxiliary cost criterion $f_{(k)}(\cdot)$ must be such that the following *auxiliary minimization problem*

6   Minimize $f_{(k)}(z)$ subject to $R_k z = c_k$ and $Q_k z \leq d_k$.

has an optimal solution whenever it has a feasible solution.

In addition, the cost criterion must be such that problem (6) can be treated by means of a standard finite-step algorithm either to obtain an optimal solution or to determine that an optimal solution does not exist. Note that this second property of $f_{(k)}(\cdot)$ is really needed only at the nonoptimal times $k < \hat{k}$. Thus, once we have computed the optimal time $\hat{k}$, we may change the form of the cost function, so that $f_{(\hat{k})}$ does not necessarily give rise to a problem, of form (6), which can be solved by finite-step algorithms.

In this book we have discussed only two classes of problems solvable by finite-step algorithms, linear programming problems and quadratic programming problems. Of these two, linear programming problems may not always satisfy the first of the previously stated conditions. However, if we choose an absolute-value type of cost function which is, a priori, bounded from below, then the *resultant* linear programming problem will have the required properties. □

7 Example. One cost criterion which could be used in problem (6) is

$$f_{(k)}(z) = \sum_{i=1}^{n} \alpha_k{}^i |z^i|,$$

where the $\alpha_k{}^i$ are nonnegative constants. If the $\alpha_k{}^i$ were chosen to be unity for those z^i which correspond to components of the controls u_j and zero for all other z^i, then the cost criterion here would correspond to the *minimum-fuel* criterion and the auxiliary minimization problem would select that minimum-time control sequence which used the least amount of *fuel*.

It should be pointed out that the addition of this auxiliary cost does not significantly deteriorate the efficiency of the linear-programming approach to solving the minimum-time problem compared with its use only to find a feasible solution. □

In many ways a quadratic cost criterion seems ideally suited as a choice in forming an auxiliary minimization problem. We saw in Section 6.2 that with a minimum of assumptions we can guarantee that the quadratic programming problem will have a solution whenever a feasible solution exists. Since the algorithm given in Section 6.10 is a finite-step procedure which will either find a solution or determine that no feasible solution exists, all our conditions are met. In addition, since the quadratic programming algorithm presented in Section 6.10 does not require that we begin with a feasible solution, it is possible to use this algorithm in direct competition with algorithms which find only feasible solutions.

The reader can doubtless form many quadratic cost criteria which have practical significance. Below are a number of examples of control problem–oriented criteria which, when translated into Basic Problem format, are acceptable for use in an auxiliary minimization problem.

8 **Example.** If we choose as a criterion in problem (6)

$$\sum_{i=0}^{k-1} \langle u_i, u_i \rangle,$$

we are minimizing the *control energy*. □

9 **Example.** A criterion sometimes proposed for problem (6) in the case where the controls u_i are scalars, restricted to satisfy $|u_i| \leq 1$, and the system is completely controllable, is

10
$$\sum_{i=k-n}^{k-1} u_i^2,$$

where n is the dimension of the state vector. For a completely controllable system, the last n control variables can be manipulated so as to take the system to all points in a neighborhood of the terminal state which can be reached when these control variables are set to zero. Thus the differences between the values which these controls take in a computed optimal control sequence and the maximum and minimum values that they can assume can be considered as a reserve, available at time $k - n$ for correcting minor errors or disturbances which have caused x_{k-n} to differ from its precomputed value. Minimizing (10) corresponds, in a crude way, to maximizing this reserve. □

This concludes our general discussion of how minimum-time problems can be solved by conventional methods. In the next two sections we shall consider a linear minimum-time problem for which there is a special algorithm.

8.4 A LINEAR MINIMUM-TIME PROBLEM

Of all the minimum-time problems, the following one has received the most attention in the literature:

1 **Problem.** Given a dynamical system described by

$$x_{i+1} - x_i = A x_i + b u_i \qquad i = 0, 1, 2, \ldots,$$

with $x_i \in E^n$ and $u_i \in E^1$[†] for $i = 0, 1, 2, \ldots$ and A and b constant matrices[1] of appropriate dimensions, find a control sequence $\hat{u}_0, \hat{u}_1$,

---

[†] We could assume $u_i \in E^r$, written $r > 1$, but this would only complicate the notation without shedding additional light on the methods we are to describe.
[1] Again, for the sake of simplicity, we keep the matrices A and b constant. There is no intrinsic difficulty encountered when these matrices are actually functions of the time index i.

$\hat{u}_2, \ldots$ which takes the system from the given initial state $x_0 = \xi_0$ to the desired terminal state ξ_T in a minimum number of steps, subject to the constraint $|\hat{u}_i| \leq 1$ for $i = 0, 1, 2, \ldots$.

We assume that the matrix $I + A$ is nonsingular, and that the system is completely controllable. □

We saw in Section 5.6 that such problems (with k fixed) are best transformed into the form of the Basic Problem (1.4.1) by means of the alternative procedure. Specifically, if we solve for x_k, we obtain

$$2 \qquad x_k = (I + A)^k x_0 + \sum_{i=0}^{k-1} (I + A)^{k-1-i} b u_i.$$

Substituting ξ_0 for x_0 and setting $x_k = \xi_T$, we have

$$3 \qquad \xi_T - (I + A)^k \xi_0 = \sum_{i=0}^{k-1} (I + A)^{k-1-i} b u_i.$$

Because of the assumption that the matrix $I + A$ is nonsingular, we can rewrite (3) as

$$4 \qquad (I + A)^{-k} \xi_T - \xi_0 = \sum_{i=0}^{k-1} (I + A)^{-1-i} b u_i.$$

Now, if for $k = 0, 1, \ldots$ we define

$$5 \qquad z^i = u_{i-1} \qquad i = 1, 2, \ldots, k$$

$$6 \qquad c_k = (I + A)^{-k} \xi_T - \xi_0$$

$$7 \qquad r_j = (I + A)^{-j} b \qquad j = 1, 2, \ldots, \dagger$$

then (4) becomes

$$8 \qquad c_k = \sum_{j=1}^{k} z^j r_j,$$

and the constraints on the vector $z = (z^1, z^2, \ldots, z^k)$ are seen to be

$$9 \qquad |z^j| \leq 1 \qquad j = 1, 2, \ldots, k.$$

Obviously, problem (1) can be solved by linear-programming techniques. In addition, if we wish, we may take advantage of the nature of the sets $\Omega_k = \{z \in E^k : |z^i| \leq 1\}$ by using the bounded-variable linear programming algorithm described in Section 5.11.

We shall now consider a somewhat less direct, but rather useful

† Note that $r_1, \ldots, r_n$ are linearly independent because of the controllability assumption.

SEC. 8.4 A LINEAR MINIMUM-TIME PROBLEM

method for using linear-programming techniques to solve the minimum-time problem (1). We begin by defining an error vector.

(10) $$e \triangleq \xi_T - x_k.$$

Substituting for x_k from (2), we obtain

(11) $$e = \xi_T - (I + A)^k \xi_0 + \sum_{i=0}^{k-1} (I + A)^{k-1-i} b u_i.$$

Next, for fixed k, we attempt to minimize

(12) $$\max_{i=1,\ldots,n} |e^i|$$

As already pointed out in (5.3.7), this term may be reduced to a linear criterion by introducing a new variable y subject to the constraints

(13) $$\begin{aligned} y + e^i &\geq 0 \\ y - e^i &\geq 0 \end{aligned} \quad i = 1, 2, \ldots, n.$$

Thus the problem of minimizing (12) subject to (11) and $|u_i| \leq 1$ for $i = 0, 1, \ldots, k - 1$ is found to be equivalent to:

(14) Minimize y subject to (11), (13), and $|u_i| \leq 1$ for $i = 0, 1, \ldots, k - 1$. □

Obviously, (14) is a linear programming problem.

Clearly, if k is less than the minimum time $\hat{k}$, then $\hat{y}$, the minimum value of y for (14), will be positive, and for the minimum time it will be zero. Hence, as before, we may solve the linear minimum-time problem by solving (14) and increasing the values of k until the minimum cost $\hat{y}$ is driven to zero.

The extra benefits derived from this approach become apparent when we note that (12) defines a norm in n-space. Hence for values of k less than the minimum time the solution obtained for (14) yields a control sequence which minimizes a distance between the final state x_k and the desired state ξ_T, the distance in this case being defined by (12). In certain applications it could be quite useful to know how close the system can get to the desired final state for those times for which the final state is not actually reachable.

(15) **Remark.** Another useful error criterion is

(16) $$\sum_{i=1}^{n} |e^i|.$$

We saw in Section 5.3 that problems with this criterion can be reduced to linear-programming form. Criterion (16) has the advantage over criterion (12) that in reducing the problem to linear-programming form there is no need for additional constraints such as (13). It does require the addition of n more variables, but the number of variables is not as important as the number of constraints in linear programming. □

Obviously, as was pointed out in the preceding section, we may use

17
$$\sum_{i=0}^{k-1} u_i^2$$

as our auxiliary cost criterion, in which case the conversion to Basic Problem form is especially simple.

Alternatively, we may choose to minimize

18
$$\sum_{i=1}^{n} (e^i)^2$$

at each step k. In this case the error criterion is recognized to be the squared euclidean norm, so that we are finding the reachable point closest to the desired terminal state. This results in a simple quadratic programming problem which is easily handled by the algorithm of Section 6.10, but the dimensions of the resultant problem are larger than in the case of criterion (16), which was proposed for use with the simplex method. For this reason it may be preferable to use (16).

19 **Remark.** Clearly, we could use any other positive-definite or positive-semidefinite quadratic form in e in place of (18). □

20 **Exercise.** Show that the quadratic programming problem obtained with (18) will not, in general, have a unique solution. □

8.5 A GEOMETRIC APPROACH TO THE LINEAR MINIMUM-TIME PROBLEM

In the preceding section we considered various methods for solving the linear minimum-time problem. All these methods depended upon increasing the value of the terminal time k by 1 at each iteration. In this section we shall obtain a rule, based on geometric considerations, for eliminating certain values of k as not being possible optimal times. Clearly, such a rule helps to reduce the total amount of calculation necessary to solve the minimum-time problem.

SEC. 8.5 GEOMETRIC APPROACH TO LINEAR MINIMUM-TIME PROBLEMS

Definition. Let r_i for $i = 1, 2, \ldots$ be a sequence of n-dimensional vectors, with $r_1, r_2, \ldots, r_n$ linearly independent. For each integer $k \geq 1$ we define the compact convex set $A_k \subset E^n$ by

$$A_k = \left\{ c : c = \sum_{i=1}^{k} \alpha^i r_i, |\alpha^i| \leq 1 \right\}. \quad \square$$

Using (2), we may state the linear minimum-time problem, defined in the preceding section, in the following way:

Problem. Given the sequence of vectors $r_i \in E^n$ for $i = 1, 2, \ldots$, satisfying the conditions given in (1), and a sequence of vectors $c_k \in E^n$ for $k = 1, 2, \ldots$, find the smallest positive integer $\hat{k}$ such that $c_{\hat{k}} \in A_{\hat{k}}$. $\quad \square$

Assumption. There is an integer $k \geq 1$ such that $c_k \in A_k$. $\quad \square$

Remark. For the linear minimum-time problem, with ξ_T taken as the origin, the set A_k consists of all the initial states ξ_0 from which the origin can be reached in time k or less. $\quad \square$

In principle, the following two theorems give a test for determining whether or not $c_k \in A_k$.

Separation theorem. Suppose that A_k is a closed convex set; then a vector c_k is not an element of A_k if and only if there exists a nonzero vector η such that

$$\langle \eta, c_k \rangle > \langle \eta, a \rangle \qquad \text{for all } a \in A_k. \quad \square$$

[The proof of this theorem can be found in (A.5.28).]

Since A_k is compact, (7) is equivalent to

$$\langle \eta, c_k \rangle - \max_{a \in A_k} \langle \eta, a \rangle > 0.$$

From definition (1) of the set A_k it is easy to see that

$$\max_{a \in A_k} \langle \eta, a \rangle = \max_{\substack{|\alpha_i| \leq 1 \\ 1 \leq i \leq k}} \sum_{i=1}^{k} \langle \eta, r_i \rangle \alpha_i = \sum_{i=1}^{k} |\langle \eta, r_i \rangle|.$$

Definition. Let the function $f_{(k)} : E^n \times E^n \to E^1$ be defined by

$$f_{(k)}(x, y) = \langle x, y \rangle - \sum_{i=1}^{k} |\langle x, r_i \rangle|. \quad \square$$

Making use of (8) and (9), we arrive at the following result.

12 Theorem. The vector c_k is not an element of A_k if and only if there exists a vector η such that

13
$$f_{(k)}(\eta,c_k) > 0. \quad \square$$

The computational significance of theorem (12) is best illustrated by an example.

14 Example. Consider problem (3). Suppose $n = 2$,

$$r_1 = \begin{bmatrix} 3 \\ 1 \end{bmatrix} \quad r_2 = \begin{bmatrix} -1 \\ 0 \end{bmatrix} \quad r_3 = \begin{bmatrix} 1 \\ -1 \end{bmatrix} \quad r_4 = \begin{bmatrix} -2 \\ 1 \end{bmatrix} \quad r_5 = \begin{bmatrix} -2 \\ -2 \end{bmatrix},$$

and

$$c_k = c = \begin{bmatrix} 5 \\ 4 \end{bmatrix} \quad k = 1, 2, \ldots, 5.$$

To show that $c \notin A_1$, it is sufficient to find a vector η such that $f_{(1)}(\eta,c) > 0$. When c is quite far from A_1, as in this case, it is not unreasonable to try $\eta = c$. To test this choice, let us evaluate $f_{(1)}(c,c)$. We find that

15
$$f_{(1)}(c,c) = \langle c,c \rangle - |\langle c,r_1 \rangle| = 22;$$

therefore $c \notin A_1$. We next try $k = 2$. Note that

16
$$f_{(k+1)}(\eta,c) = f_{(k)}(\eta,c) - |\langle \eta, r_{k+1} \rangle|.$$

Thus we see that $f_{(2)}(\eta,c)$ can be obtained from $f_{(1)}(\eta,c)$ by evaluating one additional scalar product, and we also find that since $f_{(1)}(c,c)$ was large, it is reasonable to ask if $\eta = c$ will also work with $f_{(2)}$. In fact, from (16), we have

17
$$f_{(2)}(c,c) = 22 - |\langle c,r_2 \rangle| = 22 - 5 = 17.$$

Continuing in this manner, we find that

18
$$\begin{aligned} f_{(3)}(c,c) &= f_{(2)}(c,c) - |\langle c,r_3 \rangle| = 17 - 1 = 16 \\ f_{(4)}(c,c) &= 16 - 6 = 10 \\ f_{(5)}(c,c) &= 10 - 18 = -8. \end{aligned}$$

Thus by evaluating only six scalar products we have determined that c does not belong to A_1, A_2, A_3, or A_4. Note that the fact that $f_{(5)}(c,c) < 0$ does not necessarily mean that $c \in A_5$. The only implication is that the hyperplane through c with outward normal c is no longer a separating hyperplane. $\quad \square$

We now propose that the first step in an algorithm for solving problem (3) is to find the smallest integer k such that $f_{(k)}(c_k,c_k) \leq 0$. If M is the integer so obtained, then $c_k \notin A_k$ for $k = 1, 2, \ldots, M - 1$.

SEC. 8.5 GEOMETRIC APPROACH TO LINEAR MINIMUM-TIME PROBLEMS

19 Remark. To summarize the above considerations, we note that the choice of $\eta = c_k$ is natural in trying to establish that $c_k \notin A_k$ by showing that $f_{(k)}(\eta, c_k) > 0$. □

Suppose that we have an integer M such that $c_k \notin A_k$ for all $k < M$ and $f_{(M)}(c_M, c_M) \leq 0$. Then c_M may belong to A_M, and we must determine whether or not the equations and inequalities

$$20 \quad \sum_{i=1}^{M} \alpha^i r_i = c_M$$
$$|\alpha_i{}^i| \leq 1 \quad i = 1, 2, \ldots, M$$

have a solution. One method for solving this problem was discussed in Section 5.11, and we shall now use a minor modification of that procedure. Thus to solve (20) we solve the following linear programming problem:

21 Auxiliary problem. Minimize $\langle \tilde{c}_M, \xi \rangle = \sum_{i=1}^{n} \tilde{c}_M{}^i \xi^i$ subject to the constraints

$$22 \quad \sum_{i=1}^{M} \alpha^i r_i + \xi = c_M$$

$$23 \quad -1 \leq \alpha^i \leq 1 \quad i = 1, 2, \ldots, M$$
$$24 \quad \tilde{c}_M{}^i \xi^i \geq 0 \quad i = 1, 2, \ldots, n,$$

where $\tilde{c}_M{}^i = c_M{}^i$ if $c_M{}^i \neq 0$ and $\tilde{c}_M{}^i = 1$ if $c_M{}^i = 0$. □

Note that $\alpha_0 = 0$ and $\xi_0 = c_M$ are a basic solution for (22) to (24). If we solve the problem (21) by means of the bounded-variable simplex algorithm (5.11.23), then an optimal solution $(\hat{\xi}, \hat{\alpha})$ will always be obtained in a finite number of steps [observe that with ξ_0 as defined above, a valid upper bound for $\tilde{c}_M{}^i \xi^i$ would be $(c_M{}^i)^2$]. If $\langle \tilde{c}_M, \hat{\xi} \rangle = 0$, then $c_M \in A_M$ and $\hat{\alpha}^i$ for $i = 1, 2, \ldots, M$ constitute a solution to problem (3). If $\langle c_M, \hat{\xi} \rangle > 0$, then $c_M \notin A_M$, and we must increase the final time k.

At this point we can observe an additional advantage in solving problem (21) by means of the simplex algorithm. Suppose $\langle \tilde{c}_M, \hat{\xi} \rangle > 0$. Since $(\hat{\xi}, \hat{\alpha})$ is an optimal solution to the linear programming problem (21), the bounded-variable simplex algorithm yields a set of Lagrange multipliers $\psi^1, \psi^2, \ldots, \psi^m$ satisfying the necessary and sufficient conditions for optimality, theorems (3.3.5) and (3.6.1), for this problem. For problem (21) these conditions assume the following form:

$$25 \quad \sum_{i=1}^{M} \hat{\alpha}^i r_i + \hat{\xi} = c_M.$$

26) $-1 \leq \hat{\alpha}^i \leq +1$ for $i = 1, 2, \ldots, M$ and $\tilde{c}_M{}^i \hat{\xi}^i \geq 0$ for $i = 1, 2, \ldots, n$.

27) For $i = 1, 2, \ldots, M$

$$\begin{aligned}\langle\psi,r_i\rangle &\leq 0 &&\text{if } \hat{\alpha}^i = -1 \\ \langle\psi,r_i\rangle &\geq 0 &&\text{if } \hat{\alpha}^i = +1 \\ \langle\psi,r_i\rangle &= 0 &&\text{if } |\hat{\alpha}^i| < 1.\end{aligned}$$

28) For $i = 1, 2, \ldots, n$

$$\begin{aligned}-\tilde{c}_M{}^i + \psi^i &\leq 0 &&\text{if } \tilde{c}_M{}^i \hat{\xi}^i = 0 \\ -\tilde{c}_M{}^i + \psi^i &= 0 &&\text{if } \tilde{c}_M{}^i \hat{\xi}^i > 0.\end{aligned}$$

Using conditions (25) to (28), we can now prove the following theorem.

29) **Theorem.** Let $(\hat{\xi},\hat{\alpha})$ be an optimal solution to problem (21), and let ψ be a vector satisfying the necessary and sufficient conditions for optimality, (25) to (28). If $\langle \tilde{c}_M,\hat{\xi}\rangle > 0$, then

$$f_{(M)}(\psi,c_M) > 0.$$

Proof. We take the scalar product of both sides of (25) with ψ to obtain

30) $$\sum_{i=1}^{M} \langle\psi,r_i\rangle\hat{\alpha}^i + \langle\psi,\hat{\xi}\rangle = \langle\psi,c_M\rangle.$$

From (27) it follows that

31) $$\langle\psi,r_i\rangle\hat{\alpha}^i = |\langle r_i,\psi\rangle| \qquad i = 1, 2, \ldots, M,$$

and from (28), since $\hat{\xi}^i = 0$ whenever $\tilde{c}_M{}^i \hat{\xi}^i = 0$ it follows that for $i = 1, 2, \ldots, n$

32) $$\begin{aligned}\psi^i \hat{\xi}^i &> 0 &&\text{if } \tilde{c}_M{}^i \hat{\xi}^i > 0 \\ \psi^i \hat{\xi}^i &= 0 &&\text{if } \tilde{c}_M{}^i \hat{\xi}^i = 0;\end{aligned}$$

that is

33) $$\langle\psi,\hat{\xi}\rangle \geq 0.$$

Since $\langle \tilde{c}_M,\hat{\xi}\rangle > 0$ implies that $\tilde{c}_M{}^i \hat{\xi}^i > 0$ for at least one $i \in \{1,2,\ldots,n\}$, it follows that

34) $$\langle\psi,\hat{\xi}\rangle > 0.$$

Combining (30), (31), and (34), we obtain

35) $$f_{(M)}(\psi,c_M) = \langle c_M,\psi\rangle - \sum_{i=1}^{M} |\langle r_i,\psi\rangle| = \langle\psi,\hat{\xi}\rangle > 0. \quad \square$$

SEC. 8.5 GEOMETRIC APPROACH TO LINEAR MINIMUM-TIME PROBLEMS

In other words, theorem (29) states that if the solution to (21) is such that $\langle \tilde{c}_M, \hat{\xi} \rangle > 0$, then any vector ψ satisfying the necessary and sufficient conditions for optimality defines a hyperplane through c_M with the property that A_M lies strictly on one side of this hyperplane.

We now combine the procedures suggested by theorems (12) and (29) into an algorithm for solving problem (3) under assumption (4), which guarantees the existence of a solution.

36 Algorithm. Let j be an index which is used to denote the iteration number.

Step 1. Find the smallest integer $M_1 \geqq 1$ such that $f_{(M_1)}(c_{M_1}, c_{M_1}) \leqq 0$, and set $j = 1$.

Step 2. Use the bounded-variable simplex algorithm to find a vector $\hat{\xi}_j \in E^n$ and a vector $\hat{\alpha}_j \in E^{M_j}$ which minimize $\langle \tilde{c}_{M_j}, \xi \rangle$ subject to the constraints

$$\sum_{i=1}^{M_j} \alpha^i r_i + \xi = c_{M_j}$$
$$-1 \leqq \alpha^i \leqq 1 \quad i = 1, 2, \ldots, M_j$$
$$\tilde{c}_{M_j}{}^i \xi^i \geqq 0 \quad i = 1, 2, \ldots, n,$$

where $\tilde{c}_{M_j}{}^i = c_{M_j}{}^i$ if $c_{M_j}{}^i \neq 0$ and $\tilde{c}_{M_j}{}^i = 1$ if $c_{M_j}{}^i = 0$. If $\langle \tilde{c}_{M_j}, \hat{\xi}_j \rangle = 0$, then $\hat{k} = M_j$ is the (minimum-time) solution to problem (3) and $\hat{u}_0 = \hat{\alpha}_j{}^1, \hat{u}_1 = \hat{\alpha}_j{}^2, \ldots, \hat{u}_{M_j-1} = \hat{\alpha}_j{}^{M_j}$ is an optimal control sequence. If $\langle \tilde{c}_{M_j}, \hat{\xi}_j \rangle > 0$, then go to step 3.

Step 3. Let ψ_j be the multiplier vector determined by the simplex algorithm according to (5.11.24). Find the smallest integer $M_{j+1} > M_j$ such that $f_{(M_{j+1})}(\psi_j, c_{M_{j+1}}) \leqq 0$. Set $j = j + 1$ and go to step 2. □

37 Remark. In carrying out step 3 considerable computation time can be saved by noting that the terms $\langle r_i, \psi \rangle$ are common to all the $f_{(k)}$ for $k \leqq i$. Therefore, if we define

38
$$g_{(k)}(\psi) = \sum_{i=1}^{k} |\langle r_i, \psi \rangle|$$

and store the value of $g_{(k)}$ at each step, we may obtain $f_{(k+1)}(\psi, c_{k+1})$ by the operations

39 $\quad g_{(k+1)}(\psi) = g_{(k)}(\psi) - |\langle r_{k+1}, \psi \rangle|$
40 $\quad f_{(k+1)}(\psi, c_{k+1}) = \langle \psi, c_{k+1} \rangle - g_{(k+1)}(\psi).$

Hence it is necessary to evaluate only two scalar products for each value of k. ☐

It should be clear that is is not possible to predict in advance how many values of k will be skipped in steps 1 and 3. This will depend in general on the nature of the vectors r_i and c_k in a particular problem. However, if the minimum-time solution corresponds to a large value of k, it is to be expected that in most cases step 1 will eliminate a large number of nonoptimal integers. In example problems it has been found that step 3 sometimes allows many values of k to be skipped, while at other times it is of no value at all. This depends very much on the initial state. Clearly, however, even in the worst case, where no steps at all can be skipped, this algorithm will not be very much slower than the ordinary linear programming procedure, since it will differ only in the fact that the functions $f_{(k)}$ must be evaluated at every step. For those cases where many steps are skipped, it could be significantly faster.

REFERENCES

1. R. E. Kalman: Optimal Nonlinear Control of Saturating Systems by Intermittent Actions, *IRE Wescon Conv. Rec.*, part IV, 1957, pp. 130–135.
2. C. A. Desoer and J. Wing: A Minimal Time Discrete System, *IRE Trans. Automatic Control*, May, 1961, pp. 111–125.
3. L. A. Zadeh and B. H. Whalen: On Optimal Control and Linear Programming, *IRE Trans. Automatic Control*, **AC-7**:45–46 (1962).
4. H. C. Torng: Optimization of Discrete Control Systems through Linear Programming, *J. Franklin Inst.*, 278(1):28–44 (1964).

appendix A
Convexity

A.1 INTRODUCTION

The purpose of this appendix is to develop, as rapidly as possible, the minimal background in the theory of convex sets and convex functions which the authors deem necessary for an understanding of the material in this book. The development given here is not intended to be complete. Accordingly, a certain number of results will be presented without proof, and the task of either constructing proofs or looking them up in the references is left to the reader.

The presentation which follows assumes an elementary working knowledge of finite-dimensional vector spaces and linear transformations. In addition, familiarity with elementary real analysis, including continuity, the ideas of open, closed, and compact sets, and the interior and closure of sets in finite-dimensional euclidean space is assumed. *Throughout this book, E^n denotes an n-dimensional, real euclidean space.*

A.2 LINES AND HYPERPLANES

1 Definition. If $x_1, x_2 \in E^n$, then the *line through x_1 and x_2* is defined to be the set of points

$$\{x: x = x_1 + \lambda(x_2 - x_1), \lambda \text{ real}\}. \quad \square$$

2   Definition. The (closed) *line segment joining* x_1 *and* x_2 is defined to be the set of points
$$\{x: x = x_1 + \lambda(x_2 - x_1),\ 0 \leq \lambda \leq 1\}.$$
We shall sometimes denote this line segment by $x_1 x_2$. □

3   Definition. If $x = (x^1, \ldots, x^n)$ and $y = (y^1, \ldots, y^n)$ are two vectors in E^n, the *scalar product of x and y*, denoted by $\langle x, y \rangle$, is defined to be
$$\langle x, y \rangle = \sum_{i=1}^{n} x^i y^i.\quad \square$$

4   Definition. The *euclidean norm* of a vector $x = (x^1, \ldots, x^n) \in E^n$, denoted by $\|x\|$, is defined to be
$$\|x\| = \sqrt{\langle x, x \rangle}.\quad \square$$

It is assumed that the reader is already familiar with this definition and with the properties of this norm.

5   Definition. If a is any nonzero vector in E^n and b is any real number, then the set of points
$$\{x: \langle a, x \rangle = b\}$$
is called an $(n-1)$-*dimensional hyperplane* in E^n (also a *linear manifold* or a *linear variety*). In the case where $b = 0$ the hyperplane is called an $(n-1)$-*dimensional subspace* of E^n. The vector a is called the *normal to the hyperplane*. □

6   Exercise. Show that an $(n-1)$-dimensional hyperplane in E^n contains exactly $(n-1)$ linearly independent vectors. □

In general, we define a k-dimensional linear manifold in the following manner.

7   Definition. Let $a_1, \ldots, a_k \in E^n$ be linearly independent vectors, and let $b \in E^n$ be arbitrary. The set
$$X = \{x: x = b + \sum_{i=1}^{k} \alpha^i a_i,\ \alpha^i \text{ real for } i = 1, \ldots, k\}$$
is called a k-*dimensional linear manifold*. If $b = 0$ (or if $b = \sum_{i=1}^{k} \alpha^i a_i$ for some α^i), then X is called a k-*dimensional subspace*. □

8   Definition. Given any hyperplane $X = \{x: \langle a,x \rangle = b\}$, the two sets given by $\{x = \langle a,x \rangle < b\}$ and $\{x: \langle a,x \rangle > b\}$ are called *open half spaces* bounded by X. The sets $\{x: \langle a,x \rangle \leq b\}$ and $\{x: \langle a,x \rangle \geq b\}$ are called *closed half spaces* bounded by X. $\square$

A.3 CONVEX SETS

1   Definition. A nonempty set X in E^n is said to be *convex* if for any two points $x_1, x_2 \in X$ the line segment joining x_1 and x_2 is contained in X, that is, if

$$\lambda x_1 + (1 - \lambda) x_2 \in X \text{ for every } \lambda \in [0,1]. \quad \square$$

Examples of convex sets are balls (open or closed), open or closed half spaces, and hyperplanes. The whole space is convex, and a set consisting of a single point is convex.

2   Remark. To avoid having to examine continually special cases, the *empty set*, denoted by ϕ in this book, is not considered to be a convex set. $\square$

3   Theorem. Let C be a family of convex sets in E^n, that is $C = \{X_w : w \in \Omega\}$. If

$$X = \bigcap_{w \in \Omega} X_w \neq \phi,$$

then X is also convex.

Proof. Let x_1 and x_2 be arbitrary points in the (nonempty) intersection X. Then $x_1, x_2 \in X_w$ for every $w \in \Omega$. Since each X_w is convex, the line segment $x_1 x_2$ joining x_1 and x_2 belongs to X_w for every $w \in \Omega$. This implies that $x_1 x_2$ belongs to X, and thus that X is convex. $\square$

4   Definition. If X_1 and X_2 are two subsets of E^n and α and β are real numbers, then the *linear combination of the sets*, denoted by $X = \alpha X_1 + \beta X_2$, is the set

$$X = \{x: x = \alpha x_1 + \beta x_2, x_1 \in X_1, x_2 \in X_2\}. \quad \square$$

5   Theorem. If X_1 and X_2 are convex sets, then $\alpha X_1 + \beta X_2$ is a convex set. $\square$

6   Exercise. Prove theorem (5). $\square$

7   Definition.   If $X_1 \subset E^n$ and $X_2 \subset E^m$, then the *direct product* of X_1 and X_2 is denoted by $X_1 \times X_2 \subset E^{n+m}$ and is the set defined by

$$X_1 \times X_2 = \{x \in E^{n+m} \colon x = (x^1, \ldots, x^n, x^{n+1}, \ldots, x^{n+m}),$$
$$(x^1, \ldots, x^n) \in X_1, (x^{n+1}, \ldots, x^{n+m}) \in X_2\}. \quad \square$$

8   Theorem.   The direct product of convex sets is a convex set. $\square$

9   Exercise.   Prove theorem (8). $\square$

On many occasions we shall need to map convex subsets from one space into another. The following material bears on this matter.

10   Definition.   A function $f \colon X \to E^m$, with $X \subset E^n$, is said to be *linear* if for all $x_1, x_2 \in X$ and for all real scalars α_1 and α_2

$$f(\alpha_1 x_1 + \alpha_2 x_2) = \alpha_1 f(x_1) + \alpha_2 f(x_2). \quad \square$$

11   Definition.   Let $\mathcal{L} \subset E^n$ be an $(n-1)$-dimensional subspace, and let η be a normal to this subspace. If X is any subset of E^n, the *projection of X onto $\mathcal{L}$* is defined to be the set

$$X' = \{z \colon z \in \mathcal{L}, z + \alpha\eta \in X \text{ for some real } \alpha\}. \quad \square$$

12   Exercise.   Show that X' as defined above is the image of X under the linear transformation

$$f(x) = x - \frac{\langle \eta, x \rangle}{\langle \eta, \eta \rangle} \eta. \quad \square$$

13   Theorem.   If $X \subset E^n$ is a convex set and $f \colon X \to E^m$ is a linear function, then $f(X) \triangleq \{y \colon y = f(x), x \in X\}$ is a convex set.

Proof.   Let y_1 and y_2 be elements of $f(X)$. Then there exist points x_1 and x_2 in X such that $y_1 = f(x_1)$ and $y_2 = f(x_2)$. Since X is convex,

$$\lambda x_1 + (1 - \lambda)x_2 \in X \qquad \text{for every } \lambda \in [0,1].$$

Since f is linear,

$$f(\lambda x_1 + (1 - \lambda)x_2) = \lambda f(x_1) + (1 - \lambda)f(x_2)$$
$$= \lambda y_1 + (1 - \lambda)y_2 \in f(X).$$

Thus $f(X)$ is convex. $\square$

14   Corollary.   The projection of a convex set onto a subspace is a convex set.

Proof. This is an immediate consequence of theorem (13) and exercise (12). □

15 Definition. A *convex combination* of a finite number of points $x_1, \ldots, x_k \in E^n$ is defined to be a point $x \in E^n$ satisfying

$$x = \sum_{i=1}^{k} \mu^i x_i,$$

where the μ^i for $i = 1, \ldots, k$ are real scalars such that

$$\mu^i \geqq 0 \quad i = 1, \ldots, k$$

and

$$\sum_{i=1}^{k} \mu^i = 1. \quad \square$$

With this definition it is now possible to give alternative characterization of a convex set.

16 Theorem. A set X in E^n is convex if and only if every convex combination of any finite number of arbitrary points of X belongs to X.

Proof. ⇒. If every convex combination of any finite number of arbitrary points of X belongs to X, then, in particular, every convex combination of any two points x_1 and x_2 in X belongs to X; that is,

17 $$\mu^1 x_1 + \mu^2 x_2 \in X$$

for all $\mu^1 \geqq 0$ and $\mu^2 \geqq 0$ such that $\mu^1 + \mu^2 = 1$. Solving for μ^2, we find that

$$\mu^2 = 1 - \mu^1,$$

and $\mu^2 \geqq 0$ implies that $0 \leqq \mu^1 \leqq 1$. Thus we may rewrite (17) as

$$\mu^1 x_1 + (1 - \mu^1) x_2 \in X \quad \text{for all } 0 \leqq \mu^1 \leqq 1.$$

Hence X is convex.

⇐. The proof in this direction proceeds by induction. The convex combination of any two points in X is in X, by the definition of convexity. Now, suppose that the statement is correct for convex combinations of k points in X. Let $x_1, \ldots, x_k, x_{k+1}$ be arbitrary points in X. Form

18 $$x = \mu^1 x_1 + \cdots + \mu^k x_k + \mu^{k+1} x_{k+1},$$

with $\mu^i \geqq 0$ for $i = 1, \ldots, k+1$ and $\sum_{i=1}^{k+1} \mu^i = 1$. If $\mu^{k+1} = 1$, then

$x = x_{k+1} \in X$, and we are done. Therefore suppose that $\mu^{k+1} < 1$. Then

$$\sum_{i=1}^{k} \mu^k = 1 - \mu^{k+1} > 0.$$

We may rewrite (18) as

19
$$x = \sum_{i=1}^{k} \mu^i \left(\frac{\mu^1}{\sum_{i=1}^{k} \mu^i} x_1 + \cdots + \frac{\mu^k}{\sum_{i=1}^{k} \mu^i} x_k \right) + \mu^{k+1} x_{k+1}.$$

Let $\lambda^j = \mu^j / \sum_{i=1}^{k} \mu^i$ and $\lambda = \sum_{i=1}^{k} \mu^i \geq 0$. Note that $\lambda^j \geq 0$ and that $\sum_{j=1}^{k} \lambda^j = 1$. Substituting in (19), we obtain

$$x = \lambda \left(\sum_{j=1}^{k} \lambda^j x_j \right) + (1 - \lambda) x_{k+1}.$$

By the induction hypothesis,

$$\sum_{j=1}^{k} \lambda^j x_j = y \in X,$$

and hence

$$x = \lambda y + (1 - \lambda) x_{k+1}.$$

By definition, $0 \leq \lambda \leq 1$, so $x \in X$. □

There are certain points in a convex set which cannot be expressed as a convex combination of other points in the set. These points are given a special name.

20 **Definition.** A point x belonging to a convex set $X \subset E^n$ is said to be an *extreme point* of X if it cannot be expressed as a convex combination of other points of X. □

The set consisting of all the convex combinations of a given finite set of points is of special interest and is given a special name.

21 **Definition.** The *convex hull* of a finite number of points $x_1, \ldots, x_k$ in E^n is defined to be the set

$$\{x : x = \sum_{i=1}^{k} \mu^i x_i, \mu^i \geq 0, \sum_{i=1}^{k} \mu^i = 1\}$$

and will be denoted by co $\{x_1, \ldots, x_k\}$. Such a set is called a *convex polyhedron* or a *convex polytope*. The extreme points of a convex polytope are called *vertices* and are a subset of the points $x_1, \ldots, x_k$. ☐

22 Definition. In the particular case where $k = n + 1$ and the vectors $x_2 - x_1, x_3 - x_1, \ldots, x_{n+1} - x_1$ are linearly independent, the convex hull of $x_1, x_2, \ldots, x_{n+1}$ is called a *simplex*. [The extreme points of a simplex are the defining vectors $x_1, \ldots, x_n$.] ☐

The condition that the vectors $x_2 - x_1, \ldots, x_{n+1} - x_1$, with $x_1, \ldots, x_n$ vertices of a simplex, be linearly independent is equivalent to requiring that any scalars $\mu^2, \mu^3, \ldots, \mu^{n+1}$ satisfying

23
$$\mu^2(x_2 - x_1) + \mu^3(x_3 - x_1) + \cdots + \mu^{n+1}(x_{n+1} - x_1) = 0$$

satisfy

$$\mu^2 = \mu^3 = \cdots = \mu^{n+1} = 0.$$

If we rewrite (23) as

24
$$-\left(\sum_{i=2}^{n+1} \mu^i\right) x_1 + \mu^2 x_2 + \cdots + \mu^{n+1} x_{n+1} = 0$$

and define

$$\mu^1 \triangleq -\sum_{i=2}^{n+1} \mu^i,$$

it becomes clear that we obtain the following condition for the linear independence of a set of vectors $x_2 - x_1, x_3 - x_1, \ldots, x_{n+1} - x_1$.

25 Theorem. The vectors $x_2 - x_1, \ldots, x_{n+1} - x_1$ are linearly independent if and only if, whenever $\mu^1, \ldots, \mu^{n+1}$ are real numbers such that

26
$$\mu^1 x_1 + \mu^2 x_2 + \cdots + \mu^{n+1} x_{n+1} = 0$$
$$\mu^1 + \mu^2 + \cdots + \mu^{n+1} = 0,$$

then

$$\mu^1 = \mu^2 = \cdots = \mu^{n+1} = 0. \quad \square$$

27 Exercise. Prove theorem (25). ☐

The following theorem and its corollary are used in proving the Fundamental Theorem (2.3.12).

28 Theorem. If S is a simplex in E^n with vertices $x_1, \ldots, x_{n+1}$, then any point in E^n can be written uniquely in the form

$$x = \sum_{i=1}^{n+1} \mu^i x_i,$$

with

$$\sum_{i=1}^{n+1} \mu^i = 1.$$

Proof. It is clear from the statement of the theorem that we wish to determine if the set of simultaneous equations in $\mu^1, \ldots, \mu^{n+1}$,

29
$$\mu^1 x_1 + \cdots + \mu^{n+1} x_{n+1} = x$$
$$\mu^1 + \cdots + \mu^{n+1} = 1,$$

has a unique solution for all x in E^n. That a solution exists and is unique follows immediately from theorem (25), since the matrices on the left-hand sides of (26) and (29) are identical. ☐

30 Corollary. If x belongs to the simplex S, then the μ^i in theorem (28) also satisfy $\mu^i \geqq 0$ for $i = 1, \ldots, n$.

Proof. This follows from the uniqueness of the μ^i and the fact that every point in S can be written as some convex combination of its vertices. ☐

31 Definition. The scalars $\mu^1, \mu^2, \ldots, \mu^n$ in theorem (38) are known as the *barycentric coordinates* of x relative to $x_1, \ldots, x_{n+1}$. ☐

So far we have defined the *convex hull* of a finite number of points. It is also possible to define the convex hull of an arbitrary subset of E^n.

32 Definition. The *convex hull* (*convex closure*) of a subset A of E^n is the intersection of all convex sets containing A. We shall denote the convex hull of A by co A. ☐

33 Remark. Note that co A is a convex set, by theorem (3), since co $A \neq \phi$. ☐

34 Exercise. Show that when A consists of a finite number of points, definitions (21) and (32) are consistent. ☐

Exercise (34) may be generalized, leading to an alternative characterization of the convex hull of an arbitrary subset of E^n.

Theorem. The convex hull of A consists of all the convex combinations of finite subsets of A; that is,

$$\operatorname{co} A = \{x: x = \sum_{i=1}^{k} \mu^i a_i, \sum_{i=1}^{k} \mu^i = 1, \mu^i \geq 0, k$$

a positive integer, $a_i \in A\}$. □

Exercise. Prove theorem (35). □

We next take up the topic of the dimension of a convex set.

Definition. The *(linear) dimension* of a convex set $X \subset E^n$ is the largest integer $k \leq n$ such that there exist $k + 1$ vectors $x_1, \ldots, x_{k+1} \in X$ with the property that $x_2 - x_1, \ldots, x_{k+1} - x_1$ are linearly independent. A convex set in E^n will be called *full dimensional* if its dimension is n. □

It is consistent with this definition to consider the dimension of a convex set consisting of a single point to be zero. Using the same convention, we consider a single point to be a zero-dimensional linear manifold. The following theorem should be interpreted in the light of these conventions.

Theorem. A k-dimensional convex set can be contained in a k-dimensional linear manifold. This manifold is uniquely determined. □

Exercise. Prove theorem (38). □

We conclude this section with certain topological properties of convex sets.

Definition. The *closure* of a set $A \subset E^n$, denoted by $\bar{A}$, is the intersection of all closed sets containing A. □

Definition. The *interior* of a set $A \subset E^n$, denoted by $\dot{A}$, is the union of all open sets contained in A. □

Remark. Definition (40) is equivalent to saying that $\bar{A}$ consists of A and all limit points of A. Definition (41) is equivalent to saying that $\dot{A}$ consists of all the points of A about which it is possible to construct a ball of some positive radius which is entirely contained in A. □

Definition. The boundary of a set $A \subset E^n$, denoted by ∂A, is defined to be the set

$$\partial A = \{x: x \in \bar{A}, x \notin \dot{A}\}. \quad \Box$$

44 Theorem. The closure of a convex set is a convex set.

Proof. Let $A \subset E^n$ be a convex set, and let $\bar{A}$ be its closure. Let $x, y \in \bar{A}$ be arbitrary points. Then, by remark (42), there exist sequences $\{x_i\}$ and $\{y_i\}$ with $x_i, y_i \in A$ for each i and with

$$x_i \to x \qquad y_i \to y.$$

Let $z = \lambda x + (1 - \lambda)y$, with $0 \leq \lambda \leq 1$. Form $z_i \triangleq x_i + (1 - \lambda)y_i$ for $i = 1, 2, \ldots$. Clearly, $z_i \in A$ for each i and $z_i \to z$; hence $z \in \bar{A}$. □

45 Theorem. The interior of a convex set is either empty or convex. □

46 Exercise. Prove theorem (45). □

47 Theorem. If X is a convex set and $\dot{X}$ is nonempty, then

$$\bar{\dot{X}} = \bar{X} \quad \text{and} \quad \dot{\bar{X}} = \dot{X}. \quad \square$$

48 Exercise. Prove theorem (47). □

49 Definition. Let X be a k-dimensional convex subset of E^n. The *relative interior* of X is defined to be the interior of X relative to the k-dimensional linear manifold $\mathcal{L}(X)$ containing it; that is, x belongs to the relative interior of X if there exists an open ball $B(x)$ about x such that $\mathcal{L}(X) \cap B(x) \subset X$. □

50 Remark. The relative interior of any convex set is always nonempty and hence convex. □

A.4 CONVEX CONES

In the development of necessary conditions for optimality in Chapter 2, convex cones play a crucial role. In this section we shall develop some of their more important properties.

1 Definition. A *cone* C in E^n is a set of points such that if $x \in C$, then $\alpha x \in C$ for every $\alpha \geq 0$. □

2 Remark. It is obvious from definition (1) that the origin is always a member of every cone. The origin is called the *vertex* of the cone. □

The definition of a cone given above requires that the cone contain the origin. However, it is often convenient to talk about sets which are *translates* of a cone.

3 Definition. A *cone C with vertex* x_0 is defined to be a set of points C such that

$$C - \{x_0\} = \{z \colon z + x_0 \in C\}$$

is a cone. ☐

It is clear that such a cone need not contain the origin.

4 Definition. A *ray*, or *half-line*, is a cone of the form $\{x \colon x = \alpha c$, where c is a fixed vector, $\alpha \geqq 0\}$. ☐

We may on occasion translate a ray in the same manner as we did a cone, and in such a case we refer to it as a *ray emanating from* x_0.

5 Definition. A cone C is called a *convex cone* if C is a convex set. ☐

6 Theorem. A cone C is a convex cone if and only if $x_1 + x_2 \in C$ whenever $x_1, x_2 \in C$.

Proof. $\Rightarrow$. Suppose that C is convex. Let $x_1, x_2 \in C$ be arbitrary. Then

$$\tfrac{1}{2} x_1 + \tfrac{1}{2} x_2 \in C,$$

by convexity, and

$$x_1 + x_2 = 2(\tfrac{1}{2} x_1 + \tfrac{1}{2} x_2) \in C$$

because C is a cone.

$\Leftarrow$. Let $y_1, y_2 \in C$ be arbitrary. Let $\lambda \in [0,1]$ also be arbitrary, and let

$$y = \lambda y_1 + (1 - \lambda) y_2.$$

Since $\lambda \in [0,1]$, $\lambda \geqq 0$, and $(1 - \lambda) \geqq 0$, $\lambda y_1 \in C$ and $(1 - \lambda) y_2 \in C$. Therefore

$$y = [\lambda y_1 + (1 - \lambda) y_2] \in C,$$

by hypothesis, and C is convex. ☐

7 Definition. The cone C *generated* by a set X is defined to be the set of points

$$C = \{y \colon y = \lambda x, \, x \in X, \, \lambda \geqq 0\}. \quad \square$$

If the set X is convex, then the following theorem holds.

8 Theorem. The cone C generated by a convex set X is a convex cone. ☐

9 Exercise. Prove theorem (8). □

In the particular case where the set X is a convex polyhedron, we make the following definition.

10 Definition. The cone generated by a convex polyhedron is called a *convex polyhedral cone*. □

11 Theorem. If $C_1, C_2 \subset E^n$ are convex cones, then the linear combination $\alpha C_1 + \beta C_2$ is also a convex cone for all real α and β.

Proof. $\alpha C_1 + \beta C_2$ is a convex set, by theorem (2.6). Therefore we need only prove that $\alpha C_1 + \beta C_2$ is a cone. Let $x \in \alpha C_1 + \beta C_2$ be arbitrary; then $x = \alpha x_1 + \beta x_2$ for some $x_1 \in C_1$ and $x_2 \in C_2$. Let $\lambda \geq 0$ be arbitrary; then

$$\lambda x = \lambda(\alpha x_1 + \beta x_2) = \alpha(\lambda x_1) + \beta(\lambda x_2).$$

Since C_1 and C_2 are cones, $\lambda x_1 \in C_1$ and $\lambda x_2 \in C_2$. Hence $\lambda x \in \alpha C_1 + \beta C_2$. □

12 Definition. A ray contained in a cone is called a *boundary ray* of the cone if every point on the ray is a boundary point of the cone. □

13 Exercise. Suppose that $\bar{x} \neq 0$ is a boundary point in a cone C with vertex x_0. Show that every point on the ray $\{x: x = \lambda(\bar{x} - x_0), \lambda \geq 0\}$ is a boundary point of C. □

14 Remark. It is clear that in E^2 a convex cone which is neither the whole space nor the singleton $\{0\}$ has exactly two boundary rays, and, except for the case where the cone is a line, the cone consists of the sector bounded by these rays.[1] It is also clear that this sector must subtend an angle less than or equal to 180° for the cone to be convex. □

15 Definition. A ray of a cone is called an *interior ray* of the cone if every point on the ray other than the vertex is an interior point of the cone. □

16 Exercise. Show that if one point on a ray, distinct from the origin, is an interior point of a cone, then every point on this ray other than the origin is an interior point of the cone. □

[1] We consider a cone consisting of a single ray to be the degenerate case where the two boundary rays coincide.

The following theorem is an essential tool in the examination of convex sets.

17 Theorem. If $K \subset E^n$ is a convex cone and $K \neq E^n$, then there exists a nonzero vector $a \in E^n$ such that $\langle a, x \rangle \leq 0$ for every $x \in K$; that is, K can be contained in a closed half space.

Proof. The proof will be by induction on the dimension of the space. The proof is trivial for $n = 1$, and the case of $n = 2$ follows immediately from remark (14).

Suppose the theorem is true in E^{n-1}, with $n - 1 \geq 2$. We shall show that it must also be true in E^n. We do this by supposing the contrary to be true. An immediate consequence of this supposition and the induction hypothesis is that the projection of K onto any $(n - 1)$-dimensional subspace $\mathcal{L}$ must be the whole subspace. This follows because if there exists an $(n - 1)$-dimensional subspace $\hat{\mathcal{L}}$ with normal $\hat{\eta}$ such that the projection $\hat{K}$ of K is not all of $\hat{\mathcal{L}}$, then, by the induction hypothesis, there exists a vector $a \in \hat{\mathcal{L}}$ such that

$$\langle a, x \rangle \leq 0 \quad \text{for every } x \in \hat{K}.$$

Now, by the definition of a projection, every $y \in K$ is of the form

$$y = x + \alpha \hat{\eta},$$

where $x \in \hat{K}$ and α is some real scalar. Furthermore,

$$\langle a, \hat{\eta} \rangle = 0,$$

because $a \in \hat{\mathcal{L}}$ and $\hat{\eta}$ is a normal to $\hat{\mathcal{L}}$. Hence

$$\langle a, y \rangle = \langle a, x \rangle + \alpha \langle z, \hat{\eta} \rangle = \langle a, x \rangle \leq 0,$$

which contradicts our supposition of the nonexistence of a containing half space.

We now show that if K cannot be contained in a half space in E^n, then the intersection of K with every two-dimensional plane cannot be contained in a half plane. To see this, suppose that there exists some two-dimensional subspace $\mathfrak{M}$ such that $K' = K \cap \mathfrak{M}$ could be contained in a half plane. Let $b \in \mathfrak{M}$ be an outward normal vector to the boundary line of the half plane, and let c be a vector in the boundary line, that is, $\langle b, c \rangle = 0$, as shown in Figure 1. If $\mathcal{L}$ is the $(n - 1)$-dimensional subspace with normal c, then b must lie in $\mathcal{L}$, and the points in E^n which project onto b come entirely from $\mathfrak{M}$. Clearly, no point of K can project onto b, and hence $\mathcal{L}$ is not covered by the projection of K. This contradicts the observation of the preceding paragraph.

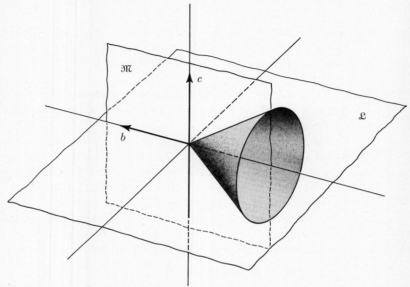

Figure 1

Now, since the theorem is true for $n = 2$, and the intersection of K with every two-dimensional plane is incapable of being contained in a half plane, it follows that the intersection of K with every two-dimensional subspace must be the whole subspace. This, of course, means that $K = E^n$, a contradiction of the original hypothesis. □

Now that we have shown the existence of at least one vector, a, with the property stated in theorem (17), it makes sense to talk about the set of all vectors having this property.

18 Definition. Let $K \subset E^n$ be a convex cone. The *polar of K* is defined to be the set

$$P(K) \triangleq \{a: \langle a,x \rangle \leq 0 \text{ for all } x \in K\}. \quad \square$$

Clearly, $0 \in P(K)$ for any convex cone K. If $K = E^n$, then $P(K) = \{0\}$. However, if $K \neq E^n$, then, by theorem (17), $P(K)$ contains at least one point other than the origin, and clearly, it also contains the ray through that point. This suggests that $P(K)$ itself may be a convex cone, which is true.

APP. A.5 SEPARATION OF SETS: SUPPORTING HYPERPLANES

19 Theorem. The polar of a convex cone is a convex cone.

Proof. Clearly, $P(K)$ is a cone, since if $a \in P(K)$, then for any $\alpha \geq 0$

$$\langle \alpha a, x \rangle = \alpha \langle a, x \rangle \leq 0 \quad \text{for all } x \in K.$$

Now let a_1 and a_2 be arbitrary elements of $P(K)$. Then

$$\langle a_1 + a_2, x \rangle = \langle a_1, x \rangle + \langle a_2, x \rangle \leq 0 \quad \text{for all } x \in K.$$

Therefore $(a_1 + a_2) \in P(K)$, and $P(K)$ is a convex cone, by theorem (6). □

A.5 SEPARATION OF SETS: SUPPORTING HYPERPLANES

The preceding sections of this appendix have introduced the fundamentals of convex sets. In this section these fundamentals are used to derive certain far-reaching properties of convex sets. The theorems which follow form the basis for many of the results contained in the body of this book. Indeed, the necessary conditions for optimality, to which approximately one-half the book is devoted, are restatements of the separation theorems of this section. The reader who has not encountered this material previously should be careful to obtain a firm grasp of the material in this section.

1 Definition. Two sets $X_1, X_2 \in E^n$ are said to be *separated* if there exists a hyperplane $\mathcal{L} = \{x : \langle a, x \rangle = b\}$ in E^n such that

$$X_1 \subset \{x : \langle a, x \rangle \leq b\}$$
$$X_2 \subset \{x : \langle a, x \rangle \geq b\}.$$

In this case the hyperplane $\mathcal{L}$ is called a *separating hyperplane*. □

Note that it is possible, as shown in Figure 1, to have two sets separated in the sense of the above definition even though their intersection is nonempty.

2 Definition. Two sets $X_1, X_2 \subset E^n$ are said to be *strictly separated* if there exists a hyperplane $\mathcal{L} = \{x : \langle a, x \rangle = b\}$ in E^n such that

$$X_1 \subset \{x : \langle a, x \rangle < b\}$$
$$X_2 \subset \{x : \langle a, x \rangle > b\}. \quad □$$

Clearly, a necessary condition for two sets to be strictly separated is that they have an empty intersection.

246 CONVEXITY

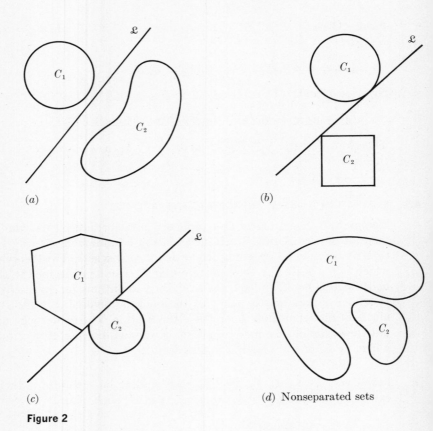

(a)

(b)

(c)

(d) Nonseparated sets

Figure 2

3 **Exercise.** Show that if $\langle a,x \rangle \leq b$ for every point x belonging to a subset X of E^n, then any interior point y of X must satisfy $\langle a,y \rangle < b$. □

A consequence of the above result is that if two sets $X_1, X_2 \subset E^n$ are separated, and these sets have nonempty interiors $\dot{X}_1$ and $\dot{X}_2$, then $\dot{X}_1$ and $\dot{X}_2$ are strictly separated.

4 **Exercise.** Show that if the relative interiors of two convex sets are separated, then the sets themselves are separated. □

The first condition we shall consider concerning the separation of sets deals with convex cones. It is this theorem which forms the basis for the necessary conditions for optimality given in theorem (2.3.12).

APP. A.5 SEPARATION OF SETS: SUPPORTING HYPERPLANES

5   **Theorem.** Consider two convex cones $K_1, K_2 \subset E^n$. If either
    a. $K_1 \cup K_2$ can be contained in an $(n-1)$-dimensional hyperplane, or
    b.   rel int $K_1 \cap$ rel int $K_2 = \phi$, where rel int K_i stands for the relative interior of K_i,
then K_1 and K_2 can be separated.

Proof. If condition (a) is satisfied, then the theorem is obviously true. Therefore, consider the case where (b) [but not necessarily (a)] holds. As observed in (3.50), rel int $K_i \neq \phi$ for $i = 1, 2$. Furthermore, it must be true that for at least one of the sets rel int K_i for $i = 1, 2$ there exists a nonzero $x \in$ rel int K_i such that $-x \notin$ rel int K_i. Otherwise we would have $0 \in$ rel int K_i for $i = 1, 2$, which would contradict (b).

Next let

$$C_i \triangleq \text{rel int } K_i \cup \{0\} \qquad i = 1, 2.$$

It is left to the reader to verify that remark (4.16) and theorem (4.6) imply that C_1 and C_2 are convex cones.

Let x be the point whose existence was shown above, and, with no loss of generality, suppose that $x \in C_1$, with $-x \notin C_1$ (clearly $x \in$ rel int K_1 and $-x \notin$ rel int K_1 imply $x \in C_1$ and $-x \notin C_1$).

Now consider the convex cone

6     $$C \triangleq C_1 - C_2.$$

That C is a convex cone follows from theorem (4.11). Suppose that $C = E^n$. Then clearly $-x \in C$, and hence

7     $$-x = x_1 - x_2 \qquad \text{for some } x_1 \in C_1, x_2 \in C_2.$$

Rearranging, we have

8     $$x_2 = x_1 + x \in C_1.$$

Since $-x \notin C_1$, we must have $x_2 \neq 0$. This implies that

9     $$x_2 \in \text{rel int } K_1 \cap \text{rel int } K_2,$$

a contradiction of (b). Therefore $C \neq E^n$. From theorem (4.17), there exists a vector $a \in E^n$ such that

10     $$\langle a, z \rangle \leq 0 \qquad \text{for all } z \in C.$$

In particular, since the origin 0 belongs to both C_1 and C_2, there are points in C of the form

$$x_1 = x_1 - 0 \qquad \text{for each } x_1 \in C_1$$

and also points of the form

11
$$-x_2 = 0 - x_2 \quad \text{for each } x_2 \in C_2.$$

Substituting into (10), we have

12 $\quad\langle a,x_1\rangle \leq 0 \quad$ for all $x_1 \in C_1$
13 $\quad\langle a,x_2\rangle \geq 0 \quad$ for all $x_2 \in C_2$.

Finally, applying exercise (4), we have

14 $\quad\langle a,x_1\rangle \leq 0 \quad$ for all $x_1 \in K_1$
15 $\quad\langle a,x_2\rangle \geq 0 \quad$ for all $x_2 \in K_2$. □

In the remainder of this section we consider some of the fundamental theorems regarding separation of sets and supporting hyperplanes to sets.

16 **Theorem.** If $C \subset E^n$ is a convex set and $x \notin C$, then there exists a hyperplane $\mathcal{L} = \{z: \langle a,z\rangle = b\}$ such that $x \in \mathcal{L}$ and $\langle a,z\rangle \leq b$ for all $z \in C$.

Proof. The set $C_1 = C - \{x\}$ is convex, since C and $\{x\}$ are convex. The origin does not belong to C_1, since $0 \in C_1$ would imply that $x \in C$. Let K be the cone generated by C_1 [see (4.7)]. K is a convex cone, by theorem (4.8). If $K = E^n$, then for any $y \neq 0$, $y \in K$ and $-y \in K$. However, by the definition of K, there must exist positive scalars α_1 snd α_2 such that $\alpha_1 y \in C_1$ and $\alpha_2(-y) \in C_1$. This, however, would mean that $0 \in C_1$, which we have already observed to be false. Hence $K \neq E^n$.

Applying theorem (4.17), we find that there exists a vector $a \in E^n$ such that

17
$$\langle a,z\rangle \leq 0 \quad \text{for every } z \in K.$$

By the definition of K, $C_1 \subset K$, and hence

18
$$z = c - x \in K \quad \text{for every } c \in C.$$

Thus

19
$$\langle a,z\rangle = \langle a,c\rangle - \langle a,x\rangle \leq 0 \quad \text{for every } c \in C.$$

Clearly, the hyperplane

20
$$\mathcal{L} = \{y: \langle a,y\rangle = \langle a,x\rangle\}$$

is the required hyperplane. □

Using theorem (16), we can prove the following theorem on the separation of convex sets.

21 Theorem. If $C_1, C_2 \subset E^n$ are disjoint convex sets, that is, $C_1 \cap C_2 = \phi$, then C_1 and C_2 are separated.

Proof. Let $C = C_1 - C_2$. Clearly, $0 \notin C$, since $0 \in C$ would imply that $x_1 = x_2$ for some $x_1 \in C_1$ and $x_2 \in C_2$, which is a contradiction. Therefore, by theorem (16), there exists a hyperplane

$$\mathcal{L} = \{z : \langle a, z \rangle = \langle a, 0 \rangle = 0\}$$

such that $\langle a, x \rangle \leq 0$ for every $x \in C$. Fix $\bar{x}_2 \in C_2$. Clearly, $\langle a, x_1 \rangle \leq \langle a, \bar{x}_2 \rangle$ for every $x_1 \in C_1$, and therefore

22 $$b \triangleq \sup_{x_1 \in C_1} \langle a, x_1 \rangle$$

must be finite. Obviously, then,

23 $$\langle a, x_1 \rangle \leq b \quad \text{for every } x_1 \in C_1.$$

We shall show by contradiction that $\langle a, x_2 \rangle \geq b$ for every $x_2 \in C_2$. Suppose that for some $x_2 \in C_2$

24 $$\langle a, x_2 \rangle < b.$$

Then there exists an $x_1 \in C_1$ such that

25 $$\langle a, x_2 \rangle < \langle a, x_1 \rangle,$$

or

26 $$\langle a, x_1 - x_2 \rangle > 0.$$

But this contradicts the fact that $\langle a, x \rangle \leq 0$ for every $x \in C$, and therefore $\mathcal{L}_1 = \{z : \langle a, z \rangle = b\}$ is the required hyperplane. $\square$

27 Exercise. Show that if C_1 is a convex cone in theorem (21), then we may always choose $b = 0$. $\square$

28 Corollary. If $C \subset E^n$ is a closed convex set and $\bar{x} \notin C$, then there exists a hyperplane $\{x : \langle a, x \rangle = b\}$ such that

$$\langle a, \bar{x} \rangle > b$$
$$\langle a, y \rangle \leq b \quad \text{for all } y \in C.$$

Proof. Since $\bar{x}$ belongs to the open complement of C, there exists a ball S of radius $r > 0$ about $\bar{x}$ such that $S \cap C = \phi$. Since a ball is convex, theorem (21) may be applied to obtain a hyperplane $\{z : \langle a, z \rangle = b\}$ such that

$$\langle a, z \rangle \geq b \quad \text{for every } z \in S$$
$$\langle a, z \rangle \leq b \quad \text{for every } z \in C.$$

However, $\bar{x}$ is an interior point of S, and hence by exercise (3), we must have $\langle a,\bar{x}\rangle > b$. □

29 **Remark.** Clearly, corollary (28) states that $\bar{x}$ and C are strictly separated, since we could consider instead the hyperplane $\{x: \langle a,x\rangle = \frac{1}{2}[b + \langle a,\bar{x}\rangle]\}$. □

In (4.18) the concept of the *polar* of a convex cone was introduced. In the following theorem and corollary we prove a very interesting and useful property of polars.

30 **Theorem.** If $C \subset E^n$ is a closed convex cone, then the polar of $P(C)$ is C itself; that is, $P(P(C)) = C$.

Proof. Since $P(C)$ is itself a convex cone, as was demonstrated in (4.19), $P(P(C))$ is well defined. Clearly, $C \subset P(P(C))$, so we need only show that $P(P(C)) \subset C$. Let $a \in P(P(C))$ be arbitrary, and suppose that $a \notin C$. Then, by corollary (28) and exercise (27), there exists an $\eta \in E^n$ such that

31 $\qquad \langle \eta,a\rangle > 0$
32 $\qquad \langle \eta,x\rangle \leq 0 \qquad$ for all $x \in C$.

By (32) and the definition of $P(C)$, $\eta \in P(C)$. Therefore, by the definition of $P(P(C))$ and the fact that $a \in P(P(C))$, we must have

33 $\qquad \langle a,\eta\rangle \leq 0,$

which contradicts (31). Therefore $a \in C$ and $C = P(P(C))$. □

The following corollary, commonly known as *Farkas' lemma*, is one of the basic tools in deriving necessary conditions for optimality for nonlinear programming problems.

34 **Corollary: Farkas' lemma.** If $a_1, \ldots, a_k$ and b are a finite set of vectors in E^n, then

35 $\qquad \langle b,x\rangle \leq 0$

for all $x \in E^n$ satisfying

36 $\qquad \langle a_i,x\rangle \leq 0 \qquad i = 1, \ldots, k$

if and only if

$$b = \sum_{i=1}^{k} \mu^i a_i,$$

with $\mu^i \geq 0$ for $i = 1, \ldots, k$.

Proof. $\Leftarrow$. The proof in this direction is trivial.

$\Rightarrow$. The theorem is also trivial if $a_i = 0$ for $i = 1, \ldots, k$, if $b = 0$, or if $k = 0$, and so we may consider the case where none of these occur. Let K be the convex polyhedral cone generated by $a_1, \ldots, a_k$, that is, the cone generated by the convex hull of $a_1, \ldots, a_k$. Clearly, K is closed. It is easily verified that the polar of K is given by

37 $$P(K) = \{x: \langle a_i, x \rangle \leq 0, i = 1, \ldots, k\}.$$

The hypothesis implies that $b \in P(P(K))$. Therefore, by theorem (30),

38 $$b \in K,$$

which is the desired result, since every element of K may be written in the form $\sum_{i=1}^{k} \mu^i a_i$ for some $\mu^i \geq 0$, with $i = 1, \ldots, k$. □

39 **Remark.** If we change (36) in the statement of corollary (34) to include $\langle a_i, x \rangle = 0$ for some $i \in \{1, \ldots, k\}$, then in the construction of b the corresponding μ^i have no specified sign. This is immediately apparent if the equality $\langle a_i, x \rangle = 0$ is rewritten as the two inequalities $\langle a_i, x \rangle \leq 0$ and $\langle -a_i, x \rangle \leq 0$. □

40 **Exercise.** Show that if $C \subset E^n$ is a closed convex cone with the property that for all nonzero vectors $x \in C$, $-x \notin C$, then there exists a vector $a \in E^n$ such that $\langle a, x \rangle < 0$ for all nonzero $x \in C$. [*Hint:* It is easy to show that any vector a in the *interior* of the polar of C must have the above property. Therefore, show that $P(C)$ has a nonempty interior by deducing that otherwise $P(P(C)) \neq C$. The interested reader may attempt to prove this result without resorting to the material in this section.] □

We conclude this introduction to convex sets with the following results on supporting hyperplanes.

41 **Definition.** A *supporting hyperplane* to a set $X \subset E^n$ at a point $\bar{x} \in \bar{X}$, the closure of X, is a hyperplane $\mathcal{L} = \{x: \langle a, x \rangle = b\}$ such that $\bar{x} \in \mathcal{L}$ and $\langle a, x \rangle \leq b$ for all $x \in X$. □

42 **Theorem.** If $C \subset E^n$ is a convex set, then there exists a supporting hyperplane to C at every boundary point of C.

Proof. If the dimension of C is less than n, then the theorem follows immediately. Therefore we suppose that C is of dimension n. It follows that C has a nonempty interior $\dot{C}$. By theorem (3.45), $\dot{C}$ is also convex. Let x be any boundary point of C. Then, since $x \notin \dot{C}$, we may use theorem (16) to obtain a hyperplane $\mathcal{L} = \{z: \langle a,x \rangle = b\}$, with the property that $\langle a,x \rangle = b$ and $\langle a,z \rangle \leq b$ for all $z \in \dot{C}$. By theorem (3.47) and exercise (4), we have that $\langle a,x \rangle \leq b$ for all $z \in C$. □

A.6 CONVEX FUNCTIONS

It is natural to study convex and concave functions in conjunction with convex sets. In our case they are of additional importance because many of the necessary conditions for optimality, developed in Chapter 3, become sufficient conditions, and because, in optimization, simplifications always result when some of the relevant functions are convex or concave.

1   **Definition.** A function $f: X \to E^1$, with X a convex subset of E^n, is said to be *convex* if for any two points x_1 and x_2 in X and any real λ, $0 \leq \lambda \leq 1$, we have

2
$$f(\lambda x_1 + (1 - \lambda)x_2) \leq \lambda f(x_1) + (1 - \lambda)f(x_2).$$

If strict inequality holds in (2) for all $x_1 \neq x_2$ in X and for all $0 < \lambda < 1$, then $f(\cdot)$ is said to be *strictly convex*. □

3   **Definition.** A function $f: X \to E^1$, with X a convex subset of E^n, is called *concave (strictly concave)* if $-f(\cdot)$ is convex (strictly convex). □

4   **Theorem.** A function $f(\cdot)$ from a convex set X into E^1 is convex if and only if for every finite number of points $x_1, \ldots, x_k$ in X

$$f\left(\sum_{i=1}^{k} \mu^i x_i\right) \leq \sum_{i=1}^{k} \mu^i f(x_i)$$

whenever $\sum_{i=1}^{k} \mu^i = 1$, with $\mu^i \geq 0$ and $i = 1, \ldots, k$. □

5   **Exercise.** Prove theorem (4). [*Hint:* Proceed as in the proof of Theorem (3.16).] □

6   **Theorem.** Let $X \subset E^n$ be a convex set. If the functions $f^i: X \to E^1$ for $i = 1, \ldots, k$ are convex and if α^i for $i = 1, \ldots, k$ are real scalars satisfying $\alpha^i \geq 0$, then the function

$$f(x) \triangleq \sum_{i=1}^{k} \alpha^i f^i(x)$$

is convex.

Proof. Let $x_1, x_2 \in X$ be arbitrary and let $\lambda \in [0,1]$ also be arbitrary. Since each $f^i(\cdot)$ is convex and $\alpha^i \geqq 0$,

$$f(\lambda x_1 + (1 - \lambda)x_2) = \sum_{i=1}^{k} \alpha^i f^i(\lambda x_1 + (1 - \lambda)x_2)$$
$$\leqq \sum_{i=1}^{k} \alpha^i [\lambda f^i(x_1) + (1 - \lambda)f^i(x_2)].$$

This inequality may be rewritten in the form

$$f(\lambda x_1 + (1 - \lambda)x_2) \leqq \lambda \sum_{i=1}^{k} \alpha^i f^i(x_1) + (1 - \lambda) \sum_{i=1}^{k} \alpha^i f^i(x_2)$$
$$= \lambda f(x_1) + (1 - \lambda) f(x_2).$$

Hence f is convex. □

9 Theorem. Let $f(\cdot)$ be a convex function from a convex set $X \subset E^n$ into E^1; then for all real α the set $\{x: f(x) \leqq \alpha\}$ is either empty or convex. □

10 Exercise. Prove theorem (9). □

Functions which have the property stipulated in theorem (9) are given a special name.

11 Definition. A function $f: X \subset E^n \to E^1$, with X convex, is said to be *quasi-convex on* X if for any real α the set $\{x: f(x) \leqq \alpha\}$ is either empty or convex. □

12 Remark. The function $f(x) \triangleq -(1 - e^{-|x|})$ mapping E^1 into E^1 is quasi-convex on E^1, but it is not convex. □

As we shall see, if a function is twice continuously differentiable, there is a relatively simple test to determine if it is convex. Before imposing any differentiability conditions, however, we prove an important property of convex functions.

13 Theorem. Let $f(\cdot)$ be a convex function from an *open* convex set $X \subset E^n$ into E^1; then f is continuous on X.

Proof. Let $x_0 \in X$ be arbitrary. Since X is open, we can construct a simplex [see (3.22)] $S \subset X$ with vertices $x_1, \ldots, x_{n+1}$ in X such that x_0 is an interior point of S. By corollary (3.30), every point in S can be written uniquely in the form

$$x = \sum_{i=1}^{n+1} \mu^i x_i,$$

where $\mu^i \geq 0$ and $\sum_{i=1}^{n+1} \mu^i = 1$. Let α be any real scalar satisfying

(15) $$\alpha > \max\{f(x_1), \ldots, f(x_{n+1})\}.$$

Since f is convex, it follows from theorem (4) that for every $x \in S \subset X$

$$f(x) = f\left(\sum_{i=1}^{n+1} \mu^i x_i\right) \leq \sum_{i=1}^{n+1} \mu^i f(x_i) < \alpha \sum_{i=1}^{n+1} \mu^i = \alpha.$$

Therefore f is bounded from above on S.

Let $B_\delta(x_0)$ be a ball centered at x_0 with radius $\delta > 0$ such that $B_\delta(x_0) \subset S$. It is possible to construct such a ball because x_0 is an interior point of S. Let $x \in B_\delta(x_0)$ be such that $0 < \|x - x_0\| < \delta$, and define the vector w to be

$$w = \frac{\delta(x - x_0)}{\|x - x_0\|}.$$

Clearly, the points $x_0 \pm w$ belong to the boundary of the ball $B_\delta(x_0)$, while the points x and x_0 lie on the line segment $(x_0 - w)(x_0 + w)$. Indeed, it is easily verified that

(16) $$x = \lambda(x_0 + w) + (1 - \lambda)x_0$$

(17) $$x_0 = \frac{x}{1+\lambda} + \frac{\lambda}{1+\lambda}(x_0 - w),$$

where

(18) $$\lambda \triangleq \frac{\|x - x_0\|}{\delta} \leq 1.$$

Since $f(\cdot)$ is convex on X and $B_\delta(x_0) \subset S$, we have

(19) $$f(x) \leq \lambda f(x_0 + w) + (1 - \lambda)f(x_0) \leq \lambda\alpha + (1-\lambda)f(x_0),$$

(20) $$f(x_0) \leq \frac{1}{1+\lambda}f(x) + \frac{\lambda}{1+\lambda}f(x_0 - w) \leq \frac{1}{1+\lambda}f(x) + \frac{\lambda\alpha}{1+\lambda},$$

where α is as defined in (15). Rearranging these inequalities, we obtain

(21) $$f(x) - f(x_0) \leq \lambda(\alpha - f(x_0))$$
(22) $$f(x) - f(x_0) \geq -\lambda(\alpha - f(x_0)),$$

which, because of the definition of λ in (18), is equivalent to

(23) $$|f(x) - f(x_0)| \leq \frac{\alpha - f(x_0)}{\delta}\|x - x_0\|.$$

Hence if $\epsilon > 0$ is given and $\|x - x_0\| \leq \min\{\delta, \delta\epsilon/[\alpha - f(x_0)]\}$, then

$|f(x) - f(x_0)| \leq \epsilon$. Therefore $f(\cdot)$ is continuous at x_0. Since x_0 was arbitrary, $f(\cdot)$ is continuous on X. □

24 Example. Let $X = \{x: 0 \leq x \leq 1\} \subset E^1$, let $f(x) = x$ if $0 \leq x < 1$, and let $f(1) = 2$. Then $f(\cdot)$ is convex on the closed convex set X, but $f(\cdot)$ is not continuous at 1. □

We now turn to properties of differentiable convex functions.

25 Theorem. Let $f(\cdot)$ be a continuously differentiable function from an open convex set $X \subset E^n$ into E^1. Then $f(\cdot)$ is convex if and only if

$$26 \quad f(x) \geq f(x_0) + \langle \nabla f(x_0), x - x_0 \rangle,$$

where

$$\nabla f(x_0) = \left[\frac{\partial f(x_0)}{\partial x^1}, \ldots, \frac{\partial f(x_0)}{\partial x^n} \right]^T,$$

for every $x, x_0 \in X$.

Proof. ⇒. Let $x_0, x \in X$ be arbitrary, let $\lambda \in (0,1)$, and let $h = (x - x_0)$. Then $x_0 + \lambda h \in X$, since X is convex, and since $f(\cdot)$ is convex on X,

$$27 \quad f(x_0 + \lambda h) \leq \lambda f(x) + (1 - \lambda) f(x_0).$$

Subtracting $\langle \nabla f(x_0), h \rangle$ from both sides of (27) and rewriting the inequality, we obtain

$$28 \quad \frac{f(x_0 + \lambda h) - f(x_0) - \lambda \langle \nabla f(x_0), h \rangle}{\lambda} \leq f(x) - f(x_0) - \langle \nabla f(x_0), h \rangle.$$

Since $f(\cdot)$ is differentiable at x_0, the left-hand side of (28) converges to zero as $\lambda \to 0^+$. Thus the right-hand side is nonnegative, which is equivalent to (26).

⇐. Suppose that (26) holds for all $x_0, x \in X$. Let $x_1, x_2 \in X$, with $x_1 \neq x_2$, and let $0 < \lambda < 1$. Set

$$29 \quad x_0 = \lambda x_1 + (1 - \lambda) x_2 \qquad h = (x_1 - x_0).$$

It is easily verified that

$$30 \quad x_2 = x_0 - \frac{\lambda}{1 - \lambda} h.$$

By (26),

$$31 \quad f(x_1) \geq f(x_0) + \langle \nabla f(x_0), h \rangle$$

$$32 \quad f(x_2) \geq f(x_0) + \langle \nabla f(x_0), h \rangle \frac{-\lambda}{1 - \lambda}.$$

Multiplying (31) by $\lambda/(1 - \lambda)$ and adding it to (32), we obtain

$$\frac{\lambda}{1 - \lambda} f(x_1) + f(x_2) \geq \left(\frac{\lambda}{1 - \lambda} + 1\right) f(x_0),$$

or

33 $\qquad \lambda f(x_1) + (1 - \lambda) f(x_2) \geq f(x_0).$

Thus we have shown that

34 $\qquad f(\lambda x_1 + (1 - \lambda) x_2) \leq \lambda f(x_1) + (1 - \lambda) f(x_2)$
$\hfill$ for all $0 < \lambda < 1$.

If $\lambda = 0$ or $\lambda = 1$, the inequality holds trivially. Therefore $f(\cdot)$ is convex. $\square$

35 **Theorem.** Let $f(\cdot)$ be a continuously differentiable function from an open convex set $X \subset E^n$ into E^1. Then $f(\cdot)$ is strictly convex if and only if

$$f(x) > f(x_0) + \langle \nabla f(x_0), x - x_0 \rangle$$

for every $x, x_0 \in X$, with $x \neq x_0$.

36 **Exercise.** Prove theorem (35). $\square$

The following results are useful in Chapters 3 and 7.

37 **Theorem.** Let $f(\cdot)$ be a convex differentiable function from an open convex set $X \subset E^n$ into E^1, and let x_0 be any point in X. Then $f(x_0) \leq f(x)$ for every $x \in X$ if and only if $\nabla f(x_0) = 0$.

Proof. $\Rightarrow$. This part of the proof follows directly from elementary calculus.
$\quad \Leftarrow$. Suppose that $\nabla f(x_0) = 0$. Since $x_0 \in X$, $f(x) \geq f(x_0)$ for all $x \in X$, by theorem (25). $\square$

38 **Theorem.** If $f(\cdot)$ is a strictly convex function from a convex set $X \subset E^n$ into E^1, then there is at most one point $x_0 \in X$ such that $f(x_0) \leq f(x)$ for every $x \in X$.

Proof. Suppose that $x_0 \neq x_1$ in X satisfy $f(x_i) \leq f(x)$ for every $x \in X$ and for $i = 0, 1$. Then, obviously, $f(x_0) = f(x_1)$. Since X is convex, $x_0 + \lambda(x_1 - x_0) \in X$ for every $0 < \lambda < 1$. Since f is strictly convex on X,

39 $\qquad f(x_0 + \lambda(x_1 - x_0)) < \lambda f(x_1) + (1 - \lambda) f(x_0) = f(x_0),$

which is in contradiction to the original hypothesis. $\square$

We conclude this section with a relatively simple test for determining whether or not a function is convex.

40 Definition. An $n \times n$ matrix Q is called *positive semidefinite* if $\langle x, Qx \rangle \geq 0$ for every $x \in E^n$. If $\langle x, Qx \rangle > 0$ for every nonzero $x \in E^n$, then Q is called *positive definite*. □

41 Exercise. Show that if Q is an $n \times n$ positive-semidefinite matrix, then Q^T and $Q + Q^T$ are positive-semidefinite matrices. □

42 Theorem. Let $f(\cdot)$ be a twice continuously differentiable function from an open convex set $X \subset E^n$ into E^1. Then $f(\cdot)$ is convex on X if and only if the $n \times n$ hessian matrix

$$43 \quad H(x) = \begin{bmatrix} \dfrac{\partial^2 f(x)}{\partial x^1 \partial x^1} & \dfrac{\partial^2 f(x)}{\partial x^1 \partial x^2} & \cdots & \dfrac{\partial^2 f(x)}{\partial x^1 \partial x^n} \\ \dfrac{\partial^2 f(x)}{\partial x^2 \partial x^1} & \dfrac{\partial^2 f(x)}{\partial x^2 \partial x^2} & \cdots & \dfrac{\partial^2 f(x)}{\partial x^2 \partial x^n} \\ \vdots & \vdots & & \vdots \\ \dfrac{\partial^2 f(x)}{\partial x^n \partial x^1} & \dfrac{\partial^2 f(x)}{\partial x^n \partial x^2} & \cdots & \dfrac{\partial^2 f(x)}{\partial x^n \partial x^n} \end{bmatrix}$$

is positive semidefinite for every $x \in X$.

Proof. ⇐. Let $x_0, x \in X$ be arbitrary, and let $h = x - x_0$. Then $(x_0 + \lambda h) \in X$ for every $\lambda \in [0,1]$, since X is convex. By Taylor's theorem,

$$44 \quad f(x) = f(x_0) + \langle \nabla f(x_0), h \rangle + \tfrac{1}{2} \langle h, H(x_0 + \lambda h) h \rangle$$
for some $\lambda \in [0,1]$.

Since $H(y)$ is positive semidefinite for every $y \in X$, the third term in (43) is nonnegative. Hence

$$45 \quad f(x) \geq f(x_0) + \langle \nabla f(x_0), h \rangle = f(x_0) + \langle \nabla f(x_0), x - x_0 \rangle$$
for every $x, x_0 \in X$.

Therefore $f(\cdot)$ is convex, by theorem (25).

⇒. The proof in this direction is by contraposition. Suppose there is an $x_0 \in X$ and an $h \in E^n$ such that $\langle h, H(x_0) h \rangle < 0$. By assumption, the real-valued function $\langle h, H(y) h \rangle$ is continuous at every point $y \in X$. Hence there is a ball $B_\delta(x_0) \subset X$ about x_0 of radius $\delta > 0$ such that $\langle h, H(y) h \rangle < 0$ for every $y \in B_\delta(x_0)$. Let $\epsilon > 0$ be chosen such that $x \triangleq (x_0 + \epsilon h) \in B_\delta(x_0)$, and set $h' = \epsilon h$. Using (44), with $h = h'$, we obtain

$$46 \quad f(x) = f(x_0) + \langle \nabla f(x_0), h' \rangle + \tfrac{1}{2} \langle h', H(x_0 + \lambda h') h' \rangle$$
for some $\lambda \in [0,1]$.

Now, $\|\lambda h'\| = \|\lambda \epsilon h\| = |\lambda| \|\epsilon h\| \leq \delta$, so that $x_0 + \lambda h' \in B_\delta(x_0)$, which implies that $\frac{1}{2}\langle h', H(x_0 + \lambda h')h'\rangle = \frac{1}{2}\epsilon^2 \langle h, H(x_0 + \lambda h')h\rangle < 0$. Thus

$$f(x) < f(x_0) + \langle \nabla f(x_0), h'\rangle = f(x_0) + \langle \nabla f(x_0), x - x_0\rangle$$

for two points $x, x_0 \in X$, and therefore f is not convex, by theorem (25). We conclude from this that if $f(\cdot)$ is convex, then $H(x)$ is positive semidefinite for every $x \in X$. $\square$

47 Corollary. Let $f: E^n \to E^1$ be defined by

$$f(x) = \langle x, Qx\rangle + \langle x, d\rangle,$$

where Q is an $n \times n$ positive-semidefinite matrix and $d \in E^n$ is a constant vector. Then $f(\cdot)$ is convex.

Proof. We observe that $\nabla f(x) = (Q + Q^T)x + d$, and so $H(x) = Q + Q^T$. By exercise (41) and theorem (42), f is convex. $\square$

We also have a test for strictly convex functions.

48 Theorem. Let $f(\cdot)$ be a twice continuously differentiable function from an open convex set $X \subset E^n$ into E^1. If the $n \times n$ hessian matrix $H(x)$ [see (43)] is positive definite for every $x \in X$, then f is strictly convex on X. $\square$

49 Corollary. Let $f(\cdot)$ be as defined in corollary (47). If the $n \times n$ matrix Q is positive definite, then f is strictly convex. $\square$

50 Exercise. Prove theorem (48) and corollary (49). $\square$

51 Remark. The converse of theorem (48) is false. Let $X = E^1$, and take $f(x) = x^4$. Then $f(\cdot)$ is strictly convex on X and twice continuously differentiable. But $H(0) = d^2 f(0)/dx^2 = 0$. $\square$

This concludes our discussion of convex sets and convex functions. For further detail, the reader is referred to standard texts on convexity.

REFERENCES

1. H. G. Eggleston: "Convexity," Cambridge Tracts in Mathematics and Mathematical Physics, Cambridge University Press, Cambridge, 1955.
2. C. Berge: "Topological Spaces," Oliver & Boyd, Ltd., Edinburgh and London, 1963.

appendix **B**
Constrained minimization problems in infinite-dimensional spaces

B.1 AN EXTENSION OF THE FUNDAMENTAL THEOREM (2.3.12)

The purpose of this appendix is to extend the Fundamental Theorem (2.3.12) to problems in linear topological spaces. As an application of the extended theorem, we shall derive the *Pontryagin maximum principle* [1] for fixed-time optimal control problems with continuous dynamics.

First let us formulate the equivalent of the Basic Problem (1.4.1) in an infinite-dimensional space.

1 The Basic Problem. Let $\mathcal{L}$ be a linear topological space. Given a continuous function $f(\cdot)$ from $\mathcal{L}$ into the reals, a continuous function $r(\cdot)$ from $\mathcal{L}$ into E^m, and a subset Ω of $\mathcal{L}$, find a vector $\hat{z} \in \Omega$, satisfying $r(\hat{z}) = 0$, such that for all $z \in \Omega$ which satisfy $r(z) = 0$

2 $f(\hat{z}) \leq f(z)$. $\square$

As before, we shall call any $\hat{z}$ with the above properties an optimal solution.

Observe that in formulation (1.4.1) we specified that the functions $f(\cdot)$ and $r(\cdot)$ are continuously differentiable. However, we could not do this in problem (1) because differentiability is not a well-defined concept in a linear topological space. Consequently, to make a straightforward extension of theorem (2.3.12) possible, we shall need to *stipulate* the existence of a continuous linear map from $\mathcal{L}$ into E^{m+1} with properties similar to those of the jacobian $\partial F(\hat{z})/\partial \hat{z}$, which we used in Chapter 2. The most convenient place for this stipulation seems to be in a modified definition of a conical approximation.

3 **Definition.** A convex cone $C(\hat{z},\Omega)$ will be called a *conical approximation* to the set Ω at $\hat{z} \in \Omega$, with respect to the functions $f(\cdot)$ and $r(\cdot)$, if there exist continuous linear functions $f'(\hat{z})(\cdot)$ and $r'(\hat{z})(\cdot)$ from $\mathcal{L}$ into E^1 and E^m, respectively, such that for any finite collection $\{\delta z_1, \delta z_2, \ldots, \delta z_p\}$ of linearly independent vectors in $C(\hat{z},\Omega)$ there exist an $\epsilon > 0$ and a continuous function $\zeta(\cdot)$ from co $\{\hat{z}, \hat{z} + \epsilon\, \delta z_1, \ldots, \hat{z} + \epsilon\, \delta z_p\}$ into Ω (with ϵ and ζ possibly depending on $\hat{z}$ and $\delta z_1, \delta z_2, \ldots, \delta z_p$) which satisfy

4
$$\lim_{\beta \to 0} \frac{1}{\beta} |f(\zeta(\hat{z} + \beta\, \delta z)) - f(\hat{z}) - f'(\hat{z})(\beta\, \delta z)| = 0$$

and

5
$$\lim_{\beta \to 0} \frac{1}{\beta} \| r(\zeta(\hat{z} + \beta\, \delta z)) - r(\hat{z}) - r'(\hat{z})(\beta\, \delta z)\| = 0$$

uniformly for $\delta z \in$ co $\{0, \epsilon\, \delta z_1, \epsilon\, \delta z_2, \ldots, \epsilon\, \delta z_p\}$. □

We are now ready to extend theorem (2.3.12).

6 **Theorem.** If $\hat{z}$ is an optimal solution to the Basic Problem (1) and $C(\hat{z},\Omega)$ is a conical approximation to Ω at $\hat{z} \in \Omega$ with respect to the functions $f(\cdot)$ and $r(\cdot)$, then there exists a nonzero vector $\psi = (\psi^0, \psi^1, \ldots, \psi^m)$ in E^{m+1}, with $\psi^0 \leq 0$, such that

7
$$\langle \psi, F'(\hat{z})(\delta z) \rangle \leq 0 \quad \text{for all } \delta z \in \bar{C}(\hat{z},\Omega),$$

where we define $F'(\hat{z})(\delta z)$ to be $(f'(\hat{z})(\delta z), r'(\hat{z})(\delta z))$.

Proof. Let us introduce the notation $F = (f,r)$; that is, $F \colon \mathcal{L} \to E^{m+1}$. We shall now essentially duplicate the steps of theorem (2.3.12).

APP. B.1 AN EXTENSION OF THE FUNDAMENTAL THEOREM (2.3.12)

Thus let $K(\hat{z}) \subset E^{m+1}$ be the cone defined by

(8) $$K(\hat{z}) = F'(\hat{z})(C(\hat{z},\Omega)).$$

The cone $K(\hat{z})$ is convex because $C(\hat{z},\Omega)$ is convex and $F'(\hat{z})(\cdot)$ is linear. Examining the statement of the theorem, we see that it implies that the cone $K(\hat{z})$ and the ray

(9) $$R = \{y: y = \beta(-1,0, \ldots ,0), \beta > 0\} \subset E^{m+1}$$

must be separated by a hyperplane with normal $\psi = (\psi^0, \psi^1, \ldots , \psi^m)$ such that

(10) $\langle \psi, y \rangle \leq 0$ for all $y \in K(\hat{z})$
(11) $\langle \psi, y \rangle \geq 0$ for all $y \in R$.

To obtain a contradiction, let us suppose that $K(\hat{z})$ and R are not separated. Then the dimension of $K(\hat{z})$ must be $m + 1$ and R must be in the interior of $K(\hat{z})$. We can therefore find a simplex S in $K(\hat{z})$ with vertices 0 and $\delta y_1, \delta y_2, \ldots , \delta y_{m+1}$ such that

(12)   There exists a vector $\delta y_0 \in R$ which lies in the interior of S;

(13)   There exists a set of vectors $\delta z_1, \delta z_2, \ldots , \delta z_{m+1}$ in $C(\hat{z},\Omega)$ satisfying

$$F'(\hat{z})(\delta z_i) = \delta y_i \quad i = 1, 2, \ldots , m + 1$$

for which the associated $\epsilon > 0$ in definition (3) may be taken to be 1.

Thus, with the choice indicated, there exists a continuous function ζ which maps co $\{\hat{z}, \hat{z} + \delta z_1, \hat{z} + \delta z_2, \ldots , \hat{z} + \delta z_{m+1}\}$ into Ω. [Note that the δz_i are linearly independent because the δy_i are linearly independent and $F'(\hat{z})(\cdot)$ is linear.]

Now for $0 < \alpha \leq 1$ let $\Sigma_\alpha \subset S$ be a closed ball with center $\alpha \, \delta y_0$ and radius αr, with $r > 0$, where δy_0 is the point indicated in (12). To produce the counterpart of the map $G_\alpha(\cdot)$ in (2.3.22), we let X be the continuous linear map[1] from E^{m+1} into $\mathcal{L}$ defined by

(14) $$X(\delta y_i) = \delta z_i \quad i = 1, 2, \ldots , m + 1,$$

where the δz_i are the elements defined in (13). Then we define the map $G_\alpha(\cdot)$ from $\Sigma_\alpha - \{\alpha \, \delta y_0\}$ into E^{m+1} by

(15) $$G_\alpha(x) = F(\zeta(\hat{z} + X(\alpha \, \delta y_0 + x))) - (F(\hat{z}) + \alpha \, \delta y_0).$$

---

[1] The continuity of X follows directly from the basic axioms defining a linear topological space.

Making use of (4) and (5), we find that

16 $G_\alpha(x) = F'(\hat{z})(X(\alpha\ \delta y_0 + x)) - \alpha\ \delta y_0 + o(X(\alpha\ \delta y_0 + x))$,

where $o(X(\alpha\ \delta y_0 + x))/\alpha \to 0$ as $\alpha \to 0$ uniformly for $x \in (\Sigma_\alpha - \{\delta y_0\})$.

Now, for any $\delta y \in S$, $F'(\hat{z})(X(\delta y)) = \delta y$, and hence (16) becomes

17 $G_\alpha(x) = x + o(X(\alpha\ \delta y_0 + x))$.

Since the composition $o \cdot X$ is continuous, we invoke the Brouwer fixed-point theorem, exactly as in the proof of theorem (2.3.12) following expression (2.3.24), to conclude that there exists an $\alpha^* \in (0,1]$ and a point $x^* \in (\Sigma_{\alpha^*} - \{\alpha^*\ \delta y_0\})$ such that

18 $G_{\alpha^*}(x^*) = 0$.

But (13), (15), and (18) imply that for $z^* = \zeta(\hat{z} + X(\alpha^*\ \delta y_0 + x^*))$

19 $z^* \in \Omega$
20 $r(z^*) = 0$
21 $f(z^*) = f(\hat{z}) + \alpha^*\ \delta y_0{}^0 < f(\hat{z})$,

where $\delta y_0 = (\delta y_0{}^0, \delta y_0{}^1, \ldots, \delta y_0{}^m)$. This contradicts the optimality of $\hat{z}$, and hence the cone $K(\hat{z})$ and the ray R must be separated. Consequently, there exists a vector $\psi = (\psi^0, \psi^1, \ldots, \psi^m)$, with $\psi^0 \leq 0$, such that

22 $\langle \psi, F'(\hat{z})(\delta z)\rangle \leq 0 \qquad$ for all $\delta z \in C(\hat{z},\Omega)$.

Inequality (7) now follows from the continuity of the linear map $\langle \psi, \cdot \rangle$. □

In the next section we shall see how to find functions $f'(\hat{z})(\cdot)$, $r'(\hat{z})(\cdot)$, and ζ in a particular case, a fixed-time optimal control problem with continuous dynamics.

B.2 THE MAXIMUM PRINCIPLE

We shall now explore the applicability of the theorem (B.1.6) by using it to obtain the *Pontryagin maximum principle* for fixed-time optimal control problems.

1 **The fixed-time optimal control problem.** Consider a dynamical system described by the differential equation

2 $$\frac{dx(t)}{dt} = h(x(t), u(t)) \qquad t \in [0,T],$$

where $x(t) \in E^n$ is the state of the system at time t, $u(t) \in E^l$ is the input, or control, of the system at time t, and h is a function from $E^n \times E^l$ into E^n. Given a cost function $h^0(\cdot,\cdot)$ from $E^n \times E^l$ into the reals, a constraint function $g(\cdot)$ from E^n into E^m, an initial state x_0, a constraint set $U \subset E^l$, and a final time $T > 0$, find a measurable, essentially bounded control $\hat{u}(\cdot)$ mapping $[0,T]$ into E^l and a corresponding trajectory $\hat{x}(\cdot)$, determined by (2) on the interval $[0,T]$, satisfying

3 $\quad\quad\hat{u}(t) \in U \quad$ for all $t \in [0,T]$
4 $\quad\quad\hat{x}(0) = x_0$
5 $\quad\quad g(\hat{x}(T)) = 0$

such that for all essentially bounded, measurable controls $u(\cdot)$ defined on $[0,T]$, with corresponding trajectories $x(\cdot)$ determined by (2) and satisfying (3) to (5),

6 $\quad\quad\int_0^T h^0(\hat{x}(t),\hat{u}(t))\, dt \leq \int_0^T h^0(x(t),u(t))\, dt. \quad\square$

Let us make the following assumptions:

7 **Assumption.** The functions $h^0(\cdot,\cdot)$ and $h(\cdot,\cdot)$ are continuously differentiable in x and continuous in u. $\quad\square$

8 **Assumption.** The function $g(\cdot)$ is continuously differentiable and its $m \times n$ jacobian matrix $\partial g(x)/\partial x$ has rank m for all $x \in \{x' : g(x') = 0\}$. $\quad\square$

We shall now transcribe the optimal control problem (1) into the form of the Basic Problem (B.1.1). Let P_1 and P_2 be the $1 \times (n + 1)$ and $n \times (n + 1)$ projection matrices defined by

9 $\quad\quad P_1 = [1,0,0,\ldots,0,0]$

$$P_2 = \begin{bmatrix} 0 & 1 & 0 & \cdots & \cdots & 0 \\ 0 & 0 & 1 & 0 & \cdots & 0 \\ \cdots & \cdots & \cdots & \cdots & \cdots & \cdots \\ 0 & \cdots & \cdots & 0 & 1 & 0 \\ 0 & \cdots & \cdots & \cdots & 0 & 1 \end{bmatrix}.$$

Thus, given a vector $z = (z^0,x)$ in E^{n+1}, with $x \in E^n$, we have, $P_1 z = z^0$ and $P_2 z = x$. Next let $\mathbf{h}(\cdot,\cdot)$ be a map from $E^{n+1} \times E^l$ into E^{n+1}, defined by

10 $\quad\quad \mathbf{h}(z,u) = (h^0(P_2 z,u), h(P_2 z, u))$,

where $z = (z^0, x^1, x^2, \ldots, x^n)$.

In this notation problem (1) becomes:

11 Problem. Minimize $P_1 z(T) = z^0(T)$ subject to

$$g(P_2 z(T)) = g(x(T)) = 0 \quad \text{12}$$

and to $z(\cdot)$ satisfying the differential equation

$$\frac{d}{dt} z(t) = \mathbf{h}(z(t), u(t)) \quad t \in [0, T], \quad \text{13}$$

with $z(0) = (0, x_0)$, for some measurable, essentially bounded function $u(\cdot)$ mapping $[0, T]$ into U. □

Now let us define the functions $f(\cdot)$ and $r(\cdot)$ and the set Ω as follows:

14 Definition. $f(z) = P_1 z(T), \quad r(z) = g(P_2 z(T))$. □

15 Definition. Let Ω be the set of all absolutely continuous functions $z(\cdot)$, from $[0, T]$ into E^{n+1}, satisfying (13), with $z(0) = (0, x_0)$, for some measurable essentially bounded function $u(\cdot)$ from $[0, T]$ into U. □

With these definitions, the transcription of problem (1) into the form of the Basic Problem (B.1.1) is complete except for the definition of the linear space $\mathcal{L}$ and a demonstration that $f(\cdot)$ and $r(\cdot)$ are continuous.

Since the set Ω consists of absolutely continuous functions, it might seem natural to take $\mathcal{L}$ to be the space of all continuous functions from $[0, T]$ into E^{n+1}. However, since we wish to use a conical approximation consisting of piecewise-continuous functions (first constructed by Pontryagin et al. [1] in the original proof of the maximum principle), we find it necessary to embed Ω into a larger linear topological space, which we now define.

Let $\mathcal{U}$ be the set of all upper-semicontinuous real-valued functions defined on $[0, T]$, and let $\mathcal{S} = \mathcal{U} - \mathcal{U}$. Clearly, $\mathcal{S}$ is a linear space. We now define the space $\mathcal{L}$ to be the cartesian product $\mathcal{S}^{n+1} = \mathcal{S} \times \mathcal{S} \times \cdots \times \mathcal{S}$, with the relativized pointwise topology [3], that is, the topology constructed from the subbase consisting of the family of all subsets of the form $\{z(\cdot) \in \mathcal{S}^{n+1}: z(t) \in N\}$, where t is a point in $[0, T]$ and N is an open set in E^{n+1}. With this topology, a net $z_s \in \mathcal{L}$, with $s \in A$ a directed set, converges to a function $z^* \in \mathcal{L}$ if and only if $z_s(t) \to z^*(t)$ for every $t \in [0, T]$.

To show that maps from $\mathcal{L}$ into E^{n+1} and from $\mathcal{L}$ into $\mathcal{L}$ are continuous, we shall need the following two simple results, which we draw from Kelley [3].

16 Definition. Let $t \in [0,T]$ be arbitrary. Then the *evaluation function* $e_t: \mathcal{L} \to E^{n+1}$ is defined by

17 $\qquad e_t(z) = z(t).$ □

18 Proposition. For every $t \in [0,T]$ the evaluation function e_t is continuous.

Proof. Let z_s be any net in $\mathcal{L}$ which converges to a function z^*. Then $z_s(t) \to z^*(t)$, and hence $e_t(z_s) \to e_t(z^*)$, which proves that e_t is continuous. □

19 Proposition. A function g mapping a linear topological space $\mathfrak{M}$ into $\mathcal{L}$ is continuous if and only if the composition $e_t \cdot g$ is continuous for every $t \in [0,T]$.

Proof. Suppose g is continuous; then, since e_t is continuous, the composition $e_t \cdot g$ is continuous.

Now suppose that $e_t \cdot g$ is continuous for all $t \in [0,T]$ but g is not continuous. Then there must be a net ξ_s in $\mathfrak{M}$ such that ξ_s converges to ξ^* in $\mathfrak{M}$ but $g(\xi_s)$ does not converge to $g(\xi^*)$. Hence there must be at least one $t \in [0,T]$ such that $g(\xi_s)(t)$ does not converge to $g(\xi^*)(t)$. But this contradicts the assumption that $e_t \cdot g$ is continuous for all $t \in [0,T]$. □

20 Proposition. The functions $f: \mathcal{L} \to E^1$ and $r: \mathcal{L} \to E^m$ defined by (14) are continuous.

Proof. Since $f = P_1 \cdot e_T$, $r = g \cdot P_2 \cdot e_T$, P_1, P_2, g, and e_T are continuous, and the compositions of continuous maps are continuous, the proposition must be true. □

21 Definition. We shall refer to the problem of minimizing $f(z)$ subject to $r(z) = 0$ and $z \in \Omega \subset \mathcal{L}$, with f, r, and Ω as defined in (14) and (15), as the *optimal control problem* in *basic form*. □

Note that with the optimal control problem cast in basic form the control $u(\cdot)$ in (13) acts as a *parameter*, and therefore (13) is considered as a parameterized family of differential equations.

Let $\hat{z}(\cdot)$ be a given optimal solution to the optimal control problem in basic form, corresponding to the optimal control $\hat{u}(\cdot)$. We shall now construct a conical approximation at $\hat{z}(\cdot)$ to the set Ω defined in (15), with respect to the map $F = (f,r)$, where $f(\cdot)$ and $r(\cdot)$ are as defined in (14).

22 Definition. Let I be the set of all points in $(0,T)$, the interior of $[0,T]$, such that for every $s \in I$ and any absolutely continuous function $z: [0,T] \to E^{n+1}$

23
$$\lim_{\alpha \to 0} \frac{1}{\alpha} \left[\int_{s+\alpha p}^{s+\alpha q} \mathbf{h}(z(t), \hat{u}(t))\, dt - \alpha(q-p)\mathbf{h}(z(s), \hat{u}(s)) \right] = 0,$$

where p and q are arbitrary real numbers. □

It can be deduced from theorem 41.2 in [2] that the set $[0,T]/I \triangleq \{s \in [0,T]: s \notin I\}$ has measure zero, that is, that almost all points in $[0,T]$ are in I.

Now, let $\Phi(t,s)$ be the $(n+1) \times (n+1)$ matrix which satisfies the linear differential equation

24
$$\frac{d}{dt}\Phi(t,s) = \frac{\partial \mathbf{h}(\hat{z}(t), \hat{u}(t))}{\partial z} \Phi(t,s) \qquad t \in [s,T],$$

with $\Phi(s,s)$ equal to the $(n+1) \times (n+1)$ identity matrix, and $\hat{z}(\cdot)$, $\hat{u}(\cdot)$ the optimal pair under consideration. Then for any $s \in I$ and $v \in U$ we define the functions $\delta z_{s,v}$ by

25
$$\delta z_{s,v}(t) = \begin{cases} 0 & 0 \leq t < s \\ \phi(t,s)\,[\mathbf{h}(\hat{z}(s),v) - \mathbf{h}(\hat{z}(s),\hat{u}(s))] & s \leq t \leq T, \end{cases}$$

and we claim that the set

26
$$C(\hat{z},\Omega) = \left\{ \delta z \in \mathfrak{L}:\ \delta z = \sum_{i=1}^{k} \alpha^i\, \delta z_{s_i, v_i} \right\},$$

with $\alpha^i \geq 0$, $s_i \in I$ and $v_i \in U$ for $i = 1, 2, \ldots, k$ and k arbitrary but finite, is a conical approximation to the set Ω defined by (15) at the optimal solution $\hat{z}$. The fact that this is a natural candidate for a conical approximation will become apparent from what follows.

To establish our claim, we must show that $C(\hat{z},\Omega)$ is a convex cone and then find functions $\zeta(\cdot)$, $f'(\hat{z})(\cdot)$, $r'(\hat{z})(\cdot)$, and $o(\cdot)$ which satisfy the assumptions of definition (B.1.3). We shall now carry out these steps.[1]

27 **Proposition.** The set $C(\hat{z},\Omega)$ defined in (26) is a convex cone.

Proof. Let $\delta z' = \sum_{i=1}^{k'} \alpha^{i\prime}\, \delta z_{s_{i'}, v_{i'}}$ and $\delta z'' = \sum_{i=1}^{k''} \alpha^{i\prime\prime}\, \delta z_{s_{i''}, v_{i''}}$ be any two elements in $C(z,\Omega)$. Then, for any $\lambda \geq 0$, $\lambda\, \delta z' \in C(\hat{z},\Omega)$, by inspection, and for any $\lambda \in [0,1]$, $\delta z_\lambda \triangleq \lambda\, \delta z' + (1-\lambda)\, \delta z''$ is seen

---

[1] The reader who is not interested in the details of this demonstration may proceed directly to (71), where the maximum principle is established.

APP. B.2 THE MAXIMUM PRINCIPLE 267

to be of the form

(28) $$\delta z_\lambda = \sum_{j=1}^{k} \alpha^j(\lambda)\, \delta z_{s_j, v_j},$$

where $k = k' + k''$, $\alpha^j(\lambda) = \lambda \alpha^{j'}$, $s_j = s'_j$ and $v_j = v'_j$ for $j = 1, 2, \ldots, k'$, and $\alpha^{j+k'}(\lambda) = (1 - \lambda)\alpha^{j''}$, $s_{j+k'} = s''_j$, and $v_{j+k'} = v''_j$ for $j = 1, 2, \ldots, k''$. Hence $\alpha^j(\lambda) \geq 0$, k is finite and δz_λ is in $C(\hat{z}, \Omega)$; that is, $C(\hat{z}, \Omega)$ is a convex cone. □

Let $\delta z_1, \delta z_2, \ldots, \delta z_p$ be any set of linearly independent elements in $C(\hat{z}, \Omega)$, and let $C = \mathrm{co}\ \{0, \delta z_1, \delta z_2, \ldots, \delta z_p\}$. We shall now choose an $\epsilon > 0$ and construct a continuous map ζ from $\{\hat{z}\} + \epsilon C$ into Ω. It will be seen that the choice of ϵ is closely connected with the way ζ is constructed.

First, each of the δz_j chosen above must be of the form

(29) $$\delta z_j = \sum_{i=1}^{k_j} \alpha_j{}^i\, \delta z_{s_i{}^j, v_i{}^j} \qquad j = 1, 2, \ldots, p.$$

Now let us order the points $s_i{}^j$ linearly. After renumbering the $s_i{}^j$ and the $v_i{}^j$ accordingly, we find that there are an integer k, points $s_1 \leq s_2 \leq s_3 \leq \cdots \leq s_k$ in I, corresponding points $v_1, v_2, \ldots, v_k$ in U, and nonnegative scalars $\beta_i{}^j$, with $i = 1, 2, \ldots, k$ and $j = 1, 2, \ldots, p$, such that (29) becomes

(30) $$\delta z_j = \sum_{i=1}^{k} \beta_j{}^i\, \delta z_{s_i, v_i} \qquad j = 1, 2, \ldots, p.$$

Since the δz_j are linearly independent, every vector $\delta z \in C$ has a unique representation of the form

$$\delta z = \sum_{j=1}^{p} \lambda^j\, \delta z_j,$$

with $\lambda^j \geq 0$ and $\sum_{j=1}^{p} \lambda^j \leq 1$; that is,

(31) $$\delta z = \sum_{j=1}^{p} \lambda^j \left(\sum_{i=1}^{k} \beta_j{}^i\, \delta z_{s_i, v_i} \right)$$
$$= \sum_{i=1}^{k} \left(\sum_{j=1}^{p} \lambda^j \beta_j{}^i \right) \delta z_{s_i, v_i}$$
$$= \sum_{i=1}^{k} \delta t_i(\lambda)\, \delta z_{s_i, v_i},$$

where

(32) $$\delta t_i(\lambda) \triangleq \sum_{j=1}^{p} \lambda^j \beta_j{}^i \qquad i = 1, 2, \ldots, k.$$

Let Λ be the simplex in E^p defined by

(33) $$\Lambda = \left\{\lambda \in E^p : \sum_{i=1}^{p} \lambda^i \leq 1, \lambda^i \geq 0\right\}.$$

Then it is clear from the above that there is a one-to-one correspondence between the elements δz of C and the elements λ of Λ. Furthermore, it can be shown that this correspondence is bicontinuous. The continuity of the $\delta t_i(\lambda)$ is self-evident from their definition.

Now let $\gamma > 0$ be small but arbitrary; we shall fix an upper bound for γ later. As in [1], we shall construct the map $\zeta(\cdot)$ by solving differential equation (13), with a "perturbed" control which depends on $\delta z \in C$. First let us define these perturbations.

Let $\delta z \in C$, and let the corresponding δt_i be as defined in (32). Now let

(34) $$l_i = \begin{cases} -(\delta t_i + \delta t_{i+1} + \cdots + \delta t_k) & \text{if } s_i = s_k \\ -(\delta t_i + \delta t_{i+1} + \cdots + \delta t_j) & \text{if } s_i = s_{i+1} = \cdots \\ & \quad = s_j < s_{j+1}, j < k, \end{cases}$$

and let

(35) $$I_i = \{t : s_i + \gamma l_i < t \leq s_i + \gamma(l_i + \delta t_i)\}, \quad i = 1, 2, \ldots, k,$$

where $\gamma > 0$ is the scalar introduced above. It is clear that there exists a positive number γ' such that for all $0 < \gamma \leq \gamma'$ the intervals I_i are disjoint and contained in $[0,T]$. (Note that I_i may be empty.)

For any fixed $\gamma \in (0,\gamma']$ we define the perturbed control $u_{\gamma,\lambda}(\cdot)$ on $[0,T]$ by the relation,[1] with v_i as in (31),

(36) $$u_{\gamma,\lambda}(t) = \begin{cases} \hat{u}(t) & \text{if } t \notin \bigcup_{i=1}^{k} I_i \\ v_i & \text{if } t \in I_i. \end{cases}$$

(37) **Definition.** For $\gamma \in (0,\gamma']$ and $\lambda \in \Lambda$ let $z_{\gamma,\lambda}(\cdot)$ be the solution of (13) corresponding to $u(\cdot) = u_{\gamma,\lambda}(\cdot)$ and satisfying $z_{\gamma,\lambda}(0) = (0,x_0)$. ☐

Since $u_{\gamma,\lambda}(\cdot)$ is an essentially bounded measurable function satisfying $u_{\gamma,\lambda}(t) \in U$ for all $t \in [0,T]$, it is clear that $z_{\gamma,\lambda}(\cdot)$ is in Ω.

---

[1] When some $s_i = s_{i+1}$ and $\delta t_i = \delta t_{i+1} = 0$, (36) defines $u_{\gamma,\lambda}(\cdot)$ almost everywhere (it may not be uniquely defined at some of the s_i). However, this is entirely adequate for our purposes.

Referring to a theorem on the continuous dependence on parameters of the solutions of a differential equation (theorem 69.1 of reference [2]), we find that there must exist a $\gamma'' > 0$ such that $z_{\gamma,\lambda}(t)$ is a continuous function of γ and λ for all $\gamma \in [0,\gamma'']$, all $\lambda \in \Lambda$, and all $t \in [0,T]$. Let $\epsilon = \min\,[\gamma',\gamma'']$, and let $\zeta(\cdot)$, mapping co $\{\hat{z}, \hat{z} + \epsilon\,\delta z_1, \ldots, \hat{z} + \epsilon\,\delta z_p\}$ into Ω, be defined by

38
$$\zeta(\hat{z} + \delta z) = z_{\epsilon,\lambda},$$

with λ defined by the relation

$$\frac{1}{\epsilon}\,\delta z = \sum_{j=1}^{p} \lambda^i\,\delta z_j.$$

Then, since the parameter λ is continuous in δz and $z_{\epsilon,\lambda}$ is continuous in λ (because of the continuous dependence of $z_{\epsilon,\lambda}(t)$ on λ for every $t \in [0,T]$ and because of proposition (19)), it is clear that ζ is a continuous map.

To complete our proof we have to exhibit the linear continuous functions $f'(\hat{z})(\cdot)$ and $r'(\hat{z})(\cdot)$ and show that relations (B.1.4) and (B.1.5) are satisfied. We shall do this by first showing that for $(\hat{z} + \beta\,\delta z) \in \text{co } \{\hat{z}, \hat{z} + \epsilon\,\delta z_1, \ldots, \hat{z} + \epsilon\,\delta z_p\}$, with $0 \leq \beta \leq 1$,

39
$$\zeta(\hat{z} + \beta\,\delta z)(T) = \hat{z}(T)$$
$$+ \beta\epsilon \sum_{i=1}^{k} \delta t_i(\lambda)\Phi(T,s_i)[\mathbf{h}(\hat{z}(s_i),v_i) - \mathbf{h}(\hat{z}(s_i),\hat{u}(s_i))] + o(\beta\,\delta z(T)),$$

where $o(\beta\,\delta z(T)/\beta \to 0$ as $\beta \to 0$, and the $\delta t_i(\lambda)$ are determined by (31) and (32); that is,

40
$$\zeta(\hat{z} + \beta\,\delta z)(T) = \hat{z}(T) + \beta\,\delta z(T) + o(\beta\,\delta z(T)).$$

Obviously, this requires certain results in differential equations, which we now digress to consider.

41 **Lemma.** Let $u(\cdot)$ be any admissible control, i.e., a measurable, essentially bounded function defined on $[0,T]$ with range in U. Let $z(\cdot)$ defined on $[0,T]$ be the absolutely continuous solution of (13) corresponding to this $u(\cdot)$ and satisfying the "initial" condition $z(s) = z_s$ for $s \in [0,T]$. Let $z'(\cdot)$ defined on $[0,T]$ be the absolutely continuous solution of (13), also corresponding to this $u(\cdot)$ but satisfying the "initial" condition $z'(s) = z_s + \beta\xi + o(\beta)$, where ξ is independent of β and $o(\beta)/\beta \to 0$ as $\beta \to 0$. Then

42
$$z'(t) = z(t) + \beta\,\delta z(t) + o'(\beta,t) \qquad \text{for all } t \in [s,T],$$

where $o'(\beta,t)/\beta \to 0$ as $\beta \to 0$ uniformly for $t \in [s,T]$, and for almost all $t \in [s,T]$, δz satisfies the differential equation

43
$$\frac{d}{dt}\delta z(t) = \frac{\partial \mathbf{h}(z(t),u(t))}{\partial z} \delta z(t) \qquad \delta z(s) = \xi.\quad \square$$

The gist of this lemma is that the solutions of (13) are Frechet differentiable with respect to initial conditions. The proof can be found in [2], theorem 69.4.

44 **Lemma.** For any $\beta \in (0,\epsilon]$ and $\lambda \in \Lambda$ let $z_{\beta,\lambda}(\cdot)$ be defined as in (37), with $\epsilon > 0$ defined as for (38). Then for $i = 1, 2, \ldots, k$ and $t \in I_i$ [with $\gamma = \beta$ in (35)]

45
$$\mathbf{h}(z_{\beta,\lambda}(t),v_i) = \mathbf{h}(\hat{z}(t),v_i) + a_i(t,\beta,\lambda),$$

where $a_i(t,\beta,\lambda) \to 0$ as $\beta \to 0$ uniformly for $t \in I_i$ and $\lambda \in \Lambda$, and $\hat{z}$ is the optimal solution under consideration.

Proof. Using the theorem on the continuous dependence of solutions of differential equations on parameters (see [2], theorem 69.1), we can show that $z_{\beta,\lambda}(t) \to \hat{z}(t)$ as $\beta \to 0$ uniformly for $t \in [0,T]$ and $\lambda \in \Lambda$, and hence the lemma must hold. $\square$

We shall now finish establishing (39) in two steps. First we shall evaluate the increments in $z_{\epsilon,\lambda}$ over the intervals I_i, and then we shall make use of lemma (41) to compute the propagation of these disturbances.

46 **Lemma.** For any $0 < \beta \leq \epsilon$, with $\epsilon > 0$ defined as for (38), and $\lambda \in \Lambda$, let $z_{\beta,\lambda}(\cdot)$ be as defined in (37). Then for any interval I_i for $i = 1, 2, \ldots, k$ [with $\gamma = \beta$ in (35)] the increment in $z_{\beta,\lambda}(t)$ on I_i is given by

47
$$z_{\beta,\lambda}(s_i + \beta(l_i + \delta t_i(\lambda))) - z_{\beta,\lambda}(s_i + \beta l_i)$$
$$= \beta \,\delta t_i \,\mathbf{h}(\hat{z}(s_i),v_i) + o(\beta\lambda),$$

where $o(\beta\lambda)/\beta \to 0$ as $\beta \to 0$ uniformly for $\lambda \in \Lambda$.

Proof. This lemma follows directly from lemma (44) and the fact that for all $s \in [0,T]$, all p and q real, all $v \in U$, and every absolutely continuous function $z: [0,T] \to E^{n+1}$ we have

$$\lim_{\alpha \to 0} \frac{1}{\alpha}\left[\int_{s+\alpha p}^{s+\alpha q} \mathbf{h}(z(t),v)\,dt - h(z(s),v)\right] = 0.\quad \square$$

APP. B.2 THE MAXIMUM PRINCIPLE 271

48 **Theorem.** For $0 \leq \beta \leq 1$ let $\beta \, \delta z$ be an arbitrary point in co $\{0, \epsilon \, \delta z_1, \epsilon \, \delta z_2, \ldots, \epsilon \, \delta z_p\}$, with $\epsilon = \min[\gamma', \gamma'']$ defined as before, and let $\lambda \in \Lambda$ correspond to δz according to (31). Then for $t \in [s_k, T]$

49
$$z_{\epsilon,\beta\lambda}(t) = z_{\beta\epsilon,\lambda}(t) = \hat{z}(t) + \beta \, \delta z(t) + o(\beta \, \delta z(t), t),$$

where $o(\beta \, \delta z(t), t)/\beta \to 0$ as $\beta \to 0$ uniformly for $\delta z \in$ co $\{0, \epsilon \, \delta z_1, \epsilon \, \delta z_2, \ldots, \epsilon \, \delta z_p\}$ and $t \in [s_k, T]$.

Proof. We begin by noting that δz must be of the form

50
$$\begin{aligned}\delta z(t) &= \epsilon \sum_{i=1}^{p} \lambda^i \, \delta z_i(t) \\ &= \epsilon \sum_{i=1}^{k} \delta t_i(\lambda) \, \delta z_{s_i, v_i}(t) \\ &= \epsilon \sum_{i=1}^{k} \Phi(t, s_i)[\mathbf{h}(\hat{z}(s_i), v_i) - \mathbf{h}(\hat{z}(s_i), \hat{u}(s_i))] \, \delta t_i(\lambda).\end{aligned}$$

We prove the theorem by induction on the integer k. Thus suppose $k = 1$. Invoking (47), we get

51
$$z_{\beta\epsilon,\lambda}(s_1) = \hat{z}(s_1 - \beta\epsilon \, \delta t_1(\lambda)) + \beta\epsilon \, \delta t_1(\lambda)\mathbf{h}(\hat{z}(s_1), v_1) + o(\beta\lambda),$$

and since (23) is satisfied by s_1,

52
$$\hat{z}(s_1 - \beta\epsilon \, \delta t_1(\lambda)) = \hat{z}(s_1) - \beta\epsilon \, \delta t_1(\lambda)\mathbf{h}(\hat{z}(s_1), \hat{u}(s_1)) + o(\beta),$$

where $o(\beta)/\beta \to 0$ as $\beta \to 0$. Consequently, we find that

53
$$z_{\beta\epsilon,\lambda}(s_1) = \hat{z}(s_1) + \beta\epsilon \, \delta t_1(\lambda)[\mathbf{h}(\hat{z}(s_1), v_1) \\ - \mathbf{h}(\hat{z}(s_1), \hat{u}(s_1))] + \tilde{o}(\beta\lambda),$$

where $\tilde{o}(\beta\lambda)/\beta \to 0$ as $\beta \to 0$ uniformly for $\lambda \in \Lambda$. Hence, by lemma (41), for $t \in [s_1, T]$

54
$$z_{\beta\epsilon,\lambda}(t) = \hat{z}(t) + \beta\epsilon \, \delta t_1(\lambda)\Phi(t, s_1)[\mathbf{h}(\hat{z}(s_1), v_1) \\ - \mathbf{h}(\hat{z}(s_1), \hat{u}(s_1))] + o(\beta\lambda, t),$$

where $o(\beta\lambda, t)/\beta \to 0$ as $\beta \to 0$ uniformly for $\lambda \in \Lambda$. Comparing with (50) and recalling that there is a bicontinuous one-to-one correspondence between $\delta z(\cdot)$ and λ, we find that for $t = T$ (54) becomes

55
$$z_{\beta\epsilon,\lambda}(T) = \hat{z}(T) + \beta \, \delta z(T) + o'(\beta \, \delta z(T)),$$

where $o'(\beta \, \delta z(T))/\beta \to 0$ uniformly for all $\delta z \in$ co $\{0, \epsilon \, \delta z_1, \epsilon \, \delta z_2, \ldots, \epsilon \, \delta z_p\}$. Thus for $k = 1$ and $0 \leq \beta \leq 1$

56
$$\zeta(\hat{z} + \beta \, \delta z)(T) = \hat{z}(T) + \beta \, \delta z(T) + o(\beta \, \delta z(T)).$$

Now suppose that (49) holds for $k = q - 1$ in (50), with $q > 1$. Let j be a positive integer such that $\cdots \leqq s_{j-1} \leqq s_j < s_{j+1} = s_{j+2} = \cdots = s_q$; that is, $s_i = s_q$ for $i = j+1, j+2, \ldots, q-1$ and $s_i < s_q$ for all $i \leqq j$. Then, by the induction hypothesis,

(57)
$$z_{\beta\epsilon,\lambda}(s_q + \beta\epsilon l_{j+1}) = \hat{z}(s_q + \beta\epsilon l_{j+1})$$
$$+ \beta\epsilon \sum_{i=1}^{j} \delta t_i(\lambda) \, \Phi(s_q + \beta\epsilon l_{j+1}, s_i)[\mathbf{h}(\hat{z}(s_i), v_i)$$
$$- \mathbf{h}(\hat{z}(s_i), \hat{u}(s_i))] + o(\beta\lambda, s_q + \beta\epsilon l_{j+1}).$$

Making use of (47) and the fact that the intervals $I_{j+1}, I_{j+2}, \ldots, I_q$ adjoin one another, we get

(58)
$$z_{\beta\epsilon,\lambda}(s_q) = z_{\beta\epsilon,\lambda}(s_q + \beta\epsilon l_{j+1})$$
$$+ \beta\epsilon \sum_{i=j+1}^{q} \delta t_i(\lambda) \, \mathbf{h}(\hat{z}(s_i), v_i) + o(\beta\lambda).$$

Since (23) is satisfied at the points $s_{j+1}, s_{j+2}, \ldots, s_q$,

(59)
$$\hat{z}(s_q + \beta\epsilon l_{j+1}) = \hat{z}(s_q) - \beta\epsilon \sum_{i=j+1}^{q} \delta t_i(\lambda) \, \mathbf{h}(\hat{z}(s_i), \hat{u}(s_i)) + o(\beta\lambda),$$

where $o(\beta\lambda)/\beta \to 0$ as $\beta \to 0$ uniformly for $\lambda \in \Lambda$.

Expanding $\beta\epsilon \, \delta t_i(\lambda) \, \Phi(s_q + \beta\epsilon l_{j+1}, s_i)$ for $i = 1, 2, \ldots, j$ about s_q, we get

(60)
$$\beta\epsilon \, \delta t_i(\lambda) \, \Phi(s_q + \beta\epsilon l_{j+1}, s_i) = \beta\epsilon \, \delta t_i(\lambda) \, \Phi(s_q, s_i) + o(\beta\lambda),$$

where $o(\beta\lambda)$ is an $(n+1) \times (n+1)$ matrix such that $(1/\beta)o(\beta\lambda)$ tends to the zero matrix as $\beta \to 0$ uniformly for $\lambda \in \Lambda$. Combining (57) to (60), we obtain

(61)
$$z_{\beta\epsilon,\lambda}(s_q) = \hat{z}(s_q) + \beta\epsilon \sum_{i=j+1}^{q} \delta t_i(\lambda) \, [\mathbf{h}(\hat{z}(s_i), v_i) - \mathbf{h}(\hat{z}(s_i), \hat{u}(s_i))]$$
$$+ \beta\epsilon \sum_{i=1}^{j+1} \delta t_i(\lambda) \, \Phi(s_q, s_i)[\mathbf{h}(\hat{z}(s_i), v_i) - \mathbf{h}(\hat{z}(s_i), \hat{u}(s_i))] + o(\beta\lambda),$$

where $o(\beta\lambda)/\beta \to 0$ as $\beta \to 0$ uniformly for $\lambda \in \Lambda$. We now invoke lemma (41) to obtain, for $t \in [s_q, T]$,

(62)
$$z_{\beta\epsilon,\lambda}(t) = \hat{z}(t) + \beta\epsilon \sum_{i=1}^{q} \delta t_i(\lambda) \, \Phi(t, s_i)[\mathbf{h}(\hat{z}(s_i), v_i)$$
$$- \mathbf{h}(\hat{z}(s_i), \hat{u}(s_i))] + o(\beta\lambda, t),$$

where we have used the fact that $\Phi(t, s_q)\Phi(s_q, s_i) = \Phi(t, s_i)$ and $o(\beta\lambda, t)/\beta \to 0$ as $\beta \to 0$ uniformly for $\lambda \in \Lambda$ and $t \in [s_q, T]$. Substituting from (50) into (62), we get, for $t \in [s_q, T]$,

(63) $$z_{\beta\epsilon,\lambda}(t) = \hat{z}(t) + \beta\,\delta z(t) + o(\beta\,\delta z(t), t),$$

where $o(\beta\,\delta z(t), t)/\beta \to 0$ as $\beta \to 0$ uniformly for $t \in [s_q, T]$ and $\delta z(\cdot) \in \text{co}\,\{0, \epsilon\,\delta z_1, \epsilon\,\delta z_2, \ldots, \epsilon\,\delta z_p\}$, and where we have made use of the bicontinuous one-to-one correspondence between $\delta z(\cdot)$ and λ.

Since we have shown that (63) holds for $k = 1$, and that if it is true for $k = q - 1$, then it is also true for $k = q$, with $q = 1, 2, \ldots$, the theorem is proved. □

It now follows from the definitions of $\zeta(\cdot)$ [see (38)] and from (63) that for every $\beta\,\delta z(\cdot)$ in co $\{0, \epsilon\,\delta z_1, \epsilon\,\delta z_2, \ldots, \epsilon\,\delta z_p\}$, with $0 \leq \beta \leq 1$,

(64) $$\zeta(\hat{z} + \beta\,\delta z)(T) = \hat{z}(T) + \beta\,\delta z(T) + o(\beta\,\delta z(T)),$$

where $o(\beta\,\delta z(T))/\beta \to 0$ as $\beta \to 0$ uniformly for $\delta z \in \text{co}\,\{0, \epsilon\,\delta z_1, \epsilon\,\delta z_2, \ldots, \epsilon\,\delta z_p\}$. Recalling definition (14), we find that for any $\beta\,\delta z \in \text{co}\,\{0, \epsilon\,\delta z_1, \epsilon\,\delta z_2, \ldots, \epsilon\,\delta z_p\}$ and $0 \leq \beta \leq 1$

(65) $$f(\zeta(\hat{z} + \beta\,\delta z)) = P_1 e_T \zeta(\hat{z} + \beta\,\delta z) \\ = P_1(\hat{z}(T) + \beta\,\delta z(T) + o(\beta\,\delta z(T))).$$

Since $f(\hat{z}) = P_1(\hat{z}(T))$, we find that

(66) $$\lim_{\beta \to 0} \frac{1}{\beta}\,(f(\zeta(\hat{z} + \beta\,\delta z)) - f(\hat{z}) - \beta\,\delta z(T)) = 0$$

uniformly for $\delta z \in \text{co}\,\{0, \epsilon\,\delta z_1, \epsilon\,\delta z_2, \ldots, \epsilon\,\delta z_p\}$. Hence if we define $f'(\hat{z}): \mathcal{L} \to E^1$ by

(67) $$f'(\hat{z})(\delta z) = P_1 e_T\,\delta z = P_1\,\delta z(T),$$

we see that (B.1.4) is satisfied, since the linearity and continuity of $f'(\hat{z})(\cdot)$ are obvious.

Similarly, for any $\delta z \in \text{co}\,\{0, \epsilon\,\delta z_1, \epsilon\,\delta z_2, \ldots, \epsilon\,\delta z_p\}$ and $0 \leq \beta \leq 1$

(68) $$r(\zeta(\hat{z} + \beta\,\delta z)) = g(P_2(\hat{z}(T) + \beta\,\delta z(T) + o(\beta\,\delta z(T)))).$$

Since $g(\cdot)$ is continuously differentiable, (68) can be expanded as

(69) $$r(\zeta(\hat{z} + \beta\,\delta z)) = g(P_2\hat{z}(T)) + \frac{\partial g(P_2(\hat{z}(T)))}{\partial x}\,\beta P_2\,\delta z(T) \\ + o(\beta\,\delta z(T)),$$

where $o(\beta\,\delta z(T))/\beta \to 0$ as $\beta \to 0$ uniformly for $\delta z \in \text{co}\,\{0,\,\epsilon\,\delta z_1,\,\epsilon\,\delta z_2,\,\ldots,\,\epsilon\,\delta z_p\}$. Therefore if we define $r'(\hat{z}): \mathfrak{L} \to E^m$ by

70
$$r'(\hat{z})(\delta z) = \frac{\partial g(P_2(\hat{z}(T)))}{\partial x} P_2\,\delta z(T),$$

we find that (B.1.5) is satisfied, since the linearity and continuity of $r'(z)(\cdot)$ are obvious.

Hence we have shown that the set $C(\hat{z},\Omega)$ defined in (26) is indeed a conical approximation to Ω at $\hat{z}$, with respect to the maps $f(\cdot)$ and $r(\cdot)$.

We can now establish the Pontryagin maximum principle for fixed-time problems.

71 Theorem: maximum principle. If $\hat{u}(\cdot)$ is an optimal control for problem (1) and $\hat{x}(\cdot)$ is a corresponding optimal trajectory, then there exist a scalar $p^0 \leq 0$ and a costate trajectory $p(\cdot)$, from $[0,T]$ into E^n, with $(p^0,p(t)) \neq 0$, such that

72
$$\frac{d}{dt}p(t) = -p^0\left[\frac{\partial h^0(\hat{x}(t),\hat{u}(t))}{\partial x}\right]^T$$
$$- \left[\frac{\partial h(\hat{x}(t),\hat{u}(t))}{\partial x}\right]^T p(t) \quad \text{for } t \in [0,T],$$

73
$$p(T) = \left[\frac{\partial g(\hat{x}(T))}{\partial x}\right]^T \psi \quad \text{for some } \psi \in E^m,$$

and for every $v \in U$ and almost all $t \in [0,T]$

74
$$p^0 h^0(\hat{x}(t),\hat{u}(t)) + \langle p(t),h(\hat{x}(t),\hat{u}(t))\rangle \geq p^0 h^0(\hat{x}(t),v) + \langle p(t),h(\hat{x}(t),v)\rangle.$$

Proof. Let us consider the optimal control problem (1) in basic form [see definition (21)]. If $\hat{z}(\cdot) = (\hat{z}^0(\cdot),\hat{x}(\cdot))$ is an optimal solution corresponding to an optimal control $\hat{u}(\cdot)$, then, by theorem (B.1.6), there exists a vector $\psi = (p^0,\psi)$ in E^{m+1} such that

75
$$p^0 f'(\hat{z})(\delta z) + \langle \psi,r'(\hat{z})(\delta z)\rangle \leq 0 \quad \text{for all } \delta z \in C(\hat{z},\Omega),$$

with $C(\hat{z},\Omega)$, $f'(\hat{z})(\cdot)$, and $r'(\hat{z})(\cdot)$ as defined in (26), (67), and (70), respectively. Substituting for $f'(\hat{z})(\cdot)$ and $r'(\hat{z})(\cdot)$ in (75) and choosing a $\delta z \in C(\hat{z},\Omega)$ of the form

76
$$\delta z = \delta z_{s,v},$$

with $s \in I$ and $v \in U$, arbitrary, we obtain for (75)

(77)
$$p^0 \langle P_1^T, \Phi(T,s)[\mathbf{h}(\hat{z}(s),v) - \mathbf{h}(\hat{z}(s),\hat{u}(s))] \rangle$$
$$+ \langle \psi, \frac{\partial g(x(T))}{\partial x} P_2 \Phi(T,s)[\mathbf{h}(\hat{z}(s),v) - \mathbf{h}(\hat{z}(s),\hat{u}(s))] \rangle \leq 0.$$

Rearranging terms yields

(78)
$$\left\langle \Phi^T(T,s) \left\{ p^0 P_1^T + P_2^T \left[\frac{\partial g(\hat{x}(T))}{\partial x} \right]^T \psi \right\}, \quad [\mathbf{h}(\hat{z}(s),v) - \mathbf{h}(\hat{z}(s),\hat{u}(s))] \right\rangle \leq 0.$$

Now let $\mathbf{p}: [0,T] \to E^{n+1}$ be defined by

(79)
$$\mathbf{p}(t) = \Phi^T(T,t) \left\{ p^0 P_1^T + P_2^T \left[\frac{\partial g(\hat{x}(T))}{\partial x} \right]^T \psi \right\},$$

and let us write $\mathbf{p}(t) = (p^0(t), p(t))$, with $p(t) = (p^1(t), p^2(t), \ldots, p^n(t)) \in E^n$. Then, by inspection of (79), (24), and (10), we obtain

(80)
$$p^0(T) = p^0 \qquad p(T) = \left[\frac{\partial g(\hat{x}(T))^T}{\partial x} \right] \psi$$

and

(81)
$$\frac{d}{dt} \mathbf{p}(t) = - \left[\frac{\partial \mathbf{h}(\hat{z}(t),\hat{u}(t))}{\partial z} \right]^T \mathbf{p}(t).$$

Expansion of (81) in components yields

(82)
$$\frac{d}{dt} p^0(t) = 0$$

(83)
$$\frac{d}{dt} p(t) = - p^0 \left[\frac{\partial h^0(\hat{x}(t),\hat{u}(t))}{\partial x} \right]^T - \left[\frac{\partial h(\hat{x}(t),\hat{u}(t))}{\partial x} \right]^T p(t).$$

If we now substitute $p(t)$ into (78), we get (74). □

The maximum principle can also be extended to free-time problems and to problems in which the function h depends on t as well as on x and u. These extensions are quite simple and are left as an exercise for the reader.

This concludes our discussion of optimization problems in infinite-dimensional spaces.

REFERENCES

1. L. S. Pontryagin, V. G. Boltyanskii, R. V. Gamkrelidze, and E. F. Mishchenko: "The Mathematical Theory of Optimal Processes," Interscience Publishers, Inc., New York, 1962.

2. E. J. McShane: "Integration," Princeton University Press, Princeton, N.J., 1944.
3. J. L. Kelley: "General Topology," D. Van Nostrand Company, Inc., Princeton, N.J., 1955.
4. R. V. Gamkrelidze: On Some Extremal Problems in the Theory of Differential Equations with Applications to the Theory of Optimal Control, *SIAM J. Control*, **3**:106–128 (1965).
5. L. W. Neustadt: An Abstract Variational Theory with Applications to a Broad Class of Optimization Problems, I. General Theory, II. Applications, *SIAM J. Control*, **4**:505–527 (1964), **5**:90–137 (1967).
6. N. O. Da Cunha and E. Polak: Constrained Minimization Under Vector-valued Criteria in Linear Topological Spaces, in A. V. Balakrishnan and L. W. Neustadt (eds.), "Mathematical Theory of Control," pp. 10–25, Academic Press, Inc., New York, 1966.

Glossary of Symbols

I. GENERAL CONVENTIONS

1. E^n denotes the euclidean space of ordered n-tuples of real numbers. Elements of E^n are denoted by lowercase letters, and when it is necessary to show the components of a vector x in E^n, it is done as follows: $x = (x^1, x^2, \ldots, x^n)$. When an n-tuplet $(x^1, x^2, \ldots, x^n)$ is a vector in E^n, it is always treated as a *column* vector in matrix multiplications; i.e., it is transposed into an $n \times 1$ matrix, but the transposition symbol is omitted.

2. When a vector z or $\mathbf{x}$, in E^{m+1} is constructed from a vector $x = (x^1, x^2, \ldots, x^m)$ in E^m by adding one component to the elements of x, and we wish to indicate this fact, then we depart from the notation in item 1, above and write the components of z or x as follows: $z = (z^0, x^1, x^2, \ldots, x^m)$ or $\mathbf{x} = (x^0, x^1, x^2, \ldots, x^m)$.

3. The scalar product in E^n is defined by $\langle x,y \rangle = \Sigma_{i=1}^{n} x^i y^i$; the euclidean norm is defined by $\|x\| = \sqrt{\langle x,x \rangle}$.

4. g or $g(\cdot)$ denotes a function, with the dot standing for the undesignated variable; $g(x)$ denotes the value of g at the point x. To indicate that the domain of a function g is A and that its range is B, we write $g: A \to B$. Assuming that $g: E^n \to E^m$, we write g in expanded form as follows: $g = (g^1, g^2, \ldots, g^m)$, so that $g(x) = (g^1(x), g^2(x), \ldots, g^m(x))$ is the image of x and is a vector in E^m. (The components of g are real-valued functions).

5. A function $g: E^n \to E^m$ is said to be *affine* if for all x in E^n $g(x) = b + \tilde{g}(x)$, where b is a fixed vector in E^m and $\tilde{g}: E^n \to E^m$ is a linear function.

6. $\dfrac{\partial g(\hat{x})}{\partial x}$ (or $\partial g(\hat{x})/\partial x$) denotes the jacobian matrix of g at $\hat{x}$. Assuming that $g: E^n \to E^m$, the jacobian is an $n \times m$ matrix whose ijth element is $\partial g^i(\hat{x})/\partial x^j$.

7. $\nabla f(\hat{x})$ denotes the gradient at $\hat{x}$ of a function $f: E^n \to E^1$. It is always treated as a column vector and is seen to be the transpose of the $1 \times n$ jacobian matrix $\partial f(\hat{x})/\partial x$. Consequently, for any y in E^n the scalar product of $\nabla f(\hat{x})$ with y satisfies $\langle \nabla f(\hat{x}), y \rangle = \dfrac{\partial f(\hat{x})}{\partial x} y$. For emphasis, we shall sometimes write the latter as $\left\langle \dfrac{\partial f(\hat{x})}{\partial x}, y \right\rangle$.

8. Superscript minus one (-1) denotes the inverse of a matrix, e.g., Q^{-1}.

9. Superscript capital T denotes the transpose of a matrix, e.g., R^T.

10. Superscript perpendicular ($\perp$) denotes the orthogonal complement of a subspace, e.g., $\mathfrak{N}^\perp$.

II. GENERAL SYMBOLS AND ABBREVIATIONS

$A \triangleq B$	A equals B by definition; denotes
$A \Rightarrow B$	A implies B
$A \Leftarrow B$	A is implied by B
$A \supset B$	A contains B
$A \subset B$	A is contained in B; A is a subset of B
$A \cup B$	union of A and B
$A \cap B$	intersection of A and B
$A \times B$	cartesian product of A and B
$\{x:P\}$	set of x's having property P

$x \in A$	x is an element of A
$x \notin A$	x does not belong to A
$A + B =$	$\{x: x = y + z, y \in A, z \in B\}$ (linear combination of sets)
$A/B =$	$\{x: x \in A, x \notin B\}$ (difference of sets)
$\dot{A}$	interior of A
$\bar{A}$	closure of A if $A \subset E^n$; complement of A if $A \subset \{1,2,\ldots,m\}$
∂A	boundary of A
co A	convex hull of A
$\overline{co}$ A	closure of convex hull of A
rel int A	relative interior of A
(a,b)	open interval $\{t: a < t < b\}$
$(a,b]$	semiclosed interval $\{t: a < t \leq b\}$
$[a,b]$	closed interval $\{t: a \leq t \leq b\}$
$\langle \cdot, \cdot \rangle$	scalar product (dots stand for undesignated variables)
$\cdot \rangle \langle \cdot$	dyad (for $x \in E^n$ and $y \in E^m$, $x \rangle \langle y$ is the $n \times m$ matrix xy^T)
$\| \cdot \|$	euclidean norm
max (a,b)	max $(a,b) = a$ if $a \geq b$, max $(a,b) = b$ if $b > a$
$\max_i$	maximum over i
$x \geq y$	for $x,y \in E^n$, $x \geq y$ if $x^i \geq y^i$ for $i = 1,2,\ldots,n$
sgn	signum function: sgn $x = 1$ for $x > 0$, sgn $x = -1$ for $x < 0$, sgn $x = 0$ for $x = 0$
sat	saturation function: sat $x = x$ for $|x| \leq 1$, sat $x =$ sgn x for $|x| > 1$
$\square$	end discussion (proof)

III. SYMBOLS WITH SPECIAL MEANING

$C(\hat{z},\Omega)$	conical approximation to Ω at $\hat{z}$
$IC(\hat{z},\Omega)$	internal cone to Ω at $\hat{z}$
$RC(\hat{z},\Omega)$	radial cone to Ω at $\hat{z}$
$TC(\hat{z},\Omega)$	tangent cone to Ω at $\hat{z}$
$TC_s(\hat{z},\Omega)$	sequential tangent cone to Ω at $\hat{z}$
f	cost function
r	equality constraint function
$F = (f,r)$	
q^0	cost function
q	inequality constraint function
Ω, Ω'	constraint sets
I	active inequality constraint index set
J	basis indicator index set
x_i	state of dynamical system at time i
u_i	system input at time i
f_i	dynamics function
f_i^0	incremental cost function
$\mathfrak{X}$	trajectory (of states)
$\mathfrak{U}$	control sequence

X_i, X_i', X_i''	state space constraints at time i
U_i	control constraint at time i
$\mathbf{x}_i$	phase (augmented state) at time i
$\mathbf{X}$	trajectory (of phases)
q_i	inequality constraint function (state space)
g_i	equality constraint function (state space)
k	duration of discrete optimal control process

Index

Active constraint, 55, 60, 61
   index set, 60, 61, 108, 126, 180
Adjoint equation, 45, 46, 79
Admissible controls, 5
Admissible trajectory, 5
Algorithms:
   for computing a feasible direction, 184
   for computing adjacent extreme points, 114
   for computing initial feasible solution:
     linear programming problems, 111
     nonlinear programming problems, 177
   for free-time problems with penalty on time, 216

Algorithms: (cont'd)
   for general discrete optimal control problems (method of feasible directions), 192
   for linear control problems, 99, 101
   for linear minimum-time problem, 229
   for linear programming problems:
     the bounded-variable simplex algorithm, 127
     the perturbed simplex algorithm, 119
     the simplex algorithm, 117
   for nonlinear programming problems:
     exterior penalty function methods, 206

Algorithms: (cont'd)
  for nonlinear programming problems: (cont'd)
    gradient projection methods, 200, 204
    interior penalty function methods, 211
    methods of feasible directions, 188
  for quadratic control problems, 131, 140, 144, 145, 148–151, 154–156
  for quadratic programming problems:
    an extension of the simplex algorithm for quadratic programming, 166
    the simplex algorithm for quadratic programming, 160
  for unconstrained problems:
    Newton's method with variable step size, 199
    steepest descent, 197
Alternative transcription:
  for linear control problem, 100
  for minimum-energy problem, 144
  for minimum-quadratic-cost problem, 151
  for quadratic control problem, 132
Arrow, K. J., 55, 73

Basic problem, 9, 14, 15, 38, 259
Basic solution, nondegenerate, 110, 126
Basis-indicator set, 111, 126
Beale, E. M. L., 176
Berge, C., 258
Boltyanskii, V. G., 48, 275
Boundary of set, 239
Bounded-variable simplex algorithm, 127
Brouwer fixed-point theorem, 30

Canon, M. D., 3, 12, 47
Charnes, A., 119, 129
Closure of set, 239
Concave functions, 252
Conditions:
  for $p^0 < 0$, 82, 83, 96, 146
  for $\psi^0 < 0$, 22, 37, 38, 55, 65, 66
Cone, 240–242
  generated by a set, 241
  polar, 244
Conical approximation:
  of the first kind, 24

Conical approximation: (cont'd)
  for infinite-dimensional problems, 260
  of the second kind, 24
Constraint qualification, 55
  weakened, 22, 37, 38, 55, 65, 66, 82, 83, 96
Constraints:
  active, 60, 61
    index set, 60, 61, 108, 126, 180
  control, 5
  inactive, 60, 61
    index set, 60, 61, 108, 126
  state-space, 5
  trajectory, 5
Control approach:
  to minimum-energy problem, 145
  to minimum quadratic cost problem, 148
Controllable, completely, 140, 222
Convex combination of points, 235
Convex cone, 240
Convex functions:
  optimality conditions with, 70–72, 96, 256
  properties of, 252–258
Convex hull, 236, 238, 239
Convex polyhedral cone, 242
Convex polyhedron, 237
  representation theorem for, 135
Convex programming problem, 177
  with affine constraints, 196
Convex set, 233
  directionally convex, 84
Cophase vector, 44–76
Cost function, 9, 19
Costate vector, 78, 91, 274
Cottle, R. W., 163, 176
Courant, R., 205, 213
Cullum, C. D., 3, 12, 47
Curry, H. B., 197, 213

Da Cunha, N. O., 275
Dantzig, G. B., 110, 163, 176
Davidon, W. C., 206, 213
Degeneracy, 110, 115, 126, 158, 160, 168
  resolution of, 119
Derived problem:
  exterior penalty functions, 207
  interior penalty functions, 211
  quadratic programming, 158, 163

INDEX

Desoer, C. A., 230
Dimension of a convex set:
  full, 239
  linear, 239
Direct product, 234
Directional convexity, 84
Directionally convex optimal control problem, 86
Discretization, 2, 3

Eggleston, H. G., 258
Equality constraint function, 9, 19
Evaluation function, 265
Extreme point, 236
  in linear programming, 108
  in quadratic programming, 159

Farkas' lemma, 250
Feasible directions, 185
  jamming, zigzagging, 187
  method of, 188
  in optimal control, 192
Feasible solution, 10, 15, 111, 176
Fiacco, A. V., 213
Fleming, W. F., 74
Fletcher, R., 206, 213
Fritz John conditions, 70
Fromowitz, S., 67, 72, 74
Fundamental theorem, 27, 260

Gamkrelidze, R. N., 48, 275
Goldman, A. J., 66, 74, 135, 176
Gradient projection, 204
  special case, 200
Graves, L. M., 48

Half spaces, 233
  open, closed, 233
Halkin, H., 3, 48, 84
Hamiltonian, 45, 79, 92, 274
Hessian matrix, 36, 257
Heuristic optimality condition, 23
Hoffman, A. J., 119, 129
Holtzman, J. M., 84
Hurewicz, W., 48
Hurwig, L., 55, 73

Hyperplane, 232
  separating, 245, 247, 250
  supporting, 251

Initial boundary constraints, 76
Initial feasible solution, 111, 177
Interior, 239
  relative, 239
Internal cone, 61, 180

Jacobian matrix, 19, 42
John, F., 70, 73
Jordan, B. W., 48, 97

Kalman, R. E., 230
Kelley, T. L., 264, 276
Kuhn, H. W., 15, 48, 54, 55, 73
Kuhn-Tucker conditions, 15, 55

Lagrange multiplier, 50
  rule, 51
Lemke, C. E., 163, 176
Line, 231, 232
Linear combination of sets, 233
Linear control problem, 99
Linear function, 234
Linear manifold, 232
Linear minimum-time problem, 221
  algorithm for, 229
Linear programming problems:
  bounded-variable form, 125
    basic solution, 126
    basis indicator set, 126
    simplex algorithm for, 127
  canonical form, 105
    basic solution, 110
    basis indicator set, 111
    conditions for optimality, 107
    nondegeneracy assumption, 115
    resolution of degeneracy, 119
    simplex algorithm for, 117
  general form, 98

McCormic, G. P., 213
McShane, E. J., 273
Mangasarian, O. L., 67, 74

Maximum principle, 46, 81, 84, 91, 274
Messerli, E. J., 37, 48
Minimum-energy problem, 46, 140, 154
Minimum-fuel problem, 93, 101
Mishchenko, E. F., 48, 275
Motzkin's lemma, 68

Necessary conditions:
  for basic problem:
    finite-dimensional case, 15, 20, 27, 35, 38, 85
    infinite-dimensional case, 260
  for continuous optimal control problems, 274
  for discrete optimal control problems, 45, 78, 80, 81, 91, 96
  for linear programming problems, 107
  for nonlinear programming problems, 51, 52, 54, 56, 64–67, 69, 70, 182
  for problems without differentiability assumptions, 38
  for quadratic programming problems, 139, 157
Negative cost ray, 20
Neustadt, L. W., 3, 48, 276
Newtons' method with variable step size, 199
Nondegeneracy assumption:
  for linear programming, 115
  for quadratic programming, 168
Nondegenerate basic solution, 110, 126
Nonlinear programming problem, 49

Optimal control problems:
  continuous fixed time:
    in basic problem form, 265
    general form, 2, 262
  discrete fixed time:
    canonical form, 8
    general form, 5, 40, 76, 86
    linear control problems, 99, 101–105
    minimum energy, 46, 140, 154
      fuel, 93, 101
      quadratic cost, 149, 153, 155
      quadratic regulator, 148
    Tchebyshev approximation, 104
  discrete free time:
    bounded final time, 217
    general form, 6, 214

Optimal control problems: (cont'd)
  discrete free time: (cont'd)
    minimum time, 218, 221
    with penalty on time, 215–217
Optimal solution, 10, 15
Optimality conditions (see Necessary conditions; Sufficient conditions)

Penalty function:
  exterior, 206, 209, 210
  interior, 211
Perturbed linear programming problem, 120
Phase, 40
Polak, E., 3, 12, 37, 47, 97, 213, 276
Polar convex cone, 244
Pontryagin, L. S., 48, 259, 264, 274, 275
Pontryagin maximum principle, 259, 274
Positive definite, 257
Positive semidefinite, 257
Powell, M. J. D., 206, 213
Projection, orthogonal, 200
Projection matrix, 202
Propoi, A. J., 97
Pseudoconvex functions, 72
  optimality conditions with, 72

Quadratic programming problems:
  algorithms for solution of, 160, 166
  canonical form, 156
  derived problem, 158, 163
  existence of solution to, 132
  general forms, 130, 132
  nondegeneracy assumption, 168
  optimality conditions for, 139, 157
  uniqueness of solution to, 137
  usable basic solution, 158
Quadratic regulator problem, 148
Quasi-convex functions, 72
  optimality conditions with, 72

Radial cone, 20
Rosen, J. B., 97, 201, 213

Separation of convex sets, 245
  strict, 245
Sequential tangent cone, 17, 18

Simplex, 237
Simplex algorithm:
  for bounded variable problem, 127
  for canonical problem, 117
  for degenerate problem, 119
  for quadratic programming problem, 160
    further extension, 166
Solution:
  feasible, 10, 15
  optimal, 10, 15
Star-shaped set, 67
State, augmented, 40
Steepest descent, 197
Sufficient conditions, 70–72, 107, 139, 157

Tangent cone, 15
  sequential, 17, 18
Terminal boundary constraints, 76
Torng, H. C., 230

Transversality conditions, 79, 92
Tucker, A. W., 15, 48, 54, 55, 73

Usable basic solution, 158
  nondegenerate, 158
Uzawa, H., 55, 73

Van Slyke, R. M., 213
Vertex, 240

Wallman, H., 48
Whalen, B. H., 230
Wing, J., 230
Wolfe, P., 159, 176, 187, 213

Zadeh, L. A., 230
Zangwill, W., 213
Zoutendijk, G., 179, 213

Undergraduate Texts in Mathematics

Readings in Mathematics

Editors
S. Axler
F.W. Gehring
P.R. Halmos

Springer
*New York
Berlin
Heidelberg
Barcelona
Budapest
Hong Kong
London
Milan
Paris
Tokyo*

Graduate Texts in Mathematics
Readings in Mathematics

Ebbinghaus/Hermes/Hirzebruch/Koecher/Mainzer/Neukirch/Prestel/Remmert: *Numbers*
Fulton/Harris: *Representation Theory: A First Course*
Remmert: *Theory of Complex Functions*

Undergraduate Texts in Mathematics
Readings in Mathematics

Anglin: *Mathematics: A Concise History and Philosophy*
Anglin/Lambek: *The Heritage of Thales*
Bressoud: *Second Year Calculus*
Hämmerlin/Hoffmann: *Numerical Mathematics*
Isaac: *The Pleasures of Probability*
Samuel: *Projective Geometry*

W.S. Anglin
J. Lambek

The Heritage of Thales

With 23 Illustrations

 Springer

PENSACOLA JR. COLLEGE LRC

W.S. Anglin
J. Lambek
Department of Mathematics
  and Statistics
McGill University
Montreal, Quebec
Canada H3A 2K6

Editorial Board

S. Axler
Dept. of Mathematics
Michigan State University
East Lansing, MI 48824
USA

F.W. Gehring
Dept. of Mathematics
University of Michigan
Ann Arbor, MI 48109
USA

P.R. Halmos
Dept. of Mathematics
Santa Clara University
Santa Clara, CA 95053
USA

Mathematics Subject Classification (1991): 01-01, 01A05

Library of Congress Cataloging-in-Publication Data
Anglin, W.S.
    The heritage of Thales / W.S. Anglin and J. Lambek.
        p. cm. — (Undergraduate texts in mathematics. Readings in
    mathematics)
    Includes bibliographical references and index.
    ISBN 0-387-94544-X (hc : alk. paper)
    1. Mathematics—History. 2. Mathematics—Philosophy. I. Lambek,
Joachim. II. Title. III. Series.
QA21.A535 1995
510′.9—dc20   95-19695

Printed on acid-free paper.

© 1995 Springer-Verlag New York, Inc.
All rights reserved. This work may not be translated or copied in whole or in part without the written permission of the publisher (Springer-Verlag New York, Inc., 175 Fifth Avenue, New York, NY 10010, USA), except for brief excerpts in connection with reviews or scholarly analysis. Use in connection with any form of information storage and retrieval, electronic adaptation, computer software, or by similar or dissimilar methodology now known or hereafter developed is forbidden.
The use of general descriptive names, trade names, trademarks, etc., in this publication, even if the former are not especially identified, is not to be taken as a sign that such names, as understood by the Trade Marks and Merchandise Marks Act, may accordingly be used freely by anyone.

Production managed by Natalie Johnson; manufacturing supervised by Jeffrey Taub.
Photocomposed copy prepared from the authors' LaTeX file.
Printed and bound by R.R. Donnelley & Sons, Harrisonburg, VA.
Printed in the United States of America.

9 8 7 6 5 4 3 2 1

ISBN 0-387-94544-X Springer-Verlag New York Berlin Heidelberg

Preface

This is intended as a textbook on the history, philosophy and foundations of mathematics, primarily for students specializing in mathematics, but we also wish to welcome interested students from the sciences, humanities and education. We have attempted to give approximately equal treatment to the three subjects: history, philosophy and mathematics.

History
We must emphasize that this is not a scholarly account of the history of mathematics, but rather an attempt to teach some good mathematics in a historical context. Since neither of the authors is a professional historian, we have made liberal use of secondary sources. We have tried to give references for cited facts and opinions. However, considering that this text developed by repeated revisions from lecture notes of two courses given by one of us over a 25 year period, some attributions may have been lost. We could not resist retelling some amusing anecdotes, even when we suspect that they have no proven historical basis. As to the mathematicians listed in our account, we admit to being colour and gender blind; we have not attempted a balanced distribution of the mathematicians listed to meet today's standards of political correctness.

Philosophy
Both authors having wide philosophical interests, this text contains perhaps more philosophical asides than other books on the history of mathematics. For example, we discuss the relevance to mathematics of the pre-Socratic philosophers and of Plato, Aristotle, Leibniz and Russell. We also have

presented some original insights. However, on some points our opinions diverge; so, in a spirit of compromise, we have agreed to excise some of our more extreme views. Some of these divergent opinions have been expressed in Anglin [1994] and Lambek [1994].

Mathematics

One of the challenges one faces in offering a course on the history and philosophy of mathematics is to persuade one's colleagues that the course contains some genuine mathematics. For this reason, we have included some mathematical topics, usually not treated in standard courses, for example, the renaissance method for solving cubic equations and an elementary proof of the impossibility of trisecting arbitrary angles by ruler and compass constructions. We have taken the liberty of presenting many mathematical ideas in modern garb, with the hindsight inspired by more recent developments, since a presentation faithful to the original sources, while catering to the serious scholar, would bore most students to tears.

In Part I we deal essentially with the history of mathematics up to about 1800. This is because thereafter mathematics tends to become more specialized and too advanced for the students we have in mind. However, we make occasional excursions into more modern mathematics, partly to relieve the tedium associated with a strictly chronological development and partly to present modern answers to some ancient questions, whenever this can be done without overly taxing the students' ability.

In Part II we deal with some selected topics from the nineteenth and twentieth centuries. In that period, mathematics became rather specialized and made spectacular progress in different directions, but we confine attention to questions in the foundations and philosophy of mathematics.

The more universal aspects of mathematics are sketched briefly in the last five sections. We introduce the language of category theory, which attempts a kind of unification of different branches of mathematics, albeit at a very basic and abstract level.

Acknowledgements

The authors wish to acknowledge partial support from the Social Sciences and Humanities Research Council of Canada, from the Natural Sciences and Humanities Research Council of Canada and from the Quebec Department of Education.

We wish to express our sincerest thanks to Matthew Egan for his undaunted dedication in typing and editing and to Mira Bhargava, Henri Darmon and Ramona Behravan for their conscientious reading and criticism of the manuscript.

W. S. Anglin and J. Lambek

Contents

Preface	v
0 Introduction	1
PART I: History and Philosophy of Mathematics	5
1 Egyptian Mathematics	7
2 Scales of Notation	11
3 Prime Numbers	15
4 Sumerian-Babylonian Mathematics	21
5 More about Mesopotamian Mathematics	25
6 The Dawn of Greek Mathematics	29
7 Pythagoras and His School	33
8 Perfect Numbers	37
9 Regular Polyhedra	41
10 The Crisis of Incommensurables	47
11 From Heraclitus to Democritus	53

12 Mathematics in Athens	59
13 Plato and Aristotle on Mathematics	67
14 Constructions with Ruler and Compass	71
15 The Impossibility of Solving the Classical Problems	79
16 Euclid	83
17 Non-Euclidean Geometry and Hilbert's Axioms	89
18 Alexandria from 300 BC to 200 BC	93
19 Archimedes	97
20 Alexandria from 200 BC to 500 AD	103
21 Mathematics in China and India	111
22 Mathematics in Islamic Countries	117
23 New Beginnings in Europe	121
24 Mathematics in the Renaissance	125
25 The Cubic and Quartic Equations	133
26 Renaissance Mathematics Continued	139
27 The Seventeenth Century in France	145
28 The Seventeenth Century Continued	153
29 Leibniz	159
30 The Eighteenth Century	163
31 The Law of Quadratic Reciprocity	169
PART II: Foundations of Mathematics	173
1 The Number System	175
2 Natural Numbers (Peano's Approach)	179
3 The Integers	183

4	The Rationals	187
5	The Real Numbers	191
6	Complex Numbers	195
7	The Fundamental Theorem of Algebra	199
8	Quaternions	203
9	Quaternions Applied to Number Theory	207
10	Quaternions Applied to Physics	211
11	Quaternions in Quantum Mechanics	215
12	Cardinal Numbers	219
13	Cardinal Arithmetic	223
14	Continued Fractions	227
15	The Fundamental Theorem of Arithmetic	231
16	Linear Diophantine Equations	233
17	Quadratic Surds	237
18	Pythagorean Triangles and Fermat's Last Theorem	241
19	What Is a Calculation?	245
20	Recursive and Recursively Enumerable Sets	251
21	Hilbert's Tenth Problem	255
22	Lambda Calculus	259
23	Logic from Aristotle to Russell	265
24	Intuitionistic Propositional Calculus	271
25	How to Interpret Intuitionistic Logic	277
26	Intuitionistic Predicate Calculus	281
27	Intuitionistic Type Theory	285

28 Gödel's Theorems	**289**
29 Proof of Gödel's Incompleteness Theorem	**291**
30 More about Gödel's Theorems	**293**
31 Concrete Categories	**295**
32 Graphs and Categories	**297**
33 Functors	**299**
34 Natural Transformations	**303**
35 A Natural Transformation between Vector Spaces	**307**
References	311
Index	321

0
Introduction

Remarks on prehistory

Long before written records were kept, people were concerned with the seasons, important in agriculture, and the sky, which permitted them to read off the passage of time. Everyone knows that the *year* is the time it takes the sun to complete its orbit about the earth. (Copernicus notwithstanding, mathematical readers will see nothing wrong with placing the origin of the coordinate system at the center of the earth.) Also, a *month* is supposed to be the time it takes the moon to go around the earth; at least, this was the case before the lengths of the months were laid down by law. But what about the *week*? Theological explanations aside, it is the smallest period, longer than a day, that can be easily observed by looking at the sky: the time it takes the moon to pass from one phase to another, from new moon to half moon, from half moon to full moon, etc.

The days of the week are named after the sun, the moon and the five planets visible to the naked eye: Mars (French *mardi*), Mercury (French *mercredi*), Jupiter (French *jeudi*), Venus (French *vendredi*) and Saturn (English *Saturday*). The English *Tuesday, Wednesday, Thursday* and *Friday* are named after the Teutonic deities which supposedly correspond to the Roman gods after whom the planets were named.

In Hindu astronomy there are nine planetary deities, the *graha*. In addition to the seven associated with the days of the week, there are two others, *rahu* and *kebu*, alleged to be associated with the so-called 'nodes'. These are the points where the orbits of the sun and the moon, when traced out on the firmament, intersect. (See Freed and Freed [1980].) The importance of the nodes is that an eclipse of the sun or the moon can only occur when both sun and moon fall on the nodes, to within 10°; according to an ancient rule of thumb, this happens once in about 18.6 years.

At Stonehenge in England there is an imposing prehistoric monument, dating from about 2,500 BC. The huge standing stones of the monument were presumably used to sight the points on the horizon where the sun and the moon, and perhaps Venus, rise and set at certain dates (Hawkins [1965]). They are surrounded by a circle of 56 holes in the ground, and Fred Hoyle [1977] has proposed the ingenious hypothesis that these were used as a calendar and to calculate the dates of possible eclipses.

According to Hoyle, the idea was to move a *sun marker* two holes in 13 days, a *moon marker* two holes each day, and two *nodal stones* three holes per year. The sun marker would thus complete an orbit in 364 days; the discrepancy could be fixed by appropriate adjustments at midsummer and midwinter. The moon marker would complete an orbit in 28 days, that is, four weeks. Of course, this should really be 29.5 days, so adjustments might have to be made each full moon and each new moon. The nodal stones would take 56/3 years to perform a complete orbit. On those occasions when both sun marker and moon marker were about to catch up with the nodal stones, the presiding priest could risk predicting an eclipse.

Foreword on history

Even so-called 'primitive' societies may be engaged in some fairly sophisticated mathematical activities, for example, the calculations involved in kinship descriptions. (How many students can tell on the spot what exactly is a *second cousin three times removed*?) For interested readers, we recommend two recent books: *Africa Counts* by Claudia Zaslovsky and *Ethnomathematics* by Marcia Ascher.

Mathematics, as usually conceived, begins with the development of agriculture in the river valleys of Egypt, Iraq, India and China. If we pay more attention to the Near East than to the Far East, this is because the former has provided us with more accessible records and because modern mathematics can be traced back directly to it. We possess written records concerning the state of mathematics in Egypt and Mesopotamia (Iraq) from as early as about 2000 BC. Around 500 BC, mathematical knowledge spread to the Greek world. This included not only modern Greece, but also the coast of Asia Minor (modern Turkey) and Magna Grecia (southern Italy and Sicily). About 300 BC, the center of mathematics moved from Athens to Alexandria in Egypt, where it was to remain for the next 800 years; for it was in Alexandria that all the books were kept.

Around 500 AD, mediterranean civilization finally came to a stop, perhaps because of the repeated impact of epidemic diseases. About 800 AD, mathematics in the Alexandrian tradition resurfaced in India, which had a long mathematical tradition of its own. The Arabs, aided by translations of Greek texts, developed and transmitted mathematical knowledge from India back to the mediterranean area and ultimately to Europe. During the so-called 'renaissance', mathematics flourished in Italy and, aided by the Chinese invention of printing, spread to Western and Central Europe. Of course, today mathematics is being pursued in all the industrial countries of the world.

Introduction to the number system

The historical and pedagogical development of the number system goes somewhat like this:

$$\mathbf{N}^+ \to \mathbf{Q}^+ \to \mathbf{R}^+ \to \mathbf{R} \to \mathbf{C} \to \mathbf{H} \ .$$

Here $\mathbf{N}^+$ is the set of positive integers, the *numbers* used for counting, known to all societies. $\mathbf{Q}^+$ is the set of positive rationals, namely, *quotients* of positive integers, surely known to all agricultural civilizations. At one time, they were believed to exhaust all the numbers, until the Pythagoreans discovered that the diagonal of a square was not a rational multiple of its side. We use $\mathbf{R}^+$ to denote the positive *reals*; these were certainly used effectively by Thales, though the Greeks originally tended to regard them as ratios of geometric quantities. A formal treatment, anticipating the nineteenth century definition by Dedekind, was first given by Eudoxus in Athens. The transition from $\mathbf{R}^+$ to $\mathbf{R}$, the set of all reals, positive, zero and negative, took place in India and may be ascribed to Brahmagupta. The set $\mathbf{C}$ of *complex* numbers was first considered by Cardano to describe the intermediate steps in solving a cubic equation with real coefficients and three real solutions. The set $\mathbf{H}$ of *quaternions* is named after their inventor William Hamilton, who may have been preceded by Olinde Rodrigues and perhaps even by Carl Friedrich Gauss.

Most of the advances in the development of the number system may have been motivated by the desire to solve equations. Thus, the equations $2x = 3$, $x^2 = 2$, $x + 1 = 0$ and $x^2 + 1 = 0$ led to the successive introduction of $\mathbf{Q}^+, \mathbf{R}^+, \mathbf{R}$ and $\mathbf{C}$, respectively. However, all polynomial equations with complex coefficients do have complex solutions, so the introduction of quaternions requires a different justification. They were motivated by the desire to pass from the plane, describable by complex numbers, to three or four dimensions.

Part I
Topics in the History and Philosophy of Mathematics

1
Egyptian Mathematics

The Greeks believed that mathematics originated in Egypt. As to the reason for this, opinion was divided. Aristotle thought that mathematics was developed by priests, 'because the priestly class was allowed leisure' (*Metaphysics* 981b 23-24). Herodotus believed that the annual flooding of the Nile necessitated surveying to redetermine field boundaries, and thus led to the invention of geometry. In fact, Democritus referred to Egyptian mathematicians as 'rope stretchers'. It may be of interest to note that the Egyptians themselves believed that mathematics had been given to them by the god Thoth. Our only original sources of information on the mathematics of ancient Egypt are the Moscow Mathematical Papyrus and the Rhind Mathematical Papyrus.

The Moscow Papyrus dates from 1850 BC, about the time the Bible dates the life of the patriarch Abraham. In 1893 it was acquired by V. S. Golenishchev and brought to Moscow (Gillings [1972], p. 246). Problem 14 of this papyrus is by far the most interesting. It is the computation of a *truncated pyramid*, a square pyramid with a similar pyramid cut off its top. If a side of the base has length a and a side of the top has length b, then the volume of the truncated pyramid of vertical height h is

$$V = \frac{h}{3}(a^2 + ab + b^2).$$

This is exactly the formula used by the Egyptians. Note that, if $b = 0$, we get the formula for the volume of the complete pyramid.

The Rhind Mathematical Papyrus seems to be based on an earlier work. It was written by one Ahmose in 1650 BC, about the time when, accord-

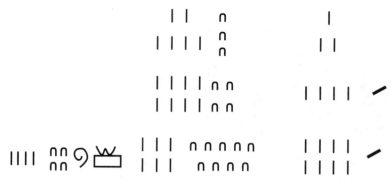

FIGURE 1.1. Rhind Papyrus

ing to the Bible, Joseph was governor of Egypt. Alexander Henry Rhind acquired it in Luxor, Egypt in 1858; the British Museum bought it from his estate in 1865. Complete photographs of the papyrus can be found in *The Rhind Mathematical Papyrus* edited by G. Robins and C. Shute.

The Rhind Papyrus opens by promising the reader 'a thorough study of all things, insight into all that exists, knowledge of all obscure secrets'. It is a bit of a letdown to find that it is, in fact, a sequence of solved problems in elementary mathematics, a sort of Schaum's outline for aspiring scribes. These scribes had to calculate how many bricks were needed to build a ramp of a certain size, how many loaves of bread were required to feed the labourers, and so on.

Problem 32 of the papyrus is an exercise in multiplication written as in Figure 1.1.

Transcribing this into modern notation, we have

```
 12         1
 24         2
 48         4      /
 96         8      /
144  =  the sum of the checked entries.
```

Clearly, this is a calculation to show that $12 \times 12 = 144$, using the fact that $12 = 4 + 8$.

By doubling and adding, the Egyptians were able to multiply any two natural numbers – without having to memorize multiplication tables! Sometimes they used a different, yet equivalent, method, as illustrated by the following multiplication of 70 by 13:

```
 70        13        /
140         6
280         3        /
560         1        /
910  =  sum of checked entries.
```

We let the reader figure out why this works. The method of repeated doubling can also be used for division. In the following example, we divide

184 by 17 (stopping at 136, as the next double exceeds 184):
```
   17   1
   34   2  /
   68   4
  136   8  /
```
The Egyptians would first check off the last row and subtract 136 from 184, obtaining 48. They would then check off the row containing 34, the highest multiple of 17 less than 48. Since $48 - 34 = 14$ is less than 17, they would now add up all the entries in the second column with check marks beside them: $2 + 8 = 10$. This gives the answer: the quotient is 10 and the remainder is 14.

In carrying out these divisions, the Egyptians sometimes interspersed doubling with multiplication by 10 (their language expressed numbers in the base 10, just as ours does). For example, Problem 69 in the Rhind Mathematical Papyrus is to calculate the number of 'ro' of flour in each loaf, if 1120 ro of flour is made into 80 loaves. In other words, we are asked to divide 1120 by 80:
```
   80            1
  800           10   /
  160            2
  320            4   /
```
sum of checked numbers = 14.

The Egyptians also knew how to extract square roots and how to solve linear equations. They used the hieroglyph h much as we use the letter x for the unknown. They used the formula $(\frac{4}{3})^4 r^2$ for the area of a circle (which implies 3.16 as an approximation to π) and they did some interesting work with arithmetic progressions. For example, Problem 64 of the Rhind Papyrus is to find an arithmetic progression with 10 terms, with sum 10, and with common difference $1/8$.

In using fractions, the Egyptians were hampered by a curious tradition. They insisted on expressing all fractions (except $2/3$) as the sum of distinct *unit* fractions of the form $1/n$, n being a positive integer. Thus $2/9$ would be written as $1/6 + 1/18$ and $19/8$ as $2 + 1/4 + 1/8$. Even $2/3$ is sometimes written as $1/2 + 1/6$.

For us it is easy to divide $5/13$ by 12, but for the Egyptians this was a substantial problem. To help with such problems, they had a table listing unit fraction decompositions for fractions of the form $2/n$, with n an odd positive integer. This table (found in the Rhind Papyrus) gives $2/13$ as $1/8 + 1/52 + 1/104$. Since $5 = (2 \cdot 2) + 1$, Ahmose would write

$$\begin{aligned} 5/13 &= 2(1/8 + 1/52 + 1/104) + 1/13 \\ &= 1/4 + 1/26 + 1/52 + 1/13. \end{aligned}$$

From this he would obtain

$$(5/13)/12 = 1/48 + 1/312 + 1/624 + 1/156.$$

Actually, any faction of the form $2/(2m+1)$ can be expressed as a sum of the unit fractions $1/(m+1)$ and $1/(m+1)(2m+1)$. Not that the Egyptians always followed this recipe; for example, Ahmose wrote $2/45 = 1/30 + 1/90$.

Recently, Paul Erdös proposed the following problem: show that, if n is an odd integer greater than 4, then $4/n$ can be written as a sum of three distinct unit fractions. The problem has not yet been solved. (See Mordell, p. 287.)

Exercises

1. Derive the formula for the volume of a truncated pyramid from that of a pyramid.

2. Explain why the above method for multiplying 70×13 works.

3. Find two ways of writing $1/4$ as the sum of two distinct unit fractions.

4. If m is a positive integer, show that $4/(4m+3)$ can be written as the sum of three distinct unit fractions.

2
Scales of Notation

The ancient Egyptian language belongs to the Hamito-Semitic group of languages. Like the Indo-European group, it contains a system of counting by tens, undoubtably arising from the habit of counting using one's fingers. The notation used for writing numbers is also clearly based on the scale of ten. For some reason, standard French departs from decimal nomenclature; it expresses 97 as 4 times 20 plus 17. This seems odd, since it was the French who introduced the decimal system for weights and measures.

Some African languages express numbers in the scale of five. One may express natural numbers in any scale b, where b is an integer greater than 1, since every natural number is uniquely expressible in the form

$$a = a_0 + a_1 b + a_2 b^2 + a_3 b^3 + \cdots + a_n b^n,$$

where $0 \leq a_i < b$ (for $i = 0, 1, \ldots, n$). We write this more briefly as

$$a = (a_n a_{n-1} \cdots a_2 a_1 a_0)_b.$$

If there is no doubt which scale is in use, the subscript b may be dropped.

The Egyptians had a number system based on the scale of ten, but, as we saw above, they often worked with scale two: to multiply by 12, Ahmose expressed 12 as $4 + 8$, that is, $2^2 + 2^3$, or $12 = (1100)_2$. The Egyptians also took $b = 7$ in some of their calculations (Gillings, p. 227), since there are seven *palms* in a *cubit*. They had no symbol for zero; instead they used special symbols for different powers of ten.

The binary scale (with $b = 2$) shows up in the Chinese *Book of Changes* (1200 BC), a system of divination in which each six place binary number represents some concept. The digit 1 was associated with the male 'yang', and the digit 0 with the female 'yin'. The number $34 = (100010)_2$ was supposed to represent 'progress and success'. The binary scale also shows up in the Hindu classification of meters in verse, about 800 BC. Finally, it is of course used in the modern computer. The digit 1 is represented by a current, and the digit 0 by the absence of a current. Number scales are often found in recreational mathematics, as in the following three problems.

Six Weight Problem: A balance is a weighing apparatus with a central pivot, a beam, two scales and a set of counter-weights that are placed in one of the scales. Suppose we have some flour and we want to be able to put it into bags weighing anywhere from one to sixty-three kilograms. How can this be done using just six counter-weights?

Answer: Weights of 1, 2, 4, 8, 16 and 32 kilograms will allow you to weigh any integral load from 1 to $32 + 16 + 8 + 4 + 2 + 1 = 63$ kilograms.

Four Weight Problem: This time, suppose we are allowed to put weights on either scale. How can we weigh bags under 42 kilograms using only four weights?

Answer: Choose weights of 1, 3, 9 and 27 kilograms, since any integer a can be written uniquely in the form

$$a = a_0 + a_1 3 + a_2 3^2 + \cdots + a_n 3^n,$$

where each a_i is one of -1, 0 or 1.

The Game of Nim: This so-called Chinese game is played by two opponents, who take turns removing matches from several piles according to the following rules:

1. A player must remove at least one match in a turn.
2. A player may remove any number of matches from a single pile in a turn.

The player who removes the last match wins. Find a strategy for winning this game.

Answer: Express the number of matches in each pile in the scale of two and write these binary numbers one below the other. If, when it is your turn, you can arrange it so that each column adds up to an even number, then you can do the same in every subsequent turn and you will win the game.

As an example, suppose there are three piles containing 7, 5 and 3 matches. It is your turn. In binary notation, the piles contain the following number of matches:

```
1 1 1
1 0 1
  1 1
```
To make the number of 1's in each column even, you take a match from the first pile, leading to
```
1 1 0
1 0 1
  1 1
```
Your opponent takes 2 matches, say, from the third pile, leaving
```
1 1 0
1 0 1
    1
```
You take two matches from the first pile, yielding
```
1 0 0
1 0 1
    1
```
Your opponent then removes all the matches from the first pile, say, resulting in
```
1 0 1
    1
```
You now take 4 matches from the first pile, leaving just one match in each pile. Your opponent has to take one of them, and you win by taking the last match.

What is going on in the Six Weight Problem? It is easily seen that the following three statements are equivalent:

1. Every integral load less than 64 kg can be weighed uniquely with 6 weights: 1, 2, 4, 8, 16 and 32 kg.

2. Every natural number less than 64 can be expressed uniquely as the sum of distinct powers of 2.

3. Every natural number less then 64 can be written uniquely in the scale of 2 with up to 6 digits, each 0 or 1.

A direct proof is quite easy, but a proof in the spirit of the 18th century is more interesting. In preparation for this proof, let us look at the following multiplication:

$$(1+x^2)(1+x^3)(1+x^5)(1+x^7) = 1+x^2+x^3+2x^5+2x^7+x^8+x^9+2x^{10}+\ldots$$

Why are the coefficients of x^5, x^4 and x^3 equal to 2, 0 and 1, respectively? Because $5 = 2 + 3$ can be written as the sum of some of 2, 3, 5 and 7 in two ways (the first sum consists of one term only), 4 cannot be written as the sum of some of 2, 3, 5 and 7 at all, and 3 can only be written as the sum of these numbers in one way, the sum having one term. In general, the coefficient of x^n will be the number of ways in which n can be written as the sum of distinct members of the set $\{2, 3, 5, 7\}$.

Now let us replace this set of numbers by the set $\{1, 2, 4, 8, 16, 32\}$ and consider

$$(1+x)(1+x^2)(1+x^4)(1+x^8)(1+x^{16})(1+x^{32}) = \sum_{i=0}^{\infty} f(n)x^n.$$

Then $f(n)$ is the number of ways in which n can be written as the sum of distinct powers of 2, up to 32. Clearly, $f(n) = 0$ when $n \geq 64$. What if $n < 64$? The left-hand side can also be written

$$\frac{1-x^2}{1-x} \cdot \frac{1-x^4}{1-x^2} \cdot \frac{1-x^8}{1-x^4} \cdot \frac{1-x^{16}}{1-x^8} \cdot \frac{1-x^{32}}{1-x^{16}} \cdot \frac{1-x^{64}}{1-x^{32}} = \frac{1-x^{64}}{1-x}$$

$$= 1 + x + x^2 + \cdots + x^{63} = \sum_{n=0}^{63} x^n.$$

Hence $f(n) = 1$ if $n < 64$.

Suppose, instead of stopping with x^{32}, we form the *infinite* product $\prod_{n=0}^{\infty}(1 + x^{2^n})$. Then we can show similarly that *every* natural number can be written in the scale of 2.

Exercises

1. Write out a scale 7 multiplication table.

2. Show how to convert a scale 10 numeral to a scale 7 numeral.

3. Give a proof, in the spirit of the 18th century, that every natural number can be written uniquely in the scale of 3. (Hint: form the infinite product $(1+x+x^2)(1+x^3+x^6)(1+x^9+x^{18}) \cdots$ and evaluate it in two different ways.)

4. Likewise, show that any integer can be written uniquely as $\sum_{k=0}^{n} a_k 3^k$, where $a_k = -1, 0$ or 1.

3
Prime Numbers

It would be impossible to write a history of mathematics without mentioning prime numbers, and it would be improper to give an account of prime numbers without going into the history of mathematics. Prime numbers enter into almost every branch of mathematics; they are as fundamental as they are ubiquitous. Their history can be used as a framework for a history of mathematics generally. In this chapter, we take a brief look at the fascinating subject of primes.

The Egyptians might have written
$$\frac{4}{5} = \frac{1}{2} + \frac{1}{4} + \frac{1}{20}.$$
From this, it follows that
$$\frac{4}{10} = \frac{1}{4} + \frac{1}{8} + \frac{1}{40}$$
and that
$$\frac{4}{15} = \frac{1}{6} + \frac{1}{12} + \frac{1}{60}.$$
The moral to be drawn from this is that, to express a/b as a sum of unit fractions, it suffices to consider the case when b cannot be factored into smaller numbers. An integer greater than 1 which cannot be factored into numbers, all of which are smaller than the original integer, is called *prime*. The first few primes are
$$2, 3, 5, 7, 11, 13, 17, 19, 23, 29, 31, 37, \ldots.$$

3. Prime Numbers

Note that a positive integer is *prime* if and only if it has exactly two positive integer divisors.

Early on, people noticed that a pile of small stones can sometimes be arranged in a rectangle and sometimes it cannot. Thus, although we do not have any record of this, the Egyptians probably knew the difference between composite and prime numbers. Indeed, it is not impossible that some Egyptian scribe may have noticed that, if every proper fraction of the form $4/p$, with p prime, and greater than 3, can be expressed as a sum of three distinct unit fractions, then every proper fraction of the form $4/n$, with n any positive integer greater than 4, can be so expressed. (See the problem of Erdös, mentioned in Chapter 1.)

It was the Greeks who first *proved* that the number of primes is infinite. A proof is found in Euclid's *Elements* (300 BC).

Euclid's Lemma (Book VII Proposition 31):

Every integer $n > 1$ is divisible by some prime number.

Proof: Among the divisors of n which are greater than 1, let p be the smallest. Then p has no divisors other than 1 and p — any other divisor of p would be a divisor of n as well — and hence p is prime.

Euclid's Theorem (Book IX Proposition 20):

Given any finite list of primes $p_1, p_2 \ldots p_k$, there is a prime not on this list.

Proof: Consider the number $n = p_1 p_2 \cdots p_k + 1$. Clearly, n is not divisible by any of the primes on the list; for, upon dividing n by p_i, we get remainder 1. From the lemma we know that n does have a prime factor q (possibly n itself). Hence there is a prime, namely, q, which is not on the list. QED.

In Proposition 14 of Book IX, Euclid proved that, if n is a square-free positive integer (that is, one with no square factor other than 1), then n has a factorization into primes which is unique (if you list the prime factors in order of increasing size). However, it was not until 1801 that the unique factorization was formally proved for *any* positive integer n. This was done by Carl Friedrich Gauss (1777–1855) in his *Disquisitiones Arithmeticae*. Although mathematicians used the unique factorization theorem long before 1801, and although almost any one of them could have found a proof for it, Gauss was the first person actually to sit down and do so. Perhaps the other mathematicians considered the theorem too obvious to be worth proving. One way to prove that every positive integer greater than 1 has a unique factorization into primes is as follows.

Proof of the Unique Factorization Theorem:

Let n be the smallest positive integer, if there is one, which has 2 (or more) factorizations into primes:

$$n = pqr \ldots = p'q'r' \ldots .$$

We assume the primes are written in nondecreasing order. By minimality of n, $p \neq p'$ (or we could cancel off the p's and get a smaller number with

two factorizations). Without loss of generality, we may suppose that $p' < p$. Hence
$$p' < p \leq q \leq r \leq \ldots . \qquad (*)$$
Since n is not prime, $n \geq p^2$, and hence $n > pp'$. By minimality of n, $n - pp'$ has a unique factorization. Both p and p' are factors of $n - pp'$ (since $n - pp' = p(qr \cdots - p') = p'(q'r' \cdots - p)$) and hence, for some positive integer z, $n - pp' = pp'z$. This gives $qr \cdots - p' = p'z$, so that p' is a factor of $qr \cdots$. Since $qr \cdots < n$, $qr \cdots$ has a unique factorization into primes. Thus p' is one of $q, r \ldots$. But this contradicts $(*)$ above. For another proof, see Part II, Chapter 15.

Like Euclid, Eratosthenes of Cyrene (230 BC) worked at the University of Alexandria. He suggested a method for making a list of all prime numbers, which is called the 'sieve of Eratosthenes'. His method is as follows: write down all the positive integers greater than 1; cross out all multiples of 2 other than 2, cross out all multiples of 3 other than 3 which have not been crossed out yet, etc. In the end, the numbers not crossed out from a complete list of primes.

People often wonder whether there is a simple formula representing prime numbers. For example, $f(x) = x^2 - x + 41$ is prime for all integer values of x from 0 to 40. While this might convince a physicist that $f(x)$ is always prime, unfortunately $f(41) = 41^2$.

In 1743, Christian Goldbach observed that a polynomial
$$f(x) = a_0 + a_1 x + a_2 x^2 + \cdots + a_n x^n$$
with integer coefficients $a_0, a_1, \ldots, a_n$ cannot represent primes only, that is, the integers $f(0), f(1), f(2), \ldots$ are not all prime.

Indeed, if $f(0) = p$, then $f(kp)$ is clearly a multiple of p for all integers k. But, as k tends to infinity, so does the absolute value of $f(kp)$. Hence, for some value of k, $f(kp)$ will be a proper multiple of p and therefore not prime.

It therefore came as a great surprise to the mathematical community when, in 1970, Yuri Matiyasevič formed a polynomial $f(x, y, z, \ldots .)$ with integer coefficients, but in several variables, such that, when positive integers are chosen for $x, y, z, \ldots$, one gets all the prime numbers and only the prime numbers as positive values of the polynomial. We shall say more about this in Chapter 21 on Hilbert's Tenth Problem in Part II.

In 1830 (in *Théorie des Nombres* Vol. II, p. 65), A. M. Legendre noted that, if $\pi(x)$ is the number of primes less than or equal to x, then $\pi(x)$ is approximately equal to $x/(\log_e x - 1.08366)$, where $e = \lim_{n \to \infty} (1 + \frac{1}{n})^n$ is the base of the natural logarithm (Chapter 26). We shall write $\log x$ and assume the base to be e. He was not able to prove this. In 1896, two mathematicians working independently proved that
$$\lim_{x \to \infty} \frac{\pi(x)}{x/\log x} = 1 .$$

These two mathematicians were the Frenchman Jacques Hadamard (1865–1963) and the Belgian Charles Jean de la Vallée Poussin (1866–1962). The result they proved is called the *Prime Number Theorem*. It implies that the nth prime is approximately equal to $n \log n$. For, if we let p_n be the nth prime, the equation implies that n is roughly equal to $p_n/\log p_n$, so that

$$p_n \approx n \log p_n \approx n \log(n \log p_n) \approx n \log n,$$

since $n \log \log p_n$ can be neglected in comparison with $n \log p_n \approx p_n$.

It was Goldbach who conjectured that every even number greater than 2 is a sum of two primes. This conjecture has not yet been proved or disproved. However, in 1937, the Russian mathematician I. M. Vinogradov made some progress towards proving Goldbach's Conjecture, by showing that every odd integer greater than, say, $10^{10^{10}}$ (or some similar bound) is a sum of three prime numbers. Some progress in the Goldbach conjecture was recently made by the Chinese mathematician Chen Jing-Run. He proved that every sufficiently large (say $> 10^{10^{10}}$) even number has the form $p+q$, where p is prime and q is either prime or the product of two primes. During the so-called 'cultural revolution' in the sixties this kind of mathematics was frowned upon in China for being far removed from any conceivable application to industry or agriculture. Because he stubbornly stuck to his esoteric research at the risk of neglecting his teaching, Chen Jing-Run was discriminated against during the reign of the so-called 'gang of four' and may have lost his academic position. After the overthrow of the gang of four, he was rehabilitated and even declared a 'hero of the revolution'.

At the moment (1995), one of the 'hot topics' in prime number theory is cryptography. In its simplest form, the idea is this: the cipher key is a product $n = pq$ of two large primes, typically having 50 to 80 digits each. Knowing n is enough to encode messages, but decryption requires knowledge of the factorization. The integer n is made public (hence the term 'public key') so that everyone can use the code to encipher messages. Security is maintained, because only the intended recipient knows the key, namely, the factorization pq, necessary to carry out the decryption. The basis for this scheme is that it takes a very long time to factor products of large primes and the war may well be over before the enemy succeeds in doing so. (Try to factor the relatively small product 1,315,685,447, and you will see that the enemy does not have an easy task.)

At the moment, much research is being done to find refinements of the above idea, refinements that are at once economical and secure for those who want to send secret messages. Much research is also being done to find ways of using computers to factor very large numbers, and thus break the codes based on the above idea.

Exercises

1. Find the 25 primes less than 100 and express 100 as the sum of two primes.

2. Prove that there exist 1,000 consecutive positive integers none of which is prime. (Hint: start with $1001! + 2$.)

3. Prove that there are infinitely many prime numbers of the form
$$4m - 1.$$
(Hint: consider $n = 4q_1 q_2 q_3 \cdots q_k - 1$, where the q_i are primes of the form $4m - 1$, and show that not all prime divisors of n can be of the form $4m + 1$.)

4
Sumerian-Babylonian Mathematics

The Sumerians were a people of unknown linguistic affinity, who lived in the southern part of Mesopotamia (Iraq), and whose civilization was absorbed by the Semitic Babylonians around 2000 BC. Babylonian culture reached its peak in about 575 BC, under Nebuchadnezzar, but most of the mathematical achievements we shall discuss in this chapter and in Chapter 5 are much older, going back as far as 2000 BC — about the time when the biblical patriarch Abraham was said to have been born in the Sumerian city of Ur.

As we shall see, Mesopotamian mathematics is quite impressive. However, we should remember that, like the ancient Egyptians, the Mesopotamians never gave what we would call 'proofs' for their results; the first people to do so were the Greeks.

In representing numbers up to (and including) 59, the Sumerians and Babylonians used a decimal system. For example, they wrote 35 as follows, where we have approximated the original cuneiform figures by ours:

$$<<< \quad YYY$$
$$YY$$

On the other hand, 60 is again denoted by Y, and so is 60^2, as well as 60^{-1}, 60^{-2}, etc. It is usually clear from the context which is meant. Here are some further examples:

$$< << \quad = \quad 30, \text{ or } 30/60 = 1/2;$$

$$< YY = 12, \text{ or } 1/5;$$

$$Y << \begin{matrix} Y & Y & Y \\ & Y & \end{matrix} = 84, \text{ or } 7/5.$$

The Babylonian use of scale 60 was taken over into Greek astronomy around 150 BC by Hipparchus of Nicaea and it is still used today in measuring time and angles. To remove ambiguities in the above three examples, we would write

$$30° \text{ or } 30',$$

$$12° \text{ or } 12',$$

$$1°24' \text{ or } 1'24''.$$

The scale 60, or *sexagesimal system*, was also employed for weights of silver: 60 shekels = 1 mina; 60 minas = 1 talent. The prophet Ezekiel, living in Babylon, wrote in 573 BC:

> The Lord Yahweh says this: ... Twenty shekels, twenty-five shekels and fifteen shekels are to make one mina (*Ezekiel* 45:9–12).

The later Babylonians even introduced a symbol for zero:

$$Y \lessgtr \begin{matrix} Y & Y & Y \\ & Y & \end{matrix} = 60^2 + 4 = 3604.$$

Ptolemy (150 AD) replaced this symbol by a small circle, probably from the Greek word 'ouden', meaning 'nothing'.

In order to divide, the Babylonians made use of the fact that $a/b = a \cdot b^{-1}$. To this end, they constructed tables of inverses, like the one given in Table 4.1 (taken from Neugebauer [1969]). Note that the scribe did not list the inverses of any integers having a prime factor other than 2, 3 or 5. It seems he was afraid of repeating sexagesimals!

The Babylonians also had tables of squares, cubes, square roots, cube roots, and even roots of the equations

$$x^2(x \pm 1) = a.$$

Their method for extracting square roots is sometimes called *Heron's method* after Heron of Alexandria (60 AD), who included it in his *Metrica*. Let a_1 be a rational number between $\sqrt{a}$ and $\sqrt{a}+1$, where a is a positive non-square integer; let $a_{n+1} = (a_n + a/a_n)/2$; then $a_n \to \sqrt{a}$ as $n \to \infty$. Indeed, if $e = a_1 - \sqrt{a}$, we have $0 < e < 1$ and

$$0 < a_{n+1} - \sqrt{a} < 2\sqrt{a}(e/2\sqrt{a})^{2^n}$$

(see Exercise 4). As $n \to \infty$, this tends to 0.

4. Sumerian-Babylonian Mathematics

b	1/b	b	1/b
2	30'	16	3'45"
3	20'	18	3'20"
4	15'	20	3'
5	12'	24	2'30"
6	10'	25	2'24"
8	7'30'	27	2'13"20"
9	6'40'	30	2'
10	6'	32	1'52"30"
12	5'	36	1'40"
15	4'	40	1'30"

TABLE 4.1. Mesopotamian table of inverses (scale 60)

For example, if $a = 2$, $a_1 = 3/2$, then $a_2 = 17/12$ and $a_3 = 577/408$. In sexagesimal notation, $577/408 = 1°24'51''10'''35''''\ldots$. The fourth approximation $a_4 = 665857/470832$, which is $1°24'51''10'''7''''\ldots$ in sexagesimal notation. The difference between a_4 and $\sqrt{2}$ is less than

$$2\sqrt{2}\left(\frac{3/2 - \sqrt{2}}{2\sqrt{2}}\right)^{2^4} < 10^{-23}.$$

The Babylonian tablet YBC7289, dating from about 1600 BC, gives $\sqrt{2}$ as $1°24'51''10'''$.

Exercises

1. Write 5000 in the Babylonian manner. (You may use our degrees, minutes and seconds.)

2. Let a/b be a proper, reduced fraction (with a and b positive integers). Let $e_1 = 60a/b$ and $e_{n+1} = 60(e_n - [e_n])$ – where $[e_n]$ is the greatest integer less than or equal to e_n. Prove that the Babylonian sexagesimal expansion for a/b is

$$(.[e_1][e_2][e_3]\ldots)_{60}.$$

3. Express $1/7$ as a repeating sexagesimal.

4. Prove by mathematical induction that

$$0 < a_{n+1} - \sqrt{a} < 2\sqrt{a}(e/2\sqrt{a})^{2^n}.$$

5. Use the Babylonian method to find $\sqrt{3}$ to within 60^{-10}.

6. Let a/b be a proper, reduced fraction (with a and b positive integers). Prove that a/b has an infinitely repeating sexagesimal expansion if and only if b has a prime factor which does not divide 60.

5
More about Mesopotamian Mathematics

In *Science Awakening I*, B. L. van der Waerden quotes the beginning of 'AO8862', a Babylonian clay tablet going back to about the same time as the Rhind Papyrus:

> Length, width. I have multiplied the length and the width, thus obtaining the area. Then I added to the area, the excess of the length over the width: 183 was the result. Moreover, I have added the length and the width: 27. Required length, width and area.
>
> 27 and 183, the sums; 15 the length; 180 the area; 12 the width;
>
> One follows this method:
>
> $27 + 183 = 210$, $2 + 27 = 29$.
>
> Take one half of 29 (this gives $14\frac{1}{2}$),
>
> $14\frac{1}{2} \times 14\frac{1}{2} = 210\frac{1}{4}$,
>
> $210\frac{1}{4} - 210 = \frac{1}{4}$.
>
> The square root of $1/4$ is $1/2$.
>
> $14\frac{1}{2} + \frac{1}{2} = 15$, the length;
>
> $14\frac{1}{2} - \frac{1}{2} = 14$, the width.
>
> Subtract 2, which has been added to 27, from 14, the width. 12 is the actual width. I have multiplied the length 15 by the width 12.
>
> $15 \times 12 = 180$, the area;
>
> $15 - 12 = 3$;
>
> $180 + 3 = 183$.

What is going on here? In modern notation, we would write x and y for length and width, respectively. The problem is to find a solution for the simultaneous equations

$$xy + (x - y) = 183 \text{ and } x + y = 27.$$

The answer is given as $x = 15$ and $y = 12$. The scribe's method is this: consider

$$xy + x - y + x + y = x(y + 2) = 210.$$

Putting $y' = y + 2$, we have $xy' = 210$. On the other hand, adding the factors of 210, we get

$$x + y' = x + y + 2 = 29;$$

hence $\frac{1}{2}(x + y') = \frac{1}{2}(29) = 14\frac{1}{2}$;

hence $\dfrac{x^2 + 2xy' + y'^2}{4} = (14\frac{1}{2})^2 = 210\frac{1}{4}$;

hence $\dfrac{x^2 - 2xy' + y'^2}{4} = 210\frac{1}{4} - 210 = \frac{1}{4}$ (the so-called discriminant);

hence $\dfrac{x - y'}{2} = \frac{1}{2}$.

Adding and subtracting $\frac{1}{2}(x + y')$ and $\frac{1}{2}(x - y')$, we get $x = 14\frac{1}{2} + \frac{1}{2} = 15$ and $y' = 14\frac{1}{2} - \frac{1}{2} = 14$. Note that 14 is not really the width; but $y = y' - 2 = 14 - 2 = 12$ is. The scribe then computes the area and checks his work. The scribe did not consider the possibility $x = 14, y + 2 = 15$, which gives the second solution $x = 14$, $y = 13$. He did not know how to take the negative square root of $\frac{1}{4}$.

The Babylonians could solve many kinds of equations, including: $ax = b$, $x^2 \pm ax = b$, $x^3 = a$, $x^2(x+1) = a$. They could also solve simultaneous equations having the following forms:

$$x \pm y = a, \quad xy = b;$$

$$x \pm y = a, \quad x^2 + y^2 = b.$$

They even managed to solve the following pair of equations:

$$x^3\sqrt{x^2 + y^2} = 3,200,000; \quad xy = 1200. \qquad (*)$$

As we saw just above, the Babylonians knew that

$$a^2 - b^2 = (a+b)(a-b).$$

They also knew that

$$(a+b)^2 = a^2 + 2ab + b^2.$$

5. More about Mesopotamian Mathematics

Like the Egyptians, the Babylonians built pyramids, or *ziggurats*. If each story of a ziggurat consists of a square platform measuring $1 \times m \times m$, then the volume of a ziggurat with a base of length n is

$$(1 \times n \times n) + \cdots + (1 \times 2 \times 2) + (1 \times 1 \times 1) = 1^2 + 2^2 + 3^2 + \cdots + n^2.$$

The Babylonians knew that the formula for this sum is

$$n(n+1)(2n+1)/6,$$

a result also known to Pythagoras, but perhaps first proved by Archimedes.

According to the biblical story of the Tower of Babel, there was once an attempt to build a ziggurat 'with its top reaching heaven'. Perhaps the people behind this project thought that there was only a finite distance between heaven and earth, or perhaps they thought that they could calculate the sum of $1^2 + 2^2 + 3^2 + \cdots$, not realizing that the series diverges.

A remarkable fact about ancient Babylonian mathematics is that it included not just the so-called theorem of Pythagoras, but a theory of 'Pythagorean triangles'. (A *Pythagorean triangle* is a triple (x, y, z) of positive integers such that $x^2 + y^2 = z^2$, and thus x, y and z are sides of a right angled triangle.) From a clay tablet called 'Plimpton 322' (dating from 1900–1600 BC), we can deduce that the Babylonians used a result of which the following is a modern version:

> Suppose u and v are *relatively prime positive integers*, that is, integers whose greatest common divisor is 1. Assume that not both are odd and that $u > v$. Then, if $a = 2uv, b = u^2 - v^2$ and $c = u^2 + v^2$, we have $\gcd(a, b, c) = 1$ and $a^2 + b^2 = c^2$.

Included in Plimpton 322 is the triangle (13500, 12709, 18541), which is generated by taking $u = 125$ and $v = 54$.

The converse of the above theorem is also true. That is, if a, b and c are relatively prime positive integers, with a even, such that $a^2 + b^2 = c^2$, then there are relatively prime positive integers u and v, not both odd, such that $a = 2uv, b = u^2 - v^2$ and $c = u^2 + v^2$. It is not impossible that the Babylonians knew this, but the earliest record we have of this result is in the solutions of Problems 8 and 9 of Book II of the *Arithmetica* of Diophantus (250 AD). Indeed, since Diophantus explained his ideas in terms of special cases, it is correct to say that the first explicit, rigorous proof of the converse of the Babylonian theorem was given only in 1738, by C. A. Koerbero (Dickson [1971], Vol. II).

According to a tablet found in 1936 in Susa, an ancient city in what is now Iran, the Babylonians sometimes used the value $3\frac{1}{8}$ for π. At other times, they seem to have been satisfied with $\pi \approx 3$. It has been suggested that this Babylonian usage is behind 1 *Kings* 7:23–24:

> He [Hiram of Tyre] made the basin of cast metal, ten cubits from rim to rim, circular in shape and five cubits high; a cord

thirty cubits long gave the measurement of its girth. Under its rim and completely encircling it were gourds; they went around the basin over a length of thirty cubits.

But perhaps the basin was hexagonal and not circular!

Exercises

1. Consider the simultaneous equations $xy + x - y = a$ and $x + y = b$, where a and b are given integers. What is a necessary and sufficient condition on a and b so that x and y will be integers?

2. Solve the simultaneous pair ($*$) (from a Susa tablet).

3. Prove by mathematical induction that
$$1^2 + 2^2 + 3^2 + \cdots + n^2 = n(n+1)(2n+1)/6.$$

4. Rabbi Nehemiah (150 AD) was unhappy with the idea that the Bible had used a very inaccurate value for π and he suggested that 'the diameter of 10 cubits included the walls of the basin, while the circumference excluded them.' Assuming that he was right, and assuming that the Bible used a perfectly accurate value for π, how wide was the wall (or rim) of the basin?

5. Prove that if a triangle has sides of lengths a, b and c, and if $a^2 + b^2 = c^2$, then the triangle is right angled.

6. Prove the Babylonian theorem for Pythagorean triangles.

7. Prove the converse of the Babylonian theorem for Pythagorean triangles.

8. In 1901, L. Kronecker gave the first proof that all positive integer solutions of $a^2 + b^2 = c^2$ are given without duplication by $a = 2uvk$, $b = (u^2 - v^2)k$, $c = (u^2 + v^2)k$, where u, v and k are positive integers such that $u > v$, u and v are not both odd, and u and v are relatively prime. Prove Kronecker's theorem.

9. The 15 Pythagorean triangles in Plimpton 322 have angles which approximate the 15 whole number angles from 44° to 58° inclusive. Find a Pythagorean triangle, with relatively prime sides, one of whose angles is within 2/5° of 47°. (Hint: see Anglin [1988].)

6
The Dawn of Greek Mathematics

The ancient Greek world was not confined to what we now call Greece, but extended to Ionia (western Turkey) in the east, southern Italy and Sicily in the west, and later to Alexandria (Egypt) in the south. Not surprisingly, Greek philosophy and mathematics began in Ionia, where the influence of the older civilizations of the east (e.g., Babylon) was greatest. Later, political events caused many Greeks to emigrate from Ionia to Italy, and this became the center of intellectual life for a while. After the war between a Greek coalition and the Persians ended in the defeat of the latter (490 BC), philosophy and mathematics flourished in Athens. Ultimately, after the founding of Alexandria (331 BC), it was there that most of the major scientific developments took place until about 500 AD.

The first Greek mathematician and philosopher is Thales of Miletus (600 BC). According to Proclus, Thales visited Egypt and brought back the knowledge of geometry from there. He may also have been influenced by Indian thought via Persia. He is said to have predicted the solar eclipse which occurred over the Near East in May of 585 BC. To do this, he may have made use of observations which the Babylonians had accumulated over many centuries.

Plato repeats a story about Thales being an absent-minded professor, who was so preoccupied with celestial matters that he did not observe what was in front of his feet and once fell into a well (*Theaetetus* 174a). According to other anecdotes, however, Thales could turn his mind to practical matters when necessary. He constructed an almanac and he used the theory of similar triangles to calculate the distance of ships from shore and the height of pyramids. To impress his business minded fellow citizens, he once cornered the market in olive oil and, incidentally, made himself rich.

Thales's name is associated with a number of elementary theorems in geometry:

1. a circle is bisected by a diameter;

2. the base angles of an isosceles triangle are equal;

3. when two lines intersect, vertically opposite angles are equal;

4. the angle-angle-side congruence theorem;

5. the angle subtended by a diameter of a circle is a right angle (that is, if A, B, C are points on a circle and AC is a diameter, then $\angle ABC$ is a right angle).

Theorem 5 is called *Thales's theorem*. To prove this he also had to know the following:

6. the sum of the angles in a triangle is equal to two right angles (or, as we now say, in slavish imitation of the Babylonians, 180°).

All of these theorems must have been known empirically by the Egyptians and Babylonians. The reason they are associated with Thales is not that he discovered them, but that he was the first to prove them. This was the essential difference between pre-Greek and Greek mathematics: the Greeks established the logical connections among their results; they gave the first abstract proofs in mathematics.

As a philosopher, Thales is known for his statement that everything is made of water. How should we interpret this, and what is its relevance to mathematics?

As we look around us, we observe that there are two kinds of things: those that can be counted, such as pebbles and cows, and those that can only be measured, such as butter and water. This physical distinction between 'discrete' and 'continuous' is reflected on a linguistic level: it is perfectly correct to say 'one cow, two cows' but it sounds rather odd to say 'one butter, two butters', to put it mildly (however, see Exercise 5). We call the former sort of nouns *count nouns*, and the latter *mass nouns*. To some extent, this distinction is a convention. For example, in modern English, we can count peas but not rice, while a hundred years ago, 'pease' was not a plural but a mass noun. (A hundred years from now, 'rice' may be the plural of 'rouse'.)

A question which physicists are still working on is this: is the material universe ultimately countable — consisting of discrete, unconnected fragments — or is the material universe ultimately continuous — that is, should it be understood in terms of a connected continuum? If the first, how do we explain the unity of nature, how do we understand the continuity of change and motion? If the second, how do we explain the diversity of nature, how do we understand the individuality of distinct, single objects?

This issue was addressed by more than one Greek thinker. As we shall see, Pythagoras and Democritus took the view that reality is basically discrete. They then tried to understand apparently continuous entities in terms of discrete entities (e.g., lengths as ratios of whole numbers). Thales, on the other hand, took the view that 'all is water'. In other words, the material universe is best understood in terms of a single substance, namely, water. (Here we are using the word 'substance' not in the Aristotelian sense of 'individual entity', but in the more common sense of 'material having uniform properties'.) Thales had undoubtedly noticed that ice and steam are both forms of water, but we do not know why he picked water as the fundamental substance. (It has been suggested by Marxist historians that this was so because his city Miletus was a *maritime* power.) What is important is not that Thales overlooked the possibility of, say, there being 90 different substances, but that he raised a fascinating problem about the universe, which has not been resolved to this day.

Other Ionian philosophers agreed with Thales that there was a single substance, but not that it was water. Anaximenes of Miletus (550 BC) identified the primal substance as air. Heraclitus of Ephesus (500 BC) held that everything is made of fire. Anaximander was a follower and compatriot of Thales, who like Thales, took the view that the universe is best understood in terms of a single substance. Unlike Thales, he did not think this substance was water. He thought it was something he called the *Infinite*. The Infinite could take on the forms of earth, water, air, and fire. Today we might refer to solid, liquid, gas and energy, respectively.

Exercises

1. Let ABC be a triangle, and let d be a straight line through A parallel to BC. Assuming that the 'alternate angles are equal', prove that the sum of the angles of ABC equals two right angles.

2. Prove the Theorem of Thales, using Exercise 1 and the theorem that the base angles of an isosceles triangle are equal.

3. Prove the converse of Thales's Theorem: if A, B and C are points on a circle and $\angle ABC$ is a right angle then AC is a diameter.

4. How would you measure the height of a pyramid (or tree, for that matter), using similar triangles?

5. Some nouns like 'rice' are definitely count nouns. We cannot say 'two rices'. Other nouns are more problematic. 'Whisky' is normally a mass noun, but 'two whiskies, please' is perfectly good English (meaning two glasses of whisky). Some languages have many fewer count nouns than English. For example, in Indonesian it is incorrect to say 'two

cows'; you have to say 'two tails of cow', as we might say 'two head of cattle'. (It is amusing to note that our mass noun 'cattle' is itself ultimately derived from the Latin word 'caput' meaning 'head'.) Write an essay on the distinction between count nouns and mass nouns and its relevance to mathematics.

7
Pythagoras and His School

Pythagoras (570–500 BC) was born in Samos, a Greek island off the coast of what is now Turkey. According to ancient sources (Iamblichus, Porphyry and Diogenes Laërtius), he traveled and studied in the Persian empire, which extended then from northern Greece to the Indus Valley and included ancient Mesopotamia. We know (Plimpton 322) that the Babylonians understood what is now called the 'theorem of Pythagoras', although the latter may have given the first proof. Pythagoras may have learned the theory of 'Pythagorean triangles' from the Babylonians.

According to the above mentioned sources, Pythagoras also studied under the Zoroastrian priests, the so-called 'Magi'. However, judging from his belief in reincarnation and his vegetarianism, it is more likely that he was influenced by Hindu tradition. Even his mathematics has an Indian flavour.

About 525 BC, Pythagoras emigrated to Croton (modern Crotone) in southern Italy, where he founded a society, half-way between a political party and a religious cult, which came to be known as the 'Pythagorean Brotherhood.' Some members of this society were admitted to an inner circle consisting of the so-called 'mathematicians'.

The word 'mathematics' was in fact introduced by Pythagoras. The first part of this word is an old Indo-European root, related to the English word 'mind'. The modern meaning of 'mathematics' is due to Aristotle.

Whereas Thales had claimed that 'all is water', Pythagoras taught that 'all is number'. For Pythagoras this implied that everything could be understood in terms of whole numbers and their ratios. In particular, he implicitly expected that every line segment was a whole number or a ratio of whole numbers (in terms of a given unit length). It seems that the dis-

7. Pythagoras and His School

FIGURE 7.1. Pythagorean star and the fourth triangular number

covery of the irrationality of the diagonal of the square of side 1 was made by his followers and that Pythagoras himself was not aware of this.

In his philosophy, Pythagoras reserved a special place for the number 10. He called it the 'divine number', noting that 10 is a triangular number and realizing that the five-pointed 'Pythagorean star' (Figure 7.1) has 10 vertices.

The Pythagoreans ascribed all their mathematical discoveries to Pythagoras, but there is not, in fact, a single theorem which we can safely credit to him. For example, in his preface to the *Introductio Arithmetica*, written by a Pythagorean, Nichomachus of Gerasa (100 AD), Iamblichus (300 AD) credits Pythagoras with a knowledge of the amicable pair 220 and 284. (Two natural numbers are *amicable* if each is the sum of the proper divisors of the other.) However, we have no way of knowing for certain whether amicable numbers had been recognized as early as 500 BC. Yet, according to a famous anecdote, when someone challenged his slogan 'all is number' by asking 'then what is friendship?', Pythagoras replied that friendship is as 220 is to 284.

Leaving behind the shadowy figure of the Master, let us review the accomplishments of his followers. Although they were primarily a religious and political group, they did a fair amount of work in arithmetic, geometry, astronomy and music – the four subjects later forming the medieval *quadrivium*. (In the university curriculum of the Middle Ages, these subjects were meant to follow the 'trivial' subjects: grammar, rhetoric and logic.)

Theorem of Pythagoras

The Pythagoreans are probably responsible for the proof found in Euclid's *Elements*, Book I, Proposition 47. They also found a proof of the converse of the theorem of Pythagoras.

Means

They examined the relationships between the following means:
arithmetic ($\frac{1}{2}(a+b)$), geometric ($\sqrt{ab}$) and harmonic ($2ab/(a+b)$).

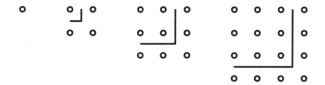
FIGURE 7.2. The sequence of squares

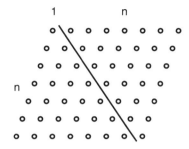
FIGURE 7.3. Triangular numbers

Perfect Numbers

They found a formula giving perfect numbers. See Chapter 8.

Regular Solids

They discovered the dodecahedron. See Chapter 9.

Irrationality of $\sqrt{2}$

They discovered that the square root of 2 is not rational. They used the integer solutions of $x^2 - 2y^2 = \pm 1$ to find approximations to it. See Chapter 10.

Figurative Numbers

They found proofs for several algebraic relations by means of studying figurative numbers. For example, looking at the sequence of squares, expressed in terms of 'arrays of pebbles' (Figure 7.2), they noticed that $n^2 + (2n + 1) = (n + 1)^2$ and hence $1 + 3 + 5 + \cdots + (2n - 1) = n^2$:

Fitting two 'triangular numbers' into a parallelogram, they noticed that the nth triangular number is $\frac{1}{2}n(n+1)$.

Looking at the sequence of triangular numbers, expressed in terms of pebble arrays, they realized that the difference between the $(n + 1)$th and nth triangular number is just $n + 1$. From this they concluded that $1 + 2 + 3 + \cdots + n =$ the nth triangular number $= \frac{1}{2}n(n+1)$. See Figure 7.3.

The study of figurative numbers is alive and well today. For example, recently some very advanced mathematics was used to show, for the first time, that there are exactly six triangular numbers that are products of three consecutive integers (See Tzanakis and de Weger [1989].)

Exercises

1. How might a Pythagorean have derived the fact that the angle at the tip of his star is 36°?

2. Check that 220 and 284 are amicable. If 12,285 is one member of an amicable pair, find the other.

3. Prove that the sum of the first n cubes is the square of the sum of the first n numbers. (Use mathematical induction or construct a square figure with sides of length $1 + 2 + \cdots + n$ and divide it into figures whose areas are the first n perfect cubes.)

8
Perfect Numbers

The Pythagoreans were interested in perfect numbers, that is, numbers such as 6 and 28, which are equal to the sum of their proper divisors. They may also be described as numbers which are amicable with themselves. Nowadays we usually speak about the sum of all the divisors of a positive integer n, including n itself. If $\sigma(n)$ denotes this sum, then n is *perfect* if and only if $\sigma(n) = 2n$. As the culmination of Book IX of the *Elements* (300 BC), Euclid proved that any positive integer of the form

$$n = 2^{m-1}(2^m - 1)$$

is perfect, whenever $2^m - 1$ is prime. This fact had probably been discovered by the Pythagoreans.

Proof of Proposition IX 36 (Perfect Number Theorem):

If $p = 2^m - 1$ is prime, then the divisors of $n = 2^{m-1}p$ are

$$1,\ 2,\ 2^2,\ \ldots,\ 2^{m-1},\ p,\ 2p,\ 2^2 p,\ \ldots,\ 2^{m-1}p$$

(thanks to unique factorization). The sum of these divisors is thus

$$\begin{aligned}\sigma(n) &= (1 + 2 + 2^2 + \cdots + 2^{m-1})(1 + p) \\ &= (2^m - 1)(1 + p) \\ &= 2(2^{m-1}(2^m - 1)) = 2n.\end{aligned}$$

Even though $2^m p$ is not square-free, Euclid did have a rigorous proof of the special case of the unique factorization theorem which is used in the

2	107	9689
3	127	9941
5	521	11213
7	607	19937
13	1279	21701
17	2203	23209
19	2281	44497
31	3217	86243
61	4253	132049
89	4423	216091

TABLE 8.1. Values of m making $2^m - 1$ prime

above proof, and he also had a rigorous proof for the formula for the sum of a geometric progression (IX 35).

An integer of the form $2^m - 1$ can only be prime if m is prime. For if $m = ab$ with $a, b > 1$, we have the factorization

$$2^{ab} - 1 = (2^a - 1)((2^a)^{b-1} + (2^a)^{b-2} + \cdots + 2^a + 1)$$

into two factors greater than 1. The converse is not true. Although 11 is prime, $2^{11} - 1$ is not; for $2^{11} - 1 = 2047 = 23 \times 89$.

Primes of the form $2^m - 1$ are called *Mersenne* primes, after Father Marin Mersenne (1588–1648). In the preface of his *Cogitata Physico-Mathematica* (1644), Mersenne correctly stated that the first 8 perfect numbers are given by the values $m = 2, 3, 5, 7, 13, 17, 19$ and 31. He also claimed that $2^{67} - 1$ is prime, and hence $2^{66}(2^{67} - 1)$ is perfect. Here he was wrong. In 1903, Frank Nelson Cole gave a lecture which consisted of two calculations. First Cole calculated $2^{67} - 1$. Then he worked out the product

$$193,707,721 \times 761,838,257,287.$$

He did not say a word as he did this. The two calculations agreed, and Cole received a standing ovation. He had factored $2^{67} - 1$ and proved Mersenne wrong.

Edouard Lucas (1842–1891) found a very efficient way of testing whether $2^m - 1$ is prime. Let $u_1 = 4$ and $u_{n+1} = u_n^2 - 2$. Thus $u_2 = 14, u_3 = 194$ and $u_4 = 37,634$. If $m > 2$ then $2^m - 1$ is prime just in case $2^m - 1$ is a factor of u_{m-1}. For example, since $2^5 - 1$ is a factor of 37,634 it follows that $2^5 - 1$ is prime (and hence $2^4(2^5 - 1) = 496$ is perfect). (For an elementary proof of Lucas's Theorem, see Sierpinski [1964].)

Thanks to Lucas's test — and the modern computer — we know, since about 1985, that $2^m - 1$ is prime when m has the 30 values given in Table

8. Perfect Numbers 39

8.1. The Greeks knew just the first four Mersenne primes and Mersenne discovered eight more. Before 1950, only the first 12 Mersenne primes were known. Then, with the help of ever more powerful computers, 18 more came to light. (Even after writing this, we learned of three more, corresponding to $m = 110,503$, $m = 756,839$, and $m = 858,433$. The last of these Mersenne primes has 258,716 digits.) We still do not know whether there are infinitely many Mersenne primes. Nor do we know if there are any odd perfect numbers. What we do know is that every even perfect number has the form given by Euclid. This was first proved by Leonhard Euler (1707–1783). His proof was as follows.

Proof that every even perfect number has Euclid's form:

Suppose n is an even perfect number. Then we can write it in the form $2^{m-1}q$ with q odd and $m, q > 1$. Each divisor of n has the form $2^r d$ where $0 \leq r \leq m - 1$ and d is a divisor of q. Therefore

$$\sigma(n) = (1 + 2 + \ldots + 2^{m-1})\sigma(q) = (2^m - 1)\sigma(q).$$

Since n is perfect, $2^m q = \sigma(n) = (2^m - 1)\sigma(q)$.

Since $2^m - 1$ is odd, 2^m must divide $\sigma(q)$, say $\sigma(q) = 2^m k$, hence $q = (2^m - 1)k$. Among the divisors of q are q itself and k. These are different, since $m > 1$, and their sum is $2^m k$, which is the sum of all the divisors of q. Therefore, q has exactly two divisors and so is prime, hence $k = 1$ and $q = 2^m - 1$.

Perfect numbers are of interest not only as a challenge to computer programmers, they also play role in religious mysticism. For example, following Philo of Alexandria, Augustine writes in the *City of God*:

> Six is a number perfect in itself, and not because God created all things in six days; rather, the converse is true. God created all things in six days because this number is perfect, and it would have been perfect even if the work of six days did not exist.

In a recent book on Sufi mysticism, it is stated that 6 is the first 'complete' number and 28 is the second. Evidently, 'complete' here means 'perfect'.

Apparently, the Pythagoreans knew only one amicable pair of numbers. Although Euler found 60 such pairs, the second smallest pair $(1184, 1210)$ was only discovered in 1866 by Nicolo Paganini.

The Arabic mathematician Thabit ibn Qurra (826–901) gave a general procedure for discovering many amicable pairs, analogous to Euclid's procedure for discovering perfect numbers. See Exercise 5.

8. Perfect Numbers

Exercises

1. Prove that every even perfect number, except 6, is the sum of the first 2^k odd cubes, for some k.

2. Show that, if m and n are relatively prime positive integers, then $\sigma(mn) = \sigma(m)\sigma(n)$.

3. Show that, if p is prime, $\sigma(p^k) = (p^{k+1} - 1)/(p - 1)$.

4. Obtain a formula for $\sigma(n)$ in terms of the prime factorizations of n.

5. Prove the result of Thabit ibn Qurra: if $p = 3 \times 2^{t+1} - 1, q = 3 \times 2^t - 1$ and $r = 9 \times 2^{2t+1} - 1$ are odd prime numbers, then $m = 2^{t+1}pq$ and $n = 2^{t+1}r$ are amicable.

6. Find two amicable pairs with the help of the above procedure.

9
Regular Polyhedra

The Pythagoreans knew that there are three ways to tile a plane (e.g., a bathroom floor) using congruent regular polygons. Indeed, since once can dissect a polygon with p sides into $p - 2$ triangles, the sum of the angles of such a polygon is $(p-2)180°$. Thus each angle of a regular 'p-gon' is $(p-2)180°/p$. If q such angles meet at a point, then

$$q(p-2)180°/p = 360°,$$

which may be simplified to yield the Egyptian problem: $1/2 = 1/p + 1/q$. Since p and q are integers greater then 2, we must have one of the following three possibilities:

p	q
3	6
4	4
6	3

The first of these gives the tiling with equilateral triangles, the second the tiling with squares, and the last the tiling with regular hexagons. No other regular polygon can be used to tile the plane.

A polyhedron is *regular* if its faces are congruent regular polygons and if the same number of faces meet at each vertex. Five regular polyhedra are the following:

- the cube, bounded by 6 squares, with 3 edges at each vertex,
- the tetrahedron, bounded by 4 equilateral triangles, with 3 edges meeting at each vertex,

9. Regular Polyhedra

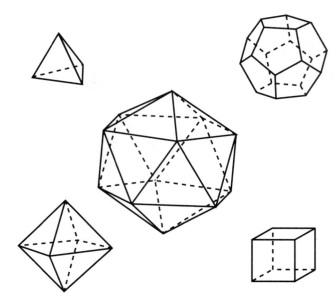

FIGURE 9.1. The Pythagorean solids

- the octahedron, bounded by 8 equilateral triangles, with 4 edges at each vertex,

- the icosahedron, bounded by 20 equilateral triangles, with 5 edges at each vertex,

- the dodecahedron, bounded by 12 regular pentagons, with 3 edges at each vertex (see Figure 9.1).

In the *Timaeus* (53-58), Plato explains the composition of the physical universe in terms of these five regular polyhedra. The cube is associated with earth, the tetrahedron with fire, the octahedron with air, the icosahedron with water and the dodecahedron with the whole cosmos. Plato explains the boiling of water by means of a 'chemical equation', which we might write as

$$F_4 + W_{20} \to 2A_8 + 2F_4.$$

That is, fire, with 4 faces, combines with water (20 faces) to produce 2 air atoms (each with 8 faces) and 2 fire atoms (each with 4 faces). Note that the numbers balance.

Such theorizing is very much in the spirit of the Pythagorean teaching that 'all is number'. Indeed, historians attribute the theory of the *Timaeus* to Pythagoras himself (Guthrie [1987]). It seems, however, that Pythagoras himself may not have known about the dodecahedron. According to one account, Hippasus (470 BC) was expelled from the Pythagorean or-

p	q	1/p + 1/q	E	F	V	Polyhedron
3	3	2/3	6	4	4	tetrahedron
3	4	7/12	12	8	6	octahedron
4	3	7/12	12	6	8	cube
3	5	8/15	30	20	12	icosahedron
5	3	8/15	30	12	20	dodecahedron

TABLE 9.1. The regular polyhedra

der because, having discovered the dodecahedron, he failed to ascribe his discovery to Pythagoras.

When Plato proposed that the creator, whom he called the 'demiurge', used the regular polyhedra when forming the universe, he may not have been that far off. The tetrahedron, cube and octahedron can be found in nature as crystals. The octahedron, icosahedron and dodecahedron occur as the skeletons of certain radiolarians (a type of microscopic sea animal).

Are there only five regular polyhedra? Yes. In fact, a proof of this is found at the end of Euclid's Elements (300 BC). This proof is based on the fact that if q regular p-gons meet at a vertex, then the sum of the q angles in the q faces is less than 360°. This is proved rigorously in Proposition 21 of Book XI of the *Elements*, but it can be seen intuitively by imagining someone cutting the q edges and flattening the angle. For example, the three angles at the vertex of a cube clearly add up to 270°, which is less than 360°.

In general, for a regular polyhedron whose faces are regular p-gons, with q faces meeting at each vertex,

$$q(p-2)180°/p < 360°,$$

which may be simplified to yield $1/2 < 1/p + 1/q$. We can easily see that there are just five possibilities for p and q, as in Table 9.1. (In the table, E is the number of edges the polyhedron has, F the number of faces, and V the number of vertices.)

We have not as yet explained how the numbers E, F and V in our table are calculated. It so happens that these numbers are related by a simple formula, even when we consider an arbitrary polyhedron, regular or not. Here are some examples:

	F	V	E
cube	6	8	12
tetrahedron	4	4	6
pyramid	5	5	8
prism	5	6	9

9. Regular Polyhedra

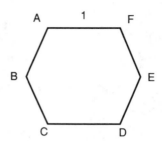

FIGURE 9.2. Cross-section of an icosahedron

We note that in each example

$$F + V - E = 2.$$

This is in fact a general rule, valid for all polyhedra. While it may first have been noted by Descartes, it was only proved by Euler and is known as *Euler's formula*. We shall give a proof in Chapter 30 of Part I.

Now suppose we are looking at a regular solid in which each face has p edges and in which q edges meet at each vertex. It follows immediately that

$$pF = 2E, \qquad qV = 2E.$$

Substituting $F = 2E/p$ and $V = 2E/q$ into Euler's formula, we obtain

$$2E/p + 2E/q - E = 2,$$

and, after dividing by $2E$,

$$1/p + 1/q - 1/2 = 1/E.$$

This allows us to calculate E from p and q, and then F and V.

The ancient Greeks were fascinated by the five regular solids. Without the help of trigonometry or calculus, they managed to prove all their basic properties. Book XIII of the *Elements* (300 BC) is devoted to showing that, for each of these five solids, there is a sphere passing through all its vertices. In each of the five cases, Book XIII calculates the ratio of the side of the regular polyhedron to the radius of this 'circumscribing' sphere.

For example, if one cuts an icosahedron in half, cutting along a side, the resulting cross-section is as in Figure 9.2. AF and CD are edges; AC and DF are diagonals in the regular pentagons formed by the sides of the icosahedron. (You may have to construct an icosahedron to see this.) Thus, if AF and CD each have unit length, AC and DF each have length $\frac{1}{2}(1 + \sqrt{5})$. The diameter of the circumscribing sphere is CF, which is the hypotenuse of the right triangle with sides CD and DF. Thus $CF^2 = 1^2 + (\frac{1}{2}(1 + \sqrt{5}))^2$, and hence the radius of the sphere is $\frac{1}{2}\sqrt{\frac{1}{2}(5 + \sqrt{5})}$.

Exercises

1. Construct a dodecahedron, e.g., by taping together 12 identical regular pentagons cut out of cardboard.

2. Show that the radius of a sphere passing through the vertices of a dodecahedron with side 1 is $(\sqrt{3} + \sqrt{15})/4$.

3. Show that the volume of a dodecahedron of side 1 is $(15 + 7\sqrt{5})/4$.

4. Given a polyhedron, not necessarily regular, in which exactly 3 edges meet at each vertex. Show that $V = 2K$, $E = 3K$ and $F = K + 2$ for some positive integer K.

5. Under the condition of the previous exercise, if F_p is the number of faces with p sides, show that
$$\sum_p (6-p) F_p = 12.$$

6. If all faces of a polygon are hexagons or pentagons and if three edges meet at each vertex, prove that the number of pentagons is twelve. (There are molecules of such a shape with twenty hexagons, called 'buckyballs', a form of carbon called 'buckminster fullerene'. See Chung and Sternberg [1993].)

10
The Crisis of Incommensurables

Two lengths a and b are said to be *commensurable* if there exist positive integers p and q such that $a/b = p/q$. When the Pythagoreans claimed that all things are numbers, they probably meant to imply that all pairs of lengths are commensurable. They were aware of the fact that, if a vibrating string is divided into two parts, of lengths a and b, so that a melodious tone is produced, then a and b are commensurable.

Unfortunately for the Pythagoreans, they soon discovered that the diagonal of a square is not commensurable with its side. A simple proof of this is found in Aristotle's *Prior Analytics* 41a23-30. Let $ABCD$ be a square, say of side $AB = 1$. By the Theorem of Pythagoras, the diagonal AC measures $\sqrt{2}$. Suppose $\sqrt{2} = AC/AB = p/q$, where p and q are positive integers. We may assume, without loss of generality, that p and q are relatively prime. In particular, they are not both even. Now $p^2 = 2q^2$, so that p^2 is even. As the Pythagoreans well knew, the square of an odd number is odd and the square of an even number is even. Thus, from the fact that p^2 is even, it follows that p is even. Putting $p = 2r$, we have $2q^2 = (2r)^2$, hence $q = 2r^2$. But this means that q is even as well, contradicting the fact that p and q are relatively prime. Thus, the assumption that AC and AB are commensurable must be false. Today we would express this result by saying that $\sqrt{2}$ is irrational.

The Pythagoreans tried to keep this discovery a secret, as it seemed to undermine their whole philosophy. Some say that it was Hippasus, whom we met before, who leaked the secret, and that he drowned in a shipwreck as a punishment for having done so. It seems that Hippasus was the Trotsky of the Pythagorean society.

10. The Crisis of Incommensurables

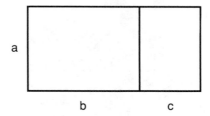

FIGURE 10.1. The distributive law

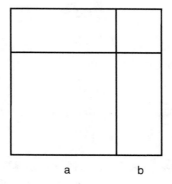

FIGURE 10.2. Binomial expansion $(a+b)^2 = a^2 + 2ab + b^2$

The Greeks did not have infinite decimals. They did not know how to handle a number like $\sqrt{2}$ in an arithmetical or algebraic fashion the way we do now, although it has recently been claimed that they could represent real numbers by continued fractions. They did, however, know that $\sqrt{2}$ was a length, and they turned to geometry for an understanding of it. The problem of incommensurables was one of the reasons that they preferred to do what we would call algebra in a geometric manner.

For example, the distributive law $a(b+c) = ab + ac$ was thought of as an addition rule for areas of rectangles, as in Figure 10.1.

Euclid put it thus:

> If there are two straight lines, and one of them be cut into any number of segments whatever, the rectangle contained by the two straight lines is equal to the rectangles contained by the uncut straight line and each of the segments (*Elements* II 1).

The law $(a+b)^2 = a^2 + 2ab + b^2$ is illustrated in Figure 10.2. We shall refrain from putting this law into words.

However, as a third example, we again quote Euclid:

> If a straight line be cut into equal and unequal segments, the rectangle contained by the unequal segments of the whole to-

10. The Crisis of Incommensurables

gether with the square on the straight line between the points of section is equal to the square on the half (*Elements* II 5).

This is equivalent to the identity $(a+b)(a-b) = a^2 - b^2$.

The Pythagoreans found ways of approximating $\sqrt{2}$ as closely as could be desired by rational numbers. Using our modern algebraic notation, we can express their method as follows.

If $x^2 - 2y^2 = \pm 1$, with x and y positive integers, then x^2 is approximately equal to $2y^2$, so that x/y is approximately equal to $\sqrt{2}$. More precisely,

$$(x/y - \sqrt{2})(x/y + \sqrt{2}) = x^2/y^2 - 2 = \pm 1/y^2$$

so that

$$x/y - \sqrt{2} = \pm 1/(y^2(x/y + \sqrt{2})).$$

Since $x/y + \sqrt{2} > 1$, it follows that

$$|x/y - \sqrt{2}| < 1/y^2.$$

Thus, if we can find positive integer solutions of $x^2 - 2y^2 = \pm 1$ with y sufficiently large, then we can find rational approximations to $\sqrt{2}$ as close as we please.

To find positive integers x and y such that $x^2 - 2y^2 = \pm 1$, the Pythagoreans proceeded as follows. Putting

$$a_1 = 1, \qquad b_1 = 1$$

and defining inductively

$$a_{n+1} = a_n + 2b_n, \quad b_{n+1} = a_n + b_n,$$

they obtained the following table:

n	a_n	b_n	a_n/b_n
1	1	1	1
2	3	2	3/2
3	7	5	7/5
4	17	12	17/12

etc., in which the last column contains successive approximations to $\sqrt{2}$.

Indeed, it is not difficult to prove by mathematical induction that

$$a_n^2 - 2b_n^2 = (-1)^n.$$

This is surely true when $n=1$, so suppose it holds for n. Then

$$\begin{aligned}
a_{n+1}^2 - 2b_{n+1}^2 &= (a_n + 2b_n)^2 - 2(a_n + b_n)^2 \\
&= a_n^2 + 4a_nb_n + 4b_n^2 - 2a_n^2 - 4a_nb_n - 2b_n^2 \\
&= -a_n^2 + 2b_n^2 \\
&= -(-1)^n = (-1)^{n+1}.
\end{aligned}$$

10. The Crisis of Incommensurables

Thus our proof by mathematical induction is complete.

This method, but not the above proof, is explained verbally by Proclus in commenting on a passage in Plato's *Republic*. Today, it is easy to obtain explicit formulas for the numbers a_n and b_n. First, one proves by mathematical induction that

$$a_n + b_n\sqrt{2} = (1 + \sqrt{2})^n .$$

Replacing the square root by its negative, one obtains

$$a_n - b_n\sqrt{2} = (1 - \sqrt{2})^n .$$

Therefore,

$$a_n = \frac{1}{2}((1+\sqrt{2})^n + (1-\sqrt{2})^n),$$

$$b_n = \frac{1}{2\sqrt{2}}((1+\sqrt{2})^n - (1-\sqrt{2})^n).$$

Although the Pythagoreans did not know it, they had actually found all solutions of the equations $x^2 - 2y^2 = \pm 1$ in positive integers. Suppose, for example, $x^2 - 2y^2 = 1$. Let n be the largest natural number such that $(1+\sqrt{2})^n \leq x + y\sqrt{2}$, then

$$(1+\sqrt{2})^n \leq x + y\sqrt{2} < (1+\sqrt{2})^{n+1}.$$

Multiplying this by $(1-\sqrt{2})^n = a_n - b_n\sqrt{2}$ and assuming that n is even, we obtain

(1) $$1 \leq (x + y\sqrt{2})(a_n - b_n\sqrt{2}) < 1 + \sqrt{2}.$$

Taking reciprocals of this, we get

(2) $$-1 \leq (-x + y\sqrt{2})(a_n + b_n\sqrt{2}) < 1 - \sqrt{2}.$$

Adding (1) and (2) and dividing by $2\sqrt{2}$, we obtain

$$0 \leq ya_n - xb_n < 1/\sqrt{2}.$$

Since $ya_n - xb_n$ is a whole number, it must be 0, hence $ya_n = xb_n$. Now we know that x and y are relatively prime, and so are a_n and b_n. It easily follows that $x = a_n$ and $y = b_n$, where n is even.

If n is odd or if $x^2 - 2y^2 = -1$, we proceed similarly.

Exercises

1. Prove that the decimal expression of $\sqrt{2}$ is not ultimately periodic.

2. Prove that the following numbers are not rational: $\sqrt{3}$, $\sqrt[3]{2}$ and $\log_{10} 2$.

3. If a, b, c and d are integers and $a + b\sqrt{2} = c + d\sqrt{2}$, show that $a = c$ and $b = d$.

4. Solve the following equations for positive integers:
$$x^2 - 4y^2 = 1, \quad x^2 - 3y^2 = 1.$$

11
From Heraclitus to Democritus

Heraclitus of Ephesus (in western Turkey) flourished about 500 BC, Parmenides of Elea (in southern Italy) about 480 BC, Zeno of Elea about 460 BC, Empedocles in Sicily about 440 BC, Democritus of Abdera (in north-eastern Greece) about 420 BC.

In the *Metaphysics* (986b4-8), Aristotle tells us that the Pythagoreans had a list of opposites: one, many; finite, infinite; male, female; etc. It was perhaps this list which led Heraclitus to his view that everything that happens is the result of a struggle between opposites. He proclaimed that all change is the result of strife.

Heraclitus believed that everything is in flux. It was he who asserted that one cannot step into the same river twice. Not surprisingly, he thought the fundamental substance was fire, and declared that all matter can be transformed into fire (and vice versa), just as all goods can be exchanged for gold. Did he anticipate the modern discovery that mass can be transformed into energy?

Heraclitus has had a great deal of influence on the twentieth century, largely through the nineteenth century Prussian philosopher Hegel. Influenced by Heraclitus, Hegel taught that the universe is a sort of debating society in which 'thesis' and 'antithesis' are forever struggling to produce a 'synthesis'. Marx adopted this philosophy, giving it a materialistic slant, and the views of Heraclitus ended up forming part of the official doctrine of Marxist governments, now much in decline.

Heraclitus has had less influence on logic. On one occasion he expressed his doctrine of continual change by saying that the river we step into both is and is not the same. Yet, in most logical systems, any statement of the

11. From Heraclitus to Democritus

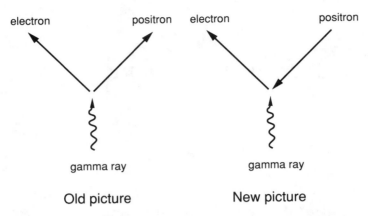

FIGURE 11.1. Positron as an electron travelling backwards in time

form 'p and not p' is regarded to be false. Hegelians sometimes adopt a similar mode of speech, claiming that 'a is not always equal to a'. Needless to say, this doctrine has not been applied to mathematics.

Yet Marxist philosophers try to understand not only history, but also mathematics in terms of a dialectic process. According to Lenin, subtraction is the antithesis of addition, yielding arithmetic as a synthesis, and integration is the antithesis of differentiation, the synthesis being calculus. Quite recently, the American mathematician Lawvere has suggested that a foundation of mathematics be built on a dialectic process in which the striving opposites are so-called 'adjoint functors', but this concept is too technical to be explained here.

Parmenides took the view opposite to that of Heraclitus, proclaiming that nothing changes, that change is an illusion: from the point of view of the 'goddess', the past and the future are all there at the same time. This is a bit like the view of the modern physicist and his four-dimensional space-time, in which the ever-changing events are replaced by unchanging world-lines.

Richard Feynman has recently shown that this way of viewing the universe allows one to give a more elegant and instructive explanation of certain fundamental processes. For example, to explain how an electron and a positron annihilate each other, giving rise to a γ ray, we may take it that the positron is an electron traveling backwards in time, having been deflected with a γ ray splitting off. Simultaneous pair creation is explained similarly. One may even speculate that there is only one electron in the universe. See Figure 11.1.

Zeno was a disciple of Parmenides. He produced four arguments attempting to prove that motion is impossible, his so-called 'paradoxes'. What he really showed was that if you do not allow infinite processes, what we now call 'limits' in mathematics, then you cannot use mathematics to analyze motion. These arguments, found in Aristotle's *Physics* 239b5-240a18 and 233a21-31, are the following.

1. A point moving from 0 to 1 on the number line first covers a distance of $1/2$, then a distance of $1/4$, then a distance of $1/8$, and so on. After n steps, it has covered a total distance of $1/2 + 1/4 + \ldots + 1/2^n = 1 - 1/2^n$. From the fact that there is no n such that $1 - 1/2^n = 1$, Zeno concluded that the point will never reach 1. In other words, motion from 0 to 1 is impossible.

   Today we get around Zeno's difficulty with the help of the notion of 'limit'. But this is a fairly sophisticated concept; according to nineteenth century mathematicians, the meaning of '$\lim_{n\to\infty} f(x) = a$' is as follows: for every real $\epsilon > 0$, there exists a natural number k such that, for all $n > k$, $|f(n) - a| < \epsilon$. While the ancients may have had an intuitive notion of what is meant by a limit, the rigorous definition was surely beyond them.

2. Achilles and the tortoise are engaged in a race along a measured line. Achilles starts at 0, but the tortoise is given a head start, at 1, since Achilles runs twice as fast as the tortoise. Thus, when Achilles arrives at the point 1, the tortoise is at $1 + 1/2$; when Achilles arrives at the point $1 + 1/2$, the tortoise is at $1 + 1/2 + 1/4$, and so on. In general when Achilles gets to $2 - 1/2^{n-1}$, the tortoise is at $2 - 1/2^n$, just a little bit ahead. From this Zeno concludes that Achilles can never catch up with the tortoise. If it looks like Achilles catches up with the tortoise, that only means that motion is an illusion.

   The modern solution to this paradox is the same as the solution to paradox (1): $\lim_{n\to\infty}(2 - 1/2^{n-1}) = \lim_{n\to\infty}(2 - 1/2^n)$.

3. Since, at any instant, a flying arrow is in exactly one place, Zeno infers that, at that instant, it is motionless. (Indeed, if you took a high-speed photograph of the arrow, it would look as if it was perfectly still.) One is tempted to conclude that the speed of the arrow is 0, that it is not really moving.

   The argument seems to be that the speed $dx/dt = 0$ because $dx = 0$. But this would only follow if $dt \neq 0$. If we assume that an interval of time is an actually infinite set of (equal) instants, then each instant has zero duration, and $dt = 0$. But then we could equally well infer that $dx/dt = 17$, since $0 = 17 \times 0$.

   Zeno's argument works only if you assume that time is basically discrete, there being some smallest, finite 'quantum' of time (e.g., $1/2^{100}$ seconds). During each quantum of time, the arrow would be motionless, for if it moved, say, from 0 to 1, there would be a time before it got to 1/2, and a time after, and the quantum of time would be divisible into two parts. Hence it really would not be moving. If there were some n such that every interval of time exceeds $1/2^n$ seconds, then Zeno would be right: motion is an illusion.

In writing the speed as dx/dt, we used the notation of the 17th century philosopher and mathematician Leibniz. He conceived of dx and dt as *infinitesimals* and thought of dx/dt as their actual ratio. The great Newton essentially shared this view, even though his notation was different. Infinitesimals were believed to be quantities which are infinitely small, yet unequal to zero. This idea was attacked by the 18th century philosopher Berkeley as being absurd, and 19th century mathematicians agreed with him. They redefined the ratio dx/dt as

$$\frac{dx}{dt} = \lim_{\delta t \to 0} \frac{\delta x}{\delta t},$$

where δx and δt are not necessarily small.

It was only in the middle of the 20th century that Abraham Robinson pointed out that infinitesimals may be introduced by fiat, just like the square root of -1, as *their existence does not lead to a contradiction*. Indeed, consider the following infinite collection of inequalities:

$$0 < dx, \; dx < 1, \; dx < 1/2, \; dx < 1/3, \; \ldots. \qquad (*)$$

Suppose we can derive a contradiction from this infinite collection. Now, it is generally agreed that a mathematical proof can have only a finite number of steps, this being part of the very definition of *proof*. Therefore, the proof of a contradiction from the assumptions $(*)$ can only mention a finite number of them, say the last being $dx < 1/n$. But this finite collection of assumptions does not lead to a contradiction, as $dx = 1/(n+1)$ satisfies all of them.

4. In Zeno's fourth argument there are three rows of people:

$$\begin{array}{ccccc} & A & A & A & A \\ & B & B & B & B & \to \\ \leftarrow & C & C & C & C \end{array}$$

The A's are stationary, the B's are moving to the right and the C's are moving to the left, at the same speed. Now suppose it takes a B one instant of time to pass an A, then it will take him half an instant of time to pass a C. It follows, that there is no such thing as an *instant*, if by that we mean an indivisible 'quantum of time': time is infinitely divisible, it is a substance.

In summary, we may not agree with Zeno that there is no motion, but we must credit him with probing into the very foundation of mathematics and physics.

Empedocles is important in experimental science and cosmology. He demonstrated that air is a substance by pushing an inverted tumbler into a tub of water; the water did not rush in to fill the apparent vacuum. He recognized not only the four traditional substances, earth, water, air and fire, but postulated two other: love and strife. He believed that, as long as love prevails, the cosmos contracts, but when strife prevails, it expands. Thus he seems to have anticipated modern astronomers in realizing that our universe is in an expanding phase. As his last experiment, he leaped into a volcano to demonstrate his immortality.

Democritus finally did away with substances and replaced them with atoms. These were assumed to be physically but not geometrically, indivisible. They were indestructible and constantly moving, the space between them being empty. The number of atoms was infinite, but they differed in shape and size. In the 19th century, people thought they had identified the atoms of Democritus; but then, in the early 20th century, Rutherford showed that the so-called atoms were divisible after all. It might have been wiser to reserve the word 'atom' for our electrons and quarks, if these indeed turn out to be the ultimate constituents of matter.

Democritus was a determinist, he believed that everything that happens must happen. He wrote books on geometry, which have been lost. He is said to have emphasized the importance of proofs and to have discovered (or rediscovered) how to calculate the volume of a pyramid or cone by taking one third of the base area times the height. It is curious that Marx wrote his doctoral dissertation on Democritus and not on Heraclitus.

12
Mathematics in Athens

After an alliance of Greek cities defeated the Persians in 490 BC, Athens became, for over a hundred years, a great center of civilization. There are things about it we do not like: Athens exacted tribute from its allies, and the leisurely life of its leading citizens was based upon slavery. Nonetheless, one can safely say that the degree of civilization achieved by the Athenians around 400 BC has rarely, if ever, been surpassed in the history of the world. Because we must confine our attention to mathematics, we shall touch on only one of the many areas in which cultural development took place.

One of the first mathematicians who worked in Athens was the sophist Hippias (420 BC), from Elis on the west coast of Greece. In a dialogue sometimes ascribed to Plato, we hear Socrates (469–399 BC) teasing Hippias about his mathematics:

> Socrates: And tell me, Hippias, are you not a skilful calculator and arithmetician?
>
> Hippias: Yes, Socrates, assuredly I am.
>
> Socrates: And if someone were to ask you what is the sum of 3 multiplied by 700, you would tell him the true answer in a moment, if you pleased?
>
> Hippias: Certainly I should.
>
> Socrates: Is not that because you are the wisest and ablest of men in these matters?
>
> Hippias: Yes.

Hippias discovered a curve called the *quadratrix*, which can be used for trisecting an arbitrary angle, and also for constructing a square equal in area to a given circle. He described the quadratix as follows: imagine that side AD of the unit square $ABCD$ moves down at a rate of 1 unit per second towards the side BC (on the 'bottom' of the square). Imagine that side AB rotates about B at a rate of 1/4 revolution per second towards BC, so that, after 1 second, both AD and AB coincide with BC. At any time t ($0 \leq t \leq 1$), the two moving sides meet at a point P. The set or *locus* of these points P is the quadratrix.

In terms of our modern analytic geometry and trigonometry, we would put it this way: the point P has coordinates

$$((1-t)/\tan(90°(1-t)), \ 1-t)$$

so the equation of the quadratrix is $y = x\tan(90°y)$, with $0 \leq y$.

To divide an angle of, say, 60° into 3 equal parts, it is enough to place it in standard position, with its vertex on the origin, and one arm along the positive x axis. If the other arm meets the quadratrix at (a,b), we find the point P where $y = b/3$ meets the quadratrix. The angle between the line joining P to the origin and the positive x axis is 20°.

Furthermore, as y tends to 0, $x = y/\tan(90°y)$ tends to $2/\pi$, and so the quadratrix can be used to 'square the circle'.

This highly ingenious method was criticized — by Plato, it seems — on the grounds that it is more elegant to use only straight lines and circles in the solution of mathematical problems. One ought to carry out geometric constructions using only a ruler (for drawing straight lines) and a compass (for drawing circles); using a quadratrix was considered to be cheating.

However, as Pierre Wantzel (1814–1848) was the first to prove, it is not possible to trisect an arbitrary angle using only straight lines and circles. One has to use some other tool — such as the quadratrix. Hippias was right and Plato was wrong; although by insisting on ruler and compass constructions he raised an interesting and challenging problem, which we shall discuss in Chapter 14.

In order to understand the Greek contribution to the beginnings of analysis, it is important to know how they attacked, and finally solved, the problem of the area of the circle. Antiphon the sophist (425 BC) was one of the early Athenian mathematicians who worked on this problem. He suggested that the area of the circle be calculated in terms of the regular polygons inscribed in it. (The *regular m-gon* is the polygon with m equal sides and angles.)

Using the assumption that the area of the union of pairwise disjoint sets equals the sum of their areas, it is not hard to show that an inscribed square takes up more than 1/2 the area of a circle, and an inscribed regular octagon takes up more than 3/4 of the area of the circle. Indeed, as the ancient Greeks realized (Euclid's *Elements* XII 2), one can use what we call 'mathematical induction' to show that an inscribed regular 2^n-gon takes up more than $1 - 1/2^{n-1}$ of the area of a circle.

If we inscribe a regular 2^n-gon in a circle, its longest diagonals are diameters of that circle. The Greeks knew that the area of a regular 2^n-gon is proportional to the square on its longest diagonal – a result which follows from the fact that the area of a triangle with given angles is proportional to the square of its longest side – and from this it follows that, insofar as a circle is like a regular 2^n-gon, its area is proportional to the square on its diameter. Indeed, somewhat later than Antiphon, Eudoxus (408 – ca. 355 BC) gave a rigorous proof that the area of a circle is, in fact, proportional to the square on its diameter.

Antiphon boldly claimed that a circle simply *is* a regular polygon (with a large number of sides). In making this claim, Antiphon entered a lively discussion, started by Zeno (450 BC) and others, about whether space is continuous or discrete. If space is discrete, then there is some minimum area e. If n is so large that $1/2^{n-1}$ of the area of the circle is less than e, then an inscribed regular 2^n-gon, in taking up more than $1 - 1/2^{n-1}$ of the area, actually takes up *all* the area.

Another early Athenian mathematician who worked on the geometry of the circle was Hippocrates, who came from the Greek island of Chios, near present-day Turkey. (He is not to be confused with the physician, famous for his oath, who came from Cos.) Hippocrates, it is said, had been swindled in business and came to Athens about 430 BC to recover his property through legal action. The case dragged on, and Hippocrates used the time to study philosophy and supported himself by teaching geometry.

Hippocrates was responsible for much of the material in Books III and IV of Euclid's *Elements*. He called the square of a quantity 'dynamos', hence our 'power'. He pioneered the custom of reducing one theorem to another and may have been one of the first to use the method of *reductio ad absurdum* in mathematics.

He was also the first to find the precise area of a region bounded by curves, as we shall now see. Construct semicircles on three sides of a right triangle. By the converse of the theorem of Thales, the semicircle on the hypotenuse passes through the vertex at the right angle. The semicircles on the other two sides of the right triangle are supposed to lie outside the triangle. (See Figure 12.1.) The areas included in the two smaller semicircles, but not in the semicircle on the hypotenuse, are called *lunes* (after the crescent moon).

Hippocrates argued as follows. If the vertices of the triangle are A, B and C, with the right angle at C, then $AC^2 + CB^2 = AB^2$ (by the theorem of Pythagoras). Since the area of a circle is proportional to the square on its diameter, the area of a semicircle is likewise proportional to the square on

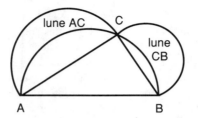

FIGURE 12.1. Lunes of Hippocrates

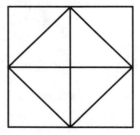

FIGURE 12.2. Doubling the square

the diameter. Therefore the sum of the areas of the semicircles on AC and CB equals the area of the semicircle on AB. Subtracting the areas where the semicircles overlap, we may conclude that the sum of the areas of the lunes equals the area of the right triangle. That is, since semicircle AC plus semicircle CB equals semicircle AB, it follows that

$$\text{lune } AC + \text{lune } CB = \text{triangle } ABC = \tfrac{1}{2} BC \times AC.$$

Hippocrates also contributed to the problem of 'doubling the cube'. This was the problem of determining the length x such that $x^3 = 2$, preferably using only the geometry of straight lines and circles.

Legend has it that, during the plague of 430 BC, the Athenians consulted the oracle of Delos for help. The oracle replied that they should double the altar of Apollo, a marble cube, in size. When the plague refused to abate, the oracle explained that the Athenians had doubled the edges of the cube, not its volume. Although the Athenians did not succeed with this task, at least not according to the methods acceptable to Plato, the plague seems to have stopped anyway.

Hippocrates noted that one could double the volume of a cube, with edge one unit in length, if one could find quantities x and y such that $1/x = x/y = y/2$; for then x^3 would equal 2. However, he did not succeed in constructing these quantities in a way that satisfied Plato (who wanted to use only straightedge and compass).

12. Mathematics in Athens 63

Socrates (469–399 BC) was the mentor of Plato. As Plato portrays him in the dialogue *Meno*, Socrates claimed that all knowledge is recollection. In an argument with Meno on the nature of virtue, Socrates bet him that he could make his slave 'remember' a geometric construction and its proof. Asked to double the unit square, the slave, who was completely ignorant of geometry, first offered to double its side, but was soon led to admit his error. Then Socrates got him to look at the figure of a square with the midpoints of its four sides all joined to each other (Figure 12.2) and soon persuaded him to 'remember' that it is the square on the diagonal of the inner square which has double its area.

As a young man, Plato (429–349 BC) was a disciple of Socrates. After the latter's death, Plato travelled to Africa, where he visited Heliopolis, now a suburb of Cairo, and Cyrene in Lybia. There he studied with Theodorus, who had proved the irrationality of the square roots of the nonsquare integers less than 18. Plato also went to Italy and became acquainted with Archytas (428–347 BC), the head of the Pythagorean school. Archytas also had 'doubled the cube', but he did so by going beyond the geometry of straight lines and circles.

Plato returned to Athens in about 380 BC and founded the famous Academy. At the entrance of this school was the inscription: *let no one ignorant of geometry enter here.*

The importance of Plato in the history of mathematics is due not so much to any mathematical contribution of his own as to the influence he exerted on others. It was he who insisted that a 'proper' solution involve no curves other than the circle (*Timaeus* 34a). It was he who emphasized the importance of clear definitions and postulates. Finally, Plato strongly encouraged people to study mathematics because he believed that this study would help them become wise and therefore virtuous. The five Platonic solids, or regular polyhedra, were not discovered by Plato, but he discussed them in the *Timaeus*.

Plato had a brilliant student, Theaetetus, who died in battle in 369 BC, and to whom he dedicated a dialogue. It was Theaetetus who showed that the square root of a natural number is irrational if and only if the natural number is not a square (*Theaetetus* 147c-148b). Theatetus also studied the regular polyhedra, and worked on the theory of proportion. According to van der Waerden [1985], Theaetetus was responsible for Books X and XIII of Euclid's *Elements*.

The most important Athenian mathematician at this time was Eudoxus of Cnidus, another small Greek island near modern Turkey. Eudoxus lived from 408 to 355 BC, and distinguished himself in astronomy, medicine, geography and philosophy – as well as mathematics. Like Plato, he studied astronomy in Heliopolis, and mathematics with Archytas in Tarentum (in what is now southern Italy). As a young man, Eudoxus studied in Plato's

Academy, commuting on foot from Piraeus, the harbour district. Later he engaged in a philosophical controversy with Plato; it seems that Eudoxus anticipated the Epicurean position that humans strive to maximize pleasure minus pain.

In mathematics, Eudoxus was responsible for Books V and XII of Euclid's *Elements*. Book V deals with the theory of proportion. Today, might define the proportion 'a is to b as c is to d' (written $a : b :: c : d$) as an equation $a/b = c/d$, and we would say that the proportion held just in case $ad = bc$. However, this presupposes our theory of the real number field. It presupposes that we already have some way of understanding what it is to multiply two irrational numbers. Eudoxus was starting from scratch. He could not use multiplication to define proportion because it was in terms of proportion that he defined multiplication. Eudoxus used his theory of proportion to prove the basic laws of multiplication, such as commutativity and associativity. The definition on which he based his development of the number system was the following:

$a : b :: c : d$ if and only if, for all positive integers p and q,

$$pa > qb \text{ if and only if } pc > qd,$$

and likewise with $>$ replaced by $<$.

If we take a/b and c/d to be positive real numbers, this statement asserts:

- the set of rationals above $a/b =$ the set of rationals above c/d
- the set of rationals below $a/b =$ the set of rationals below c/d,

thus anticipating the modern definition of real numbers due to Dedekind.

Actually, Eudoxus assumed that a, b, c and d are geometric quantities. For example, a and b could be arcs of circles and c and d could be angles. This is another reason why he did not write $ad = bc$; for how do you multiply an arc by an angle?

One particular ratio Eudoxus was interested in arose from the following problem; to divide a segment AB by a point H so that $AB/AH = AH/HB$, that is, the whole is to the larger part as the larger part is to the smaller. Taking $AH = x$ and $HB = 1$, we obtain the quadratic equation $x^2 - x - 1 = 0$, so that, upon discarding the negative solution, we find $x = (1 + \sqrt{5})/2$. This, or sometimes its reciprocal $(-1 + \sqrt{5})/2$, is known as the *golden section*.

Eudoxus also gave a proof that the area of a circle is proportional to the square on its diameter (Euclid's *Elements* XII 2), by inscribing regular polygons with 2^n sides in both circles and taking n sufficiently large.

Finally, we should mention Menaechmus (350 BC), another student of Plato's. Menaechmus discovered the conics — the ellipse, hyperbola and parabola — and used them to 'double the cube'. Using modern analytic geometry, we can express his solution to the 'Delian problem' in the following simple fashion: the parabolas $y = \frac{1}{2}x^2$ and $x = y^2$ intersect at a point whose y coordinate is the cube root of 2. In his 'doubling of the cube', Menaechmus did not stick to straight lines and circles. Indeed, as already mentioned, in 1837, Pierre Wantzel showed that it is not possible to obtain a segment of length equal to the cube of root of 2, using only the geometry of straight lines and circles. We shall give a proof of this in Chapter 14.

Exercises

1. Prove that the inscribed regular 2^n-gon takes up more than $1 - 1/2^{n-1}$ of the area of the circle.

2. If the diameter of a circle is d, prove that the area of the inscribed regular 2^n-gon is $2^{n-3}d^2 \sin(\pi/2^{n-1})$.

3. Prove the theorem of Theaetetus that a natural number has an irrational square root if and only if it is not a perfect square.

4. Using the definition of proportion given by Eudoxus, show that $a : b :: c : d$ if and only if $d : c :: b : a$.

13
Plato and Aristotle on Mathematics

Plato (427–347 BC) believed that the objects in the universe fall into two very different classes, the material and the immaterial. A chair or an ox belongs to the class of material things. A soul or a number belongs to the class of immaterial things. The drawing of a square belongs to the material realm but the square itself belongs to the immaterial realm. Plato says of the students of geometry that they

> make use of the visible forms and talk about them, though they are not thinking of them but of those things of which they are a likeness, pursuing their inquiry for the sake of the square as such and the diagonal as such, and not for the sake of the image of it which they draw (*Republic* 510d).

For Plato, the class of material things is characterized by change, uncertainty, ignorance and imperfection. The drawing of a square can be erased and it is doubtful whether its angles are each exactly 90° or whether its sides are perfectly straight.

On the other hand, the class of immaterial things is characterized by their constancy and perfection and by our certain knowledge of them. The square 'as such' has sides which remain perfectly straight forever. Its properties can be deduced with infallible rigour. We can know with absolute certainty that its diagonals are equal.

Scientists understand change and motion in the universe in terms of unchanging formulas or laws. Plato had a similar outlook (*Timaeus* 52-58). Moreover, he stressed that the formulas or laws have an existence of their own, independent of the material universe.

13. Plato and Aristotle on Mathematics

According to Plato, mathematical objects are not the only immaterial objects. Other immaterial objects are God, goodness, courage and the human soul (*Republic* 380d-383c). However, the best way to begin to know the immaterial realm is to do mathematics. One is to study number theory 'for facilitating the conversion of the soul itself from the world of generation to essence and truth' (*Republic* 525c). One is to study geometry 'to facilitate the apprehension of the idea of good' (*Republic* 525e).

Plato believes that the truths of mathematics are absolute, necessary truths. He believes that, in studying them, we shall be in a better position to know the absolute, necessary truths about what is good and right, and thus be in a better position to become good ourselves.

Platonism as a philosophy of mathematics is the view that at least the most basic mathematical objects (e.g., real numbers, Euclidean squares) actually exist, independently of the human mind which conceives them. Their properties are discovered, not created.

Aristotle (384–322 BC) was a student of Plato, but he disagreed with him about the nature of mathematics. In Book XIII of the *Metaphysics*, Aristotle asserts that

> conclusions contrary alike to truth and to the usual views follow, if one is to suppose the objects of mathematics to exist thus as separate entities (*Metaphysics* 1077a).

For Aristotle, a word like 'two' is not a noun designating an abstract object but rather an adjective describing a concrete object (the *two* yard ladder, a *two* year period).

Whereas Platonism is quite compatible with the view that there are actually infinite lines and sets with an infinite numbers of elements, Aristotle is a staunch finitist. He would have rejected Cantor's 'aleph-null' (*Metaphysics* 1084a); he would have rejected infinitesimals (*Physics* 266b); he did reject infinite sets and infinite magnitudes (*Physics* III). For Aristotle, the geometer can have as much as he needs of an infinite line but he cannot have the whole line in its infinite totality.

Under the influence of Plato, Aristotle formulated a principle according to which every (mathematical) statement is either true or false, but he had his doubts when it came to applying this principle to the everyday temporal world. He wondered whether a statement like 'there will be a sea battle tommorrow' is either true or false (*On Interpretation* 9). How can it be true if the battle may not occur? How can it be false if the fight is a real possibility?

Like the view of the 20th century 'intuitionists', Aristotle's view is human-centered. The reality of numbers has to do, not with some alien heaven, but with the way we describe our surroundings. The infinite must be rejected because we humans work in a finite way. The truth about certain propositions may be left in abeyance if it is, for the moment, inaccessible to human beings.

Aristotle also had something to contribute to the old problem that had divided the Ionic philosophers from those based in Italy: whether the universe is made up of substances or atoms, whether things should be measured or counted. He pointed out that when we talk about a loaf of bread or a glass of wine, bread and wine are measured, but loaves and glasses are counted. He asserted that we measure *matter* by counting its *forms*.

Exercise

How might Aristotle answer Zeno's arguments against motion?

14
Constructions with Ruler and Compass

Ancient Greek mathematicians were haunted by three problems:
I   doubling the cube, that is, finding the cube root of 2;
II trisecting any given angle, say an angle of 60°
    (of course some angles are easily trisected, for example, one of 90°);
III squaring the circle,
    that is, constructing a square equal in area to that of a given circle.

Assorted solutions to these problems were proposed at various times, but these did not conform to the rules of the game, presumably laid down in Plato's Academy, that *only constructions with ruler and compass* be admitted. (Actually, we only know that Pappus attributed these rules to Plato more than 600 years later.) Moreover, the ruler could only be used for joining two points and the compass could only be used for drawing a circle with a given point as center and a given segment as radius. The reader will have no difficulty in carrying out the following constructions with ruler and compass:

(a) to bisect a given angle;

(b) to find the right bisector of a given segment;

(c) to draw a line through a given point parallel to a given line;

(d) to construct an equilateral triangle.

If we adopt a given segment as our unit of length, we can represent any positive real number by a segment, actually by the ratio of this segment to

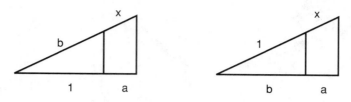

FIGURE 14.1. Finding ab and a/b

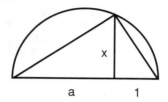

FIGURE 14.2. Constructing root of a

the unit segment. With the help of ruler and compass, the Greeks were able to perform the following arithmetical operations on positive real numbers: adding, subtracting (the smaller from the larger), multiplying, dividing and extracting square roots. The first four of these are called *rational* operations.

Indeed, for addition and subtraction this is obvious. To find $x = ab$ one considers the proportion

$$x : b = a : 1$$

and to find $x = a/b$ one considers the proportion

$$x : 1 = a : b.$$

In both cases, the problem is that of finding the fourth proportional to three given lengths, which can easily be done, following Thales, with the aid of similar triangles, using only ruler and compass constructions. See Figure 14.1.

To find $x = \sqrt{a}$, one looks at the proportion

$$1 : x = x : a.$$

Here the problem is that of finding the *mean* proportional of two given lengths, which the Greeks solved ingeniously by ruler and compass constructions, as illustrated by Figure 14.2, which exhibits the semicircle on a segment of length $a + 1$.

To attack problem **II**, the trisection problem, we would nowadays use trigonometry. First note that we can construct an angle if and only if we can construct its cosine. If $\theta = 60°/3 = 20°$, then $\cos 3\theta = \cos 60° = 1/2$, as is seen from Figure 14.3.

14. Constructions with Ruler and Compass

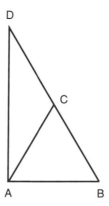

FIGURE 14.3. The cosine of 60°

In Figure 14.3, ABC is an equilateral triangle and $CD = AC$. It easily follows that the angle at B is 60°, the angle at D is 30° and the angle BAD is 90°, hence $\cos 60° = AB/BD = 1/2$.

Now

$$\begin{aligned}
\cos 3\theta &= \cos(2\theta + \theta) = \cos 2\theta \cos \theta - \sin 2\theta \sin \theta \\
&= (\cos^2 \theta - \sin^2 \theta) \cos \theta - 2 \sin \theta \cos \theta \sin \theta \\
&= \cos^3 \theta - 3 \sin^2 \theta \cos \theta = \cos^3 \theta - 3(1 - \cos^2 \theta) \cos \theta \\
&= 4 \cos^3 \theta - 3 \cos \theta.
\end{aligned}$$

Thus we want to solve the equation $8 \cos^3 \theta - 6 \cos \theta = 1$. Putting $2 \cos \theta = u$, we obtain the cubic equation

$$u^3 - 3u - 1 = 0 \, .$$

The question is therefore whether a solution of this cubic equation can be expressed in terms of rational operations and square roots. We shall return to this problem in the next chapter. First let us look at a problem which the Greeks were able to solve by their methods.

One of the highlights in Euclid's *Elements* is the construction of a regular pentagon, equivalently, that of an angle of $360°/5 = 72°$. Today we would attack this problem too with the help of trigonometry; for an elegant argument we may even invoke complex numbers. Let $\theta = 72°$, then $5\theta = 360°$, hence, by de Moivre's Theorem,

$$\begin{aligned}
(\cos \theta + i \sin \theta)^5 &= \cos 5\theta + i \sin 5\theta \\
&= \cos 360° + i \sin 360° \\
&= 1 + i0 \\
&= 1.
\end{aligned}$$

14. Constructions with Ruler and Compass

Thus, we wish to solve the equation $z^5 = 1$, that is,

$$(z-1)(z^4 + z^3 + z^2 + z + 1) = 0,$$

where $z = \cos\theta + i\sin\theta$. Note that $z^{-1} = \cos\theta - i\sin\theta$, so that

$$2\cos\theta = z + z^{-1} = u$$

say. Clearly, $z = 1$ is not a satisfactory solution, so our problem reduces to solving

$$z^4 + z^3 + z^2 + z + 1 = 0.$$

Dividing by z^2, we write this as

$$z^2 + z + 1 + z^{-1} + z^{-2} = 0.$$

Now $z + z^{-1} = u$, hence $z^2 + 2 + z^{-2} = u^2$, so that $z^2 + z^{-2} = u - 2$. The equation to be solved then becomes

$$u^2 + u - 1 = 0.$$

Discarding the negative solution, we find

$$u = \frac{-1 + \sqrt{5}}{2},$$

a number which we recall from Chapter 12 as the golden section.

Of course this is not how the Greeks attacked the problem, as they did not know about complex numbers, and, at the time of Euclid, had not yet invented trigonometry. Nonetheless, our analysis shows that they could construct the angle θ, since $2\cos\theta = (-1 + \sqrt{5})/2$ involves only rational operations and square roots. Indeed, when Euclid constructs a regular pentagon in Book IV, Proposition 11, by what looks to us like a rather complicated method, he makes use (via Proposition 10) of the earlier construction of the golden section in Book II, Proposition 11 (taken up once more in Book VI, Proposition 30). The Greeks could of course construct squares and regular hexagons, but one problem they did leave open was the following:

IV constructing a regular heptagon, that is, a seven sided figure.

By the same method we just employed for constructing a regular pentagon, we can see that this problem reduces to the following cubic equation in $u = 2\cos(360°/7)$:

$$u^3 + u^2 - 2u - 1 = 0.$$

14. Constructions with Ruler and Compass

Although the ancient Greeks did not know this, the only arithmetical operations that can be carried out with ruler and compass constructions are rational operations and square root extractions and, of course, combinations of such. To understand why this is so, we have to make use of analytic geometry, which was only developed in the seventeenth century by René Descartes.

His pioneering idea was to represent every point in the plane by a pair of real numbers (x, y) and to observe, conversely, that every such pair represents a point. Unlike the Greeks, we need not confine x and y to be positive: if we use the modern rectangular coordinate system, x is negative in the second and third quadrant, y is negative in the third and fourth quadrant. We can now say that a straight line consists of all points (x, y) satisfying an equation of the form

$$ax + by + c = 0,$$

where a, b and c are given real numbers, and a circle consists of all points (x, y) satisfying an equation of the form

$$x^2 + y^2 + dx + ey + f = 0.$$

Now what happens when we perform the following operations:

1. join two given points,
2. draw a circle with given center and radius,
3. intersect two straight lines,
4. intersect a circle and a straight line,
5. intersect two circles?

(1) Suppose the given points are (x_1, y_1) and (x_2, y_2). Then we easily see that the straight line has the equation

$$(y_1 - y_2)x + (x_2 - x_1)y + (x_1 y_2 - x_2 y_1) = 0,$$

in other words, an equation of the form

$$ax + by + c = 0,$$

where a, b and c are expressed in terms of the given quantities x_1, y_1, x_2, y_2 by means of the operations of addition, subtraction and multiplication.

(2) Suppose the center is (α, β) and the radius is ρ, then the equation of the circle is

$$(x - \alpha)^2 + (y - \beta)^2 = \rho^2,$$

that is,
$$x^2 + y^2 - 2\alpha x - 2\beta y + \alpha^2 + \beta^2 - \rho^2 = 0.$$

Thus the equation of the circle is of the form
$$x^2 + y^2 + dx + ey + f = 0,$$

where d, e and f are again expressed in terms of the given quantities α, β and ρ by means of addition, subtraction and multiplication.

(3) To find the intersection of two given straight lines, we must solve the pair of equations:
$$ax + by + c = 0,$$
$$a'x + b'y + c' = 0.$$

We obtain the solution
$$x = -\frac{cb' - c'b}{ab' - a'b}, \quad y = -\frac{ac' - a'c}{ab' - a'b}.$$

(It is assumed that $ab' - a'b \neq 0$, otherwise the two lines are parallel or even coincide.) Again we see that the new quantities x and y are obtained from the given quantities a, b, c, a', b' and c' by means of the rational operations, including division.

(4) To find the intersection of a circle and a straight line, we must solve the pair of equations:
$$ax + by + c = 0,$$
$$x^2 + y^2 + dx + ey + f = 0.$$

Assuming, for example, that $b \neq 0$, we get
$$y = -\frac{a}{b}x - \frac{c}{b}$$

from the first equation. When we substitute this into the second equation, we obtain a quadratic equation:
$$Ax^2 + Bx + C = 0,$$

where A, B and C are expressed by means of the rational operations in terms of the given quantities a, b, c, d, e and f. In particular, $A = 1 + a^2/b^2 > 0$. Finally, solving for x, we obtain:
$$x = \frac{-B \pm \sqrt{B^2 - 4AC}}{2A}.$$

Here x is expressed not only by means of the rational operations, but also requires a square root. Note that, if $B^2 - 4AC = 0$, the line is tangent to the circle and, if $B^2 - 4AC < 0$, it does not meet the circle at all.

(5) To find the intersection of two circles, we must solve the pair of equations:
$$x^2 + y^2 + dx + ey + f = 0,$$
$$x^2 + y^2 + d'x + e'y + f' = 0.$$
By subtracting, we may here replace the second equation by the linear equation
$$(d - d')x + (e - e')y + f - f' = 0,$$
so the situation is the same as that already treated under (4) above. The last equation represents the straight line passing through the two points of intersection of the given circles, or, if the given circles merely touch, their common tangent. Of course, it may also happen that the two circles have no common points at all.

We had seen earlier that geometric constructions with ruler and compass allow us to carry out the rational operations and to extract square roots. We have now shown the converse: combinations of rational operations and square roots are the only arithmetical operations which can be carried out in this way. That is, when we perform ruler and compass constructions (1) to (5), we only get lengths that can be expressed in terms of the operations $+$, $-$, $\times$, $/$, and $\sqrt{}$. To solve problems **I** to **IV** by ruler and compass constructions is thus equivalent to expressing the real numbers $\sqrt[3]{2}$, $\cos 20°$, $\sqrt{\pi}$ and $\cos(360°/7)$ by rational operations and square roots.

Exercises

1. Show how to carry out the constructions (a) to (d) in the text, using ruler and compass only.

2. Carry out a ruler and compass construction of the golden section $(-1 + \sqrt{5})/2$.

3. To construct a regular heptagon one has to find the angle $\theta = 360°/7$. Show that $u = 2\cos\theta$ satisfies the cubic equation
$$u^3 + u^2 - 2u - 1 = 0.$$

15
The Impossibility of Solving the Classical Problems

The ancient Greeks were unable to solve problems **I** to **IV** using ruler and compass constructions, for a good reason: it cannot be done. Concerning problem **III**, this was shown only in 1882 by C.L.F. Lindemann (1852–1939), who proved that π is not an algebraic number, which implies, in particular, that $\sqrt{\pi}$ cannot be constructed by rational operations and square roots. His method was based on an earlier proof by Hermite, who had shown that $e = \lim_{n\to\infty}(1+\frac{1}{n})^n$ is not algebraic.

Problems **I**, **II** and **IV** have one thing in common: they can all be expressed by cubic equations, namely,

$$u^3 - 2 = 0, \quad u^3 - 3u - 1 = 0, \quad u^3 + u^2 - 2u - 1 = 0.$$

(For the last see Exercise 3 of Chapter 14.) First let us make sure that they have no rational solutions. (For the first equation we already know this.)

Lemma 15.1. *If $a_n x^n + a_{n-1} x^{n-1} + \cdots + a_1 x + a_0 = 0$ is a polynomial equation with integer coefficients, then any rational solution is of the form p/q, with p a factor of a_0 and q a factor of a_n. In particular, when $a_n = 1$, any rational solution will be an integer and a factor of a_0.*

Proof: Let p/q be a rational solution, where $q \neq 0$ and $\gcd(p,q) = 1$. Putting $x = p/q$ in the equation and multiplying by q^n, we obtain

$$a_n p^n + a_{n-1} p^{n-1} q + \cdots + a_0 q^n = 0.$$

Since p divides all the terms except possibly the last, it must divide the last term also. Since $\gcd(p,q) = 1$, it follows from the unique factorization into primes that p divides a_0. Similarly, q divides a_n.

It follows from the lemma that the only possible rational solutions of the equation $u^3 - 2 = 0$ are $u = \pm 1$ or $u = \pm 2$, all four of which are quickly seen not to be solutions. Similarly, the only possible solutions of the second and third equations are $u = \pm 1$, which are also seen not to work. We may conclude that solutions to these three equations cannot be expressed by means of rational operations alone, but there remains the possibility that they can be expressed with the help of square roots.

Over the years, many people have tried their hand at problems **I** and **II**. For example, the much envied Casanova worked on doubling the cube and, even today, there are still many determined angle trisectors. However, in 1837, Pierre Wantzel (1814–1848) showed that none of $\sqrt[3]{2}$, $2\cos 20°$, and $2\cos(370°/7)$ can be expressed in terms of rational operations and square roots and, therefore, that problems **I**, **II** and **IV** cannot be solved by ruler and compass constructions.

To give a simple exposition of why this is so, we shall introduce a more modern concept, that of a field. For our purposes, a *field* is a set of numbers, real or complex, which contains the number 1 and which is closed under the rational operations. (There are other fields, but we shall not need them here.) In particular, the rationals form a field $\mathbf{Q}$ and so does $\mathbf{Q}[\sqrt{2}]$, the set of all numbers of the form $a + b\sqrt{2}$, where a and b are rational. More generally, if F is any field, then so is $F[\sqrt{c}]$, where c is a given element of F, by which we understand the set of all numbers of the form $a + b\sqrt{c}$ with $a, b \in F$.

It is clear that $F[\sqrt{c}]$ is closed under addition, subtraction and multiplication. To show that it is also closed under division, we assume that $\sqrt{c}$ is not in F, otherwise there would be nothing to prove, and that $a + b\sqrt{c} \neq 0$. We calculate

$$\frac{1}{a+b\sqrt{c}} = \frac{1}{a+b\sqrt{c}} \times \frac{a-b\sqrt{c}}{a-b\sqrt{c}} = \frac{a-b\sqrt{c}}{a^2-b^2c} = \frac{a}{a^2-b^2c} + \frac{-b}{a^2-b^2c}\sqrt{c},$$

which is again of the form $a' + b'\sqrt{c}$ with $a', b' \in F$. A small argument is necessary to check that $a^2 - b^2 c \neq 0$.

We can now say that a real number u is constructible with ruler and compass, equivalently, expressible by rational operations and square roots, if and only if there exists a sequence of fields

$$\mathbf{Q} = F_0 \subseteq F_1 \subseteq \cdots \subseteq F_n$$

such that $F_{k+1} = F_k[\sqrt{c_k}]$ with $c_k \in F_k$ and $u \in F_n$.

Proposition 15.2. *Suppose $f(x) = x^3 + a_2 x^2 + a_1 x + a_0$ is a cubic polynomial with coefficients in a field F. Suppose further that the equation $f(x) = 0$ has a solution in $F[\sqrt{c}]$ with $c \in F$. Then it already has a solution in F.*

Proof. Let $x_1 = a + b\sqrt{c}$ be the given solution with $a, b \in F$. Then

$(x_1 - a)^2 = b^2 c$, hence $x_1^2 + px_1 + q = 0$ with $p, q \in F$. Dividing $f(x)$ by the polynomial $x^2 + px + q$, we obtain

$$f(x) = (x^2 + px + q)(x + d) + (ex + f),$$

where the quotient is $x + d$ with $d \in F$ and the remainder is $ex + f$ with $e, f \in F$. Since $f(x_1) = 0$ and $x_1^2 + px_1 + q = 0$, we deduce that $ex_1 + f = 0$. If $e \neq 0$, then $x_1 = -f/e \in F$ and we need look no further. If $e = 0$, then also $f = 0$, hence $x + d$ is a factor of $f(x)$. But then $f(-d) = 0$ and so $x_2 = -d \in F$ is the required solution.

Corollary 15.3. *If a number expressible by rational operations and square roots satisfies a cubic equation with rational coefficients, then this equation must have a rational solution.*

Proof: Suppose the cubic equation $f(x) = 0$ has no rational solution. Then, by Proposition 15.2 with $F = \mathbf{Q}$, it has no solution in $\mathbf{Q}[\sqrt{c_1}]$ with $c_1 \in \mathbf{Q}$. Again, by the Proposition with $F = \mathbf{Q}[\sqrt{c_1}]$, it has no solution in $\mathbf{Q}[\sqrt{c_1}][\sqrt{c_2}]$ with $c_2 \in \mathbf{Q}[\sqrt{c_1}]$. Continuing in this way, we see that it has no solution in $\mathbf{Q}[\sqrt{c_1}] \cdots [\sqrt{c_n}]$ for any n, where $c_k \in \mathbf{Q}[\sqrt{c_1}] \cdots [\sqrt{c_{k-1}}]$ for $1 < k \leq n$. Thus it has no solution expressible by rational operations and square roots.

Since the cubic equations

$$u^3 - 2 = 0, \quad u^3 - 3u - 1 = 0, \quad u^3 + u^2 - 2u - 1 = 0$$

have no rational solutions, as we verified earlier, we can now infer from Corollary 15.3 that they have no solutions expressible in terms of rational operations and square roots. In view of Chapter 14, we may therefore conclude that problems **I**, **II** and **IV** cannot be solved using only ruler and compass constructions. In summary:

Theorem 15.4. *It is impossible to double a cube, to trisect an arbitrary angle or to draw a regular heptagon by ruler and compass constructions.*

We have seen that the Greeks were able to draw regular polygons with 3 or 5 sides, but not with 7 sides. The question arises, for which primes p is it possible to construct a regular p-gon using ruler and compass only? Carl Friedrich Gauss showed that this is possible whenever p is a prime of the form $2^n + 1$ and Wantzel proved the converse. Gauss was so pleased with his discovery that he wanted a regular 17-gon inscribed on his tombstone. His request was not carried out, but a regular 17-gon was inscribed on a monument to Gauss in Braunschweig, Germany.

Odd prime numbers of the form $2^n + 1$ are called 'Fermat primes', after Pierre de Fermat (1601–1665). It is easy to prove that $2^n + 1$ cannot be prime unless n has the form 2^k.

To see this, one makes use of the identity

$$x^b + 1 = (x+1)(x^{b-1} - x^{b-2} + \cdots + 1)$$

for odd b. If $n = ab$ and b is odd, put $x = 2^a$ and infer that $2^a + 1$ is a factor of $2^n + 1$. Unless $b = 1$, it follows that $2^a + 1$ is an odd proper divisor of $2^n + 1$ different from 1, so $2^n + 1$ cannot be prime. Thus, if $2^n + 1$ is prime, n cannot have any odd proper divisors other than 1, hence n must be a power of 2.

Fermat was under the impression that $2^{2^k} + 1$ is prime for all natural numbers k. Euler later found that $2^{2^5} + 1$, a ten digit number, is divisible by 641. An easy, though tricky, way of seeing this is as follows:

$$\begin{aligned} 2^{2^5} &= 16 \times 2^{28} = (641 - 5^4)2^{28} \\ &= 641m - (5 \times 2^7)^4 \\ &= 641m - (641 - 1)^4 \\ &= 641m - (641n + 1) \\ &= 641(m-n) - 1, \end{aligned}$$

where m and n are integers, hence $2^{2^5} + 1$ is a multiple of 641.

At the moment, the only known Fermat primes are 3, 5, 17, 257, and 65,537, corresponding to $k = 0, 1, 2, 3$ and 4, respectively. Not surprisingly, two people in the 19th century tried to break records by actually constructing regular polygons with 257 and 65,537 sides.

For $k = 5, 6, \ldots, 22$, it is known that $2^{2^k} + 1$ is composite.

It follows from the work of Gauss and Wantzel that a regular polygon with m sides can be constructed by ruler and compass if and only if

$$m = 2^k p_1 \cdots p_l,$$

where $k \geq 0$ and $p_1, \cdots, p_l$ are distinct Fermat primes.

Exercises

1. If $a, b \in F$ but $\sqrt{c} \notin F$, show that $a + b\sqrt{c}$ is either 0 or possesses an inverse in $F[\sqrt{c}]$.

2. If $d \in F[\sqrt{c}]$, show that any element of $F[\sqrt{c}][\sqrt{d}]$ satisfies an equation of degree 4 with coefficients in F.

3. Explain how the Greeks could construct regular polygons with 15 and 60 sides.

4. If n is a positive integer, show that an angle of n degrees can be constructed with ruler and compass if and only if n is a multiple of 3.

16
Euclid

The city of Alexandria, on the mediterranean coast of Egypt, was founded by Alexander the Great in 332 B.C., who brought Greeks, Egyptians and Jews to settle there. One of his generals, Ptolemy I, made Alexandria the capital of his kingdom and founded a dynasty consisting of a long line of rulers, also named 'Ptolemy' and ending with the reign of the famous queen Cleopatra, who picked the wrong side in a Roman civil war.

Ptolemy established a university in Alexandria, called the 'Museum', which was soon to acquire a library holding more than 600,000 papyrus scrolls. For well over 600 years, Alexandria was to be the mathematical and scientific center of the world, with only some schools of philosophy surviving in Athens, although, after the extinction of the Ptolemaic line with Cleopatra, Alexandria was ruled by Rome. It was ultimately conquered by the Arabs in 641 AD.

The first chair of mathematics at the Museum was occupied by Euclid (330 to 275 BC), said to have been a student of a student of Plato. Apart from a couple of anecdotes, we know little about his life, and some ancient authors even thought he was a committee, like the 20th century Nicolas Bourbaki. According to one anecdote, Euclid told the impatient king that 'there is no royal road to learning'. According to another, he gave a small coin to a student who demanded to know the practical value of the lectures he had been attending.

Euclid wrote a number of books, on optics, music, astronomy etc., but his fame rests on the *Elements*, a collection of 13 so-called books (which we would now call chapters), which presented the foundations of all the mathematics known in his day. Nothing like this was to be published again, until the middle of the 20th century, when Nicolas Bourbaki issued a collection of books that purported to cover the elements of all the mathematics we study now.

None of the theorems contained in the 13 books can with certainty be ascribed to Euclid himself. It is believed that the Pythagoreans, including Archytas, were responsible for much of what appears in Books I, II, VI, VII, VIII, IX and XI and that Hippocrates was behind Books III and IV. For Books V and XII we are to thank Eudoxus, and Books X and XIII are said to be based on the work of Theaetetus.

However, the logical organization of the *Elements* is undoubtedly Euclid's contribution. Its success can be measured by the fact that, after more than 2,000 years, it was still used as a textbook in British schools. Moreover, throughout the ages, its structure was often imitated. Thomas Aquinas used a similar axiomatic presentation in his *Summa*, Newton's *Principia* is written in the style of the *Elements* and Spinoza's *Ethics* follows its logical arrangement. Undoubtedly the *Elements* has been the most influential scientific textbook in history.

Euclid's grandiose plan was to deduce all of mathematics from a small number of initial definitions and assumptions. The assumptions are subdivided into *axioms*, dealing with mathematics in general, and *postulates*, dealing with geometry in particular.

His treatment illustrated the ideal described by Aristotle at the beginning of his *Posterior Analytics*: sure knowledge is obtained by the rigorous deduction of the consequences of basic truths. To Euclid, these basic truths were either definitions or basic assumptions, largely assertions of unique existence. Let us take a closer look at his definitions, axioms and postulates.

The *Elements* begins with a list of 23 definitions, of which we will mention the first four:

1. A *point* is that which has no parts.

2. A *line* is length without width.

3. The extremities of a line are points.

4. A *straight line* is a line which lies evenly with the points on itself.

These statements are not definitions in the modern sense, though they make it clear that a point has no extension, that a line is not necessarily straight and that it is of finite length. Today we prefer to regard points and straight lines as undefined primitive concepts and leave the definition of curved lines to more advanced mathematics. The obscurity of Definition 4 may be due to the translation.

Euclid's axioms are intended to apply to all of mathematics, not just to geometry. A typical axiom asserts: 'If equals are added to equals, their sums are equal.' One cannot quarrel with this statement, though today we might derive it from axioms of equality and the view of addition as an operation.

Euclid lists five postulates, which we shall now state and comment upon.

I. To draw a straight line from any point to any other point.

Presumably this means that there exists a unique straight line joining two distinct given points. Thus, a 'straight line' cannot be interpreted as referring to a great circle on a sphere, as there are many great circles joining two antipodes, e.g., the meridians passing through the two poles on the globe. The way to get around this objection is to *identify* antipodal points; one then obtains *elliptic geometry*, which also satisfies Postulate I.

II. To produce a finite straight line continuously in a straight line.

Here 'continuously' is usually interpreted to imply 'indefinitely', thus ruling out not only spherical, but also elliptic geometry.

III. To describe a circle with any center and any distance [as radius].

Like Postulate I, this is a construction, or unique existence statement, the word 'circle' having previously been defined, in Definition 15, as 'a plane figure contained by one line [i.e. curve] such that all the straight lines falling upon it from one point among those lying within the figure are equal to one another.' It would appear that by a circle Euclid means not just its circumference but also its interior.

IV. That all right angles are equal to one another.

The status of this assertion as a postulate is rather dubious, and it has been argued, already in antiquity, that it should be listed as an axiom instead.

V. That, if a straight line falling on two straight lines makes the interior angles on the same side less than two right angles, the two straight lines, if produced indefinitely, meet on that side on which are the angles less than two right angles.

This is the most famous of Euclid's postulates and it is to his credit that he recognized its significance. It will be discussed at length in Chapter 17.

It is on these definitions and assumptions that Euclid plans to erect his impressive edifice of logical deductions. Here is how he begins:

Proposition 1.
On a finite straight line to construct an equilateral triangle.

In his proof he considers a segment AB and constructs circles with centers A and B and radius AB. He then considers the point C in which the two circles intersect and goes on to show that $\triangle ABC$ is equilateral.

This proof falls short of modern standards of rigour. In general, two circles may meet in two points, touch at one point, or not meet at all. In the present situation, they do, in fact, meet in two points; but this does not follow from Euclid's explicit assumptions.

Book I concludes with proofs of the Theorem of Pythagoras and its converse. Euclid is careful to show that there is a square on the hypotenuse, before discussing its properties. This is interesting, in view of Legendre's later proof that the existence of such a square implies Euclid's Postulate V.

To prove the Theorem of Pythagoras, Euclid uses a theory of area. Nowadays we are tempted to define the area of a rectangle as 'length times width'. This presupposes a theory which explains what it means to multiply two irrationals. Euclid approached the question of area from a more elementary point of view. He began with the idea that two polygons have the same area if they first can be dissected into triangles which can be reassembled, as in a jigsaw puzzle, to form a polygon exactly like the second polygon. It is only in Book VI, after Euclid has presented Eudoxus's theory of irrationals, that the length times width formula is justified.

However, in Book II, Euclid gives geometric treatments of certain basic algebraic identities, such as $a(b+c) = ab + ac$, using the areas of rectangles to handle products. He also gives a proof of a statement equivalent to what we now call the Law of Cosines.

Book III discusses the basic properties of the circle. Euclid goes to great length to give rigorous proofs. For example, in spite of the fact that it is 'obvious from the diagram', Euclid offers a demonstration of the fact that the points on a chord of a circle lie in the interior of the circle. Euclid is not always successful in his attempt at rigour, but it is clear that he does understand the need for it.

Book IV gives constructions for various regular polygons. It culminates with a treatment of the regular 15-gon. This achievement remained unsurpassed until 1796, when Carl Friedrich Gauss (1777–1855) found a construction for the regular 17-gon.

In Book V, Euclid uses Eudoxus's definitions of proportion to deduce an arithmetic for line segments. The 'commutativity of multiplication' is the subject of Proposition 16.

In Book VI, Euclid uses the material of Book V to derive the basic properties of similar triangles. The book concludes with the theorem that the length of a circular arc is proportional to the angle it subtends at the center of the circle. In talking about 'arclength', Euclid is implicitly presupposing the 'completeness' of the plane.

Books VII to IX present some elementary theorems of number theory. Included are proofs for Euclid's Algorithm (VII 2), the unique factorization of square-free integers (IX 14), the infinitude of primes (IX 20), the formula for the sum of a geometric progression (IX 35), and the formula for even perfect numbers (IX 36).

Book X is occupied with what we might call 'field extensions of degree 4 over rationals'. Euclid is interested in knowing when an expression like $\sqrt{7 + 2\sqrt{6}}$, which looks like it has 'degree 4' is actually equal to an expression like $1 + \sqrt{6}$, which involves only one 'layer' of square roots.

Book XI derives the basic theorems of solid geometry. A 'cone' is defined in terms of the revolution of a right triangle. A 'cube' is 'a solid figure contained by six equal squares'. Proposition XI 21 says that 'any solid angle is contained by plane angles [whose sum is] less than four right angles'. This proposition is used at the end of Book XIII to show that there are at most 5 regular polyhedra.

Book XII is the masterpiece of Eudoxus. Without the help of calculus, he manages to give a rigorous treatment of the volumes of the pyramid, cone and sphere.

Book XIII is the apex of the *Elements*. For each of the five regular polyhedra, Euclid derives the ratio of its side to the radius of the sphere in which it is inscribed. Although Euclid failed to give a complete theory of regular polygons – for example, the construction of the regular 17-gon is missing – he succeeded in giving a complete theory of regular polyhedra.

Euclid's *Elements*, or watered down versions of it, was used for over 2,000 years in universities and schools to teach not only geometry but also rigorous thinking. Not long after World War II, a reaction against this program set in and educators decided that geometry was not the appropriate place for training in logic. Anyway, they argued, Euclid was not rigorous enough and Hilbert's rigorous treatment (Chapter 17) was too cumbersome. So geometry was swept away in favour of 'New Mathematics'. The French mathematician Dieudonné, one of the founding members of the Bourbaki group, suggested that linear algebra should replace what he contemptuously called 'the theory of the triangle'.

Exercises

1. True or false? If two triangles have the same area, you can cut one of them up into little triangles, which can then be placed side by side to form a triangle congruent to the second triangle. Give a reference or a reason for your answer.

2. How did Euclid construct the regular 15-gon?

17
Non-Euclidean Geometry and Hilbert's Axioms

The parallel postulate V, Euclid's fifth postulate, seems less natural or convincing than the others. Ever since Euclid's time, people have felt that it ought to be deducible from Euclid's other postulates I to IV or from some logically equivalent set of axioms.

Noteworthy attempts to prove Euclid's fifth postulate were made by Proclus (410–85 AD), Saccheri (1667–1733), Thibault (1775–1822), and many others. We now know that these attempts were doomed to fail. Postulate V is independent of I to IV and one of Euclid's contributions to mathematics was his implicit recognition of this fact by presenting V as an axiom.

Given 'absolute geometry' (that is, the geometry based only on postulates I to IV) there are a number of important statements equivalent to the parallel postulate:

- Through a point not on a line there is exactly one line parallel to that line — Playfair.

- Every segment is a side of a square (with four right angles) — Legendre.

- Not every pair of similar triangles is congruent — Wallis.

- Every triangle has a circumcircle — Legendre.

- There is at least one triangle whose angle sum is 180° — Legendre.

(Heath notes that the first of these is due to Proclus.)

17. Non-Euclidean Geometry and Hilbert's Axioms

The search for a proof of the parallel postulate led to the discovery of many such equivalent statements, but each one was felt to be insufficiently 'self-evident' or 'basic' to count as a proper Euclidean axiom. What was really wanted was a deduction of postulate V from postulates I to IV alone.

Gauss may have been the first person to suspect the truth. In a letter to Franz Taurinus, written in 1824, Gauss says that he is sure that the parallel postulate cannot be proved.

Consider the alternative postulate:

(**H**) Through any point not on a line, there are at least two lines through that point and parallel to that line.

If we replace Euclid's parallel postulate by (H), we get the axioms of 'hyperbolic geometry'. It seems that Gauss believed hyperbolic geometry to be consistent.

The first person to publish results in hyperbolic geometry was the Russian N. I. Lobachevsky (1793–1856), of the University of Kasan, in 1829. In the same year, essentially the same results were discovered independently by the Hungarian J. Bolyai.

It was not until 1868 that it was proved that postulates I to IV do not imply postulate V. In that year, E. Beltrami (1835–1900) gave a Euclidean model for hyperbolic geometry. This showed that, if hyperbolic geometry contained any logical contradiction (for example, the assertion that both (H) and V are true), then that contradiction could be translated into a contradiction in Euclidean geometry. Since, presumably, there is no inconsistency in Euclidean geometry, there is none in hyperbolic geometry either.

In 1882, in the first article ever published in *Acta Mathematica*, Henri Poincaré (1854–1912) gave a sketch of a second Euclidean model for hyperbolic geometry. This model goes as follows. We interpret 'point' as 'point of the Cartesian plane in the interior of the unit circle $x^2 + y^2 = 1$'. We interpret a 'line' to mean either a 'diameter of the unit circle (minus endpoints)', or else 'a circular arc in the interior of the unit circle and orthogonal to it'. (Two arcs are 'orthogonal' if they intersect at right angles.)

'Betweenness' is defined in the obvious way. Segment equality is defined as follows. Let AB be a 'segment', that is, part of a diameter or orthogonal arc. Let A^e be the endpoint of that diameter or orthogonal arc which is on A's side of the diameter or arc. Let B^e be the other endpoint. Let

$$d(AB) = (AB^e/BB^e)(BA^e/AA^e),$$

where the segments on the right of this equation are ordinary Euclidean segments. Then two Poincaré segments AB and CD are 'equal' if and only if $d(AB) = d(CD)$.

The 'angle' between two Poincaré lines is the Euclidean angle between their tangents through the point where the lines meet. Angle equality is defined in the usual way.

17. Non-Euclidean Geometry and Hilbert's Axioms

Given the usual definition of the 'circle', it turns out, somewhat surprisingly, that 'circles' in the Poincaré sense are ordinary circles. It is just that their Poincaré centers are not where you would expect.

Using Steiner's geometry of 'inversion', one can prove that, under this interpretation, postulates I to IV are theorems in Euclidean geometry. However, postulate V, so interpreted, is not; through the center of the unit circle, for example, one can draw many 'lines' which do not meet a given orthogonal arc.

If V followed from I to IV, then V interpreted in the Poincaré sense, would both hold and not hold. We would have a statement about Euclidean lines and circles (related to the unit circle) which was both provable and disprovable relative to I to IV. Thus, assuming that I to IV are consistent, so are I to IV together with (H).

The reader can find the details of the 'proof by inversion' on pages 402 to 407 of volume 1 of Eves [1963]. Note that, although Eves uses logarithms in his proof, the reasoning is just the same if one drops them. (Eves uses logarithms because he wants the smallest 'distance' to be 0, not 1, and because of some results he aims to derive in a later section of his book.)

Nowadays people consider not only hyperbolic, but also 'elliptic geometry'. This was developed by Riemann in 1854, but is not to be confused with the more general 'Riemannian geometry', which we shall discuss below. In elliptic geometry, the straight lines are finite, and there are *no* parallels. A 'point' is like a pair of points on a sphere, and a 'line' is like a great circle on that sphere. Unfortunately, elliptic geometry does not satisfy postulate II, according to the way we interpreted 'continuously'.

The attitude of modern mathematicians is that one can vary the postulates of Euclid at will, constructing as many different geometries as one wishes. In the 19th century, this was a radical idea. People thought of Euclid's axioms as necessary truths about space, and hence truths which underlay the whole of astronomy and physics.

A modern physicist uses whichever geometry suits his purposes. According to the general theory of relativity, space-time is a four-dimensional Riemannian geometry, but with its curvature varying from place to place, depending on the local density of matter. The sum of the angles of a triangle might be two right angles (as in Euclidean geometry) if one was in a vacuum; however, if matter were present, the angle sum would differ from two right angles (as in non-Euclidean geometry), on account of the bending of light rays under the gravitational influence of that matter. Ideas such as these would have amazed mathematicians living in the early part of the 19th century.

We have seen in the last chapter that Euclid's postulates were not really adequate to describe the system he had in mind. Surprisingly, it was only in 1899 that Hilbert gave a completely adequate axiomatic description of three-dimensional Euclidean space. Since Hilbert required 21 postulates, or 'axioms' as he preferred to call them, we shall only state some of them here to give the flavour of his work.

Hilbert deals with the following undefined concepts: point, line, plane, incidence (between points and lines, between points and planes, between lines and planes), order (a ternary relation of 'betweenness' for three collinear points) and congruence (a binary relation between 'segments', which are themselves defined in terms of betweenness).

He lists seven axioms of incidence. For example, the first two can be combined to say this:

'Given two distinct points A and B there is a unique line a such that A lies on a and B lies on a.'

He lists five axioms of order. For example, the first of these says this:

'If B is between A and C then B is between C and A.'

More significant is his fifth axiom of order:

'If A, B, and C are three non-collinear points, and if a is a line which meets the segment AB, then a also meets the segment AC or the segment BC.'

He lists six axioms of congruence; for example the second one says

'If $AB \equiv A'B'$ and $AB \equiv A''B''$ then $A'B' \equiv A''B''$.'

He lists two axioms of continuity. The first of these is the so-called axiom of Archimedes: 'If e and f are geometric quantities and $e \neq 0$, then there is a natural number n such that $ne > f$.'

The second is his controversial axiom of completeness. He only reluctantly added it to the French translation of his lecture notes when it became apparent that otherwise one still could not deduce Euclid's Proposition 1. It was later simplified by Bernays as follows:

'No points can be added to a straight line so that all other postulates remain valid.'

Exercises

1. How does one use the parallel postulate to show, in Euclidean geometry, that every triangle has a circumcircle?

2. Show that, in the Poincaré model, there is exactly one line through any two distinct points. That is, prove, in Euclidean geometry, that, given any two points in the interior of a circle, there is exactly one other circle which goes through those points and is orthogonal to the first circle.

3. 'The true geometry is the one which is the simplest and most beautiful.' Write a short essay on this statement, saying something about the relation between the simple, the beautiful and the true.

18
Alexandria from 300 BC to 200 BC

The school of mathematics established by Euclid in Alexandria produced some first rate mathematicians in the third century BC Among them were the following:

- Aristarchus of Samos, 310 – 250 BC,
- Archimedes of Syracuse, 287 – 212 BC,
- Apollonius of Perga, 260 – 190 BC,
- Eratosthenes of Cyrene, 275 – 195 BC.

Aristarchus came from Samos, the same Greek island Pythagoras came from. He gave an interesting application of mathematics to astronomy. Let SEM be the triangle whose vertices are the sun (S), the earth (E) and the moon (M) (Figure 18.1). Aristarchus noted that when the moon is at its first quarter, the angle SME is a right angle. This is why we see exactly half of the part of the moon's surface that faces the earth. When the moon is in its first quarter, one can see the sun and the moon together in the sky, at the same time. Thus Aristarchus was able to measure the angle SEM. He found it to be 29/30 of a right angle. (A more accurate value is 0.9981 of a right angle.) Constructing a right triangle with an acute angle of 29/30 of a right angle — there is a ruler and compass construction for this — Aristarchus found that the ratio of its short side to its hypotenuse is about 1/19. He concluded that the distance from the earth to the sun is about 19 times greater than the distance from the earth to the moon. Had his

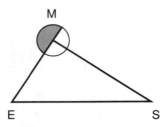

FIGURE 18.1. Relative distances of the sun and moon

measurement of the angle SEM been correct, he would have found that SE is about 400 times SM.

His calculation would have been easier had he used trigonometry, which was only developed a century later. If $\angle SEM = 87°$, then $\angle ESM = 3° = \pi/60$ radians, hence $EM/ES = \sin(\pi/60) \approx \pi/60 \approx 1/19$, as the sine of a small angle is approximately equal to that angle when expressed in radian measure.

Since the apparent sizes of the sun and the moon are approximately equal, as is seen during a solar eclipse, their actual diameters are in the same ratio as their distance from the earth.

By looking at the shadow cast by the earth upon the moon during a lunar eclipse one may also compare the size of the moon with that of the earth. (Since the sun is far away, the size of the earth is approximately the same as that of its shadow.) Aristarchus found

$$\frac{\text{diameter of the earth}}{\text{diameter of the moon}} \approx 7.$$

The actual figure is about 4. According to Plutarch, Aristarchus also proposed the hypothesis that 'the earth moves in an oblique circle about the sun at the same time as it turns around its axis'. It seems that Copernicus suppressed his acquaintance with the work of Aristarchus!

Although Archimedes is assumed to have studied in Alexandria, his productive life was spent in Syracuse. We shall leave him to Chapter 19.

Apollonius (260–190 BC) came from Perga in the south of what is now Turkey. He wrote a treatise on conics which contained 400 propositions. These were arranged in eight books, four of which survived in the original Greek, and three of which survived in Arabic translation. We do not read this treatise anymore, because we feel we can do the same things more easily using analytic geometry.

According to the modern definition of a *conic section*, it is the set of all points P in the plane such that P's distance from a fixed point, called the *focus*, bears a constant ratio to its distance from a fixed line, called the *directrix*. This ratio is called the *eccentricity*. Apollonius did not give this definition and it is doubtful whether he was aware of the eccentricity, which is used to classify conic sections as follows: ellipses have eccentricity < 1 (in particular, a circle has eccentricity 0), parabolas have eccentricity $= 1$, and hyperbolas have eccentricity > 1.

Apollonius also wrote a treatise on 'Tangencies' in which he showed how to give a ruler and compass construction for a circle tangent to three given circles.

Eratosthenes of Cyrene (in North Africa) became the chief librarian at Alexandria. He was interested in many things: philosophy, poetry, history, philology, geography, astronomy and mathematics. We have already mentioned his sieve for constructing the list of primes. Eratosthenes also invented the Julian calendar, with every fourth year containing an extra day, and he calculated the size of the earth. Perhaps the reason his students called him 'beta' (the second letter of the Greek alphabet) was that, although he studied many different things, he never considered himself the leading expert in any one field. It is reported that, in his old age, Eratosthenes went blind and committed suicide by starvation.

Eratosthenes's greatest achievement was the measurement of the circumference of the earth. Eratosthenes correctly assumed that, since the sun is so far from the earth, those of its rays which hit the earth can be regarded as parallel. (Here he used the result of Aristarchus.) Eratosthenes knew that Syene (present day Aswan) is almost exactly on the Tropic of Cancer, that is, at noon on midsummer's day (June 21), the sun is directly overhead, as could be witnessed from the bottom of a well. Eratosthenes observed that at Alexandria, at noon on midsummer's day, the sun was $360°/50$ from the point directly overhead. He argued that this same angle was subtended at the center of the earth by the arc joining Alexandria to Syene, which is due south of Alexandria. According to Euclid (theorem VI 33), the length of an arc of a circle is proportional to the angle it subtends at the center. So all Eratosthenes had to do was to measure the distance from Alexandria to Syene. This he found to be 5,000 stadia, a stadium being the length of the famous Olympic track. Eratosthenes concluded that the circumference of the earth is to 5,000 stadia as $360°$ is to $360°/50$, and hence the circumference is $5,000 \times 50 = 250,000$ stadia (Figure 18.2). As the Olympic stadium is about 180 meters, this would make the circumference of the earth about 45,000 kilometers.

Eratosthenes's calculation of the circumference of the earth is remarkably accurate. The correct value is almost exactly 40,000 km; in fact, the *kilometer* was originally defined as $1/40,000$ of the circumference of the earth. Had Columbus known this, he might never have set out on his journey or called the inhabitants of the New World 'Indians'.

18. Alexandria from 300 BC to 200 BC

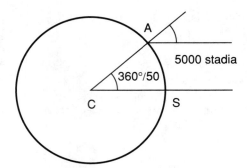

FIGURE 18.2. Circumference of the Earth in stadia

Exercises

1. Obtain the equation of a conic section with focus $(a, 0)$, directrix the y axis and eccentricity e.

2. Suppose you know the actual size of the moon. What is a simple way of finding its distance from the earth — without using anything Eratosthenes could not have used.

19
Archimedes

Archimedes (287–212 BC) was the greatest applied mathematician and physicist before Newton. Many stories are told about him. One story relates that, while he was taking a bath, Archimedes suddenly discovered a simple way of determining the ratio of gold to silver in a gold-silver alloy. Elated by his discovery, he leapt from the bath, and ran through the streets of Syracuse, shouting 'eureka!' (which means 'I have found it!'). Unfortunately, he had forgotten to get dressed.

It was no accident that Archimedes made his discovery in a bath. Suppose you have an object made of gold and silver weighing m ounces. Suppose you wish to determine the number x of ounces of gold which the goldsmith has put into it. If g is the density of gold and s is the density of silver, the volume of the object is $(x/g) + (m-x)/s$. What Archimedes realized was that, by immersing the object in a rectangular bath tub, and observing the increase in water level, you can easily determine its volume v. Solving for x in the equation
$$v = (x/g) + (m-x)/s,$$
you obtain the mass of gold in the object. Thanks to his mathematics, Archimedes was able to tell his friend, King Hieron of Syracuse, whether the goldsmith had cheated the king by charging him for pure gold, while, in fact, using a certain percentage of silver for his crown. Newton, the first modern physicist to surpass Archimedes, centuries later, was to perform a similar task in unmasking counterfeiters.

When Syracuse was besieged by a Roman army, Archimedes constructed various machines to help defend his city. As well as catapults and crossbows, Archimedes designed devices which dropped huge stones on the Roman ships. He even constructed a crane which could lift a ship from the water, and drop it back in, stern-first. When Syracuse finally fell (212 BC), the Roman general Marcellus gave orders to bring Archimedes to him unharmed. These were not obeyed. Archimedes, it seems, was slain by an unknown soldier. There are various accounts of this story; the best known is that by Plutarch in his biography of Marcellus. Who today would know of Marcellus were it not for Archimedes?

Archimedes wrote on many subjects: the circle, the parabola, the spiral, the sphere, the cylinder, arithmetic, mechanics, statics and hydrostatics. One of his more interesting books, the *Method*, was rediscovered only in 1906.

By considering a regular 96-gon inscribed in a circle, Archimedes showed that $\pi < 3\frac{1}{7}$. By considering a regular 96-gon circumscribing a circle, he showed that $3\frac{10}{71} < \pi$. He was aware that one could calculate π to any desired accuracy by letting the number of sides of the regular polygon tend to infinity.

He proved that the area of a circle is πr^2 and that the volume of a sphere is $\frac{4}{3}\pi r^3$. He knew how to calculate the area bounded by a parabola and a chord, the area of a sector of a spiral, the volume of an ellipsoid of revolution, the volume of a segment of a sphere, the centroid of a hemisphere and, perhaps most remarkably, the volume common to two equal right circular cylinders intersecting at right angles. All these are calculus problems and, indeed, Archimedes was using what we would now call the technique of 'integration'.

As an example of Archimedes's mathematics, let us see how he proved that the area of a circle is πr^2. He started with the following assumptions and theorems:

1. Circles and circle segments have areas.

2. The area of a set of pairwise disjoint triangles and circle segments equals the sum of the areas of those triangles and circle segments; thus if we dissect a circle into triangles and circle segments, the area of the circle is the sum of the areas of the triangles and circle segments into which it has been dissected; also the area of the circle is greater than the sum of the areas of any proper subset of those triangles and circle segments.

3. Given any circle, there is a straight line segment which is longer than the perimeter of any polygon inscribed in the circle, and shorter than the perimeter of any polygon circumscribing the circle; this is the 'circumference' of the circle. (The first person to give the name π to the circumference of the circle with unit diameter was not Archimedes, but William Jones, in 1706.)

4. Given any areas e and f, there is a natural number m such that $me > f$; (this assumption is found at the beginning of Archimedes's 'On the Sphere and the Cylinder', so it is often called the 'Axiom of Archimedes'; however, it is also found at the beginning of Book V of Euclid's *Elements* and, even earlier, at 266b in Aristotle's *Physics*).

5. A regular 2^n-gon inscribed in a circle takes up more than $1 - 1/2^{n-1}$ of its area; a regular 2^n-gon circumscribed about a circle has an area less than $1 + 1/2^{n-2}$ of that of the circle.

6. The area of a circle is proportional to its diameter squared (see Euclid's *Elements* XII 2).

Using these assumptions and theorems, Archimedes derived the formula for the area of a circle by first obtaining two contradictions:
(A) Suppose the circle has area greater than that of a right triangle T whose legs equal the radius and circumference of the circle.
By (4) and (5) we can find a natural number n such that

circle area − inscribed regular 2^n-gon area < circle area − area of T

and hence

area of T < 2^n-gon area.

Let AB be a side of the inscribed regular 2^n-gon, and ON a perpendicular from the center O of the circle to AB (with N being the midpoint of AB). Then ON is less than the radius of the circle. Using (3), we have

$$\begin{aligned} 2^n\text{-gon area} &= 2^n(\tfrac{1}{2}AB \cdot ON) \\ &= \tfrac{1}{2}(2^n AB)ON \\ &< \tfrac{1}{2}\text{circumference} \times \text{radius} \\ &= \text{area of } T. \end{aligned}$$

Contradiction. Thus (A) must be rejected.

(B) Suppose the circle has area less than T.
By (4) and (5) there is a natural number n such that the area of T > circumscribed regular 2^n-gon area. However, if AB is a side of the circumscribing regular 2^n-gon, then, by (3),

$$\begin{aligned} 2^n\text{-gon area} &= 2^n(\tfrac{1}{2}AB \times \text{circle radius}) \\ &> \tfrac{1}{2}\text{circumference} \times \text{radius} \\ &= \text{area of } T. \end{aligned}$$

Contradiction. Thus (B) must be rejected.

Since (A) and (B) must be rejected, it follows from Aristotle's 'Law of the Excluded Middle' that (C) the area of the circle equals that of a right triangle whose legs equal the radius and circumference of that circle. In other words,

$$\text{circle area} = \tfrac{1}{2} \text{ circumference} \times \text{radius}.$$

The expression on the right-hand side is of course πr^2 (see (6) above).

Archimedes's proof of this formula was the culmination of about two hundred years of previous work on the circle, beginning with Antiphon (425 BC).

Archimedes, working in Syracuse, would communicate his results to the mathematicians back in Alexandria. He became annoyed when, suspiciously often, they claimed that they had made the same discoveries. To fool them, Archimedes included some false results in a book on the sphere and the cylinder, but history does not reveal the outcome. To challenge the mathematicians at Alexandria, Archimedes posed the following problem (which we have 'translated' into modern algebraic notation).

The sungod had a herd of cattle consisting of w white bulls, g grey bulls, b brown bulls and s spotted bulls, as well as w', g', b' and s' cows of matching colours. What was the total number of bulls and cows if

$$s = w - 5g/6 = g - 9b/20 = b - 13w/42,$$
$$w' = 7(g+g')/12, \quad g' = 9(b+b')/20,$$
$$b' = 11(s+s')/30, \quad s' = 13(w+w')/42,$$

and $w + g$ is a square and $b + s$ is a triangular number?

It is a curious triumph of tradition that Archimedes used the Egyptian method for representing the fractions appearing in this problem as sums of reciprocals of positive integers. Aside from the last two restrictions, we have here seven equations in eight unknowns, which cannot be solved by algebraic methods alone. However, if one is looking for positive integer solutions, such problems are called 'Diophantine', after the mathematician Diophantus, who will appear about 500 years after Archimedes. We sketch how a solution may proceed, though the reader will have to fill in many details.

From the equations

$$s = w - 5g/6 = g - 9b/20 = b - 13w/42$$

we find that, for some positive integer m,

$$w = 2226m, \quad g = 1602m, \quad b = 1580m, \quad s = 891m.$$

From the next four equations, we find that there is some natural number k such that

$$m = 4657k, \quad w' = 7,206,360k, \quad g' = 4,893,246k,$$

$$b' = 3,515,820k, \quad s' = 5,439,213k.$$

If $w + g$ is a square then $4(957)(4657)k$ is a square. Since 4657 is prime and 957 is the product of two distinct primes, it is easy to show that this occurs only if k has the form $(957)(4657)t^2$, where t is an integer. If $b+s$ is triangular, then $2471m$ has the form $\frac{1}{2}n(n+1)$. Thus $8(2471)(4657)(957)(4657)t^2$ has the form $4n^2 + 4n$. In other words, to find an integer solution to Archimedes's Cattle Problem, we have to find an integer solution to

$$(2n+1)^2 - 8(2471)(957)(4657^2)t^2 = 1.$$

The mathematicians at Alexandria were not able to solve this problem. Indeed, it was not solved until 1965, when H. C. Williams, R. A. German and C. R. Zarnke used a computer to generate the 206,545 digit answer. The answer was published for the first time in 1980–1981. The reader can find the full 206,545 digit answer printed in Harry L. Nelson's article 'A Solution to Archimedes' Cattle Problem', *Journal of Recreational Mathematics* 13, pp. 164–176.

It is impossible to do full justice here to Archimedes's important contributions to physics. Let us only mention that he developed the theory of the lever and investigated the properties of floating bodies.

Exercises

1. Let a and b be positive real numbers. Archimedes proved that $x^3 - ax^2 + (4/9)a^2 b$ has a positive root if and only if $a > 3b$. Do the same.

2. If, in a cube of side 1, two cylinders, each of diameter 1, are constructed so that their axes are perpendicular, show that the volume common to these cylinders is $2/3$.

3. Prove that a regular 2^n-gon circumscribed about a circle has an area less than $1 + 1/2^{n-2}$ of that of the circle.

4. Find the least positive integer solution of $x^2 - 5y^2 = 1$.

5. Let B be a point in straight line segment AC. Construct three semi-circles with diameters AB, BC and AC, all on the same side of AC. The area which is in the semi-circle on AC but not in either of the two smaller semi-circles is called the 'arbelos' (or 'shoe-maker's knife'). Archimedes found the area of the arbelos, in terms of AB and BC. Do the same.

6. Justify the existence of the numbers m and k in the above solution of the Cattle Problem.

7. If Archimedes were alive today, would he have a moral obligation to help his country design nuclear weapons or would he have a moral obligation not to help them design nuclear weapons? Support your answer with reasons related to the role of science in human history.

20
Alexandria from 200 BC to 500 AD

In this chapter we discuss the more important mathematicians who worked in Alexandria after 200 BC:

- Hipparchus of Nicea, born about 180 BC,
- Heron of Alexandria, about 60 AD,
- Menelaus of Alexandria, about 100 AD,
- Ptolemy of Alexandria, died in 168 AD,
- Diophantus, about 250 AD,
- Pappus, about 320 AD.

Less significant as mathematicians, but nonetheless important in the history of the subject are

- Nicomachus of Gerasa, about 100 AD,
- Hypatia, died in 415 AD,
- Proclus, 410 – 485 AD,
- Boethius, 475 – 524 AD.

Hipparchus came from Nicea, a town near present day Istanbul, which was to be the site of the great Church Council of 325 AD. Hipparchus made many contributions to astronomy. He calculated the duration of the year to within 6 minutes, the angle between the ecliptic and the equator, the annual precession of the equinoxes, the lunar parallax, the eccentricity of the solar orbit, etc. He knew that the moon moves only approximately in a circle with center at the earth, a better approximation being an 'epicycle'. The same was true of the sun. Hipparchus suggested that epicycles of higher orders were necessary to describe the motions of the planets.

In mathematics his great contribution was the founding of trigonometry. He drew up a table giving for each angle with vertex at the center of a circle of radius 1 the length of the chord it cuts off in the circle. For example, suppose $\angle AOB$ is $30°$, with O the center of the circle, and $OA = OB = 1$ two radii of the circle. Then the chord in question is the segment AB. This has length $31.06/60$, so, in the table of Hipparchus, we would find

$$\text{chord}(30°) = 31.06/60.$$

In modern terms, $\text{chord}(x) = 2\sin(x/2)$.

To construct this table, Hipparchus made use of formulas which we would express as follows:

$$\sin(x \pm y) = \sin x \cos y \pm \cos x \sin y$$

and

$$2\sin^2(x/2) = 1 - \cos x.$$

Heron is known for the formula, probably discovered by Archimedes, which expresses the area of a triangle in terms of the lengths a, b and c of its sides. If $s = (a+b+c)/2$, this formula gives the area of the triangle as

$$\sqrt{s(s-a)(s-b)(s-c)}.$$

We shall meet this formula again when we discuss the mathematics of India.

Menelaus was the first to study spherical trigonometry. He is also known for the following theorem, which is found in his *Spherica*.

Menelaus's Theorem:

Let ABC be a triangle. Suppose D is on the line through B and C, E is on the line through A and C, and F is on the line through A and B. Suppose that either two or none of D, E, F are on sides of the triangle. Then D, E, F are collinear if and only if $BD \cdot CE \cdot AF = CD \cdot AE \cdot BF$.

Proof: Suppose D, E, F are collinear, in line l. We may suppose l does not pass through A, B or C. Let A', B', C' be points in l such that AA', BB', CC' are all perpendicular to l. Then $BD/CD = BB'/CC'$, $CE/AE = CC'/AA'$

1	α	10	ι	100	ρ
2	β	20	κ	200	σ
3	γ	30	λ	300	τ
4	δ	40	μ	400	υ
5	ε	50	ν	500	φ
6	f	60	ξ	600	χ
7	ζ	70	o	700	ψ
8	η	80	π	800	ω
9	θ	90	q	900	

TABLE 20.1. Ptolomaic notation

and $AF/BF = AA'/BB'$. Multiplying the three equations together we obtain the result.

The converse is easy.

As his *Tetrabiblos* shows, Ptolemy was a keen believer in the superstition called astrology. In spite of this, he did for astronomy what Euclid had done for geometry and arithmetic: he wrote the definitive textbook, known by its Arabic name, the 'Almagest' or 'Greatest'. Like Hipparchus, Ptolemy gave a table of chords.

Book I of the 'Almagest' contains 'Ptolemy's Theorem': in a cyclic quadrilateral, the product of the diagonals is equal to the sum of the products of the two pairs of opposite sides. He used Greek letters to denote numbers. Curiously, however, he retained two Phoenician letters, corresponding to Latin f and q, which had actually disappeared in Greek. He also added one symbol at the end, giving him a total of 27 symbols, which allowed him to represent the numbers 1 to 9, 10 to 90 and 100 to 900. He also made use of a small circle to denote zero. In his tables, he employed the Babylonian system to denote not only angles, as we still do, but also lengths as had Hipparchus before him. Thus he wrote

$$120°0'0'' = \rho\kappa | \circ | \circ,$$

$$\text{chord } 1° = 1°2'50'' = \alpha|\beta|\nu.$$

So he took the radius to be 1.

His underlying decimal notation was based on the alphabetic code in Table 20.1. We have substituted the Latin letters for 6 and 90 and omitted the symbol for 900.

In 250 AD, in Rome, Plotinus was teaching his version of Platonism. At the same time, in Alexandria, Diophantus was writing the *Arithmetica*. This originally contained 13 books. Until 1973 we had only six of these, but then three more were discovered in an Arabic translation going back to the ninth century. (See Jacques Sesiano, *Books IV to VII of Diophantus's Arithmetica.*)

These 13 books consisted of solutions to algebraic problems. The solutions are all rational numbers. Some of the equations or systems of equations are indeterminate and often there is more than one rational solution. Diophantus, however, is usually content to give just one solution. Note that what we now call a 'Diophantine equation' is one whose unknowns are not just rational but integers. Diophantus, however, accepted any rational solution.

As an example, let us consider problem 9 of Book II. The problem is 'to divide a given number which is the sum of two squares into two other squares.' That is, given rationals a and b, find a nontrivial rational solution of
$$x^2 + y^2 = a^2 + b^2.$$
Diophantus takes the special case where $a = 2$ and $b = 3$, but his solution is easily generalized. He writes:

> Take $(x + 2)^2$ as the first square and $(mx - 3)^2$ as the second (where m is an integer), say $(2x - 3)^2$. Therefore $(x^2 + 4x + 4) + (4x^2 + 9 - 12x) = 13$, or $5x^2 + 13 - 8x = 13$. Therefore $x = 8/5$, and the required squares are $324/25$ and $1/25$.

Note that, to get the general solution, m should be any rational number.

An interesting question is whether Diophantus was aware of the algebraic rules that lay behind many of his solutions. In Book III, problem 19, Diophantus notes that 65 is a sum of two squares in two ways since 65 'is the product of 13 and 5, each of which numbers is the sum of two squares'. From this we can deduce that he knew that the product of two integers, each of which is a sum of two squares, is itself a sum of two squares, and in two ways. Did Diophantus also know the stronger proposition that
$$(a^2 + b^2)(c^2 + d^2) = (ac \mp bd)^2 + (ad \pm bc)^2 \ ?$$
Basing himself just on the remark which we have quoted from Book III, problem 19, T. L. Heath conjectured that Diophantus did know this identity (see page 105 in Heath's translation of the *Arithmetica*). The person who first published the algebraic identity, however, was Abu Jafar al-Khazin (950 AD), and, later, Fibonacci gave it in his *Liber Quadratorum* (1225 AD).

Diophantus was the first to make systematic use of a symbolic notation for algebraic expressions. He denoted + by juxtaposition, − by the symbol ⋏ and = by ι. He wrote
$$K^v \text{ for } x^3,$$
$$\Delta^v \text{ for } x^2,$$

ς for x^1 (ςς for the plural),
$\overset{o}{M}$ for x^0.

For example,
$$\Delta^v \overline{\gamma} \; \overset{o}{M} \; \overline{\iota\beta}$$

stands for $3x^2 + 12$, while

$$K^v \overline{\alpha\varsigma\varsigma\eta} \wedge \Delta^v \overline{\epsilon} \; \overset{o}{M} \; \overline{\alpha} \wr \varsigma\overline{\alpha}$$

represents the equation

$$(x^3 + 8x) - (5x^2 + 1) = x.$$

In 320 AD, the Roman Empire had its first Christian emperor, Constantine, after whom the Eastern Roman capital was named Constantinople. In Alexandria, Athanasius was defending the divinity of Jesus against Arius, who asserted that Jesus was *like* God, but not *equal* to him. (The difference between these two words in Greek consisted of one letter, the Greek letter iota. We have preserved this difference in Mathematics, when we distinguish between 'homeomorphism' and 'homomorphism'.)

Meanwhile in Alexandria, Pappus was writing his encyclopaedic *Collection* of earlier mathematical works. The school of mathematics had declined, and Pappus was its last, lone member.

The 'Theorem of Pappus' appears in Book VII of the *Collection*. It is far more important than Pappus realized. It expresses the commutativity of multiplication and is fundamental to projective geometry. Hilbert made use of it as a key theorem in his presentation of Euclidean geometry. The theorem of Pappus can be proved with the help of the theorem of Menelaus as follows.

Theorem of Pappus: Given points ABC on one line, $A'B'C'$ on another, the three points of intersection $P = BC' \cap CB'$, $Q = AB' \cap BA'$ and $R = CA' \cap AC'$ are collinear (Figure 20.1).

In stating this result, we have assumed that BC' and CB' etc., are not parallel. We shall also assume that ABC and $A'B'C'$ meet at a point X and that none of the other lines in the diagram are parallel.

Proof: Let $A'B \cap B'C = U$, $AC' \cap A'B = V$ and $B'C \cap AC' = W$. We apply Menelaus's Theorem five times to the triangle UVW. Since $A'CR$, $BC'P$, $AB'Q$, $A'B'C'$, and ABC are all collinear, we have

$$VR \cdot WC \cdot UA' = RW \cdot CU \cdot A'V,$$

$$VC' \cdot WP \cdot UB = C'W \cdot PU \cdot BV,$$

$$VA \cdot WB' \cdot UQ = AW \cdot B'U \cdot QV,$$

$$VC' \cdot WB' \cdot UA' = C'W \cdot B'U \cdot A'V,$$

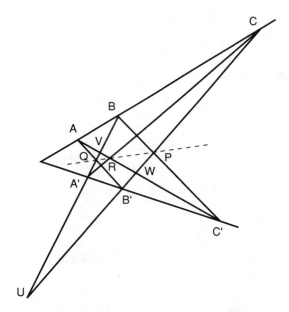

FIGURE 20.1. Pappus's Theorem

$$VA \cdot WC \cdot UB = AW \cdot CU \cdot BV.$$

Multiplying the first three equations and dividing by the product of the last two, we obtain

$$VR \cdot WP \cdot UQ = RW \cdot PU \cdot QV.$$

The result now follows by another application of Menelaus's Theorem. Note that the argument makes use of the commutativity of multiplication.

When CD and EF are parallel, the proof proceeds in a similar fashion. However, one uses, not Menelaus's Theorem, but the theory of similar triangles. We leave the details to the reader.

Among the minor mathematicians of this era was Nicomachus (100 AD) from Palestine. He was a Pythagorean, and he published a book on number theory, which is the basis for many of our speculations about the nature of 'Pythagorean' mathematics.

Hypatia (d. 415) was the daughter of Theon of Alexandria, who had put out an edition of Euclid's *Elements* and a commentary on Ptolemy's *Almagest*. Hypatia wrote commentaries on Apollonius and Diophantus.

According to Socrates Scholasticus (380–450 AD) in Chapter 15 of Book VII of his *History of the Church*, Hypatia was murdered by a mob in the course of an anti-pagan riot. This tragedy is sometimes blamed on the Christian bishop, Cyril, but there is no evidence to support this accusation. The 19th century author C. Kingsley wrote a fascinating historical novel *Hypatia*, which makes this story come to life.

Another minor mathematician at this time was Proclus. He studied in Alexandria and then worked in Athens. He wrote a commentary on the first book of Euclid, which contains valuable information about the history of Greek mathematics.

Boethius studied in Athens but lived in Rome. He is most famous for his *De Consolatione Philosophicae*, which he wrote in prison, about 525 AD His *Arithmetic* and *Geometry* were standard textbooks in the Middle Ages. Unfortunately, they contained much less mathematics than the *Elements*.

In 529 AD, the emperor Justinian closed the pagan schools of philosophy at Athens. The 'Dark Ages' of Europe had begun.

It is interesting that we do not know of many mathematicians of the period (50–500 AD) converting to Christianity. It seems that the Academy at Athens and the University at Alexandria both rejected the new religion. An interesting dialogue between reason and faith might have taken place, but, as it turned out, it was only in the later Middle Ages that thinkers, such as Aquinas (1250 AD), advanced philosophies that were influenced by the *Elements* as well as by the Bible.

When the Arabs conquered Alexandria in 641 AD, there probably was not much left of the famous library. However, according to an often repeated story, dating back to Moslem sources in the 13th century, it was the Arabs who destroyed it. Their commander, Amru, was willing to spare the library, but was dissuaded by Caliph Omar I, who argued thus: 'If the books of the Greeks confirm what is written in the Koran, they are superfluous; if they contradict the Koran, they are dangerous. In either case they should be destroyed.' The story continues, saying that the books served to heat the furnaces of the public bathhouses for six months! An almost identical story is told about another library in Persia. What was the origin of these stories? According to Bernard Lewis, in a Letter to the Editor, *The New York Review of Books* 37, Number 14 (October 27, 1990), they were invented to justify the destruction of a completely different library, a collection of Fatimid books, deemed to be heretical by the orthodox Sunni Sultan Saladin in the 12th century.

Exercises

1. Prove Heron's formula.

2. Give the details of the converse in the proof of the Theorem of Menelaus.

3. Find all the integer solutions of $x^2 + y^2 = z^2$ by using the same technique as Diophantus did in problem 9 of Book II of the *Arithmetica*. (Hint: the given equation has a solution in the integers if and only if $x'^2 + y'^2 = 1^2 + 0^2$ has a corresponding solution in rationals.)

4. Find an infinite family of rational solutions to $x^2 + y^2 = z^2$ (*Arithmetica* IV 1 (Sesiano)).

5. Prove a version of the Theorem of Pappus in case BC' and AB' are parallel.

21
Mathematics in China and India

Not much is known about the development of mathematics in China before contact with the West was established. The 'Arithmetic in Nine Sections' ('Chiu Chang Suan Shu') was written before 200 AD. Like the Rhind Papyrus, it is a list of problems and solutions. Chapter 8 shows how to solve n linear equations in n unknowns, using a method which is essentially the same as Gaussian elimination. One system which is solved is the following:

$$3x + 2y + z = 39,$$
$$2x + 3y + z = 34,$$
$$x + 2y + 3z = 26.$$

The Chinese interest in systems of linear equations was perhaps linked to their interest in magic squares. The square

$$\begin{array}{ccc} 4 & 9 & 2 \\ 3 & 5 & 7 \\ 8 & 1 & 6 \end{array}$$

was supposedly brought to humankind on the back of a tortoise from the River Lo in the days of Emperor Yu. Its 'magic' property is that all rows and columns and the two diagonals have the same sum.

A 'Chinese Remainder Problem' was solved by Sun Tsu (400 AD):

divide by 3, the remainder is 2;
divide by 5, the remainder is 3;
divide by 7, the remainder is 2;
what will be the number?

Only one of the infinitely many solutions is given, but Sun Tsu's method allowed him to give as many others as he wanted.

Tsu Ch'ung Chih (475 AD) gave upper and lower bounds for π. He used the method of Archimedes, but he obtained sharper estimates. In the 'Nine Sections of Mathematics' (1247), Ch'in Chiu Shao (Qin Jiushao) found a root of
$$x^4 - 763200x^2 + 40642560000,$$
using what is in effect Horner's method (rediscovered by William Horner in 1819). In the 'Precious Mirror of the Four Elements' (1303), Chu Shih Chieh (Zhu Shijie) gives 'Pascal's Triangle' (also known in India and later rediscovered by Pascal in 1653).

In the 16th and 17th centuries, Christian missionaries from Europe entered China, introducing Western mathematics (e.g., the theory of logarithms). Today, the Chinese contribute to all branches of mathematics. They have retained, however, a strong interest in the theory of numbers. For example, in 1985, De Gang Ma gave the first elementary solution of the Diophantine equation
$$x(x+1)(2x+1) = 6y^2.$$
(See Anglin [1995], Section 4.4.)

We have already heard about Chen Jing-run and the progress he made on the Goldbach conjecture.

The earliest Indian mathematical texts we have are the *Sulvasutras* or 'rules of the cord'. They were written sometime between 500 BC and 200 AD. They contain some elementary geometry related to the construction of altars. It is observed that $12^2 + 35^2 = 37^2$ and, presumably, the general method for constructing such 'Pythagorean triples' was known.

Mathematics in India turns up in unexpected places. Around 800 BC a certain Pingala wrote a book on the *Science of verse meters*. Syllables in Vedic poetry are distinguished to be either 'light' or 'heavy'; the former are assigned the binary digit 1, the latter the binary digit 0. Thus, with each line of verse is asssociated a number written in binary scale, which is then converted into decimal scale. Conversion in the converse direction is also discussed. (See van Nooten [1993].)

Four noteworthy Indian mathematicians were:

- Aryabhata the Elder, born in 476 AD,

- Brahmagupta, flourished in 628 AD,

- Mahavira, lived about 850 AD,

- Bhaskara, 1114–1185 AD.

In their work, ancient Indian traditions and Alexandrian mathematics seem to flow together. Yet, unlike the ancient Greeks, these mathematicians did not include many proofs in their books. Also, unlike the ancient Greeks, they wrote their mathematics books in verse! For example, Bhaskara gives the following problem:

> The square root of half the number of bees in a swarm
> Has flown out upon a jasmine bush;
> Eight ninths of the swarm has remained behind;
> And a female bee flies about a male who is buzzing inside a lotus flower;
> In the night, allured by the flower's sweet odour, he went inside it
> And now he is trapped!
> Tell me, most enchanting lady, the number of bees.

This is certainly a poetic way of asking for the solution of

$$\sqrt{x/2} + 8x/9 + 2 = x.$$

Aryabhata wrote on arithmetic and algebra, on plane and spherical trigonometry, and on astronomy. He knew the formulas for the sum $1^k + 2^k + \cdots + n^k$ for $k = 1, 2$ and 3. He solved quadratic and indeterminate equations. He was one of the first to make use of the sine function, defining $\sin A$ as (chord $2A$)/2, thus helping to simplify the addition formulas and giving trigonometry its modern form. He also produced a quite accurate sine table.

Brahmagupta (b. 598) studied the Diophantine equation $x^2 - Ry^2 = 1$ (where R is a given, positive nonsquare integer). This equation is mistakenly called 'Pell's equation', after John Pell (1610–1685), who actually had nothing to do with it. A complete solution to this equation was given by Lagrange in 1767.

Brahmagupta calculated the surface and volume of pyramids and cones, using $\sqrt{10}$ as an approximation for π. Most importantly, he was the first person to give a systematic presentation of the rules for working with zero and negative numbers, and he may have been the first to introduce negative numbers into the number system. According to *The Treasury of Mathematics*, edited by Henrietta O. Midonick, Brahmagupta's *Brahmasphuita Siddhanta* tells us that positive divided by positive, or negative divided by negative, is positive, whereas positive divided by negative, or negative divided by positive is negative. It seems, however, that Brahmagupta had some trouble dividing by zero.

One of Brahmagupta's more technical achievements was his discovery of the formula for the area of a cyclic quadrilateral with sides a, b, c and d:

$$\sqrt{(s-a)(s-b)(s-c)(s-d)},$$

where $s = (a+b+c+d)/2$, thus generalizing Heron's formula. He did not give a proof for this formula, and it seems that a careless copyist omitted the word 'cyclic'.

Following Brahmagupta, Mahavira discusses the properties of zero, stating that
$$a + 0 = a, \ a - 0 = a, \ a \cdot 0 = 0.$$
He does not quite know what to do with $a/0$, apparently believing that it equals a, unless we again blame a copyist. He also talks about negative numbers and asserts that they are not squares. He tries to give some social significance to his equations. When posing the equation
$$x/4 + 2\sqrt{x} + 15 = x,$$
he writes:

> One fourth of a herd of camels was seen in a forest. Twice the square root of that herd had moved on to the mountain slope. Three times five camels, however, were found to remain on the river bank. What is the number of that herd of camels?

Bhaskara (1114–ca. 1185) named one of his mathematics books after his daughter Lilavati. This book deals with weights and measures, the decimal notation, the operations of arithmetic, reduction of fractions to common denominators, linear and quadratic equations, arithmetic and geometric progressions, interest and discount, triangles and quadrilaterals, approximations to π, trigonometric formulas, volumes of solids, indeterminate linear equations, and combinations. It contains the earliest extant exposition of the decimal notation using the sign for zero. Concerning calculations with zero, Bhaskara gives the following rules:
$$a \pm 0 = a, \ 0^2 = 0, \ \sqrt{0} = 0, \ a/0 = \infty.$$

Some of his problems throw light on economic and social history. A female slave of age 16 is said to be worth eight oxen who have worked for two years; her price declines as she grows older.

In the first volume of his *History of Mathematics*, D.E. Smith relates that the *Lilavati* was translated into Persian in 1587 by someone called Fyzi. Fyzi claims that Bhaskara dedicated the book to his daughter in order to console her for remaining single. According to astrologers, there was but one lucky moment at which Lilavati might marry. Unfortunately, one of Lilavati's pearls fell into the water clock, and it stopped without anyone noticing. The lucky moment passed, and Lilavati missed her chance of an astrologically sound marriage. Bhaskara, accepting the advice of the astrologers, decided to give Lilavati a book of mathematics instead of a husband.

Perhaps the greatest Indian mathematician in the 20th century was Srinivasa Ramanujan (1887–1920), who was essentially self-taught and discovered many beautiful and ingenious formulas.

Exercises

1. Solve the system of three equations in three unknowns given in the Chiu Chang Suan Shu.

2. Find a 4 by 4 magic square, using the numbers from 1 to 16.

3. Find all the solutions to Sun Tsu's Chinese Remainder Problem.

4. Show that Ch'in Chiu Shao's polynomial had four linear factors.

5. Find a solution of De Gang Ma's equation (first posed by E. Lucas in 1875) with $x > 1$.

6. How many camels were there in Brahmagupta's problem?

7. How many bees were there in Bhaskara's problem?

8. If you know the sides of a cyclic quadrilateral, can you determine the radius of the circumscribing circle? Why? How?

22
Mathematics in Islamic Countries

When we speak of Arabic mathematicians, we must remember that Arabic was the common language of intellectuals in the Islamic world, just as Latin was in medieval Europe. In fact, the mathematicians may have been Turks or Persians.

During the period we are concerned with here, the intellectual center of the Arab world was Baghdad, which was founded in the 8th century and was the seat of the eastern Caliphs. It was visited by Greek and Jewish physicians from the West and by Indian and Persian scholars from the East.

Caliph al-Mamun established a 'House of Wisdom', or university, at Baghdad at the beginning of the 9th century. He ordered the translations of many Greek manuscripts, and it was thanks to his policy that many of these works were preserved. Three mathematicians associated with the House of Wisdom are

- Al-Khwarizmi, about 830,

- Thabit Ibn-Qurra (836–901),

- Al-Khayyami (Omar Khayyam) (ca. 1050–1123).

Muhammed ibn-Musa al-Khwarizmi probably came from Khorezm, which corresponds to the modern Khiva and the district surrounding it, south of the Aral Sea in central Asia. But some people deny this and say that he was merely a member of the Khwarizmi tribe, which in fact came to rule the Turko-Persian empire in 1194 AD. Our word 'algorithm' comes from a book he wrote on the use of Indian numerals with the title: 'Spoken has al-Khwarizmi ...', or, in the English translation of the Latin translation: 'Spoken has Algoritmi ...'.

Al-Khwarizmi's most important book was the *Hisab al-jabr w'al-muqabalah*, from which we get the word 'algebra'. The word 'al-jabra' means something like 'combining', as in combining the terms to solve an equation. The same root shows up in old Spanish as 'algebrista', a bone-setter, that is, one who joins together the parts of a broken bone.

Al-Khwarizmi's 'Algebra' was based on the work of Brahmagupta. Although it became extremely influential, it is not as interesting from a mathematical point of view as its fame would suggest. It contains nothing that was not known to the ancient Babylonians or ancient Greeks. There are few proofs, and one of them is woefully inadequate. This is al-Khwarizmi's 'proof' of the Theorem of Pythagoras, which only works if the right triangle is isosceles!

Al-Khwarizimi gives three approximations for π. None of them is supported by any reasoning, and al-Khwarizmi does not seem to think it matters which one is used. However, the book is extremely interesting as a source of sociological information. Many of the text problems deal with questions of inheritance according to Muslim religious law. (These problems appear to have been omitted in the Latin translation by Robert of Chester, which popularized al-Khwarizmi's work in Europe.) Here is an example:

> A man, in his illness before his death, makes someone a present of a slave girl, besides whom he has no property. Then he dies. The slave girl is worth 300 dirhams, and her dowry is 100 dirhams. The man to whom she has been presented cohabits with her. What is the legacy?

Here is how one is supposed to solve the problem:

$$300 - x - \frac{100}{300}x = 2x,$$

yielding $x = 90$ dirhams (Rosen [1831]).

Thabit Ibn-Qurra was an extraordinary polymath. He lived in Baghdad and was an active member of a neo-Pythagorean group called the Sabians. He wrote on politics, grammar, symbolism in Plato's *Republic*, smallpox, the anatomy of birds, the beam balance, the salinity of seawater, the sundial, Euclid's Parallel Postulate, cubic equations, the new crescent moon, etc.

Thabit believed there is an actual infinity (as opposed to Aristotle's potential infinity). He did work in spherical trigonometry and in what we now would call calculus. In his *Book on the Determination of Amicable Numbers*, he gave a wholly original rule:

Let n be a positive integer greater than 1. Let $p = 3 \cdot 2^n - 1$, $q = 3 \cdot 2^{n-1} - 1$ and $r = 9 \cdot 2^{2n-1} - 1$. If p, q and r are primes, then $2^n pq$ and $2^n r$ are 'amicable' (that is, the sum of the proper divisors of each equals the other).

When $n = 2$, we get the amicable pair 220 and 284. When $n = 3$ (or any multiple of 3), r is divisible by 7, and so is not prime. However, when $n = 4$, we have another amicable pair, namely, 17296 and 18416.

Al-Khayyami wrote in Arabic on astronomy and mathematics. He revised the Julian calendar, approaching our Gregorian calendar in accuracy. In a book with the same title as that by al-Khwarizmi, he developed a geometric method for finding the positive real roots of the cubic and quartic equations. The idea was this: to solve the cubic equation

$$x^3 + ax^2 + b^2 x + b^2 c = 0,$$

one intersects the hyperbola $y = bc/x + b$ with the circle $(x + \frac{1}{2}(a+c))^2 + y^2 = \frac{1}{4}(a-c)^2$ and discards the point $(-c, 0)$.

As Omar Khayyam, he wrote a famous poem in Persian, the *Rubaiyat*. Its 19th century translation by Edward Fitzgerald is still a bestseller. It expounds a rather Epicurean philosophy, claiming that the most important thing is wine and the only sure thing is death:

> Oh, threats of Hell and Hopes of Paradise!
> One thing at least is – *This* Life flies;
> One thing is certain and the rest is Lies;
> The Flower that once has blown for ever dies.

(See LXIII in Appendix 1 of E. FitzGerald's translation of the *Rubaiyat* (London: Bernard Quaritch, 1859).) Omar was not popular with the religious establishment of his day, and even now many Persians prefer the more mystical poetry of his fellow countrymen Hafiz.

One of Omar's contemporaries, Nizam-ul-Mulk, wrote in his autobiography that as a youth he made a mutual assistance pact with two fellow students in Naishapur, namely, Omar Khayyam and Hasan Ben Sabbah. He himself later became vizier to two Sultans of the Seljuk dynasty and was able to carry out his pledge by helping Omar to a yearly pension. Unfortunately, so he recounts, Hasan was not satisfied with the government post offered to him and ultimately became the head of a religious sect, the Ismailis. He surrounded himself by a group of fanatics, called the 'assassins', this word being derived from 'hashish'. The sect exists today, though without the practice of assassination; its leader is the Agha Khan.

Exercises

1. Show that 1184 and 1210 is an amicable pair not generated by Thabit's rule.

2. Find another amicable pair which is generated by Thabit's rule.

3. In Thabit's rule, prove that if n is a multiple of 3, then r is a multiple of 7.

4. Prove the following generalization of Thabit's rule: assuming that $p = (2^k + 1)2^{t+k} - 1$, $q = (2^k + 1)2^t - 1$ and $r = (2^k + 1)2^{2t+k} - 1$ are odd primes, then $a = 2^{t+k}pq$ and $b = 2^{t+k}r$ are amicable.

5. Let ABC be a triangle with $\angle A$ obtuse. Let B' and C' be in BC such that $\angle AB'B = \angle AC'C = \angle BAC$. Derive the following theorem of Thabit: $AB^2 + AC^2 = BC(BB' + CC')$.

6. Show that al-Khayyami's method will produce a real solution of the cubic equation in the text.

23
New Beginnings in Europe

Europeans only began to rouse themselves from the intellectual slumber of the Dark Ages as they came in contact with Arab civilization, mostly in Spain.

Gerbert (940–1003) had studied in Spain, where he learned the Indian numerals (but not zero). He wrote on arithmetic and geometry, which so overawed his contemporaries that they believed he had a pact with the devil. In spite of this, he became Pope and was known as Sylvester II from 999 to 1003.

One of the most difficult problems in his *Geometry* was the following: find x and y such that $x^2 + y^2 = a^2$ and $\frac{1}{2}xy = b$. This would have been an easy exercise for a Babylonian scribe!

Contemporary with Gerbert was another mathematician and churchperson, Hrotsvitha of Saxony (932–1002), who had an interest in perfect numbers.

The Englishman Adelhard of Bath (1075–1160) attended lectures at Cordova in Spain, about 1120, disguising himself as a Moslem. There he obtained a copy of Euclid's *Elements* in Arabic, which he translated into Latin. All European editions of the *Elements* were based on this translation until 1533, when the Greek original finally became available.

Abraham Ben Ezra (ca. 1095–1167), while based in Spain, travelled widely between Egypt and England. His book *Sefer ha-Mispar* explained the Hindu arithmetic, using Hebrew letters for numerals, with a zero added.

23. New Beginnings in Europe

He wrote poetry of a pessimistic nature, asserting that, if he were to trade in candles, it would always be noon.

Jordanus Nemorarius (early 13th century) wrote about triangles, circles, regular polygons, Arabic numerals, primes, perfect numbers, polygonal numbers, ratios, powers and progressions. Like Diophantus, he used letters to denote the unknowns in equations.

In his *Tractatus de numeris datis*, Jordanus discusses problems of the following sort: find x and y such that $x + y = 10$ and $x^2 + y^2 = 58$. This is the sort of problem the ancient Mesopotamians were good at.

Nicole Oresme (1320–1382) was a French bishop who wrote extensively on mathematics and gave the first proof of the divergence of the harmonic series.

In the thousand years from 400 to 1400 AD there was exactly one outstanding European mathematician, namely, Leonardo of Pisa (1180–1250), who was also known as Fibonacci. A contemporary of St. Francis of Assisi, he learned his mathematics in Algeria, where his father was a custom house official. In 1202 he published the *Liber abaci*, in which he explained the Indian system of numerals and introduced the famous 'Fibonacci sequence'

$$1, 1, 2, 3, 5, 8, 13, 21, 34, 55, \ldots,$$

where each number is the sum of the preceding two. This sequence has important applications in science and in advanced number theory.

Another book by Fibonacci is his *Liber quadratorum*. This original work in indeterminate analysis gave the first proof of the identity

$$(a^2 + b^2)(c^2 + d^2) = (ac - bd)^2 + (bc + ad)^2.$$

This is equivalent to the theorem that the product of the norms of two complex numbers equals the norm of their product. (The *norm* is the square of the absolute value.) The identity implies that, if each of two integers is a sum of two squares, then so is their product.

We recall that a numerical instance of Fibonacci's identity had already been mentioned by Diophantus (Problem 19, Book III, *Arithmetica*). An explicit statement of the identity first occurred in a commentary on Diophantus by al-Khazin. (See Anbonba [1979].)

In 1225, emperor Frederick II delayed his departure on a crusade to organize a mathematical contest. Leonardo answered all the challenges with flying colours. Two of the problems were the following:

1. Find a fraction a/b such that $(a/b)^2 \pm 5$ are both squares of fractions.

2. Solve $x^3 + 2x^2 + 10x = 20$.

Fibonacci understood negative numbers, interpreting them, in one place, as losses.

23. New Beginnings in Europe

The Fibonacci numbers (1, 1, 2, 3, 5, 8, ...) proved to be so interesting that now there is a whole journal, called the *Fibonacci Quarterly*, devoted to them. Two examples of their relevance are the following.

The seeds of a sunflower head are arranged in such a way that they form two sets of arcs, emanating from the center. The number of clockwise arcs might be 21, and the number of counterclockwise arcs might be 13. In every case, these numbers are consecutive Fibonacci numbers.

One of the great feats of 20th century number theory was Matiyasevič's proof that there is no general procedure for solving Diophantine equations. We shall examine this proof in the Section on Hilbert's Tenth Problem in Part II, Chapter 21. A crucial element in this proof was Matiyasvič's use of the Fibonacci numbers.

To conclude this section, we derive the formula for the nth Fibonacci number u_n, where $u_0 = 0$ (which we include for convenience), $u_1 = 1$ and $u_{n+2} = u_n + u_{n+1}$ for all $n \geq 0$. Put

$$U(x) = u_0 + u_1 x + u_2 x^2 + \cdots = \Sigma_{n=0}^{\infty} u_n x^n.$$

This formal power series is usually called the *generating function* of u_n, although it is not, strictly speaking, a function. We easily calculate

$$U(x)(1 - x - x^2) = u_0 + (u_1 - u_0)x,$$

since $(u_{n+2} - u_{n+1} - u_n)x^{n+2} = 0$ for all $n \geq 0$.
Hence

$$U(x) = \frac{x}{1 - x - x^2} = \frac{A}{1 - \alpha x} + \frac{B}{1 - \beta x} = A\Sigma_{n=0}^{\infty} \alpha^n x^n + B\Sigma_{n=0}^{\infty} \beta^n x^n$$

in partial fractions, where

$$(1 - \alpha x)(1 - \beta x) \equiv 1 - x - x^2,$$

$$A(1 - \beta x) + B(1 - \alpha x) \equiv x.$$

We see from the first identity that $\alpha + \beta = 1$ and $\alpha\beta = -1$, hence

$$\alpha = \tfrac{1}{2}(1 + \sqrt{5}), \quad \beta = \tfrac{1}{2}(1 - \sqrt{5}).$$

From the second identity one easily determines

$$A = \frac{1}{\alpha - \beta} = \frac{1}{\sqrt{5}}, \quad B = \frac{1}{\beta - \alpha} = \frac{-1}{\sqrt{5}}.$$

Comparing the two expansions of $U(x)$ above we obtain

$$u_n = A\alpha^n + B\beta^n = \frac{1}{\sqrt{5}}\left(\frac{1 + \sqrt{5}}{2}\right)^n - \frac{1}{\sqrt{5}}\left(\frac{1 - \sqrt{5}}{2}\right)^n.$$

Furthermore, $\beta^n/\sqrt{5}$ is close to 0 and it is not hard to show that u_n is the integer nearest $\alpha^n/\sqrt{5}$.

The number α is called the 'golden ratio'. It played a role in Euclid's construction of the regular pentagon. The above formula for u_n was discovered and proved by A. de Moivre (1730). (See Section 9.6 in Stillwell's *Mathematics and its History*.)

The Fibonacci numbers arose in connection with the following problem from the *Liber abaci*: how many pairs of rabbits will be produced in a year, beginning with a single pair, if in every month each pair bears a new pair which becomes productive from the second month on?

Exercises

1. Solve Gerbert's problem.

2. Solve Jordanus's problem.

3. Solve problem (1) on Frederick II's math contest.

4. Solve problem (2) on Frederick II's math contest.

5. Starting with the formula $u_n = (\alpha^n - \beta^n)/\sqrt{5}$, show that u_n is the integer nearest $\alpha^n/\sqrt{5}$.

6. Show that u_{n+1}/u_n tends to the golden ratio as n tends to infinity.

7. Show that u_n and u_{n+1} are relatively prime.

8. Prove that the greatest common divisor of any two Fibonacci numbers is also a Fibonacci number.

24
Mathematics in the Renaissance

Aside from the invention of the Indian numerals, and aside from the work of a few persons of talent, such as Pappus and Fibonacci, no significant advances in mathematics had taken place in the thousand years following Diophantus. In the 15th and 16th centuries there was a sudden spurt of activity, aided by the Chinese invention of printing, which reached Europe in 1450 and which carried mathematics, both pure and applied, beyond the achievements of the ancients. It is hard to overemphasize the importance of printing for the spread of mathematical knowledge. Copying mathematical texts by hand required much time and labour. In ancient times, many texts existed only in a single copy, which would be found in the library of Alexandria. This is why, for about 800 years, all mathematical activity was concentrated in one place. Now such texts were available all over the civilized world and people could learn mathematics even in such outlying places as Bohemia or Scotland. In this chapter, and in the next two chapters, we shall discuss advances in the following areas:

1. mathematical notation,

2. the theory of equations,

3. the invention of logarithms,

4. mechanics and astronomy.

(1) Mathematical notation

Johannes Regiomontanus (1436–1476) of Königsberg, then in Germany, gave the first systematic exposition of plane and spherical trigonometry, using both sines and cosines. In algebra he wrote 'res' for x, and 'census' for the square. Regiomontanus probably died of the plague, but there was a rumour that he was poisoned by the sons of a rival scholar.

Columbus took a copy of Regiomontanus's *Ephemerides* on his fourth voyage, and used its prediction of the lunar eclipse of February 29, 1504 to intimidate some hostile Indians in Jamaica.

Johannes Widman (1462–1500) of Eger, now in the Czech Republic, published a book, *Mercantile Arithmetic*, in 1489, in which the modern symbols + and − appeared for the first time.

Luca Pacioli of Italy (1445–1517) was a Franciscan monk. He used the 'res' and 'census' terminology of Regiomontanus. In 1509 he published the *Divina Proportione* (*Divine (or Golden) Ratio*), a book that was illustrated by none other than Leonardo da Vinci. There is a famous painting of Pacioli by Jacopo de' Barbari, which is now in the Museum of Naples. It shows the friar with his friend Guidebaldo standing in the presence of a dodecahedron. One of the problems solved by Pacioli was the following:

> The radius of the inscribed circle of a triangle is 4, and the segments into which one side is divided by the point of contact are 6 and 8. Determine the other sides.

Robert Recorde of England (1510–1558) was the first person to use the symbol = for equality, asserting that 'noe 2 thynges can be more equalle'. Recorde got into a tangle with the Earl of Pembroke and died in jail.

Christoff Rudolff of Germany used $\sqrt{}$ for 'radix' in 1525.

Adam Riese of Bavaria (1492–1559) published arithmetic books that went through more than a hundred editions, and established the use of the signs + and −.

Michael Stifel of Germany (1487–1567) was a monk who became an early follower of Luther. He introduced $1A$, $1AAA$, $1AAA$ for A, A^2 and A^3. He was the first to use negative integers as exponents and had a way of applying mathematics to the Bible which led him to conclude that Pope Leo X was the beast of the *Book of Revelation,* and also to prophesy the end of the world for 18 October, 1533. The peasants of the village where he was pastor believed this prophecy and spent all their money. When the world failed to end, Stifel found himself, not in heaven, but in a jail in Wittenberg.

Thomas Harriot of England (1560–1621) wrote a, aa, aaa for a, a^2, a^3 and introduced the signs > and < for strict inequality. He went to America with Sir Walter Raleigh and became a tobacco addict. In 1603 Harriot computed the area of a spherical triangle:

> Take the sum of all three angles and subtract 180 degrees. Set the remainder as numerator of a fraction with denominator 360

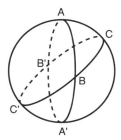

FIGURE 24.1. Area of a spherical triangle

degrees. This fraction tells us how great a portion of the hemisphere is occupied by the triangle.

To prove this, first note that, if a sphere has unit radius, then its surface is 4π or $720°$. Hence, the area between two great circles, e.g., between two meridians on the earth, one degree apart, on one hemisphere, is $720°/360 = 2°$. It follows, moreover, that the area between two great circles, on one hemisphere, separated by an angle A is $2A$.

The spherical triangle ABC is bounded by three great circles, as in Figure 24.1, where A', B' and C' are the antipodes of A, B and C respectively. We note that

$$\triangle ABC + \underline{\triangle A'BC} = 2A,$$
$$\triangle A'B'C' + \underline{\triangle A'BC'} = 2B,$$
$$\triangle ABC + \underline{\triangle ABC'} = 2C.$$

The four underlined triangles make up the visible hemisphere (Figure 24.1), namely, $360°$; hence adding the three equations we get

$$\triangle ABC + \triangle A'B'C' + 360° = 2A + 2B + 2C.$$

Now $\triangle A'B'C' = \triangle ABC$, the two being antipodal triangles. Dividing by 2, we obtain

$$\triangle ABC + 180° = A + B + C.$$

Thus the area of the spherical triangle, measured in degrees, is its *spherical excess* $A + B + C - 180°$.

(2) The theory of equations

Ever since the ancient Babylonians, people knew how to find positive, real solutions to any linear or quadratic equation. This they could do arithmetically or geometrically. Omar Khayyam (1100 A.D.) had developed a method for drawing a line segment whose length was a positive real root of a given cubic polynomial. In 1225, Leonardo of Pisa gave an arithmetical solution to $x^3 + 2x^2 + 10x = 20$. Because he used arithmetic rather than geometry, Leonardo was able to obtain an approximation to the positive root which was accurate to nine decimals places.

The first person to develop anything like a complete method for solving cubic equations — one that could in principle handle negative and imaginary roots as well as positive real roots — was alleged to have been Scipione Ferro of Bologna (1465–1526). He could solve any equation of the form $x^3 + bx = c$, giving the answers to any degree of accuracy. Ferro kept his method a secret — so that he would have an advantage over other mathematicians in mathematical contests — but, just before he died, he passed it on to one Antonio Fiore.

In 1530 Zuanne da Coi sent the following problems to Niccolo Tartaglia (1500–1557):

$$x^3 + 3x^2 = 5,$$
$$x^3 + 6x^2 + 8x = 1000.$$

Tartaglia announced that he could solve these equations and was promptly challenged by Fiore to a contest. Each contestant had to deposit a certain amount of money with a notary and propose a number of problems for his rival to solve. Whoever solved more of the problems within 30 days was to get all the money.

Tartaglia, suspecting that Fiore would pose equations of the form $x^3 + bx = c$, quickly worked out a general method for solving such equations. Indeed, Fiore's problems were of this nature, and Tartaglia was able to solve them all. He himself posed equations of the form $x^3 + ax^2 = c$, which he could solve, but which were too difficult for Fiore.

Tartaglia was born in Brescia, Italy. This town was captured by the French in 1512 and most of the inhabitants were massacred. Tartaglia's jaws were split by a soldier's sword and it was thus that he acquired his name, which means 'the stammerer'. He lectured at Verona and Venice, becoming famous through his victory over Fiore. He published a book on ballistics in 1537, in which he correctly stated that a projectile achieves its maximum range when fired at an angle of 45°. However, he gave no proof of this. In 1560 Tartaglia wrote a book on number theory, which contained some amusing puzzles, for example:

> Three couples wish to cross a river in a boat that holds only two people. How can this be done if no woman is to be left with a man unless her husband is present?

Three people wish to share the oil in a 24 ounce jar. They have empty measuring jars of capacity 5, 11 and 13 ounces. How can they divide the oil?

The later part of Tartaglia's life was embittered by a quarrel with Girolamo Cardano (1501–1576), another Italian, whose autobiography has been republished by Dover. Cardano was a famous physician in Milan and once travelled to Scotland to cure an Archbishop of asthma. He applied his mathematics to mechanics, gambling and astrology. Indeed, he might be called the discoverer of probability theory. Cardano's oldest son was executed for having poisoned his wife, and Cardano himself was imprisoned, in 1570, for heresy, for having published a horoscope of Jesus Christ. (This was heretical because it suggested that God was subject to the stars.) Cardano was freed only after he recanted. There is a story that he foretold the day of his own death, using astrology, and then felt compelled to commit suicide to make his prediction come true.

Cardano had persuaded Tartaglia to tell him the secret method for solving cubic equations. This Tartaglia did only on condition that Cardano would never reveal it. However, some years later Cardano learned of the prior work by Ferro, and he decided to publish the secret method in his *Ars Magna* (1545). Cardano gave due credit to Tartaglia for the method, but Tartaglia was upset that Cardano had broken his promise to keep it secret. Henceforth, Tartaglia would have no special advantage in mathematical contests!

Annoyed, Tartaglia challenged Cardano to a competition. The latter did not show up, being represented instead by his student, Ferrari (1522–1565). It seems that Ferrari did better than Tartaglia, and Tartaglia lost both prestige and income.

Cardano's *Ars Magna* was the best book on algebra so far. It still used geometry to prove the algebraic identity

$$(a-b)^3 = a^3 - b^3 - 3ab(a-b),$$

and it still shied away from negative numbers, listing the following equations separately:

$$x^3 + px = q, \quad x^3 = px + q, \quad x^3 + px + q = 0, \quad x^3 + q = px.$$

Nonetheless, the *Ars Magna* contained a full explanation of the cubic equation, including a treatment of imaginary numbers.

Ferarri came from Bologna in Italy. He extended the work of Tartaglia and Cardano, solving the general fourth degree equation. His solution appears in the *Ars Magna*. Ferrari became rich in the service of the Cardinal Fernando Gonzalo, but ill health forced him to retire to Bologna, in 1565, to teach mathematics. According to W.W. Rouse Ball in *A Short Account of the History of Mathematics*, p. 225, Ferrari was murdered by his sister or by her boyfriend.

Rafael Bombelli of Bologna (1526–1572) published an algebra book in 1572, in which he traced the history of the subject back to Diophantus. He discussed complex radicals at length and showed that the irreducible case of the cubic equation leads to three real roots (see Case 3 Chapter 25). He also pointed out that the ancient Greek problem of trisecting an angle was equivalent to solving a cubic equation. He wrote $\smile^n$ for x^n.

Francois Viète, also called Vieta, (1540–1603) was a French lawyer and member of parliament, but his avocation was mathematics. He wrote *In artem analyticem isagoge* in 1591, in which he applied algebra to geometry. (Hitherto people had applied geometry to algebra.) He was challenged by King Henry IV to solve a special equation of the 45th degree, and managed to give the answer in a few minutes, having noticed that the equation was satisfied by the chord of an angle of $360°/45$. He constructed the circles touching three given circles, using only Euclidean geometry, and thus recaptured an ancient construction that was probably contained in a lost book by Apollonius. Viète deciphered a Spanish code for the French. His solutions of cubic and quadratic equations were just like ours.

We close this chapter with a question about astrology. Why did many mathematicians, such as Ptolemy, Cardano and later, Kepler, waste their time and talent on astrology when a little reflection reveals that it seems unlikely to have any truth in it? Astrology is based on the unproven and implausible assumption that there is a correlation between the constellations of the stars at the time of a person's birth and his or her character and ultimate life history. Did Ptolemy, et al., only espouse the practice of astrology because it helped to supplement their incomes, or did they genuinely believe in it, as perhaps a majority of people still do today? Kepler, for one, had a cynical view of astrology, as we shall see.

Exercises

1. Professor Smith learns a secret mathematical technique from Professor Brown only because he solemnly promises Brown that he will never publish it. Later Smith discovers that this technique was long ago published, by Professor Jones, in an obscure little journal that no one ever reads. Is it morally permissible for Brown to publish this technique (giving due credit, of course, to Smith and Jones)? Support your answer with reasons.

2. Solve Pacioli's problem, given above.

3. Solve Tartaglia's puzzle about the three couples.

4. Solve Tartaglia's puzzle about the oil.

5. Let $a_1 = 1/\sqrt{2}$, $a_{n+1} = \sqrt{\frac{1}{2} + \frac{1}{2}a_n}$. Viète proved that
$$2/\pi = a_1 a_2 \cdots a_n \cdots$$
Prove this formula. (Hint: first show
$$(\sin x)/(2^n \sin(x/2^n)) = (\cos x/2)(\cos x/2^2)(\cos x/2^3) \cdots (\cos x/2^n),$$
whence
$$(\sin x)/x = (\cos x/2)(\cos x/2^2)(\cos x/2^3) \cdots$$
and, finally, let $x = \pi/2$.)

25
The Cubic and Quartic Equations

Cardano was the first person to use imaginary numbers in print. In this chapter, we shall use imaginary numbers to present what is essentially Cardano's solution to the cubic equation. We shall also give Ludovico Ferrari's solution to the fourth degree polynomial equation.

Recall that $i = \sqrt{-1}$. Recall that, if a, a', b and b' are real numbers and $a + bi = a' + b'i$, then $a = a'$ and $b = b'$ (lest $i = (a - a')/(b - b')$, a real number). Let $\omega = \frac{-1}{2} + \frac{1}{2}\sqrt{3}i$. A quick calculation shows that $\omega^2 = \frac{-1}{2} - \frac{1}{2}\sqrt{3}i$, and hence $\omega^3 = 1$.

Lemma 25.1. *Any complex number $x + iy$ can be written in the* polar form $r(\cos A + i \sin A)$, *where r is a non-negative real number and A is a real number.*

Lemma 25.2. $(r(\cos A + i \sin A))^3 = r^3(\cos 3A + i \sin 3A)$.

Proof:
$$\cos 3A = \cos^3 A - 3 \cos A \sin^2 A,$$
$$\sin 3A = 3 \cos^2 A \sin A - \sin^3 A.$$

Lemma 25.3. *The equation $z^3 = 1$ has exactly three complex solutions: $1, \omega, \omega^2$.*

Lemma 25.4. *If b is any given complex number $\neq 0$, the equation $z^3 = b$ has exactly three complex solutions. If z_1 is one solution, the others are $z_1 \omega$ and $z_1 \omega^2$.*

Lemma 25.5. *If y and p are any complex numbers, there are complex numbers u and v such that $u + v = y$ and $uv = p$.* (This is the old Babylonian problem.)

The proofs of these Lemmas are left to the reader.

The general cubic equation (after division by the leading coefficient) has the form
$$x^3 + ax^2 + bx + c = 0.$$
(We assume here that a, b, and c are real.) Putting $x = y + k$, this equation becomes
$$y^3 + (3k + a)y^2 + (\ldots)y + (\ldots) = 0.$$
We choose $k = -a/3$, so that the square term disappears. (Tartaglia was the discoverer of this trick.) The resulting equation has the form
$$y^3 - 3py - 2q = 0$$
with p and q real and where the numbers -3 and -2 are introduced for convenience only. It is this *reduced* equation which we now want to solve.

If we put $y = u + v$, then the reduced equation becomes
$$u^3 + v^3 + 3(uv - p)(u + v) - 2q = 0.$$

To make this as simple as possible, we choose $v = p/u$ (see Lemma 25.5). The equation then becomes
$$u^3 + v^3 = 2q.$$
Since, however, $u^3 v^3 = p^3$, we are back with the Babylonian problem of seeking two numbers with given sum and product. Clearly u^3 and v^3 are solutions of
$$t^2 - 2qt + p^3 = 0.$$
Therefore we have, say,
$$u^3 = q + \sqrt{q^2 - p^3},$$
$$v^3 = q - \sqrt{q^2 - p^3}.$$
Thus $y = u + v = u + p/u$, where u is a cube root of
$$q + \sqrt{q^2 - p^3}.$$

If one of these cube roots is u_1, then the others are $u_1\omega$ and $u_1\omega^2$. Let $v_1 = p/u_1$. Then $u_1\omega + p/(u_1\omega) = u_1\omega + v_1\omega^2$ and $u_1\omega^2 + p/(u_1\omega^2) = u_1\omega^2 + v_1\omega$. Hence the reduced cubic has the following three solutions only:
$$u_1 + v_1, \quad u_1\omega + v_1\omega^2, \quad u_1\omega^2 + v_1\omega.$$

To discuss these solutions in detail, we consider three cases.

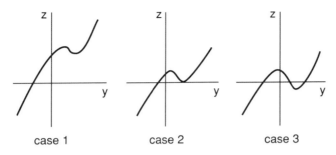

FIGURE 25.1. Cubic polynomials

Case 1. $q^2 - p^3 > 0$. Say $q^2 - p^3 = r^2$, where r is real. Let u_1 be the real cube root of $q + r$, so that v_1 is the real cube root of $q - r$ (the product $u_1 v_1$ must be real). Then the solutions are the real number $u_1 + v_1$ and the imaginary 'conjugates' $u_1 \omega + v_1 \omega^2$ and $u_1 \omega^2 + v_1 \omega$. (Note that ω^2 is the *conjugate* of ω, that is, it may be obtained from ω by replacing i with $-i$.)

Case 2. $q^2 - p^3 = 0$. Then we can take u_1 as the real cube root of q, and v_1 the same. The equation then has only two distinct roots, namely, $2u_1$ and $-u_1$ (since $\omega + \omega^2 = -1$), but we say that the latter root occurs twice.

Case 3. $q^2 - p^3 < 0$. Say $q^2 - p^3 = -r^2$, where r is a positive real number. Then $u^3 = q + ir$ and $v^3 = q - ir$. Let $u_1 = a + bi$ with a and b real. Then $(a^2 + b^2)^3 = q^2 + r^2 = p^3$, so that $a^2 + b^2 = p$ and hence $v_1 = p/u_1 = a - bi$. We calculate

$$u_1 \omega + v_1 \omega^2 = a(\omega + \omega^2) + bi(\omega - \omega^2) = -a - b\sqrt{3}.$$

Similarly, $u_1 \omega^2 + v_1 \omega = -a + b\sqrt{3}$. It is not hard to show that $a = \sqrt{p}\cos((\arctan r/q)/3)$ and $b = \sqrt{p}\sin((\arctan r/q)/3)$. (Note that, in Case 3, $p > 0$.)

If $z = y^3 - 3py - 2q$, the three cases are illustrated by the three graphs in Figure 25.1. However, it should be borne in mind that these graphs were not available in the Renaissance, as analytic geometry had not yet been invented.

The general quartic equation has the form

$$x^4 + ax^3 + bx^2 + cx + d = 0,$$

which we may write

$$x^4 + ax^3 = -bx^2 - cx - d.$$

Adding $a^2x^2/4$ to both sides of this equation, we obtain

$$\left(x^2 + \frac{1}{2}ax\right)^2 = \left(\frac{1}{4}a^2 - b\right)x^2 - cx - d.$$

In an attempt to get a perfect square on both sides of the equation, we add $t(x^2 + \frac{1}{2}ax) + \frac{1}{4}t^2$ to both sides:

$$\left(x^2 + \frac{1}{2}ax + \frac{1}{2}t\right)^2 = \left(\frac{1}{4}a^2 - b + t\right)x^2 + \left(-c + \frac{1}{2}at\right)x - d + \frac{1}{4}t^2.$$

Now $Ax^2 + Bx + C$ is a perfect square when $B^2 - 4AC = 0$. In fact, if $A \neq 0$, we can then write

$$Ax^2 + Bx + C = (\sqrt{A}x + B/(2\sqrt{A}))^2.$$

So, to get a square, it will suffice to pick t so that

$$-d + \frac{1}{4}t^2 = \frac{\left(-c + \frac{1}{2}at\right)^2}{4\left(\frac{1}{4}a^2 - b + t\right)},$$

that is,

$$t^3 - bt^2 + (ac - 4d)t + 4bd - a^2d - c^2 = 0.$$

But this is a cubic equation! In practice, this associated cubic equation can often be solved by trial and error. We only need one value of t.

We see that if x is such that $x^4 + ax^3 + bx^2 + cx + d = 0$, then there is a t, which we can determine by finding a real root of the above cubic equation, such that

$$\left(x^2 + \frac{1}{2}ax + \frac{1}{2}t\right)^2 = (\sqrt{A}x + B/2\sqrt{A})^2,$$

with $A = \frac{1}{4}a^2 - b + t$, $B = -c + \frac{1}{2}at$ and $C = -d + \frac{1}{4}t^2$.
This gives us

$$x^2 + \frac{1}{2}ax + \frac{1}{2}t = \pm(\sqrt{A}x + B/2\sqrt{A}),$$

which is a quadratic equation in x.

For several centuries people tried to find similar methods for solving equations of degree greater than 4. They failed. It was only in the 19th century that it was shown by Ruffini, Abel and Galois that the general equation of degree 5 or more cannot be solved by 'radicals' (e.g. fifth roots). Of course, special cases can be solved by radicals, for example $x^5 + x = 34$.

Exercises

1. Prove all the Lemmas in this chapter.

2. Solve the following equations, obtaining exact answers. Do not use decimal approximations. Simplify answers.

   (a) $x^3 + x^2 - 2 = 0$,
   (b) $x^3 + 9x - 2 = 0$,
   (c) $y^3 - 3y + 1 = 0$,
   (d) $y^3 - 7y - 7 = 0$,
   (e) $x^3 + 2x^2 + 10x = 20$,
   (f) $x^3 + 3x^2 = 5$,
   (g) $x^3 + 6x^2 + 8x = 1000$,
   (h) $x^4 + x^3 - 6x^2 - x + 1 = 0$,
   (i) $x^3 = 15x + 4$.

3. Why cannot a fourth degree equation have five or more distinct complex roots?

26
Renaissance Mathematics Continued

(3) The invention of logarithms

John Napier (1550–1617), a Scottish aristocrat, spent 20 years of his life on the construction of logarithms. Like Stifel, he was interested in proving that the Pope was the Antichrist (the opponent of Jesus, who is supposed to appear just prior to the latter's prophesied second coming). Napier was convinced that the world would end before 1700.

Napier describes his technique for calculating logs in his *Mirifici Logarithmorum Canonis Constructio*, which was published in 1619, two years after the author's death.

This *Construction of the Wonderful Canon of Logarithms* is tricky to read, because what Napier calls the 'logarithm' of x is actually what we call $10^7 \log_{1/e}(x/10^7)$. The book is also a bit tedious because Napier pays minute attention to error bounds. However, if we simplify and modernize Napier's presentation somewhat — as we shall do below — we obtain a very lucid piece of mathematics.

Napier's basic ideas are these. Suppose there is a particle on the negative half of the real number line, moving towards the origin at a speed proportional to its distance from the origin. At time 0, it is at -1. For any distance d, there is a time when the particle is that distance from the origin. Call this time the 'logarithm' of d.

Assume that the constant of proportionality is 1. Then $dx/dt = -x$ and $x(0) = -1$. Hence $x(t) = -e^{-t}$, and so $t = -\log_e d = \log_{1/e} d$. In this section 'log x' shall mean $\log_{1/e} x$. Note that $\log_{1/e} d$ decreases as d increases from $1/2$ to 1.

Napier used the following three ideas to calculate his tables.

1. When z is very small, $\log(1-z) \approx z$.

2. The natural number powers of $(1-10^{-m})$ — where m is a natural number — are easy to calculate. For, if $s = 1 - 10^{-m}$, then $s^2 = s(1-10^{-m}) = s - s/10^m$. Similarly, $s^3 = s^2 - s^2/10^m$. It is always just a matter of shifting the decimal point m places and subtracting. Thus Napier has

$$\begin{aligned}
1 - 10^{-5} &= 0.999\ 990\ 000\ 000\ 000, \\
(1 - 10^{-5})^2 &= 0.999\ 990\ 000\ 000\ 000 \\
&\quad -.000\ 009\ 999\ 900\ 000 \\
&= 0.999\ 980\ 000\ 100\ 000, \\
(1 - 10^{-5})^3 &= 0.999\ 980\ 000\ 100\ 000 \\
&\quad -.000\ 009\ 999\ 800\ 001 \\
&= 0.999\ 970\ 000\ 299\ 999.
\end{aligned}$$

3. Near 1, the log is smooth, and linear interpolations give excellent approximations to it.

Near -1, Napier's particle is moving at about 1 unit/second, and we can assume that $\log d = 1 - d$ for $d = 1 - 10^{-5} = 0.99999$. As in (2) above, Napier calculates $d, d^2, \ldots, d^{50}$. He finds that $d^{50} = 0.9995001225$. Hence $\log 0.9995001225 = 50 \times \log d = 50 \times 0.99999$. With linear interpolation, Napier can thus obtain a very accurate value for $\log u$, where $u = 0.9995$.

Next, using ideas similar to those in (2), Napier quickly and accurately calculates $u^2, u^3, \ldots, u^{20}$. He obtains $u^{20} = 0.990047358$. Using interpolation, he then gets a very accurate value for $\log w$, where $w = 0.99$. (Your pocket calculator will not be more accurate.)

For $a = 1, 2, \ldots, 20$ and $b = 0, 1, \ldots, 68$, Napier calculates $u^a w^b$. This gives him 1380 points between $u = 0.9995$ and $u^{20} w^{68} = 0.499860940$ for calculating logarithms in that range.

For example, $u^{19} w^{68} = 0.500110996 > 1/2 > 0.499860940 = u^{20} w^{68}$. If

$$k = \frac{\log(u^{19}w^{68}) - \log(u^{20}w^{68})}{u^{19}w^{68} - u^{20}w^{68}},$$

then k is the slope of the line joining two given points. Thus, by linear interpolation we have

$$\begin{aligned}
\log 1/2 &\approx \log(u^{20}w^{68}) + (\tfrac{1}{2} - u^{20}w^{68})k \\
&\approx 20 \log u + 68 \log w - 0.000278 \\
&\approx 0.693147.
\end{aligned}$$

This value of $\log 1/2$ is accurate to six decimal places.

Of course, it is now easy to calculate other logs. For example, if $0 < t < 1/2$, we can calculate $\log t$ by finding an integer m such that $1/2 < 2^m t < 1$. Given such an m, we have $\log t = \log(2^m t) + m \log 1/2$. The log of 2 (to our base $1/e$) is just $-\log 1/2$.

Henry Briggs (1561–1631) travelled to Edinburgh in 1615 to discuss logarithms with Napier. They agreed that there were many advantages to having logs to the base 10. In 1617 Napier died, but Briggs continued the work, publishing tables for logs to the base 10 in 1624. It was Briggs who introduced the terms 'mantissa' and 'characteristic'. The practical advantage of base 10 is of course that the logarithm of numbers such as 173, 17.3, 1.73, 0.173 and 0.0173 all have the same mantissa, which can be found in the table, while their characteristics $2, 1, 0, -1$, and -2 are seen by inspection. Briggs persuaded many of his contemporaries, including Kepler, of the importance of logarithms.

(4) **Mechanics and astronomy**

Nicholas Copernicus (1473–1543) was of Polish origin. He conjectured that the earth and the other planets move around the sun. At any rate, he said that this assumption gives a simpler explanation of what is going on. He did not assert that the planets move in circles with the sun at the center. He could only describe the orbits as epicycles, as did Ptolemy, whose observational data he used.

Simon Stevin (1548–1620) lived in the Netherlands (which included what is now Belgium). He was one of the earliest expositors of the theory of decimal fractions, which he compared to an unknown island 'having beautiful fruits, pleasant plains, precious minerals' (see page 21 in Smith's *A Source Book in Mathematics*). Stevin is perhaps best known for his *Statistics and Hydrostatics*, published in 1586. In this book he discusses the triangle of forces, resolution of forces, stable and unstable equilibrium and pressure. He was the first to advance beyond Archimedes in these subjects. He also wrote on algebra and geometry.

Galileo Galilei (1564–1642) is often regarded as the father of modern physics. There is a story that, instead of paying attention to the church service, he used his pulse to time the oscillations of a lamp swinging from the church roof. He discovered that the period of oscillation does not depend on its amplitude. This discovery was to be exploited for constructing pendulum clocks. Galileo disrupted his medical studies to devote himself to mathematics. After publishing a book on hydrostatics and centers of gravity, he was appointed professor. Everyone has heard the story about his experimenting with falling bodies on the leaning tower of Pisa. Whatever the truth of this story, he found that a falling body undergoes a uniform ac-

FIGURE 26.1. Kepler's solar system

celeration which is independent of its weight. He checked these conclusions by studying bodies sliding on inclined planes.

Galileo is most famous as an astronomer. He was the first to use the recently invented telescope to study the heavens and was rewarded by discovering, among other things, the moons of Jupiter (in 1610). These seemed to confirm the Copernican hypothesis which Galileo advocated. It is well-known that, in 1633, the Inquisition forced him to recant the Copernican view.

Johann Kepler (1571–1630) studied in Tübingen (Germany) and became a professor in Grätz (Austria). His early espousal of the Copernican system led him to seek an explanation of the distances of the various planets from the sun. His first idea was to construct an equilateral triangle with its vertices on the orbit of Saturn — it was assumed that this orbit was circular — and then to inscribe another circle in this triangle (Figure 26.1).

This second circle was supposed to be the orbit of Jupiter, and, indeed, this construction gave more or less the correct ratio for the distances of Saturn and Jupiter from the sun. Kepler tried to extend this idea by constructing a square with its vertices on the orbit of Jupiter, and inscribing a circle in this square for the orbit of Mars. Here, however, the observed facts did not confirm to the theoretical model.

Undiscouraged, Kepler then replaced the circles by spheres and the regular polygons by regular polyhedra. Remarkably, there were five intervals between the known planetary orbits to account for and five Platonic solids to explain them. Using the observational data of Ptolemy, Kepler made his model fit the facts more or less, and he published his findings in his *Cosmic Mystery*, a book which owed much to Pythagorean number mysticism. Today we do not attribute any significance to these speculations.

Kepler knew that, in order to check his theories, he had to use more recent observational data than those recorded by Ptolemy. It so happened that very accurate observations of celestial phenomena were being made by the Danish astronomer Tycho Brahe (1546–1601), who was, however, reluctant to part with his data, hoping to save them for his own theory, which was a modification of Ptolemy's.

26. Renaissance Mathematics Continued   143

The capital of the Holy Roman Empire was at that time in Prague. Tycho Brahe was called to Prague as 'imperial mathematicus' and Kepler managed to become his assistant in 1599. Brahe put Kepler to work on the orbit of Mars, which deviated from a circle more than any other planetary orbit did. Two years later, Brahe died, and Kepler succeeded him as imperial mathematicus. His job description included casting horoscopes for the emperor. Kepler expressed his views on astrology in *Tertius Interveniens*, asserting that it is all very well for philosophers to criticize the 'daughter' of astronomy, without realizing that 'the daughter must support the mother by her charms'. He pointed out that an astronomer could not make a living, unless he encouraged people in the belief that they could learn the future from the stars.

With the help of Brahe's magnificent data, Kepler was able to formulate his three laws of planetary motion:

1. Each planet describes an ellipse with the sun at one focus.

2. The line joining the sun to the planet sweeps out equal areas (bounded by the ellipse) in equal times.

3. The square of the period of revolution of each planet (its 'year') is proportional to the cube of the major axis of its orbit.

Kepler's third law came in 1619, about ten years after the first two. These three laws were later to confirm Newton's theory of universal gravitation.

After the job in Prague petered out, owing to political events and a shortage of funds in the imperial treasury, Kepler supported himself by casting horoscopes. He used mathematics even in his private life, making a careful calculation to decide which eligible woman to choose as his second wife.

There was to be one occasion when Kepler's connection to the court in Prague proved useful. When jealous neighbours had accused his mother of witchcraft, a very serious accusation in those days, he managed to get the case dismissed.

Exercises

1. How should Napier have had his particle moving to get logs to the base e?

2. Suppose we have perfectly accurate values for $\log(u^a v^b)$ and $\log(u^{a-1} v^b)$. Suppose $u^a v^b < x < u^{a-1} v^b$, and suppose we calculate the value of $\log x$ from the values of the two given logs, using linear interpolation. What is the maximum error possible?

3. You have been stranded on a desert island without your calculator. Being very bored, you calculate 2^{1000}, doubling 999 times but keeping only five significant figures (leaving a trail of zeros at the right of the calculation). You find that $2^{1000} = 1.0715... \times 10^{301}$. By what easy way can you now find $\log_{10} 2$, accurate to three decimal places?

4. Briggs calculated logs using square roots. He found $10^{1/2}, 10^{1/4}, 10^{1/8}$ etc. He then found numbers such as $10^{3/8} = 10^{1/4} 10^{1/8} = 2.37....$ This gave him the $\log_{10} 2.37....$ Use Briggs's method, together with linear interpolation, to get a value for $\log_{10} 2.37....$

5. According to Kepler's initial model, what is the ratio of the radius of Saturn's orbit to the radius of Jupiter's orbit?

27
The Seventeenth Century in France

The 17th century saw a blossoming of mathematical activity in France. Some of the important mathematicians were:

- Marin Mersenne (1588–1648),
- Gerard Desargues (1591–1661),
- René Descartes (1596–1650),
- Pierre de Fermat (1601–1665),
- Blaise Pascal (1623–1662).

Mersenne was a friar belonging to the Minim order. In the chapter on perfect numbers, we learned about primes of the form $2^n - 1$, which are named after him. His main importance lies in the fact that he corresponded with all the other French mathematicians, keeping them in touch with each other's ideas.

Desargues was the discoverer of projective geometry, the part of the geometry which deals entirely with incidence and ignores distance. Parallel lines are presumed to meet at a point 'at infinity'. In retrospect, it appears that the theorem of Pappus was also a theorem in projective geometry. Desargues's famous theorem is this:

> If two triangles, in the same plane or not, are so situated that lines joining pairs of corresponding vertices are concurrent (if

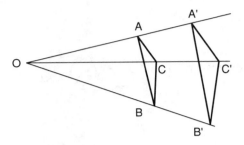

FIGURE 27.1. Desargues's Theorem

only at infinity), then the points of intersection of pairs of corresponding sides are collinear (if only on the 'line at infinity'), and conversely.

The 'parallel case' for two triangles in the same plane is pictured in Figure 27.1. According to the theorem, if the lines joining corresponding vertices meet in a single point O and if AB is parallel to $A'B'$ and AC is parallel to $A'C'$, then BC is parallel to $B'C'$.

The theorem of Desargues can be proved, in Euclidean plane geometry, using only axioms I to V of Chapter 16. This is shown by David Hilbert (1862–1943) in his *Foundations of Geometry*.

Hilbert also shows that, in the absence of the Axiom of Archimedes (mentioned in Chapters 17 and 19), the Theorem of Desargues can be used for defining multiplication within Euclidean geometry, and for proving, within that geometry, that multiplication is associative. (Hilbert also shows that the Theorem of Pappus does not depend on the Axiom of Archimedes. The Theorem of Pappus is the key ingredient in any proof of the commutativity of multiplication which does not rest on the Axiom of Archimedes.)

For a simple proof of the 'parallel case' of Desargues' Theorem, the reader may wish to consult, e.g., Ewald [1971].

Descartes was educated by the Jesuits, who kindly permitted him to spend the mornings in bed, because of his delicate health. Throughout his life, he did much of his intellectual work in bed. Being a man of 'good' family, he was supposed to choose the church or the army for his career, and he chose the latter. He first served in the army of Maurice of Orange, but transferred to that of the Duke of Bavaria when the Thirty Years War broke out. Even on military campaigns, he spent a good part of his time in bed, thinking about mathematics and philosophy. He resigned his commission in 1621, and travelled for five years, studying mathematics. In the end, he settled in Holland, where he spent twenty years in full time intellectual pursuits. In 1649 he went to Sweden at the invitation of Queen Christina. The vigorous young queen insisted on having mathematics lessons at five in the morning, and within two months the aging Descartes, who would have preferred to sleep in, caught pneumonia and died.

Descartes's first book, *Le Monde*, advanced the Copernican model of the solar system. Just as he was completing it, the Inquisition condemned the Copernican views expressed by Galileo. Descartes decided not to publish his book, writing sadly to Mersenne:

> This has so strongly affected me that I have almost resolved to burn all my manuscript, or at least show it to no one. But on no account will I publish anything that contains a word that might displease the Church.

In 1637 Descartes published his famous *Discours de la Méthode pour bien conduire sa Rasion et chercher la Verité dans les Sciences*. This 'discours' has three important appendices. Appendix I, *La Dioptrique* treats optics and the laws of refraction. Appendix II, *Les Météores*, deals with atmospheric phenomena; in particular, it offers an explanation of the shape of the rainbow. For us, the most important is Appendix III, *La Géometrie*. This is divided into three books and sets forth the principles of analytic geometry. In it we find the usual formulas for the conics, and also Descartes's 'rule of signs' (stated without proof). Descartes was thus the creator of analytic geometry, although Fermat must share equal credit.

The revolutionary idea of analytic geometry was this: points in the Euclidean plane could be represented by pairs of real numbers and, consequently, straight lines and conic sections could be described by sets of pairs (x, y) satisfying equations of the form

$$Ax + By + C = 0$$

and

$$Ax^2 + Bxy + Cy^2 + Dx + Ey + F = 0,$$

respectively. Thus geometry was reduced to algebra! It was now possible to solve some of the problems which the Greeks had left open, for instance, the problem of doubling the cube, although the actual solution had to wait a couple of centuries more.

Descartes's rule of signs asserts that the number of positive real roots of a polynomial equation $f(x) = 0$ with real coefficients is $v - 2k$, where v is the number of *variations in sign* and k is some natural number. To calculate v one writes $f(x)$ in descending powers of x, omitting all terms with zero coefficients. Then v is the number of times we change sign as we go from left to right. For example, the equation

$$(1) \qquad x^6 - 3x^5 - x^4 + 2x - 5 = 0$$

has three variations in sign, so we can tell that it has either three positive roots or only one. Sometimes we can be certain; for example, putting $x = -y$ in equation (1), we get

$$(2) \qquad y^6 + 3y^5 - y^4 - 2y - 5 = 0,$$

148   27. The Seventeenth Century in France

which has one variation in sign, hence exactly one positive root. It follows that equation (1) has exactly one negative root. Descartes's rule of signs is a special case of what logicians call *elimination of quantifiers*. Descartes wrote two other books, the *Meditations* (1641) and *Principia Philosophicae* (1644). The latter deals with physical science and proposes the theory of vortices to explain planetary motion. This theory was later refuted by Newton.

The *Meditations* contains a 'geometrical proof' of the existence of God:

> existence can no more be separated from the essence of God than the idea of a mountain from that of a valley, or the equality of its three angles to two right angles, from the essence of a triangle (Meditation V).

In other words, God exists because existence is just one of the defining properties of God.

Fermat was a councillor for the parliament of Toulouse, and only did mathematics in his spare time. He published almost nothing during his lifetime; his contributions to mathematics are contained in his correspondence (e.g., with Mersenne) and in the papers that were found after his death. Nonetheless, he is considered to be the greatest amateur mathematician of all times.

Fermat introduced his version of analytic geometry in his *Ad Locos Planos et Solidos Isagoge*. He also collaborated on probability theory with Pascal. In addition, Fermat studied tangents to curves, maxima and minima, and areas under curves, coming very close to discovering calculus. Fermat's methods were influenced by those of his contemporaries, Cavalieri and Wallis, whom we shall meet in the next Chapter.

Today Fermat is best known for his contributions to the theory of numbers. Diophantus knew that a prime number of the form $4n - 1$ is never the sum of two squares. Fermat went further and proved that every prime of the form $4n + 1$ can be written as the sum of two squares in exactly one way. He also noted that every odd prime p can be written as the difference of two squares in one and only one way.

The last statement is easy to prove. Indeed, $p = (\frac{1}{2}(p+1))^2 - (\frac{1}{2}(p-1))^2$ where $\frac{1}{2}(p+1)$ and $\frac{1}{2}(p-1)$ are both integers, since p is odd. On the other hand, if $p = x^2 - y^2 = (x+y)(x-y)$, then $x + y = p$ and $x - y = 1$ (since p, being prime, has no factors except 1 and p). This gives $x = \frac{1}{2}(p+1)$ and $y = \frac{1}{2}(p+1)$ as before.

Fermat's so-called 'Little Theorem' asserts that, if p is a prime and a is an integer which is not a multiple of p, then $a^{p-1} - 1$ is a multiple of p. There are many proofs of this theorem. One of them depends on the

Binomial Theorem (known to the Chinese centuries before Fermat):

$$(x+1)^p = x^p + \binom{p}{1}x^{p-1} + \binom{p}{2}x^{p-2} + \cdots + \binom{p}{p-1}x + 1,$$

where

$$\binom{p}{k} = \frac{p(p-1)\cdots(p-k+1)}{k(k-1)\cdots 1}$$

is an integer. If $0 < k < p$, then p is not a factor of $k(k-1)\cdots 1$. Since p is a factor of

$$\binom{p}{k}k(k-1)\cdots 1 = p(p-1)\cdots(p-k+1)$$

it follows that p is a factor of $\binom{p}{k}$ when $0 < k < p$. Hence

$$(x+1)^p = x^p + 1 + mp$$

for some integer m. If $x = 1$, we obtain $2^p = 2 + mp$, so that p is a factor of $2^{p-1} - 1$. If $x = 2$, we get $3^p = 2^p + 1 + m'p = 2 + mp + 1 + m'p = 3 + (m+m')p$, so that p is a factor of $3^{p-1} - 1$. Continuing in the same way (using mathematical induction), we see that, for all a, $a^p = a + np$ for some integer n. Thus p divides $a^p - a = a(a^{p-1} - 1)$ and the theorem follows (since p does not divide a).

More famous still is Fermat's 'Last Theorem'. This was proved in 1994 by Andrew Wiles.

What is Fermat's 'Last Theorem'? In reading Bachet's translation (from Greek into Latin) of the work of Diophantus, Fermat came across the equation $x^2 + y^2 = z^2$ with its solutions (e.g., $x = 3, y = 4$ and $z = 5$). Fermat wrote in the margin of the book that he had been able to prove that the equation

$$x^n + y^n = z^n$$

has no positive integer solutions for $n > 2$, but that the margin was too small for him to write the proof there. (Of course, $0^3 + 7^3 = 7^3$, but here one of the integers is 0; this is called a 'trivial' solution.)

It is quite possible that Fermat had a proof of the nonexistence of the positive integer solutions for $n = 3$. We still have his proof of the 'Last Theorem' for the special case $n = 4$. However, it is unlikely that he ever had a proof for the complete theorem.

Legendre disposed of the case when $n = 5$ in 1823, and Dirichlet handled the case when $n = 14$ in 1832. In 1849 Kummer made a big step forward and was able to vindicate Fermat's statement for all $n < 100$ except 37, 59 and 67. Before 1994, thanks to the help of the computer, we knew that Fermat was right for all $n < 10^8$ or so, but a proof of the general theorem still escaped us. In 1994, however, Wiles gave a complete proof, for all n.

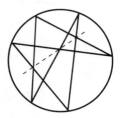

FIGURE 27.2. Pascal's Theorem

Over the centuries, research into Fermat's Last Theorem proved very fruitful. As one example, Kummer's research led to the discovery of 'ideals' and their unique factorization. As another example, Faltings's research led to advances in algebraic geometry — this as recently as 1983. In fact, Faltings came close to proving Fermat's Last Theorem, by showing that, for any $n > 2$, the equation $x^n + y^n = 1$ has at most a finite number of rational solutions.

Pascal was educated at home and forbidden to study mathematics. By the age of 12, he had rediscovered many of Euclid's theorems, so his father relented and gave him a copy of the *Elements*. At 14 Pascal began attending meetings of a group of mathematicians which included Mersenne. At 16 he wrote an essay on conic sections (with an account of the 'mystical hexagram') and, at 18, he constructed a calculating machine, one of the first computers.

At 27, Pascal abandoned mathematics to devote himself wholly to the philosophy of religion and the worship of God. At least one historian of mathematics (E.T. Bell) has taken this as a sure indication of insanity, but it is unlikely that a madman could have produced the elegant French prose or the brilliant philosophical analyses that we find in the *Pensées*, which Pascal wrote during this period. In his later life, Pascal returned to mathematics for a couple of brief periods. Pascal died at age 39. His last feat was the creation of a public transportation system, with the profits going to help the poor.

What did Pascal do as a mathematician? In 1639 (at age 16) he used some ideas of Desargues to obtain 'Pascal's Theorem' (with the 'mystic hexagram'):

Theorem 27.1. *If a hexagon is inscribed in a conic, the points of intersection of opposite sides, assumed to be nonparallel, lie in a straight line* (Figure 27.2).

Actually, this is a theorem in projective geometry, and it suffices to prove it for the case when the conic is a circle, since other conics can be obtained from the circle by projection. If one views a pair of straight lines as a degenerate conic, one obtains the Theorem of Pappus as a special case of the Theorem of Pascal. A simple proof for Pascal's Theorem in the case of the circle is found in Coxeter and Greitzer's *Geometry Revisited*.

In 1653, Pascal rediscovered what we call 'Pascal's triangle':

$$
\begin{array}{ccccccccc}
& & & & 1 & & & & \\
& & & 1 & & 1 & & & \\
& & 1 & & 2 & & 1 & & \\
& 1 & & 3 & & 3 & & 1 & \\
1 & & 4 & & 6 & & 4 & & 1 \\
& \cdot & & \cdot & & \cdot & & \cdot &
\end{array}
$$

As we noted, this goes back to the Chinese, but Pascal was the first to give clear and complete demonstrations of its basic properties (making the first explicit use of mathematical induction). He showed that the kth entry in the nth row is $\binom{n-1}{k-1}$, this being the number of ways of choosing $k-1$ out of $n-1$ things. (Today we prefer to call $\binom{n}{k}$ the number of k-element subsets of an n-element set.) Pascal showed that

$$\binom{n-1}{k-1} = \frac{(n-1)(n-2)...(n-k+1)}{(k-1)!}$$

and he showed that

$$\binom{n}{k} = \binom{n-1}{k} + \binom{n-1}{k-1}.$$

(This latter equation follows from the observation that, to choose k out of n things, we may set one of the n things aside and consider two cases: in the first case, the special thing is included in the choice, and hence there are $\binom{n-1}{k-1}$ possibilities; in the second case the special thing is excluded, leaving $\binom{n-1}{k}$ possibilities.) Finally, Pascal proved that $\binom{n}{k}$ is the coefficient of $x^k y^{n-k}$ in the binomial expansion of $(x+y)^n$.

With Fermat, Pascal developed the theory of probability, including the concept of mathematical expectation. Pascal uses probability in the *Pensées* as part of a proof that it is wiser to believe in God (Pascal, *Oeuvres Complètes*, p. 1212). This argument, called 'Pascal's wager', goes as follows: Even if the probability that a god exists is very small (but positive), if he rewards those who believe in him with eternal happiness, assumed to be of infinite value, a rational human being ought to believe in this god. (Did Pascal consider the possibility that God might punish those who adopt their beliefs in expectation of personal gain?)

27. The Seventeenth Century in France

Exercises

1. Assuming the usual theory of similar triangles (which Euclid based on the Axiom of Archimedes), prove the parallel case of Desargues's Theorem, and its converse. (The converse is: if $AB, A'B'$ and $AC, A'C'$ and $BC, B'C'$ are three pairs of parallels then AA', BB' and CC' are concurrent.)

2. Descartes showed that, roughly speaking, a curve is a conic if and only if its equation has the form
$$Ax^2 + Bxy + Cy^2 + Dx + Ey + F = 0.$$
Give examples of equations of this form which represent the following curves:
   (a) a circle,
   (b) two parallel straight lines,
   (c) a point,
   (d) the whole plane.

3. Why cannot an integer of the form $4n - 1$ be the sum of two integer squares?

4. What is the smallest positive integer which quadruples when its final (base ten) digit is shifted to the front? (For example, it is not 125 since 512 is not 4 times 125.)

5. Prove that, if $x^p + y^p = z^p$ has one solution in positive integers, then it has infinitely many such solutions.

6. Show that, if $x^p + y^p = z^p$, with p an odd prime, then p is a factor of $x + y - z$.

7. Give a proof of the Binomial Theorem.

8. Find two distinct proofs to show that the sum of the entries in any row in Pascal's triangle is a power of 2.

9. Show that the alternating (adding and subtracting) sum of the entries in any row of Pascal's triangle is 0.

10. Using mathematical induction, show that the $(n+1)$th Fibonacci number is
$$\binom{n}{0} + \binom{n-1}{1} + \binom{n-2}{2} + \cdots.$$

11. Project: Prove Pascal's mystic hexagram theorem for the case of the circle. Note that some of the points of intersection may be outside the circle. (Hint: use the Theorem of Menelaus.)

28
The Seventeenth Century Continued

Mathematics in the 17th century was by no means confined to France. Elsewhere in Europe, some of the many great mathematicians were the following:

- Bonaventura Cavalieri (1598–1647) in Italy,
- John Wallis (1616–1703) in England,
- Nicolaus Mercator (1620–1687) in Germany and England,
- Christian Huygens (1629–1695) in Holland,
- Isaac Barrow (1630–1677) in England,
- James Gregory (1638–1675) in Scotland,
- Isaac Newton (1642–1727) in England,
- Gottfried Wilhelm Leibniz (1646–1716) in Germany.

Cavalieri rediscovered Archimedes's *method of indivisibles* for calculating areas and volumes. As an example of Cavalieri's approach, let us consider the following derivation of the area of the ellipse $y = (b/a)(a^2 - x^2)^{\frac{1}{2}}$. First we plot the circle $y = (a^2 - x^2)^{\frac{1}{2}}$ on the same rectangular coordinate frame of reference. Second, we note that a vertical chord of the ellipse is just b/a of the corresponding vertical chord of the circle (i.e. the chord lying in the same straight line). Third, we think of the areas of the ellipse and the circle as somehow being the 'sums' of their chords. The ratio of ellipse chord to corresponding circle chord is always b/a, so it follows that the ratio of the area of the ellipse to the area of the circle is also b/a. Hence the area of the ellipse is $(b/a)\pi a^2 = \pi ab$.

Cavalieri's most important result is perhaps the theorem which, in our notation, reads as follows:

$$\int_0^a x^n dx = \frac{a^{n+1}}{n+1}.$$

Wallis was a professor of geometry at Oxford. He was a Royalist, and ended up as chaplain to Charles II. He also invented a method for teaching deaf mutes.

In algebra, he used negative and fractional exponents: $x^{-n} = 1/x^n$ and $x^{p/q} = \sqrt[q]{x^p}$. Using Cavalieri's methods, he calculated the area under the curve

$$y = a_0 x^0 + a_1 x^1 + \cdots + a_n x^n.$$

He also discovered, but could not give a rigorous proof for, the following expression of π as an infinite product:

$$2 \cdot \frac{2}{1} \cdot \frac{2}{3} \cdot \frac{4}{3} \cdot \frac{4}{5} \cdot \frac{6}{5} \cdot \frac{6}{7} \cdot \frac{8}{7} \cdot \frac{8}{9} \cdots.$$

Mercator calculated the area under the curve

$$y = \frac{1}{1+x} = 1 - x + x^2 - x^3 + \cdots,$$

obtaining

$$\log_e(1+x) = x - x^2/2 + x^3/3 - x^4/4 + \cdots$$

(converging for $-1 < x < 1$). (It was Gregory of St.Vincent (1584–1667) who first showed that the integral of $1/x$ is the natural log of x.) The cartographer's Mercator projection is due, not to Nicolaus Mercator, but to Gerhardus Mercator (1512–1592).

Like Pascal, Huygens made a study of the cycloid. This is the curve described by a point fixed to the rim of a wheel as it rolls along a flat surface. Huygens showed that the 'evolute' of a cycloid is again a cycloid. He also discovered (Galileo notwithstanding) that a pendulum had to swing in a cycloidal arc, and not in a circular arc, if the period of oscillation was to be strictly independent of the amplitude of the swing.

Huygen's most profound contribution to physics was the wave theory of light. Assuming that light is propagated in waves in an oscillating 'ether', he was able to explain the laws of reflection and refraction. His theory was in one respect better than Newton's corpuscular (particle) theory, which failed to explain interference phenomena. Nonetheless, Huygen's theory was

28. The Seventeenth Century Continued   155

eclipsed by Newton's for many years. It is only in modern quantum mechanics that we acknowledge that light may be viewed as waves as well as particles.

Often a promising young mathematician is denied a job in mathematics because all the places are filled by older professors who have not done a thing since the day they were given tenure. These old professors might learn a lesson from Isaac Barrow. In 1669, he resigned his professorship of geometry at Cambridge, so that the young Newton might take his place. Thereafter, Barrow devoted himself to divinity.

Gregory used the method of Wallis to obtain 'Gregory's series':
$$x = \tan x - \tfrac{1}{3}\tan^3 x + \tfrac{1}{5}\tan^5 x - \cdots.$$
Putting $x = \pi/4$, this gives
$$\pi/4 = 1 - 1/3 + 1/5 - 1/7 + \cdots,$$
an interesting but slowly converging series for π. (This tan series had also been obtained by Indian mathematicians, and is found in a book called the *Tantrasangraha-vyakhya* (c. 1530).)

Newton never knew his father, who had died before his birth. Newton was supposed to learn farming, but spent his time doing experiments and building mechanical models. Finally, his uncle relented and sent him to Cambridge, where he read Euclid, Descartes, Kepler and Wallis, and attended the lectures of Isaac Barrow.

By the time he got his B.A., Newton had discovered derivatives, which he called 'fluxions', and had established the Binomial Theorem for integer and fractional exponents (without, however, giving a rigorous proof).

Newton used the Binomial Theorem to expand certain functions $f(x)$ into power series. When he wanted to find the area under the curve $y = f(x)$ he could then apply Wallis's method, replacing x^n by $x^{n+1}/(n+1)$. At this time, there was little knowledge about the conditions under which one can treat an infinite sum in the same way as a finite sum. Indeed, there is no evidence that Newton worried about the convergence of the series in his generalized Binomial Theorem.

During the plague of 1665–66, Newton withdrew to the family farm and thought about gravitation. Back in Cambridge, he first helped Barrow with some lecture notes, and then took over his professorship in 1669. Around this time he discovered the decomposition of white light by a prism. From 1673 to 1683 he lectured on algebra and the theory of equations.

Newton's greatest contribution to knowledge was his theory of universal gravitation, which once and for all provided a rational explanation for the apparently erratic motions of the heavenly bodies. The fundamental idea was simple: the same force which causes an apple to fall must act on the moon and the planets. (Note that Newton had to reject the Aristotelian idea that an apple falls simply because it is the nature of 'earthy' things to go downwards.)

Similar ideas had previously occurred to Hooke, Huygens, Halley and Wren. These thinkers realized that Kepler's laws implied that any 'force of gravity' would have to obey an inverse square law, such as that given by Newton. What Newton did was to show that, conversely, an inverse square law implies Kepler's laws.

The *law of gravitation* asserts that, given two bodies with masses M and m, at a distance x apart, the force between them is given by

$$F = kMmx^{-2},$$

where k is a universal constant. If M is large compared with m, we may think of this force as being exerted by M on m. Since Newton had defined *force* to mean rate of change of momentum, in this case $F = -m\ddot{x}$, it follows that the acceleration $\ddot{x} = -kMx^{-2}$ of the smaller body does not depend on its mass m. This result is still valid today, even if Newton's law has to be slightly modified to conform to the general theory of relativity. If M is the mass of the earth, assumed to be concentrated at its center, and m is the mass of the apple on or near the surface, $\ddot{x}$ is practically constant, confirming Galileo's original observation.

At first Newton worked out the planetary motions from the assumption that the sun and the planets were points. However, he was not happy about this assumption, and so he did not publish his results immediately. It was only in 1685, about twenty years later, that he was able to prove that the gravitational force due to a solid sphere is the same as if the entire mass were concentrated in the center. He assumed that the density of matter at a point inside the sphere depends only on its distance from the center, this presumably being the case with all the planets in our solar system.

Having overcome this last difficulty, Newton finally published his epoch-making *Principia* in 1686. To avoid all controversy about his methods, he replaced his original arguments involving the infinitesimal calculus by classical geometrical arguments in the style of Euclid, which his contemporaries were able to understand. Unfortunately, for this very reason, today the *Principia* is difficult to read.

The second volume of the *Principia* dealt with hydrostatics and hydrodynamics. It showed that Descartes's theory of vortices did not work. Newton's theories were soon accepted everywhere. Even in France, Voltaire advocated Newton against Descartes (in 1733).

It was only in 1692 that Newton published two letters on 'fluxions', as he called derivatives. He wrote $\dot{x}, \ddot{x}$ for the first and second derivatives with respect to a parameter t (for time). He wrote o for dt and $\dot{x}o$ for dx.

28. The Seventeenth Century Continued

In 1704, in an appendix to a book on optics, he gave a study of 'fluents', his name for indefinite integrals or antiderivatives. He showed the connection between fluents and 'quadratures', that is, definite integrals. He also treated maxima and minima, tangents to curves and lengths of curves.

In 1696, Newton abandoned his Cambridge professorship for a government position in London and, three years later, he was Master of the Mint.

We defer the discussion of Leibniz to the next chapter, ending this chapter with a quotation from Newton:

> I do not know what I may appear to the world; but to myself I seem to have been only like a boy, playing on the sea-shore, and diverting myself, in now and then finding a smoother pebble, or a prettier shell than ordinary, whilst the great ocean of truth lay all undiscovered before me.

Exercises

1. Show that if the three Laws of Indices are to hold for negative and fractional exponents, Wallis's way of using such exponents is the only way. (The three Laws of Indices are $x^{m+n} = x^m x^n$, $(x^m)^n = x^{mn}$, $(xy)^m = x^m y^m$.)

2. Show that, when m is odd,
$$\int_0^{\pi/2} \sin^m x \, dx = \frac{(m-1)(m-3)\cdots 2}{m(m-2)\cdots 3 \cdot 1}.$$
Derive a similar formula for the case in which m is even. Finally, use the above two formulas to give a derivation of Wallis's product formula for π.

3. Suppose the wheel in the definition of the cycloid is rolling along the positive x axis, with the fixed point starting at the origin. What is the equation of the resulting cycloid if the radius of the wheel is 1?

4. How many terms of Gregory's series do you have to add up to get π accurate to two decimal places?

5. According to Newton's calculations, your weight on a planet of uniform density is determined only by the matter which is closer to the center than you are. Suppose that you dig down towards the center of such a planet (assumed spherical). Draw a graph showing how your weight varies as you descend towards the center. (Hint: at the center, your weight is 0.)

6. You are on a planet consisting of a spherical inner core of density 5 and radius R. This inner core is in the middle of a spherical outer core of radius $5R$ and density 1. Show that, as you dig down towards the center of this planet, your weight at first decreases, then increases, and, finally, decreases. Where does your weight start to increase?

29
Leibniz

Gottfried Wilhelm Leibniz (1646–1716) was born in Leipzig. His father died when he was six. Leibniz educated himself, using his late father's library, and entered the university in Leipzig when he was only fifteen. In 1666 he was refused the degree of Doctor of Law on the grounds that he was too young. In the same year, Leibniz conceived the idea of symbolic logic, a universal language in which all rational thinking could be expressed.

Leibniz worked as a diplomat for the Elector of Mainz. In this capacity, he went to Paris, where Louis XIV rejected his idea of attacking Egypt instead of another European country. Here Leibniz met Huygens, who introduced him to geometry and physics.

Huygens challenged Leibniz to sum the series

$$\sum_{n=1}^{\infty} \frac{2}{n(n+1)} = 1 + 1/3 + 1/6 + \cdots.$$

Leibniz solved the problem thus: $2/n(n+1) = 2(1/n - 1/(n+1))$, so the series equals $2(1 - 1/2 + 1/2 - 1/3 + 1/3 - 1/4 + 1/4 - \cdots) = 2(1 + 0) = 2$. In the 20th century, we would object to this on the grounds that Leibniz might equally well have written

$$2(2 - 3/2 + 3/2 - 4/3 + 4/3 - 5/4 + 5/4 - \cdots) = 2(2 + 0) = 4.$$

We now prefer to solve this problem by first showing that the mth partial sum of the series is $2(1 - 1/(m+1))$ and then taking the limit as $m \to \infty$.

In 1673, Leibniz visited England and became a fellow of the Royal Society. If he got his ideas about the Calculus from Newton, he never acknowledged this. Two years later, he developed his own version of the Calculus, introducing the notation we still use today. In particular, he obtained the rule for the derivative of a product that is named after him.

Leibniz's work on the Calculus appeared in 1684, before Newton had got around to publishing his results. Not surprisingly, Leibniz was accused of plagiarism by some British mathematicians, not without Newton's acquiescence, and a bitter priority battle ensued. Today, we ascribe the invention of the Calculus to both Newton and Leibniz.

Leibniz had been working as librarian for the Duke of Hannover. When the latter became king of England as George I, he left two people behind: his wife, whom he divorced and shut up in a cloister, and Leibniz, because he did not want to antagonize the British academic establishment.

Leibniz thought of dy and dx in dy/dx as 'infinitesimals'. Thus dx was an infinitely small increment in x which was yet different from 0, and dy, defined as

$$dy = f(x + dx) - f(x)$$

for a given function $y = f(x)$, was also different from 0 (unless f happened to be constant near x). For example, if $f(x) = x^2$, then $dy = (x+dx)^2 - x^2 = 2x(dx) + (dx)^2$. This represented the 'rise' of the function corresponding to the 'run' dx. Hence, the slope of the tangent was rise/run = $dy/dx = 2x + dx$, so that, at x, the tangent has slope $2x$.

The concept of the infinitesimal — also implicit in Newton's fluxions — was criticized by many, including the philosopher and bishop George Berkeley (1685–1753). How, he asked, can we divide by dx if it is 0? How can we get the slope of the tangent right, in our example $2x$, if it is not 0?

Karl Weierstrass (1815–1897) agreed that there were problems, and he responded by putting calculus on the firm footing it has today. At the moment, dy/dx is not seen as a quotient but as a limit of quotients:

$$\frac{dy}{dx} = \lim_{h \to \infty} \frac{f(x+h) - f(x)}{h}.$$

If we insist on the separate existence of dx and dy, we may put $dx = h$ and $dy = \frac{dy}{dx} h$. Weierstrass also gave a perfectly rigorous definition of a limit (the *epsilon-delta* definition) which does not depend on vague notions like 'small', 'approaches', etc.

There is also at the moment a rigorous version of the infinitesimal itself. In 1966 Abraham Robinson introduced a *nonstandard model* for real numbers in which there is an entity ξ such that

$$0 < \xi < 1, \ \ 0 < \xi < 1/2, \ \ 0 < \xi < 1/3, \ \ \ldots . \qquad (*)$$

Can this be done consistently? Yes, and the reason, roughly speaking, is as follows. We know from mathematical logic that, if a contradiction were deducible from (∗), then this contradiction would be deducible in a finite

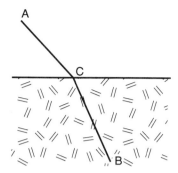

FIGURE 29.1. Refraction of light

number of steps, hence from a finite subset of (∗). Since this is not possible (assuming that the ordinary interpretation of the rational numbers is consistent), there is no contradiction.

Leibniz's infinitesimals are simple and intuitive but not rigorous; Robinson's infinitesimals are rigorous but neither simple nor intuitive. Anyone studying nonstandard analysis will discover a complex and bizarre system. On the other hand, it should be stated that nonstandard analysis has helped some mathematicians obtain new results in ordinary analysis.

Leibniz also made contributions to philosophy. He is famous for his view that this is the best of all possible universes (Could God have failed to create the best?). This view was ridiculed by Voltaire in his novel *Candide*. However, it is unlikely that Leibniz intended 'best' to mean 'happiest'. What he had in mind may have been something like the discovery of Willebrord Snell, who was able to explain the refraction of light by assuming that a ray of light minimizes the time in going from point A in one medium to point B in another. (Think of a person trying to walk from point A on a paved surface to a point B in a rough field. Since he finds it harder to walk in the field, he will not go straight from A to B, but in a broken line ACB, as shown in Figure 29.1.) Leibniz makes explicit reference to Snell in Section XXII of the *Discourse on Metaphysics*, where Leibniz discusses the 'easiest way' in which a ray of light might travel. Thus it is quite possible that by 'best' universe, Leibniz meant, among other things, 'easiest', or 'most energy-efficient' universe.

The behaviour of light rays is only one instance of the *Principle of Least Action*, which was put forward by Pierre Maupertuis (1698–1759) in 1751. As now formulated, this principle asserts that any physical process happens in such a way as to minimize the action $\int_a^b E\,dt$, where E is the difference between kinetic and potential energy and t is the time. This principle was formally established by Lagrange.

Some utilitarian philosophers, such as J.S. Mill, have claimed that even human psychology is, or ought to be, determined by something like the

Principle of Least Action. In this they followed Plato, who had Socrates argue in *The Protagoras* that the way we ought to behave is to maximize pleasure minus pain, that virtue consists in knowing when to forego a present gain in favour of a larger future gain (temperance) and when to face the danger of an immediate loss against the expectation of ultimate gain (courage). However, Socrates seems to have had second thoughts about this later. When Crito urged him to flee from his prison so as to escape the death penalty and be able to bring up his sons, Socrates refused, believing it was his moral duty to remain in jail. He replied thus:

> Whatever the popular view is, and whether the alternative is pleasanter than the present one or even harder to bear, the fact remains that to do wrong is in every sense bad and dishonourable (Plato's *Crito* 49b; see also *Laws* 707d).

Leibniz's most original contribution to philosophy is probably the system incorporated in his *Monadology*. In this work he proposes that the universe is made up of certain ultimate elements, called *monads*, which are capable of perception. Human souls are monads with memory and reason.

If Leibniz was influenced by his mathematical background in formulating this model of the universe, then he might have thought of a monad as a point together with the set of all points infinitely close to it. In fact, Robinson attributed this technical meaning to the word 'monad' in his nonstandard analysis.

On the other hand, Bertrand Russell thought of the monads as points, with arrows connecting different points (see his *Mysticism and Logic*). Each monad has a physical aspect, consisting of all arrows emerging from it, and a mental aspect, consisting of all arrows converging on it.

Leibniz asserted that a monad has no windows. (Its contact with the rest of the universe is via a 'pre-established harmony'.) This suggests that anything we know about a monad must be deducible from the 'arrows' relating it to all the other monads in the universe.

One is reminded here of the 20th century notion of a category (see Part II, Chapter 31). For instance, in the category of groups, we have as points all groups, and as arrows all group homomorphisms. If we want to know the elements of a group G, we need not look 'inside' the group at all, but only at the arrows from the free group in one generator to G.

Exercise

1. Suppose the half-plane $y > 0$ is a rough field where you would walk at u km/h, and the half-plane $y < 0$ is a paved surface where you would walk at v km/h. You are at point $(0, -6)$ and want to walk to the point (a, b). How should you go to get there fastest?

30
The Eighteenth Century

We shall only mention some of the most important mathematicians of the eighteenth century:

- Brook Taylor (1685–1731),
- Colin Maclaurin (1689–1746),
- Abraham de Moivre (1667–1754),
- Leonhard Euler (1707–1783),
- Joseph Louis Lagrange (1736–1813),
- Pierre Simon Laplace (1749–1827),
- Adrien Marie Legendre (1752–1833).

Brook Taylor, an ardent admirer of Newton, discovered the *Taylor series*

$$f(a+x) = f(a) + xf'(a) + x^2 f''(a)/2! + ...,$$

publishing it in 1715.

Colin Maclaurin, a Scotsman, is best known for the special case $a = 0$ of Taylor's series. This appeared in his *Treatise of Fluxions* (1742). In his book, Maclaurin tried to be sufficiently rigorous to answer Berkeley's objections to the Calculus, but he did not even get to the point of demonstrating conditions under which his *Maclaurin series* converges.

Abraham de Moivre was born in France, but lived in England. He is famous for his formula

$$(\cos x + i \sin x)^n = \cos nx + i \sin nx ,$$

which is easily proved, for natural numbers n, by mathematical induction. De Moivre published an important book on the theory of probability, called the *Doctrine of Chances*.

Society failed de Moivre: in spite of letters of recommendation from both Newton and Leibniz, he was never given a proper job in mathematics. He had to earn a meagre living by private tutoring and answering gamblers' questions on probability. It is said that, as he approached the end of his life, de Moivre slept fifteen minutes longer each day. When he reached a full twenty four hours, he died.

Although Leonhard Euler was Swiss, he spent part of his professional life in Berlin and most of it in St. Petersburg. Towards the end of his life he became blind, but this did not slow down his mathematical output. He found many interesting and exciting results in mathematics. Indeed, it has been said that Euler picked all the raisins out of the mathematical cake. Some of his results are the following:

1. If a convex polyhedron has V vertices, F faces and E edges, then $V + F - E = 2$. For example, a cube has 8 vertices, 6 faces and 12 edges; we have $8 + 6 - 12 = 2$. (Descartes came close to this formula, but he did not actually state it.)

2. $e^{i\pi} = -1$, where e is the 'Euler number':

$$e = \lim_{n \to \infty} (1 + 1/n)^n.$$

3. $1/1^2 + 1/2^2 + 1/3^2 + 1/4^2 + \cdots = \pi/6$.

   Euler's proof of this was not rigorous but, before Euler, no one even guessed that the sum of the series was $\pi/6$.

4. Every even perfect number has the form $2^{n-1}(2^n - 1)$ where $2^n - 1$ is prime.

5. If n is a positive integer, let $\phi(n)$ be the number of natural numbers less than or equal to n and relatively prime to it. Then, if a is a positive integer relatively prime to n, it follows that n is a factor of $a^{\phi(n)} - 1$. Fermat's Little Theorem is a corollary of this.

6. The circumcenter, orthocenter and centroid of a triangle are collinear. The line that passes through them is called the *Euler line*.

(The *circumcenter, orthocenter* and *centroid* of a triangle are the meeting points of the right bisectors, altitudes and medians, respectively.)

7. Fermat was wrong when he conjectured that all natural numbers of the form $2^{2^n} + 1$ are primes.

Euler recognized the importance of convergence in dealing with infinite series, but he did not always pay attention to it. For example, he would write
$$1/(x-1) = 1/x + 1/x^2 + 1/x^3 + \cdots$$
(which is correct when $|x| > 1$) and put $x = 1/2$ to obtain $-2 = 2 + 4 + 8 + \cdots$. He also showed a lack of rigour in employing his principle of 'conservation of form', according to which a theorem true for natural number exponents also holds for any real exponent. In this way, he obtained facile 'proofs' of the generalized Binomial Theorem, and the generalized de Moivre's Theorem.

In addition to his numerous discoveries in pure mathematics, by no means all of which have been discussed here, Euler also made important contributions to mechanics. He elaborated the Principle of Least Action. Finally, he worked out a theory of lunar motion. His collected works run to about 75 volumes.

The following proof of Euler's formula $V + F - E = 2$ was suggested by H. S. M. Coxeter.

Let O be a point in the interior of the convex polyhedron. About O as center describe a sphere which contains the polyhedron. Now imagine a source of light placed at O. Rays emanating from O will project the polyhedron onto the surface of the sphere, mapping each flat polygon onto a spherical polygon whose sides are arcs of great circles. (This idea is said to be due to the Arabic mathematician Abu'l Wafa.) Choose a point in the interior of each spherical polygon and join it to the vertices by arcs of great circles, thus dividing each spherical polygon into as many spherical triangles as it has sides. Then

$720°$ = area of sphere
= sum of areas of spherical triangles
= sum of angles of spherical triangles $- 180° \times$ number of triangles
= sum of angles at interior points +
  sum of angles at vertices $- 180° \times 2E$
= $360° \times F + 360° \times V - 360° \times E$.

Dividing by $360°$, we obtain Euler's formula.

30. The Eighteenth Century

Joseph Lagrange was born in Italy of mixed French and Italian parentage. His father lost the family fortune through speculation, but Lagrange later commented that, if it had not been for this, he might never have turned to mathematics. He was converted to mathematics through an essay by Halley.

At age 23, Lagrange was able to explain, on the basis of Newton's theory of gravitation, why the moon always shows the same face to the earth.

Having acquired an early fame, Lagrange spent 25 years in Prussia at the invitation of Frederick II. After Frederick's death, Lagrange moved to Paris, where he became a favourite of Marie Antoinette. He had mixed feelings about the Revolution, especially when his friend, the chemist Lavoisier, was guillotined, but he stuck it out. He was involved in the introduction of the decimal system for weights and measures. When people pleaded the advantages of the base 12, he would ironically defend the base 11. He became professor of mathematics at the Ecole Polytechnique.

Lagrange was a universal mathematical genius, his interests ranging from number theory to physics. Among his achievements are the following:

1. The first proof of Wilson's Theorem that, if p is a prime number, then it is a factor of $(p-1)! + 1$;

2. The first complete solution of the Diophantine equation $x^2 - Ry^2 = 1$, where R is a given nonsquare positive integer; Lagrange generalized this to give a complete treatment of Diophantine equations of the form
$$Ax^2 + Bxy + Cy^2 + Dx + Ey + F = 0,$$
where $A, B, C, D, E,$ and F are given integers;

3. The first proof that every natural number is a sum of four squares of natural numbers (e.g., $7 = 2^2 + 1^2 + 1^2 + 1^2$ and $9 = 3^2 + 0^2 + 0^2 + 0^2$);

4. A systematic theory of differential equations;

5. The *Mécanique*, which he conceived at the age of 19 but only published at 52, in which he expressed the dynamics of a rigid system by the equations
$$\frac{d}{dt}\frac{\partial T}{\partial \dot{\theta}} - \frac{\partial T}{\partial \theta} + \frac{\partial V}{\partial \theta} = 0,$$
where T is the total kinetic energy, V is the potential energy, t is the time, θ is any coordinate, and $\dot{\theta} = d\theta/dt$; Lagrange observed that his equations expressed the fact that the total *action* $\int_a^b (T - V)dt$ was minimal; to justify this observation, he had to invent the calculus of variations.

When Lagrange wrote to Euler about his results in the calculus of variations, Euler was so impressed that he withheld his own results from publication so that Lagrange could publish first. Sad to say, such unselfish acts are rare.

After his wife died in 1783, Lagrange wore himself out publishing the *Mécanique*. The excesses of the Revolution upset him and he became subject to fits of depression. From these the lonely genius was rescued by the love of a teenaged girl, Renée Le Monnier, who insisted on marrying him in 1792. For the remaining twenty years of his life, Lagrange was both happy and mathematically productive.

Laplace was the son of humble parents but ended up as a marquis under the restored Bourbons. Politically, he was an opportunist, but occasionally he stood up for his principles. Napoleon once told him, 'you have written a big book on the universe without mentioning its creator', to which Laplace replied: 'I don't need that hypothesis'.

Laplace was more of a mathematical physicist than a pure mathematician. He introduced the potential V and showed that it satisfied

$$\frac{\partial^2 V}{\partial x^2} + \frac{\partial^2 V}{\partial y^2} + \frac{\partial^2 V}{\partial z^2} = 0.$$

His greatest contribution to mathematics was the useful phrase 'it is easy to see', which peppers his *Mécanique Céleste*. In *The History of Mathematics*, David Burton reports:

> The American astronomer Nathaniel Bowditch (1773-1838), who translated four of the five volumes into English, observed, "I never came across one of Laplace's 'Thus it plainly appears' without feeling sure that I had hours of hard work before me to fill up the chasm and find out and show how it plainly appears."

Legendre was a great promoter of Euclid. He showed that the Parallel Postulate follows from the assumption that the plane contains real squares (i.e., quadrilaterals with four equal sides, each of whose angles is a right angle.) He also did work on the method of least squares.

Legendre is best known for his work in number theory. He was the first to prove that the Diophantine equation

$$x^5 + y^5 = z^5$$

has no nonzero integer solutions. He introduced the Legendre symbol $\left(\frac{n}{p}\right)$, where p is a prime and n an integer not divisible by p. He wrote $\left(\frac{n}{p}\right) = 1$ when n has the form $kp + r^2$ (with k and r integers) and $\left(\frac{n}{p}\right) = -1$ when n does not have this form.

30. The Eighteenth Century

Euler had conjectured a theorem, called the Law of Quadratic Reciprocity. Using the the Legendre symbol, it can be expressed as follows: if p and q are distinct odd primes, $\left(\frac{p}{q}\right)\left(\frac{q}{p}\right) = (-1)^{(p-1)(q-1)/4}$. Euler could not prove this, but Legendre found proofs for some special cases. It was Gauss who published the first complete proof, in 1801. Gauss later gave five other proofs of the same result.

The Legendre symbol $\left(\frac{n}{p}\right)$ can be calculated quite easily, in view of the following observation due to Euler:

$$n^{\frac{p-1}{2}} = \left(\frac{n}{p}\right) + \text{a multiple of } p.$$

Indeed, it follows from Fermat's Little Theorem that

$$(n^{\frac{p-1}{2}} - 1)(n^{\frac{p-1}{2}} + 1) = n^{p-1} - 1$$

is a multiple of p, so p divides either $n^{\frac{p-1}{2}} - 1$ or $n^{\frac{p-1}{2}} + 1$ (but not both, since it does not divide their difference). We claim that p divides the former if and only if $\left(\frac{n}{p}\right) = 1$, hence p divides the latter if and only if $\left(\frac{n}{p}\right) = -1$, from which facts the observation follows.

To see this, at least in one direction, suppose $\left(\frac{n}{p}\right) = 1$, that is, $n = r^2 + kp$ for some integers r and k. Then

$$n^{\frac{p-1}{2}} = (r^2 + kp)^{\frac{p-1}{2}} = r^{p-1} + k'p = 1 + k''p$$

for some integers k' and k'', hence p divides $n^{\frac{p-1}{2}} - 1$. The converse implication, though not difficult, is a little tricky, and we shall omit its proof.

Exercises

1. Give conditions sufficient for the convergence of the Maclaurin series.

2. Prove de Moivre's formula for positive integers n.

3. Show that e, as defined above, is bound below by 2 and above by 3.

4. Prove that the circumcenter, orthocenter and centroid of any triangle are collinear.

5. Give an example of a formula with exponents which is true when the exponents are natural numbers but not always true when the exponents are rationals.

6. Prove Wilson's Theorem.

7. Check the Law of Quadratic Reciprocity for $p = 5$ and $q = 13$.

31
The Law of Quadratic Reciprocity

In this chapter we single out a great 19th century mathematician and present one of his most elegant proofs. The reader should be warned, however, that the 19th century was so rich in mathematics that it really deserves a book of its own.

Carl Friedrich Gauss (1777–1855) was inspired to become a mathematician by his discovery of a ruler and compass construction for the regular polygon with 17 sides — this when he was only a teenager — but his gift had revealed itself much earlier: as a three year old, he had pointed out an error in his father's payroll accounts!

Gauss's first major contribution was his Number Theory book *Disquisitiones Arithmeticae*, which appeared in 1801. As well as the first construction for the regular 17-gon, it included the first proof of the Fundamental Theorem of Arithmetic (that every integer can be uniquely written as a product of primes), the first proof that every prime p has a primitive root g (meaning that no two of the numbers $1, g, g^2, \ldots, g^{p-2}$ differ by a multiple of p), the first proof that every natural number is a sum of three triangular numbers, and the first proof of the theorem featured below, namely, the Law of Quadratic Reciprocity.

Gauss was also an astronomer and a physicist. In 1807, subsequent to his calculation of the position of the asteroid Ceres, he was appointed director of the observatory at Göttingen and, in 1809, he published a book on planetary astronomy. In physics, Gauss did pioneering work in electromagnetism; the *gauss* is a unit of measure denoting magnetic intensity.

We conclude this chapter with a short proof of the quadratic reciprocity law, adapted from one of the proofs given by Gauss.

31. The Law of Quadratic Reciprocity

Let p and q be distinct odd primes and write $f(x,y) = py - qx$. Consider the region of the Cartesian plane consisting of all pairs of real numbers (x,y) such that
$$1/2 < x < p/2, \quad 1/2 < y < q/2 \,.$$
This region is subdivided into four mutually disjoint parts
$$A: f(x,y) < -q/2, \quad B: -q/2 < f(x,y) < 0 \,,$$
$$C: 0 < f(x,y) < p/2, \quad D: p/2 < f(x,y) \,.$$
Replacing x by $\frac{p+1}{2} - x$ and y by $\frac{q+1}{2} - y$, we establish a one-to-one correspondence between A and D. For any subset S of the Cartesian plane, let $L(S)$ denote the number of pairs of integers (x,y) in S; then clearly $L(A) = L(D)$. Now
$$L(A) + L(B) + L(C) + L(D) = \frac{p-1}{2} \cdot \frac{q-1}{2} \,,$$
so that
$$L(B) + L(C) = \frac{p-1}{2} \cdot \frac{q-1}{2} - 2L(A) \,.$$
Thus, the law of quadratic reciprocity can be written
$$\left(\frac{p}{q}\right)\left(\frac{q}{p}\right) = (-1)^{L(B)+L(C)} \,,$$
and this will follow if we show that
$$\left(\frac{p}{q}\right) = (-1)^{L(B)}, \quad \left(\frac{q}{p}\right) = (-1)^{L(C)}.$$
For reasons of symmetry, it will suffice to prove one of these equations, say the latter. In view of how the Legendre symbol is calculated, this may be stated in the following form, known as Gauss's Lemma:
$$q^{\frac{p-1}{2}} = (-1)^{L(C)} + \text{a multiple of } p.$$

To prove Gauss's Lemma, consider all pairs of integers (x,y) in the following region:
$$E: 1/2 < x < p/2, \quad 1/2 < y < q/2, \quad -p/2 < f(x,y) < p/2.$$
In E,
$$y - 1/2 < \frac{qx}{p} < y + 1/2,$$
so that y is the integer *closest* to $\frac{qx}{p}$; we shall write $y = g(x)$ to indicate that y is completely determined by x. Clearly, E contains C; in fact, $(x,y) \in C$ if and only if $(x,y) \in E$ and $f(x,y) > 0$.

31. The Law of Quadratic Reciprocity

As x ranges from 1 to $\frac{p-1}{2}$, and therefore y from 1 to $\frac{q-1}{2}$, the absolute value of $f(x,y) = f(x,g(x))$ takes exactly $\frac{p-1}{2}$ values. For, if $|f(x,y)| = |f(x',y')|$, where x' and y' range as x and y, then

$$q(x \pm x') = p(y \pm y'),$$

hence $x \pm x' = kp$ for some integer k, so that $x = x'$ and therefore $y = y'$. Since, in E, $|f(x,y)| < p/2$, these $\frac{p-1}{2}$ values are $1, \ldots, \frac{p-1}{2}$ and hence

$$\prod_{x=1}^{\frac{p-1}{2}} |f(x,y)| = \left(\frac{p-1}{2}\right)!$$

Finally, we calculate

$$q^{\frac{p-1}{2}} \left(\frac{p-1}{2}\right)! = \prod_{x=1}^{\frac{p-1}{2}} qx = \prod_{x=1}^{\frac{p-1}{2}} (py - f(x,g(x))).$$

This differs by a multiple of p from

$$\prod_{x=1}^{\frac{p-1}{2}} (-f(x,g(x))) = (-1)^{L(C)} \prod_{x=1}^{\frac{p-1}{2}} |f(x,g(x))| = (-1)^{L(C)} \left(\frac{p-1}{2}\right)!,$$

since $-f(x,g(x))$ will be negative if and only if $(x,g(x)) \in C$. Since p does not divide $\left(\frac{p-1}{2}\right)!$, it follows that $q^{\frac{p-1}{2}}$ differs from $(-1)^{L(C)}$ by a multiple of p. This completes the proof of Gauss's Lemma and therefore the proof of the Law of Quadratic Reciprocity.

Part II
Topics in the Foundations of Mathematics

1
The Number System

Much of 19th and 20th century mathematics is not accessible or meaningful at the undergraduate level. Still, we plan to examine some important and up to date material, including some exciting recent discoveries.

The following letters are used to represent sets of numbers:

N the natural numbers 0, 1, 2, ...;

Z the integers (Z being for 'Zahlen');

Q the rationals (Q being for 'quotients');

R the reals;

C the complex numbers $a + bi$;

H the quaternions (H being for Hamilton, the person who introduced them).

This now widely accepted notation was first proposed by N. Bourbaki, actually a slowly changing group of French mathematicians who have been engaged for half a century in writing up the *Elements of Mathematics* in a systematic and trend setting fashion. One of the founding members was the late Jean Dieudonné, to whom many of their early decisions may be attributed.

The number systems are arranged here in what mathematicians conceive to be the correct logical order, which differs from the historical one (see the Introduction). Zero was not originally considered to be a natural number and positive rationals and reals were studied before negative integers were considered.

Each of the sets after **N** 'extends' the preceding one and, with the exception of **H**, is motivated by our desire to solve equations which are otherwise unsolvable. Thus, **Z, Q, R**, and **C** are needed to solve the equations $x + 2 = 0$, $2x = 3$, $x^2 = 2$ and $x^2 + 1 = 0$, respectively.

Mathematicians often 'construct' a set on this list from the set just above it. For example, they build the rationals out of the integers by equating rationals with certain classes of pairs of integers: the fraction 2/3 is equated with $\{(2,3), (4,6), \ldots, (-2,-3), \ldots\}$. However, the list begins with the natural numbers. How were they constructed? According to Kronecker, 'God created the natural numbers, all the other numbers were made by man'. Still, as we shall see, some people have tried to construct the natural numbers also. This has led one wit to suggest that 'man created the natural numbers, all the others were Dieudonné'.

We should note that there are other kinds of numbers that do not appear on our list, such as the transfinite numbers and the infinitesimals.

What are the natural numbers? In particular, what is the number 2? We know it is not just the sign or numeral '2' (or 'II'). It is not something perishable or changing. But what is it? There is more than one answer.

The *Platonists* hold that numbers are abstract, necessarily existing objects. The number two is that Platonic 'form' or 'idea' in virtue of which things have the property of two-ness.

For the *logicists*, numbers are things which can be defined in terms of logic. For example, for Bertrand Russell, 2 is

$$\{x \mid \exists_y \exists_z (y \neq z \land x = \{y, z\})\}.$$

In other words, 2 is the set of all unordered pairs. On the other hand, von Neumann claimed that 2 is the set $\{0, 1\}$, where $1 = \{0\}$, and 0 is just the empty set. Pursuing yet another approach, Church held that 2 is the process of iteration

$$\lambda_f (f \circ f),$$

more precisely, the mapping which assigns to every function f its iterate $f \circ f$, defined by $(f \circ f)(x) = f(f(x))$.

For the *formalists*, 2 is just a class of expressions manipulated according to certain rules. Often they do not define numbers, but rather give axioms that characterize them. For example, Peano sees 2 as $SS0$, where '$SS0$' is a string of symbols which are manipulated according to certain axioms (see below).

The *intuitionists* hold that numbers are mental entities which would not exist unless people thought about them. For Brouwer, 2 is the concept which expresses the principle of 'two-ity'.

We see that each of these schools has a different view of the matter. What is the true answer? Fortunately, Professor Tournesol was able to discover this when attending a conference in France. The number 2 is a pair of platinum balls kept at room temperature in the second drawer at the Bureau of Standards in Paris.

2
Natural Numbers (Peano's Approach)

For Peano, the natural number system is a triple $(N, 0, S)$, where N is a set, 0 is an element of that set, and S is a function whose domain and codomain are that set, such that the following axioms hold for all elements x and y of **N**:

1. $Sx \neq 0$,
2. $Sx = Sy \Rightarrow x = y$,
3. when $\phi(x)$ is any proposition involving the natural number x, if
   (a) $\phi(0)$,
   (b) $\forall_{x \in N} \phi(x) \Rightarrow \phi(Sx)$,

   then $\phi(x)$ is true for any natural number x.

We define 1 as $S0$ and 2 as $SS0 = S1$, etc. (Normally, we leave out the parentheses.)

Note that (3) above is not a single axiom, but a whole 'scheme' or class of axioms, one for each ϕ. This scheme is called *mathematical induction*.

In Peano's system we define addition as follows:

$$x + 0 = x,$$
$$x + Sy = S(x + y).$$

Such definitions are called *recursive* definitions. We add 3 and 2 thus:
$3 + 2 = 3 + S1 = S(3 + 1) = S(3 + S0) = S(S(3 + 0)) = SS(SSS0) = 5.$

Multiplication is defined recursively as follows:
$$x \cdot 0 = 0,$$
$$x \cdot Sy = (x \cdot y) + x.$$

We sometimes write $(x \cdot y) + x$ as $xy + x$. Assuming that we already know how to add, we can now mutiply. Thus $3 \cdot 2 = 3 \cdot S1 = (3 \cdot 1) + 3 = (3 \cdot S0) + 3 = ((3 \cdot 0) + 3) + 3 = (0 + 3) + 3 = 3 + 3 = 6$.

Exponentiation is defined recursively as follows:
$$x^0 = 1,$$
$$x^{Sy} = (x^y) \cdot x.$$

Here even $0^0 = 1$. The reader who has already learned how to multiply will now have no difficulty in calculating 3^2. All the usual laws of arithmetic follow from the above axioms and definitions; for example:

commutativity: $x + y = y + x$, $xy = yx$;

associativity: $x + (y + z) = (x + y) + z$, $x(yz) = (xy)z$;

distributivity: $x(y + z) = xy + xz$, $(x + y)z = xz + yz$;

laws of indices: $(x^y)^z = x^{zy}$, $x^y x^z = x^{y+z}$, $x^z y^z = (xy)^z$.

We give some examples of how the above laws (and others) follow from Peano's axioms and definitions.

Lemma 2.1. $0 + x = x$.

Proof: We prove this using mathematical induction. First we show that it is true when $x = 0$: $0 + 0 = 0$ by the definition of addition (df+). Second we assume it holds for x and prove it for Sx. Our assumption that it is true for x is called the *induction hypothesis* (hyp). We have

$$\begin{aligned} 0 + Sx &= S(0 + x) & \text{(df+)} \\ &= S(x + 0) & \text{(hyp)} \\ &= Sx & \text{(df+)}. \end{aligned}$$

Since the assumption that the result holds for x implies that it holds for Sx, and since it is true for 0, we may conclude, by mathematical induction, that it is true for all natural numbers.

Lemma 2.2. $Sx + y = S(x + y)$.

Proof: Use induction on y. The result is true when $y = 0$: $Sx + 0 = Sx$ (df +) and $Sx = S(x + 0)$ (df +). If the result holds for y,

$$\begin{aligned} Sx + Sy &= S(Sx + y) & \text{(df+)} \\ &= SS(x + y) & \text{(hyp)} \\ &= S(x + Sy) & \text{(df+)}. \end{aligned}$$

2. Natural Numbers (Peano's Approach)

Theorem 2.3. $x + y = y + x$.

Proof: By induction on y. The result follows from Lemma 2.1 when $y = 0$. Supposing the result holds for y, we have

$$\begin{aligned} x + Sy &= S(x+y) &&\text{(df +)} \\ &= S(y+x) &&\text{(hyp)} \\ &= Sy + x &&\text{(Lemma 2.2).} \end{aligned}$$

Theorem 2.4. $x + (y + z) = (x + y) + z$.

Proof: Use induction on z:

$$x + (y + 0) = x + y = (x + y) + 0.$$

Supposing the result holds for z, we argue thus:

$$\begin{aligned} x + (y + Sz) &= x + S(y+z) &&\text{(df +)} \\ &= S(x + (y+z)) &&\text{(df +)} \\ &= S((x+y) + z) &&\text{(hyp)} \\ &= (x+y) + Sz &&\text{(df +).} \end{aligned}$$

Theorem 2.5. $x^z y^z = (xy)^z$.

Proof: $x^0 y^0 = 1 \cdot 1 = 1 \cdot S0 = (1 \cdot 0) + 1 = 0 + 1 = 1 = (xy)^0$, so the result is true when $z = 0$. Suppose it holds for z. Then $x^{Sz} y^{Sz} = (x^z x)(y^z y)$ and $(xy)^{Sz} = (xy)^z xy = (x^z y^z) xy$ (hyp). The result now follows by mathematical induction on z — provided we can first establish the associativity and commutativity of multiplication. This we leave to the reader.

Theorem 2.6. *If $x + y = x + z$ then $y = z$.*

Proof: By Theorem 2.3, this is true when $x = 0$. Assume it holds for x. If $Sx + y = Sx + z$ then $S(x + y) = S(x + z)$ (Lemma 2.2) and hence, by Peano's second axiom, $x + y = x + z$. By the induction hypothesis, $y = z$.

Every natural number except 0 has a predecessor. We can define a *naive predecessor* as follows:

$$\begin{aligned} P0 &= 0, \\ PSy &= y. \end{aligned}$$

Given the naive predecessor function, it is easy to define naive subtraction. Again, we use a recursive definition:

$$\begin{aligned} x \dot- 0 &= x, \\ x \dot- Sy &= P(x \dot- y). \end{aligned}$$

2. Natural Numbers (Peano's Approach)

The reason we use the sign $\dot{-}$ rather than the sign $-$ is that naive subtraction is not quite the same as ordinary subtraction. We cannot say that $1 - 3 = -2$ since -2 is not a natural number. Instead we say that $1 \dot{-} 3 = 0$. Using the above definition, we have

$$1 \dot{-} 3 = 1 \dot{-} S2 = P(1 \dot{-} 2) = P(1 \dot{-} S1) = P(P(1 \dot{-} 1)) = P(P(1 \dot{-} S0))$$
$$= P(P(P(1 \dot{-} 0))) = P(P(P(1))) = P(P(P(S0))) = P(P(0)) = P(0) = 0.$$

We define $\min(x, y)$ as $x \dot{-} (x \dot{-} y)$, and $\max(x, y)$ as $x + y \dot{-} \min(x, y)$.

Giuseppe Peano (1858–1932) published essentially this system in his *Arithmetices Principia* (1889), a book written in a language he invented.

Exercises

1. Prove that $0 \cdot x = 0$ without using the commutative law for multiplication.

2. Prove $1^x = 1$.

3. Prove $x(y + z) = xy + xz$.

4. Prove $xy = yx$.

5. Prove $x \dot{-} x = 0$.

6. Prove $Sx \dot{-} Sy = x \dot{-} y$.

7. Prove $\max(x, y) + \min(x, y) = x + y$.

8. In the *Arithmetices Principia*, Peano actually defines $x \dot{-} y$ as

$$x \dot{-} y = \begin{cases} z \text{ such that } y + z = x & \text{if there is such a } z \\ 0 & \text{otherwise} \end{cases}$$

   (a) Why can't there be two natural numbers z and w such that $y + z = x$ and $y + w = x$?

   (b) Show that Peano's definition of $\dot{-}$ is equivalent to the recursive definition given above.

3
The Integers

It is not difficult, though rather boring, to construct the integers from the natural numbers. Instead, we shall demonstrate in the next chapter how to construct the rationals from the integers, by essentially the same process. But first let us state the properties which make the set of integers into what is called an *integral domain*.

A *ring* $(R, 0, -, +, 1, \cdot)$ is a set R with operations $0, -, +, 1$, and $\cdot$, where $+$ and $\cdot$ are *binary* operations, $-$ is a *unary* operation and 0 and 1 are *nullary* operations, that is, specified elements of R, which moreover satisfy the following axioms or identities:

1. $(x + y) + z = x + (y + z)$ (associativity),
2. $x + 0 = x$,
3. $x + (-x) = 0$,
4. $x + y = y + x$ (commutativity),
5. $x \cdot 1 = x = 1 \cdot x$,
6. $(x \cdot y) \cdot z = x \cdot (y \cdot z)$ (associativity),
7. $(x + y) \cdot z = (x \cdot z) + (y \cdot z)$,
   $z \cdot (x + y) = (z \cdot x) + (z \cdot y)$ (distributivity).

We may take this opportunity to review a bit of abstract algebra. Axioms

(1) to (3) make $(R, 0, -, +)$ into a group. It is not difficult to prove that, in a group,

(2') $\qquad 0 + x = 0,$

(3') $\qquad -x + x = 0.$

A group is said to be *Abelian* if it also satisfies the commutative law (4). We have stated this as an axiom, even though it is a consequence of the remaining axioms of a ring (not of a group). The operation $\cdot$ may or may not obey the commutative law for multiplication:

(8) $\qquad x \cdot y = y \cdot x.$

If it does, we call the ring *commutative*. For example, **Z** and **Q** are commutative rings, but the ring of 2×2 matrices with entries from **Z** is not commutative.

A commutative ring is called an *integral domain* if

(9) $\qquad 0 \neq 1$ and $x \cdot y = 0$ implies $x = 0$ or $y = 0.$

It is called a *field* provided

(10) $\qquad 0 \neq 1$ and, if $x \neq 0$, there exists an element y such that $x \cdot y = 1 = y \cdot x.$

Z is an integral domain, but not a field. On the other hand, **Q** is a field as well as an integral domain. In fact, every field is an integral domain.

Exercises

1. Prove that (2') and (3') must hold in a group.

2. Prove that, in a ring, $x \cdot 0 = 0 = 0 \cdot x$ and $x \cdot (-1) = -x = (-1) \cdot x.$

3. Prove that (4) follows from the other axioms of a ring.

4. Show that every field is an integral domain.

5. In a group, if $0'$ is another element such that, for all x, $x + 0' = x$, show that $0' = 0$.

6. In a group, if $\sim$ is another unary operation such that, for all x, $x + (\sim x) = 0$, show that $\sim x = -x$.

7. Prove that, in any integral domain, we have the following *cancellation law* : if $a \cdot c = b \cdot c$ and $c \neq 0$ then $a = b$.

4
The Rationals

In this chapter we shall construct the field of rationals from the ring of integers. Our argument will use only properties (1) to (9) of the integers (Part II, Chapter 3), and hence the *same* argument will produce a field from *any* integral domain.

One's first idea is to say that a rational number a/b is a pair of integers (a, b), where $b \neq 0$. But this does not work, because then $2/3 \neq 4/6$. So we shall begin by introducing an equivalence relation on pairs of integers (a, b) such that $b \neq 0$. We define

$$(a, b) \equiv (c, d) \iff ad = bc,$$

it being assumed, of course, that b and d are nonzero. This is an *equivalence relation*, that is

$(a, b) \equiv (a, b)$ \hfill (reflexivity),

if $(a, b) \equiv (c, d)$, then $(c, d) \equiv (a, b)$ \hfill (symmetry),

if $(a, b) \equiv (c, d)$ and $(c, d) \equiv (e, f)$, then $(a, b) \equiv (e, f)$ \hfill (transitivity).

These three properties of an equivalence relation are easily checked; we shall verify only transitivity.

Given $ad = bc$ and $cf = de$, we wish to show that $af = be$. Making free

188   4. The Rationals

use of associativity and commutativity, we calculate

$$afd = adf = bcf = bde = bed.$$

As we are given that $d \neq 0$, we may use the cancellation law (Exercise 7 of Chapter 3) to infer that $af = be$, as was required to be shown.

Assuming that $b \neq 0$, we define the *ratio* (rational number) a/b as the equivalence class of (a,b), that is, as the set of all pairs (c,d), with $d \neq 0$, such that $ad = bc$. Note that

$$a/b = c/d \iff (a,b) \equiv (c,d) \iff ad = bc.$$

If $\mathbf{Q}$ is the set of such ratios, we shall obtain a field $(\mathbf{Q}, \underline{0}, -, +, \underline{1}, \cdot)$ where we have underlined $\underline{0}$ and $\underline{1}$ to distinguish them, at least temporarily, from the integers 0 and 1.

Here is how we attempt to define the operations of $\mathbf{Q}$:

$$\begin{aligned}
\underline{0} &= 0/1, \\
-(a/b) &= (-a)/b, \\
a/b + c/d &= (ad+bc)/(bd), \\
\underline{1} &= 1/1, \\
a/b \cdot c/d &= (ac)/(bd).
\end{aligned}$$

Note that, because of axiom (9) for an integral domain, $bd \neq 0$. There is no problem with $\underline{0}$ and $\underline{1}$, but we must check that the other operations are *well-defined*. We shall do so here for the operation $+$.

Thus, suppose that $a/b = a'/b'$ and $c/d = c'/d'$. We must show that

$$a/b + c/d = a'/b' + c'/d',$$

that is,

$$(ad+bc)/(bd) = (a'd' + b'c')/(b'd'),$$

that is,

$$(ad+bc)(b'd') = (bd)(a'd' + b'c').$$

The reader is invited to verify that this is indeed the case.

The ratios with operations $\underline{0}, -, +, \underline{1}$, and $\cdot$ as defined above form a field. To prove this one must check all the axioms of a field. For example, we shall check (4) and (10). To prove (4) for ratios we argue as follows:

$$\begin{aligned}
a/b + c/d &= (ad+bc)/(bd) \\
&= (da+cb)/(db) \text{ (commutativity of } \cdot \text{ for integers)} \\
&= (cb+da)/(db) \text{ (commutativity of } + \text{ for integers)} \\
&= c/d + a/b.
\end{aligned}$$

Here we made use of the commutativity of · and + for integers. To prove (10), we may assume that $a/b \neq \underline{0}$, that is, $a1 \neq b0$, that is, $a \neq 0$ (see Exercise 2 of Chapter 3). We claim that $(a/b) \cdot (b/a) = \underline{1}$. Indeed,

$$(a/b) \cdot (b/a) = (a \cdot b)/(b \cdot a) = 1/1,$$

since

$$(a \cdot b) \cdot 1 = a \cdot b = b \cdot a = (b \cdot a) \cdot 1.$$

We have thus constructed the field **Q** from the ring **Z**. More generally, the same construction leads from any integral domain to a field, called its *field of quotients*.

What is the relationship between **Q** and the ring **Z** from which it is constructed? Strictly speaking, the set **Q** does not contain the set **Z**. However, **Q** does contain a subset, consisting of the ratios of the form $a/1$, which is *isomorphic* to **Z**. To make this notion more precise, consider the mapping $h : \mathbf{Z} \to \mathbf{Q}$ such that $h(a) = a/1$. Then h is a *homomorphism*, that is, it preserves the operations of **Z**:

$$\begin{aligned} h(a+b) &= h(a) + h(b), \\ h(ab) &= h(a)h(b), \\ h(-a) &= -h(a), \\ h(0) &= \underline{0}, \\ h(1) &= \underline{1}. \end{aligned}$$

Furthermore, h is a 1-to-1 mapping (an *injection*). The set $h(\mathbf{Z})$ is thus an *isomorphic image* of **Z**. We also say that h *embeds* **Z** in **Q**. It is a mathematical convention to identify 2 and 2/1 and to say that **Z** is contained in **Q**.

Exercises

1. Show that $-(-(a/b)) = a/b$.

2. Show that $((a/b)^{-1})^{-1} = a/b$ (assuming $a, b \neq 0$).

3. Prove that $a/(b/c) = (ac)/b$.

4. Define $\max(a/b, c/d)$.

5. What definition might one use to construct the integers out of the natural numbers?

6. If b is a nonzero integer, let $b\mathbf{Z} = \{bx | x \in \mathbf{Z}\}$. Consider any mapping $f : b\mathbf{Z} \to \mathbf{Z}$ such that $f(bx + by) = f(bx) + f(by)$ for all $x, y \in \mathbf{Z}$. Show that $f(0) = 0$ and $f(-bx) = -f(bx)$. If g is any other function

4. The Rationals

like f (that is, $g : c\mathbf{Z} \to \mathbf{Z}$ for $c \neq 0$ and $g(cx+cy) = g(cx) + g(cy)$) define $f \equiv g$ to mean that $f(bcx) = g(bcx)$ for all integers x. Show that $\equiv$ is an equivalence relation and that the equivalence class $[f]$ can be taken as a definition of the quotient $f(b)/b$.

5
The Real Numbers

There are two well-known ways of constructing the reals from the rationals: the Dedekind cut approach, which goes back to Eudoxus, and the Cauchy sequence approach.

Eudoxus, a member of Plato's Academy, was interested in defining the notion of proportion for geometric quantities. What he did can be interpreted in modern terms as defining *equality* between two real numbers (ratios of geometric quantities) α and β as follows: $\alpha = \beta$ iff the set of all rationals below α is the same as the set of all rationals below β, and similarly for the sets of rationals above α and β. Dedekind exploited this idea further by defining the real number α as the pair (L, U) of sets of rationals below and above α, respectively. Such pairs can be described without mentioning α; e.g., L might be the set of all rationals x for which $x^2 < 2$ and U the set of rationals y for which $y^2 > 2$. We shall not develop this idea further, as it is discussed in many algebra books.

Analysts prefer a construction of the reals proposed by Cauchy, according to which the real number α is defined as the set of all sequences of rational numbers which converge to α. Again, this can be done without mentioning α. A *Cauchy sequence* is a sequence $\{a_n | n \in \mathbf{N}\}$ of rational numbers such that $|a_m - a_n|$ can be made as small as one likes by taking m and n sufficiently large. Two Cauchy sequences $\{a_n | n \in \mathbf{N}\}$ and $\{b_n | n \in \mathbf{N}\}$ are said to be *equivalent* if $|a_n - b_n|$ can be made as small as one likes by taking n sufficiently large. A *real number* is then defined as an equivalence class of Cauchy sequences.

An amusing construction of the real numbers, which is not so well-known, bypasses the rationals altogether and defines a real number as an equivalence class of certain mappings $f : \mathbf{Z} \to \mathbf{Z}$. Call f *almost linear* if the set of all $|f(m+n) - f(m) - f(n)|$, where $m, n \in \mathbf{Z}$, is bounded. Call two almost linear mappings $f, g : \mathbf{Z} \to \mathbf{Z}$ *equivalent* if the set of all $|f(n) - g(n)|$, where $n \in \mathbf{Z}$, is bounded. Define a *real number* to be an equivalence class of almost linear mappings $\mathbf{Z} \to \mathbf{Z}$. The real number α will be the equivalence class of the mapping f for which $f(n) = [\alpha n]$ is the greatest integer $\leq \alpha n$. Addition and multiplication of the real numbers corresponding to the almost linear mappings f and g are easily defined as the equivalence classes of $f + g$ and $f \circ g$, respectively, where $(f+g)(n) = f(n) + g(n)$ and $(f \circ g)(n) = f(g(n))$. This definition is due to Steve Schanuel.

Instead of constructing the real numbers, one may describe the field of real numbers axiomatically as a *complete ordered field*. We already know what is meant by a field, so we only have to define the words 'ordered' and 'complete'.

A field F is *ordered* if it has a subset P such that

1. $x, y \in P \Rightarrow x + y \in P$,

2. $x, y \in P \Rightarrow xy \in P$,

3. exactly one of the following holds: $x = 0$, or $x \in P$, or $-x \in P$.

Note that by (3) either 1 or -1 is an element of P. Since $(-1)(-1) = 1$, it follows from (2) that $1 \in P$. The elements of P are the *positive* elements of the field. The existence of P allows us to define an *order relation* on the field F:
$$x \leq y \Longleftrightarrow x = y \text{ or } y - x \in P.$$
From this definition we obtain the following propositions:

$x \leq x$	reflexivity,
$x \leq y$ and $y \leq z \Rightarrow x \leq z$	transitivity,
$x \leq y$ and $y \leq x \Rightarrow x = y$	antisymmetry,
$x \leq y$ or $y \leq x$	dichotomy.

It is because $\leq$ has these four properties that it is called an *order relation*. $\mathbf{Q}$ and $\mathbf{R}$ are ordered fields, but $\mathbf{C}$ is not. To see that $\mathbf{C}$ is not ordered, consider the element i. If $i \in P$ then $-1 = i \cdot i \in P$. This is impossible since $1 \in P$. If $-i \in P$ we again have that $-1 = (-i)(-i) \in P$. So neither i nor $-i$ is in P, and this contradicts (3) above.

An ordered field is *complete* if every nonempty set of positive elements has a greatest lower bound. For example, $\sqrt{2}$ is the greatest lower bound of all positive reals r such that $2 \leq r^2$. Since $\sqrt{2}$ is not rational, we may conclude from this example that $\mathbf{Q}$ is not a complete ordered field. There is really only one complete ordered field in the sense of the following:

Theorem 5.1. *Any two complete ordered fields are isomorphic.*

For the proof, see Chapter 4.3 of Birkhoff and Mac Lane [1977].

Exercises

1. Pick any of the three definitions of the reals mentioned here and prove that they form a field.

2. Prove the four properties of the order relation defined above.

6
Complex Numbers

Negative numbers are required to solve the equation $x + 2 = 1$. To solve $2x = 3$, we need the rationals. The equation $x^2 = 2$ has an irrational solution. Finally, the equation $x^2 = -1$ has an *imaginary root* called i.

Numbers of the form $a + bi$ where a and b are real, are called *complex numbers*. Complex numbers with $a = 0$ are also called *imaginary*. The complex number $a + bi$ is often associated with the point (a, b) in the Cartesian plane. The absolute value of a complex number is just its distance $\sqrt{a^2 + b^2}$ from the origin. The angle θ measured counterclockwise from the positive x axis to the line joining (a, b) to the origin is called the *angle* of the complex number $a + bi$; thus $\tan \theta = b/a$.

Complex numbers were introduced by Girolamo Cardano (1501–1576), who used them in his *Ars Magna* (1545) to solve cubic equations. Cardano tells us to multiply $5 + \sqrt{-15}$ by $5 - \sqrt{-15}$, 'putting aside the mental tortures involved' (see T. Richard Witmer's translation of the *Ars Magna*, p. 219).

There are many ways to define the complex numbers, all of them being essentially equivalent. For example, we can define them as ordered pairs of real numbers subject to the multiplication rule

$$(a, b) \cdot (c, d) = (ac - bd, ad + bc).$$

Here $(1, 0)$ plays the role of 1 and $(0, 1)$ the role of i.

Another way to introduce the 'field' of complex numbers is to say that it is the quotient ring $\mathbf{R}[x]/(x^2+1)$. The elements of this ring are equivalence classes of polynomials with real coefficients, where two polynomials are said to be equivalent if they differ by a multiple of $x^2 + 1$. Here the role of i is played by the equivalence class whose representative is the polynomial x.

Yet another way to define complex numbers is to say that they are 2×2 matrices with real entries of the form

$$\begin{pmatrix} a & b \\ -b & a \end{pmatrix}.$$

It is easy to see that the set of matrices of this form is closed under addition and multiplication. Here the role of 1 is played by the identity matrix and the role of i is played by

$$\begin{pmatrix} 0 & 1 \\ -1 & 0 \end{pmatrix}.$$

The advantage of defining complex numbers in this way is that one can use the arithmetic of matrices to give a quick proof of the fact that the complex numbers form a ring. The commutative law of multiplication still needs checking, since it does not hold for matrices in general. Moreover, the inverse of a nonzero matrix of the form

$$\begin{pmatrix} a & b \\ -b & a \end{pmatrix}$$

is just the matrix

$$\begin{pmatrix} a/k & -b/k \\ b/k & a/k \end{pmatrix},$$

where $k = a^2 + b^2$ is the determinant of the former.

Since a real number is not usually thought of as a matrix, one might well ask how the reals relate to the complex numbers if the latter are conceived as matrices. Well, the mapping $h : \mathbf{R} \to \mathbf{C}$, where

$$h(a) = \begin{pmatrix} a & 0 \\ 0 & a \end{pmatrix}$$

is a 1-to-1 homomorphism. Hence $h(\mathbf{R})$ is an isomorphic image of $\mathbf{R}$ and may be identified with $\mathbf{R}$, and we may say that $\mathbf{R}$ is a subset of $\mathbf{C}$.

Let u and v be any complex numbers. Then

$$|u + v| \leq |u| + |v|,$$

$$|uv| = |u|\,|v|.$$

If we think of the complex numbers as points in the Cartesian plane, the above inequality is just the triangle inequality of Euclid (Book I, 20). The above equality is equivalent to the algebraic identity

$$(ac - bd)^2 + (ad + bc)^2 = (a^2 + b^2)(c^2 + d^2).$$

This identity was known to al-Khazin in 950 AD. (See Part I, Chapter 23.)

6. Complex Numbers

Associating $a + bi$ with the point (a, b) in the Cartesian plane, let r be its distance from the origin, and let θ be its angle. Then

$$a + bi = r(\cos\theta + i\sin\theta).$$

If $a' + b'i$ has absolute value r' and angle θ', we may exploit the well-known addition formulas for trigonometric functions to obtain

$$(a + bi)(a' + b'i) = rr'(\cos(\theta + \theta') + i\sin(\theta + \theta')).$$

By induction on the natural number m, this formula leads to the equation

$$(r(\cos\theta + i\sin\theta))^m = r^m(\cos(m\theta) + i\sin(m\theta)),$$

which is known as de Moivre's theorem after Abraham de Moivre (1667–1754), who was the first to make use of it (Part I, Chapter 30).

Exercises

1. Using the matrix definition of complex numbers, verify that $\mathbf{C}$ is a field.

2. What is the multiplicative inverse of $5 + \sqrt{-15}$?

3. Give an algebraic proof of the triangle inequality.

4. Prove the theorem of de Moivre.

7
The Fundamental Theorem of Algebra

The negative integers, fractions, irrationals and imaginary numbers were all introduced in order to supply solutions to polynomial equations. Do we need more numbers? Is there a polynomial equation with complex coefficients with a root that is not a complex number? The answer is no, and the fact that no new numbers are required is called the Fundamental Theorem of Algebra. It was first stated by Albert Girard (1595–1632) in 1629, and proofs were given by Jean d'Alembert (1717–1783) and Carl Friedrich Gauss (1777–1855). However, John Stillwell [1989] argues that there is a flaw in Gauss's proof, and the first rigorous proof was given only after Weierstrass established the basic properties of continuous functions.

Theorem 7.1. (The Fundamental Theorem of Algebra)
Every polynomial equation with complex coefficients has complex solutions.

Proof: We assume some simple facts about continuity. For example, we assume that if a continuous closed curve which loops around the origin n times gradually changes so that it loops around the origin only $n-1$ times, then, at some time, part of it passes through the origin. Let

$$w = p(z) = z^m + c_1 z^{m-1} + \cdots + c_{m-1} z + c_m \qquad (c_m \neq 0),$$

where the c_k's are complex numbers. Without loss of generality, we may take our polynomial equation to be $p(z) = 0$. Let

$$g(z) = \frac{p(z)}{z^m} = 1 + c_1/z + \cdots + c_m/z^m.$$

Then

$$\begin{aligned}
|g(z) - 1| &= |c_1/z + \cdots + c_m/z^m| \\
&\leq |c_1/z| + \cdots + |c_m/z^m| \\
&= \frac{|c_1|}{|z|} + \cdots + \frac{|c_m|}{|z|^m} \\
&\leq m\left(\frac{\max(|c_k|)}{|z|}\right)
\end{aligned}$$

if $|z| > 1$. Geometrically, this means that, when $|z|$ is large, $g(z)$ is represented by a point close to $(1,0)$. In particular, if z moves around a large circle with centre at the origin, then $g(z)$ will move in some continuous closed curve near $(1,0)$ and not loop around $(0,0)$.

We consider two planes, one to represent $z = x + yi$, and one to represent $w = u + vi$. As z moves around a circle of radius $|z| = r$ and centre $(0,0)$, w moves in some continuous closed curve in its own plane.

Suppose that z moves around a circle with radius large enough to keep $g(z)$ from winding around $(0,0)$. How many times does w loop around $(0,0)$? Equivalently, how does the angle of w change as z moves around its large circle? In the previous chapter we noted that the angle of a product of two complex numbers is the sum of the angles of those two numbers. Since $w = z^m g(z)$, the angle of w changes by an amount equal to the change in the angle of z^m plus the change in the angle of $g(z)$. Since $g(z)$ stays close to $(1,0)$ the net change in the angle of $g(z)$ as z goes around the large circle is 0. Thus the change in the angle of w is just the change in the angle of z^m. But the angle of a product of complex numbers is equal to the sum of their angles, so the change in the angle of w is just m times the change in the angle of z. As z moves around a circle, the change in the angle of z is 2π. Hence the change in the angle of w is $2\pi m$. In other words, as z moves around the large circle, w loops around the origin m times.

Now imagine the radius of the large circle gradually decreasing to 0. When it is close to 0, w moves in a continuous closed curve close to $p(0) \neq 0$. (We are assuming that $c_m \neq 0$.) Thus w no longer loops around the origin; at some point as the radius of the large circle decreased, w moved through the origin, that is, $p(z)$ was equal to 0 and the polynomial equation had a complex number solution.

Corollary 7.2. *A polynomial of degree m with complex coefficients has exactly m linear factors.*

Proof: We have seen that $p(z) = 0$ has a complex number solution, call it z_1. If $p(z) = q(z)(x - z_1) + R$ then $R = p(z_1) = 0$, so $x - z_1$ is a factor of $p(z)$. Now $q(z)$ has degree $m-1$, and it also has a complex root. Continuing the process, we can write $p(z) = (z - z_1)(z - z_2) \cdots (z - z_m)$.

7. The Fundamental Theorem of Algebra

It is a pity that the proof of the fundamental theorem of algebra makes use of the notion of continuity, which belongs to analysis rather than to algebra. It often happens in mathematics that, when we want to prove a basic property of a certain system, we have to go outside that system. Still, the above proof can be employed constructively for solving polynomial equations. Before the advent of modern computers, there was a mechanical device which traced out the w curve for any given radius $|z| = r$. As r was gradually reduced, a bell would ring as soon as the w curve passed through the origin and a solution had been found.

Exercises

1. Let $w = z^2 - z - 2$. Graph w as z moves around the origin on a circle of radius 2.1.

2. Repeat the above exercise with a circle of radius 2.

3. Assuming that the coefficients of $p(z)$ are all real numbers, show that if $x + yi$ is a root of $p(z)$ then so is $x - yi$.

8
Quaternions

It was William Rowan Hamilton (1805–1865) who first conceived complex numbers as ordered pairs of reals, subject to the multiplication rule $(a,b)(c,d) = (ac - bd, ad + bc)$. His greatest contribution to pure mathematics, however, was his creation of an algebraic system in which the commutative law of multiplication does not hold. Just as many people before Bolyai and Lobachevsky had thought of Euclid's fifth postulate as a necessary and sacred truth, so many people before Hamilton believed that the law of commutativity for the multiplication of numbers was decreed by heaven. Hamilton discovered that there are consistent algebraic systems for which this law does not not hold. (Matrix algebra came fourteen years later.)

What Hamilton discovered were 'quaternions'. The idea dawned on him while he was strolling along the Royal Canal in Dublin in 1843.

Quaternions are numbers of the form

$$a + bi + cj + dk,$$

where a, b, c, and d are real, and $i^2 = j^2 = k^2 = ijk = -1$.

If quaternion multiplication were commutative, it would be easy to derive $ij = -ij$ and from this it would follow that $i = j = 0$, and the whole system would collapse. We must have noncommutative multiplication. In particular, $ij \neq ji$. How do we know that such entities as i, j, and k exist? There are in fact three complex matrices:

$$i_1 = \begin{pmatrix} 0 & 1 \\ -1 & 0 \end{pmatrix}, \qquad i_2 = \begin{pmatrix} 0 & i \\ i & 0 \end{pmatrix}, \qquad i_3 = \begin{pmatrix} i & 0 \\ 0 & -i \end{pmatrix}$$

such that $i_1^2 = i_2^2 = i_3^2 = i_1 i_2 i_3 = -1$, where 1 is interpreted as the identity matrix. We may define the quaternion $a = a_0 1 + a_1 i_1 + a_2 i_2 + a_3 i_3$ as the complex matrix
$$\begin{pmatrix} a_0 + ia_3 & a_1 + ia_2 \\ -a_1 + ia_2 & a_0 - ia_3 \end{pmatrix},$$
namely, $\begin{pmatrix} u & v \\ -\bar{v} & \bar{u} \end{pmatrix}$, where $u = a_0 + ia_3$ and $v = a_1 + ia_2$, the a_k being real numbers. Addition and multiplication of quaternions is just the usual matrix arithmetic. The *conjugate* of a quaternion, as distinguished from the conjugate of the complex matrix which represents it, is the quaternion
$$\bar{a} = a_0 1 - a_1 i_1 - a_2 i_2 - a_3 i_3.$$

Note that we often identify the real number a_0 with the matrix $a_0 1$. Using this identification, we have
$$a\bar{a} = a_0^2 + a_1^2 + a_2^2 + a_3^2 = \bar{a}a.$$

The *norm* $N(a)$ of a quaternion a is the product $a\bar{a}$ viewed as a real number, namely, $a_0^2 + a_1^2 + a_2^2 + a_3^2$. Note that the norm is 0 just in case the quaternion is the zero matrix. For a nonzero quaternion a, we can write
$$a(\bar{a}/N(a)) = 1 = (\bar{a}/N(a))a.$$

(Here $N(a)$ is the real number, not the matrix.) This means that nonzero quaternions have multiplicative inverses.

A *division ring* is a system of elements closed under two binary operations, addition and multiplication, such that

1. under addition, it is a commutative group,

2. under multiplication, the nonzero elements form a group, and

3. multiplication distributes over addition (on both sides).

A division ring is thus a ring (Chapter 3) in which every nonzero element has an inverse under multiplication. It satisfies all the axioms of a field, exept the commutative law of multiplication. It is sometimes called a *skewfield*. A key fact about quaternions is that they form a division ring.

It is also possible to represent quaternions by 4×4 real matrices. A cheap way to obtain such a representation is to replace i by $\begin{pmatrix} 0 & 1 \\ -1 & 0 \end{pmatrix}$ and 1 by $\begin{pmatrix} 1 & 0 \\ 0 & 1 \end{pmatrix}$ in the complex 2×2 matrix representing the quaternion. Instead we shall look at a more interesting way to obtain such a representation, which has wider implications.

The quaternion $x = x_0 + x_1 i_1 + x_2 i_2 + x_3 i_3$ gives rise to a column vector

$$[x] = \begin{pmatrix} x_0 \\ x_1 \\ x_2 \\ x_3 \end{pmatrix}.$$

Given a quaternion a, consider the mapping which assigns to every such vector $[x]$ the vector $[ax]$. Clearly, this is a linear transformation of the four-dimensional vector space over $\mathbf{R}$. Hence there will be a 4×4 matrix $A = L(a)$ such that $A[x] = [ax]$. We say that $L(a)$ *represents* the quaternion a. In particular, $I = L(1)$ is the identity matrix. For example, the matrix $I_1 = L(i_1)$ is obtained as follows: the equation

$$i_1(x_0 + x_1 i_1 + x_2 i_2 + x_3 i_3) = -x_1 + x_0 i_1 - x_3 i_2 + x_2 i_3$$

may be interpreted thus:

$$\begin{pmatrix} 0 & -1 & 0 & 0 \\ 1 & 0 & 0 & 0 \\ 0 & 0 & 0 & -1 \\ 0 & 0 & 1 & 0 \end{pmatrix} \begin{pmatrix} x_0 \\ x_1 \\ x_2 \\ x_3 \end{pmatrix} = \begin{pmatrix} -x_1 \\ x_0 \\ -x_3 \\ x_2 \end{pmatrix}.$$

Hence $I_1 = L(i_1)$ is the 4×4 matrix appearing above. The matrices $I_2 = L(i_2)$ and $I_3 = L(i_3)$ are obtained similarly. Note that

$$L(ab)[x] = [(ab)x] = [a(bx)] = L(a)[bx] = L(a)L(b)[x]$$

for all vectors $[x]$. Therefore

$$L(ab) = L(a)L(b),$$

from which it easily follows that L is an injective homomorphism from the division ring of quaternions into the ring of all 4×4 matrices over $\mathbf{R}$.

Exercises

1. Show that the conjugate of ab is $\bar{b}\bar{a}$.

2. Show that $N(\bar{a}) = N(a)$ and $N(ab) = N(a)N(b)$.

3. Prove that every quaternion satisfies a quadratic equation with real coefficients.

4. Find all quaternions whose square is 1.

5. Prove that $L(\bar{a}) = A^t$ is the transpose of $A = L(a)$ and that $(N(a))^2 = \det(A)$, the determinant of A.

8. Quaternions

6. Show that with every quaternion a one may associate another 4×4 real matrix $R(a)$ such that $R(a)[x] = [xa]$ for all quaternions x.

7. Prove that $R(ab) = R(b)R(a)$ and $L(a)R(b) = R(b)L(a)$ for all quaternions a and b.

9
Quaternions Applied to Number Theory

In this chapter we shall use integer quaternions to show that every natural number is a sum of four perfect squares. This was first proved by J. L. Lagrange in 1770 (*Oeuvres*, Vol. 3, pp. 189-201).

As a warming up exercise, note that every integer is a sum of five cubes. Indeed, let m be an integer. Since $m - m^3 = -(m-1)m(m+1)$, it follows that $m - m^3$ is divisible by both 2 and 3, and hence by 6. Thus $x = (m - m^3)/6$ is an integer. Moreover, $m = m^3 + 6x = m^3 + (x+1)^3 + (x-1)^3 + (-x)^3 + (-x)^3$, a sum of five cubes. It is not known whether every integer can be written as a sum of four cubes.

About sums of non-negative cubes, it is known that every natural number, except 23 and 239, can be written as a sum of 8 non-negative cubes. As of 1971, it was not known whether the 8 could be lowered for large positive integers (Ellison [1971], pp. 10-36).

To prove the theorem of Lagrange, we shall require the following lemma, due to Euler.

Lemma 9.1. *For every odd prime p there exist integers x and y such that*

$$x^2 + y^2 + 1 = mp,$$

where m is an integer such that $0 < m < p$.

Proof: Let x range from 0 to $\frac{1}{2}(p-1)$. The squares x^2 all leave different remainders when divided by p. For suppose x_1^2 and x_2^2 leave the same remainder. Then $(x_1 + x_2)(x_1 - x_2) = x_1^2 - x_2^2$ is a multiple of p, hence p must divide $x_1 + x_2$ or $x_1 - x_2$. Without loss of generality, we may assume that

$x_1 > x_2$. Then $x_1 \neq x_2$ and

$$0 < x_1 + x_2 < p - 1, \qquad -\frac{p-1}{2} \leq x_1 - x_2 \leq \frac{p-1}{2},$$

hence p divides neither $x_1 + x_2$ nor $x_1 - x_2$. Thus we have a contradiction and the assertion has been proved.

Similarly, we can show that, as y ranges from 0 to $\frac{1}{2}(p-1)$, the numbers $-y^2 - 1$ all leave different remainders when divided by p.

As x and y range from 0 to $\frac{1}{2}(p-1)$, the set of all x^2 thus takes on $\frac{1}{2}(p+1)$ different values and so does the set of all $-y^2 - 1$. Since there are only p possible remainders when one divides by p, the two sets must overlap; hence there exist integers x and y in the given range such that $x^2 + y^2 + 1 = mp$ is a multiple of p. Moreover,

$$1 \leq mp \leq \tfrac{1}{4}(p-1)^2 + \tfrac{1}{4}(p-1)^2 + 1 < p^2,$$

hence $1 \leq m < p$, as required.

Following Lipschitz [1886], p. 404, we define an *integer quaternion* as a quaternion with integer coefficients.

Theorem 9.2. (Lagrange)
Every natural number n is the sum of four perfect squares, that is, n is the norm of an integer quaternion.

Proof: Since the norm of the product of integer quaternions is the product of their norms and since n is a product of primes, it suffices to show that every prime is the norm of an integer quaternion. Since $2 = 1^2 + 1^2 + 0^2 + 0^2$, it suffices to prove this for odd primes. Let p be any odd prime. Then we know from Euler's lemma that there is an integer quaternion x such that $N(x) = mp$ with $0 < m < p$. Pick $m = m_0$ as small as possible with this property. We claim that $m_0 = 1$.

First let us show that m_0 cannot be even. If it is, then so is $x_0^2 + x_1^2 + x_2^2 + x_3^2$, hence also $x_0 + x_1 + x_2 + x_3$ is even. There are three cases: either all the x_i are even, or they are all odd, or exactly two are even, say x_0 and x_1. In all three cases, $x_0 \pm x_1$ and $x_2 \pm x_3$ are even, hence

$$\tfrac{1}{2}m_0 = \left(\frac{x_0 + x_1}{2}\right)^2 + \left(\frac{x_0 - x_1}{2}\right)^2 + \left(\frac{x_2 + x_3}{2}\right)^2 + \left(\frac{x_2 - x_3}{2}\right)^2$$

is the sum of four perfect squares. But $\tfrac{1}{2}m_0$ is a positive integer less than m_0, which contradicts the assumption that m_0 was chosen as small as possible.

We now know that m_0 is odd. Let z_i be the closest integer to $\frac{x_i}{m_0}$, hence $\left|\frac{x_i}{m_0} - z_i\right| < \tfrac{1}{2}$. (It cannot be equal to $\tfrac{1}{2}$, or else $m_0 = 2|x_i - m_0 z_i|$ would be even.)

Consider the integer quaternion $y = x - m_0 z$, where $z = z_0 + z_1 i_1 + z_2 i_2 + z_3 i_3$. Then
$$|y_i| = |x_i - m_0 z_i| < \tfrac{1}{2} m_0,$$
hence $N(y) < 4(\tfrac{1}{2} m_0)^2 = m_0^2$. But
$$N(y) = y\bar{y} = x\bar{x} - m_0(x\bar{z} + \bar{z}x) + m_0^2 z\bar{z}.$$

Write $x\bar{z} = w$, so $x\bar{z} + \bar{z}x = 2w_0$, where w_0 is the scalar part of w, and hence
$$N(y) = m_0 p - 2m_0 w_0 + m_0^2 N(z) = m_0 m_1,$$
where $m_1 = p - 2w_0 + m_0 N(z)$. Now $m_0 m_1 = N(y) < m_0^2$, hence $m_1 < m_0$. Consider now the integer quaternion
$$y\bar{x} = x\bar{x} - m_0 z\bar{x} = m_0 p - m_0 z\bar{x} = m_0(p - z\bar{x}).$$

Then
$$m_0 m_1 m_0 p = N(y)N(\bar{x}) = N(y\bar{x}) = m_0^2 N(p - z\bar{x}),$$
hence
$$m_1 p = N(p - z\bar{x}).$$

Since $m_1 < m_0$, this would contradict the assumption that m_0 was chosen as small as possible, unless $m_1 = 0$.

This leaves only the possibility that $m_1 = 0$, hence $N(y) = 0$, hence $y = 0$, hence $x = m_0 z$, hence $m_0 p = N(x) = m_0^2 N(z)$, hence $p = m_0 N(z)$, hence $m_0 = 1$ or $m_0 = p$. But $m_0 < p$, so $m_0 = 1$, as was required.

Exercises

1. Express 239 as a sum of nine positive cubes.

2. Show that numbers of the form $8k + 7$ cannot be expressed as sums of three perfect squares.

3. Prove that every prime number of the form $4k+1$ can be expressed as the sum of two perfect squares. (Hint: imitate the above proof using complex integers instead of integer quaternions.)

10
Quaternions Applied to Physics

If Hamilton's intention had been to apply quaternions to physics, he was stymied by the fact that physical space has only three dimensions. Writing the quaternion x as $x = x_0 + \xi$, where $\xi = i_1 x_1 + i_2 x_2 + i_3 x_3$, we call x_0 the *scalar part* and ξ the *vector part*. A 3-*vector* is then a quaternion whose scalar part is 0. Unfortunately, if we multiply ξ by another vector $\eta = y_1 i_1 + y_2 i_2 + y_3 i_3$, we obtain not a vector, but the quaternion

$$-(x_1 y_1 + x_2 y_2 + x_3 y_3) + (y_2 x_3 - x_2 y_3) i_1 + \cdots.$$

Oliver Heaviside pointed out the importance of the two separate parts of this product and wrote

$$\xi \circ \eta = x_1 y_1 + x_2 y_2 + x_3 y_3,$$

$$\xi \times \eta = (y_2 z_3 - x_2 y_3) i_1 + \cdots$$

calling them the *scalar* and *vector product*, respectively. His *vector analysis* soon replaced the use of quaternions in physics. It enabled him to give a concise formulation of Maxwell's laws of electromagnetism. Writing $\nabla = i_1 \frac{\partial}{\partial x_1} + i_2 \frac{\partial}{\partial x_2} + i_3 \frac{\partial}{\partial x_3}$ for the vector representing partial differentiation with respect to the space coordinates and letting $E = E_1 i_1 + E_2 i_2 + E_3 i_3$ and $M = M_1 i_1 + M_2 i_2 + M_3 i_3$ represent the electric and magnetic fields, respectively, Maxwell's equations may be written as follows:

$$\nabla \circ M = 0, \qquad \nabla \times E + \frac{\partial M}{c \partial t} = 0, \qquad \nabla \circ E = \rho, \qquad \nabla \times M - \frac{\partial E}{c \partial t} = \rho \frac{d\xi}{c dt},$$

where c is the velocity of light, ρ is the charge density, and $\frac{d\xi}{dt}$ the velocity of the matter bearing the electric charge.

Using the language of quaternions, albeit quaternions with complex components, these four equations may be combined into one, namely

$$\left(\frac{\partial}{c\partial t} - i\nabla\right)(M + iE) + \left(\rho + i\rho\frac{d\xi}{cdt}\right) = 0.$$

This was pointed out by Silberstein [1924] pp. 44-46; but one may wonder whether it wasn't already known to Maxwell himself, in view of his assertion that 'the invention of the calculus of quaternions is a step towards the knowledge of quantities related to space which can only be compared, for its importance, with the invention of triple coordinates by Descartes. The ideas of this calculus... are fitted to be of the greatest use in all parts of science.' (See Maxwell [1869].)

Einstein's theory of relativity made it clear that space and time should be combined into a single entity and that the expression

$$s^2 = c^2 t^2 - x^2 - y^2 - z^2$$

should be invariant under coordinate transformations passing from a stationary to a moving platform. The minus sign in this expression suggests that we are talking about the norm of a quaternion

$$x = x_0 + i\xi, \qquad x_0 = ct, \qquad \xi = i_1 x_1 + i_2 x_2 + i_3 x_3,$$

whose scalar part is real, but whose vector part is imaginary. Such a quaternion represents a point in so-called *Minkowski space*. Following Silberstein [1924], pp. 154-55, we observe that $s^2 = N(x)$ is invariant under a *Lorentz transformation*, which may itself be expressed with the help of 'biquaternions'.

A *biquaternion* $a = a_0 + i_1 a_1 + i_2 a_2 + i_3 a_3$ has complex components $a_0, \ldots, a_3$. We must here distinguish the quaternion conjugate a^t of a from its complex conjugate a^c:

$$a^t = a_0 - i_1 a_1 - i_2 a_2 - i_3 a_3,$$

$$a^c = \bar{a}_0 + i_1 \bar{a}_1 + i_2 \bar{a}_2 + i_3 \bar{a}_3.$$

If a is represented as a 4×4 matrix $L(a)$ over **C** as in Chapter 8, $L(a^t) = L(a)^t$, the transpose of $L(a)$. We call a biquaternion x *Hermitian* if $x^t = x^c$; this is what characterizes the quaternion $x = ct + i\xi$ describing a point in Minkowski space, that is, an event in space-time. A *Lorentz transformation* sends x onto pxp^{ct}, where p is a biquaternion of norm 1. Indeed,

$$(pxp^{ct})^t = p^{ct} x^t p^t = p^c x^c p^t = (pxp^{ct})^c,$$

so pxp^{ct} is also Hermitian. Moreover, $N(pxp^{ct}) = pxp^{ct}p^c x^t p^t = pxx^t p^t = N(x)N(p) = N(x)$, since $p^{ct}p^c = (p^t p)^c = N(p)^c = 1^c = 1$. Thus a Lorentz transformation preserves the norm of a Hermitian biquaternion.

Another Hermitian biquaternion which transforms like x is

$$m_0 \frac{dx}{ds} = m + im\frac{d\xi}{cdt},$$

where $m = m_0 \frac{cdt}{ds}$ is the mass of the moving particle and $m\frac{d\xi}{dt}$ is its momentum. Here m_0 is called the *rest mass*; it is assumed to be invariant under Lorentz transformations. As Einstein observed, the principle of conservation of momentum should carry with it also that of

$$m = m_0 \frac{cdt}{ds} = m_0(1 - v^2/c^2)^{-\frac{1}{2}},$$

since$(\frac{ds}{dt})^2 = c^2 - v^2$, where $v^2 = (\frac{dx_1}{dt})^2 + (\frac{dx_2}{dt})^2 + (\frac{dx_3}{dt})^2$ is the square of the velocity of the particle. If v is small compared to to c, we have

$$mc^2 = m_0 c^2 (1 - v^2/c^2)^{-\frac{1}{2}} \approx m_0 c^2 + \tfrac{1}{2} m_0 v^2,$$

which must then also be conserved. Einstein considered this to be the total energy $E = mc^2$ of the particle, consisting of the rest energy $m_0 c^2$ and the kinetic energy $\tfrac{1}{2} m_0 v^2$.

As we have seen, Maxwell's equations may be condensed into

$$\left(\frac{\partial}{c\partial t} - i\nabla\right)(M + iE) + \left(\rho + i\rho \frac{d\xi}{cdt}\right) = 0.$$

This may be written more concisely:

$$D^c F + J = 0,$$

where $D = \frac{\partial}{c\partial t} + i\nabla$ is sent by a Lorentz transformation onto $p^c D p^t$ and the so-called *six-vector* $F = M + iE$ is sent onto $p^c F p^{ct}$, so that $J = \rho + i\rho\frac{d\xi}{cdt}$ is sent onto pJp^{ct}. Thus J transforms like the mass-momentum biquaternion and may also be written as $\rho_0 \frac{dx}{ds}$, where

$$\rho = \rho_0 \frac{cdt}{ds} = \rho_0 \left(1 - \frac{v^2}{c^2}\right)^{-\frac{1}{2}}.$$

If the above rules for transforming D, F and J are adopted, it thus becomes clear that Maxwell's equations are preserved under Lorentz transformations. This fact, though without the aid of quaternions, appears to have first been noted by Poincaré.

11
Quaternions in Quantum Mechanics

What about quantum mechanics? If we adopt the representation of quaternions by 2×2 complex matrices (Chapter 8), the matrices ii_1, ii_2 and ii_3 are known as *Pauli spin matrices*, except for sign.

Now let us consider the relativistic form of Schrödinger's wave equation for the electron. This is known as the *Klein-Gordon equation* and is usually written

$$\left(\frac{\partial^2}{c^2 \partial t^2} - \nabla \circ \nabla\right)\phi = -\mu^2 \phi,$$

where $\mu = 2\pi m_0/h$ is proportional to the rest-mass m_0 of the electron, h being Planck's constant. Using biquaternion notation, we write this

$$D^c D \phi = -\mu^2 \phi.$$

It is assumed that $\phi = \phi_0 + i\phi_1$ is a complex valued function of the position x in Minkowski space.

The second order Klein-Gordon equation may be replaced by two first order equations as follows. Putting $D\phi = \mu\chi$, where $\chi = \chi_0 + i\chi_1$, we obtain

$$\mu D^c \chi = D^c D \phi = -\mu^2 \phi.$$

Hence the Klein-Gordon equation is equivalent to the following pair of equations:

$$D\phi = \mu\chi, \qquad D^c \chi = -\mu\phi.$$

Can these be combined into one first order equation?

Assume for the moment that there is an entity j such that $j^2 = -1$, $ji = -ij$, and $ji_k = i_k j$ for $k = 1, 2$ and 3. Then we have

$$\begin{aligned} D(\phi + j\chi) &= D\phi + Dj\chi \\ &= \mu\chi + jD^c\chi \\ &= \mu(\chi - j\phi) \\ &= -j\mu(\phi + j\chi). \end{aligned}$$

There is certainly no complex 4×4 matrix j which anticommutes with the complex number i. But let us pass to real 4×4 matrices and identify i_k with its first representation $L(i_k)$ (Chapter 8). We shall write j_k for $R(i_k)$, the second representation of i_k, as in Chapter 8, Exercise 7. Then

$$j_1^2 = j_2^2 = j_3^2 = j_3 j_2 j_1 = -1, \qquad j_k i_l = i_l j_k \qquad (k, l = 1, 2, 3).$$

Now replace the complex number i by the real matrix j_1 and write j_2 for j. Then the assumption made above is justified. Putting

$$\psi = \phi + j_2 \chi = \phi_0 + j_1 \phi_1 + j_2 \chi_0 + j_3 \chi_1,$$

we may write the above equation as follows:

$$D\psi + j_2 \mu \psi = 0.$$

This is essentially Dirac's equation for the electron.

It can be shown that the sixteen matrices 1, i_k, j_l, $i_k j_l$ ($k, l = 1, 2, 3$) are linearly independent (e.g., Jacobson [1980] p. 218, Theorem 4.6). Thus ψ is just an arbitrary real 4×4 matrix. However, as we may multiply Dirac's equation by the column vector $(1\,0\,0\,0)^t$, we may assume, without loss of generality, that ψ itself is a column vector $(\psi_0\,\psi_1\,\psi_2\,\psi_3)^t$ with real components. Nothing prevents us from allowing the ψ_k to be complex numbers, but, as far as the present analysis is concerned, there is no compelling reason for doing so. (However, complex values are forced upon us, as soon as we look at the electron in an electromagnetic field.)

Since a Lorentz transformation sends D onto $p^c D p^t$, we want ψ to be transformed to $p\psi$, hence $D\psi$ to $p^c D\psi$ and $j_2 \mu \psi$ to $j_2 \mu p \psi = p^c j_2 \mu \psi$, thus making the Dirac equation Lorentz invariant.

It is important to note that the biquaternions of norm 1, p and $-p$, while yielding the same Lorentz transformation, both sending x onto pxp^{ct}, induce distinct transformations on ψ, sending it to $p\psi$ and $-p\psi$, respectively. This is the mathematical reason for saying that the electron has *spin* $\frac{1}{2}$.

Exercises

1. If the biquaternion a is viewed as a real 4×4 matrix, show that its transpose is not a^t, but a^{ct}, hence a point in Minkowski space is represented not by a Hermitian, but by a symmetric matrix. A general symmetric matrix has the form

$$x = x_0 + \sum_{k,l=1}^{3} x_{kl} i_k j_l,$$

   but, for a point in Minkowski space $x_{kl} = 0$ unless $k = l = 1$, since Minkowski space has only four and not ten dimensions. (One version of *string theory* does allow for ten dimensions, presumably for quite different reasons.)

2. Maxwell predicted from his equations that electromagnetic energy is propagated in waves. Einstein used the equation $E = mc^2$ to predict that mass is convertible into energy. Dirac observed that in his equation μ might as well be replaced by $-\mu$ and predicted that the electron must have an anti-particle, now called the positron. Discuss to what extent such predictions from mathematical symbolism to physical reality are justified.

12
Cardinal Numbers

The ideas in this chapter (and Chapter 13) are due to Georg Cantor (1845–1918). Many mathematicians at first rejected Cantor's work for ideological reasons, claiming that there could be no 'actual infinity' in mathematics. Cantor found it impossible to get a decent job and, in 1884, suffered a mental breakdown from which he never fully recovered.

The *cardinal numbers* include the natural numbers $0, 1, 2, \ldots$, but go beyond them by including various kinds of infinity. We have some of the same problems in defining a cardinal number as we had with the number 2. Roughly speaking, a cardinal number tells you how many elements there are in a given set. This notion is clear enough for finite sets, but for infinite sets we need some more discussion.

Two sets are said to have the *same cardinality* if and only if there is a one-to-one correspondence between them:
$A \cong B \iff$ there is some $f : A \to B$ such that f is one-to-one and onto (or *injective* and *surjective*).
For example, $\{3, 5, 7\}$ has the same cardinality as $\{0, 1, 2\}$. Note that $\cong$ is an equivalence relation.

If n is a positive integer, we say that a set A has *cardinality* n if and only if $A \cong \{0, 1, 2, \cdots, n-1\}$, and we write $|A| = n$. The cadinality of the empty set is 0: $|\emptyset| = 0$.

A set A is *finite* when $|A|$ is a natural number. Otherwise A is *infinite*. For example, $\mathbf{N}$ is infinite.

We say a set A has cardinality $\aleph_0$ if and only if $A \cong \mathbf{N}$. Such sets are called 'countably infinite' or just 'countable'. For example, $f(m) = 2m$ is a 1-to-1 function from the set $\mathbf{N}$ of natural numbers onto the set E of even numbers. Thus the cardinality of E is $\aleph_0$ and we write $|E| = \aleph_0$. The above example is a special case of what is usually called 'Galileo's Theorem':

If a set S has cardinality $\aleph_0$, then any infinite subset of S has cardinality $\aleph_0$.

Galileo's theorem might lead one to think that there is only one infinite cardinal. However, one of Cantor's achievements was to show that the power set of any set has a higher cardinality than the set itself. By the *power set* of a set S, we mean the set $P(S)$ whose members are the subsets of S.

We write $|A| < |B|$ (A has a lower cardinality than B) if and only if $A \not\cong B$ but B has a proper subset C such that $A \cong C$.

Note that $P(S)$ always has a proper subset C, the set of singletons of S, such that $C \cong S$.

We now have Cantor's Theorem:

Theorem 12.1. *Any set A has a lower cardinality than $P(A)$.*

Proof: To obtain a contradiction, suppose there is a 1-to-1, onto function $f : A \to P(A)$. Then every subset of A has the form $f(a)$ for some $a \in A$.

Let $S = \{x \in A | x \notin f(x)\}$. This is a subset of A, so there is some $a \in A$ such that $S = f(a)$. If $a \in S$ then, by the defining property of S, $a \notin f(a) = S$, hence $a \notin S$. But then $a \notin f(a)$ and so $a \in S$. Contradiction.

If a finite set has n elements then its power set has 2^n elements. (This is because, in forming a subset, there are two choices for each of the elements in the original set: include it in the subset or leave it out.) In general, if a set has cardinality k, we use the symbol 2^k to denote the cardinality of its power set. For example, $|P(\mathbf{N})| = 2^{\aleph_0}$. By Cantor's Theorem,

$$\aleph_0 < 2^{\aleph_0} < 2^{2^{\aleph_0}} < \cdots.$$

In other words, there are an infinite number of infinities. (It was statements like these that got Cantor into trouble with Kronecker.)

Is there some subset of $P(N)$ whose cardinality is greater than $\aleph_0$ but not as great as $2^{\aleph_0}$? The hypothesis that the answer to this question is NO is called the *continuum hypothesis*. In 1940 Kurt Gödel showed that the continuum hypothesis is consistent with the usual axioms of set theory. In 1963 Paul Cohen showed that the negation of the continuum hypothesis is also consistent with the usual axioms of set theory. In other words, our basic notions about sets neither imply nor preclude the continuum hypothesis. We say that the continuum hypothesis is *independent* of the axioms of set theory.

Exercises

1. Suppose that $f : A \to B$, $g : B \to A$, $fg = 1_B$ and $gf = 1_A$ (where 1_S is the identity function on S). Prove that f is one-one and onto. Conversely, if f is one-one and onto, show that it has an inverse g.

2. Let S be the set of natural numbers of the form $2^a 3^b$ where a and b are positive integers whose gcd is 1. Use Galileo's theorem to show that $|S| = \aleph_0$.

3. Using Exercise 2, show that the set of positive rationals has cardinality $\aleph_0$.

4. Let T be the set of all sequences formed from 0's and 1's. Show that $|T| = 2^{\aleph_0}$.

5. Let T be the set of reals from 0 to 1. Show that $|T| = 2^{\aleph_0}$.

6. Prove Galileo's Theorem.

7. Prove that the set of all functions from $\mathbf{N}$ to $\mathbf{N}$ has higher cardinality than $\mathbf{N}$.

13
Cardinal Arithmetic

We begin by defining three binary operations for sets in general:

$$A \times B = \{(a,b) | a \in A \text{ and } b \in B\},$$
$$A^B = \{f | f : B \to A\},$$
$$A + B = (A \times 0) \cup (B \times 1).$$

Note that $A \times \emptyset = \emptyset$, and, if $B \neq \emptyset$, $\emptyset^B = \emptyset$. These definitions are motivated by the fact that, for finite sets A and B,

$$|A \times B| = |A| \times |B|,$$
$$|A^B| = |A|^{|B|},$$
$$|A + B| = |A| + |B|.$$

We also have the following theorem, which generalizes the results of Chapter 2. We have written 0 for $\emptyset$ and 1 for $\{\emptyset\}$.

Theorem 13.1.

1. $A + B \cong B + A$,

2. $(A + B) + C \cong A + (B + C)$,

3. $A \times B \cong B \times A$,

4. $(A \times B) \times C \cong A \times (B \times C)$,

5. $A \times (B + C) \cong (A \times B) + (A \times C)$,

6. $(A^B)^C \cong A^{C \times B}$,

7. $A^B \times A^C \cong A^{B+C}$,

8. $(A \times B)^C \cong A^C \times B^C$,

9. $A + 0 \cong A$,

10. $A \times 1 \cong A$,

11. $A \times 0 \cong 0$,

12. $A^0 \cong 1$, $A^1 \cong A$ and, if $B \neq 0$, then $0^B \cong 0$.

We shall not give a complete proof of this theorem, but, as an example, we shall prove that $(A^B)^C \cong A^{C \times B}$.

Let f be any member of $(A^B)^C$, that is, let f be any function such that $f : C \longrightarrow A^B$. Then, if c is a typical element of C, $f(c) \in A^B$. Hence, if b is a typical element of B, $(f(c))(b) \in A$. Let g be any member of $A^{C \times B}$, that is, let g be any function such that $g : C \times B \rightarrow A$. Then, if (c, b) is a typical element of $C \times B$ (with $c \in C$, $b \in B$), $g((c, b)) \in A$.

We define $F : A^{C \times B} \longrightarrow (A^B)^C$ such that $((F(g))(c))(b) = g((c, b))$. We define $G : (A^B)^C \longrightarrow A^{C \times B}$ such that $(G(f))((c, b)) = (f(c))(b)$.

Then FG is just the identity function on $(A^B)^C$. For $(FG)(f) = F(G(f))$ and this equals f just in case, for any $c \in C$, $(F(G(f)))(c) = f(c)$. Moreover, the last equation is true just in case, for any $b \in B$, $((F(G(f)))(c))(b) = (f(c))(b)$. Now $((F(G(f)))(c))(b) = (G(f))((c, b))$, by the definition of F. Furthermore, by the definition of G, $(G(f))((c, b)) = (f(c))(b)$ as required.

Again, GF is just the identity function on $A^{C \times B}$. For $(GF)(g) = G(F(g))$ and this equals g just in case, for any $(c, b) \in C \times B$, $(G(F(g)))((c, b)) = g((c, b))$. By the definition of G, $(G(F(g)))((c, b)) = ((F(g))(c))(b)$, and, by the definition of F this equals $g((c, b))$ as required.

Hence F is one-one and onto (Chapter 12, Exercise 1). Applying the function F is called *currying* in Computer Science, after the logician Haskell B. Curry.

Here are some hints for proving the remaining 11 parts of the above theorem, skipping a few unnecessary parentheses.

(1) An element of the left-hand side must be of the form $(a, 0)$ or $(b, 1)$, where $a \in A$ and $b \in B$. Let $F(a, 0) = (a, 1)$ and $F(b, 1) = (b, 0)$. Find the inverse G of F.

(3) An element of the left-hand side has the form (a, b), where $a \in A$ and $b \in B$. Let $F(a, b) = (b, a)$ and find the inverse G of F.

(5) An element of the left-hand side has the form $(a, (b, 0))$ or $(a, (c, 1))$, where $a \in A$, $b \in B$ and $c \in C$. Let $F(a, (b, 0)) = ((a, b), 0)$, $F(a, (c, 1)) = ((a, c), 1)$ and find the inverse G of F.

(7) An element of the left-hand side has the form (f,g), where $f : B \longrightarrow A$ and $g : C \longrightarrow A$. Define $F(f,g)$ by stipulating $F(f,g)(b,0) = f(b)$ and $F(f,g)(c,1) = g(c)$. An element of the right-hand side has the form $h : B+C \longrightarrow A$. Define $G(h)$ as the pair $(G(h)_0, G(h)_1)$, where $G(h)_0(b) = h(b,0)$ and $G(h)_1(c) = h(c,1)$. Show that G is the inverse of F.

(8) An element of the left-hand side has the form $h : C \longrightarrow A \times B$. From this we obtain two functions $h_0 : C \longrightarrow A$ and $h_1 : C \longrightarrow B$ such that $h(c) = (h_0(c), h_1(c))$. Let $F(h) = (h_0, h_1)$. An element of the right-hand side is a pair (f,g), where $f : C \longrightarrow A$ and $g : C \longrightarrow B$. Let $G(f,g)(c) = (f(c), g(c))$ and show that G is the inverse of F.

(12) An element of A^B is a function $f : B \longrightarrow A$, that is, a subset f of $A \times B$ such that, for every $b \in B$, there exists a unique $a \in A$ such that $(a,b) \in f$. Show that there is exactly one function $\emptyset \longrightarrow A$, that the functions $\{\emptyset\} \longrightarrow A$ are in one-to-one correspondence with the elements of A, and, finally that, when $B \neq \emptyset$, there is no function $B \longrightarrow \emptyset$.

We can now define equivalent binary operations for cardinals, so that

$$|A| + |B| = |A + B|,$$
$$|A| \times |B| = |A \times B|,$$
$$|A|^{|B|} = |A^B|.$$

For example, $\aleph_0 + 2 = \aleph_0$ since $\mathbf{N} + \{0,1\} \cong \mathbf{N}$.

Exercises

1. Prove the remaining eleven parts of the above theorem.

2. Prove that $|A| \times (|B| + |C|) = (|A| \times |B|) + (|A| \times |C|)$.

3. Prove that $\aleph_0 + \aleph_0 = \aleph_0$.

4. Prove that $\aleph_0 \times 2 = \aleph_0$.

5. Prove that $\mathbf{N} + \mathbf{N}$ and $\mathbf{N} \times \mathbf{N}$ have the same cardinality as $\mathbf{N}$.

6. Prove that $\aleph_0^2 = \aleph_0$.

7. Prove that $|A| \times |A| \times |A| = |A|^3$.

8. Simplify $2^{\aleph_0} \times 2^{\aleph_0}$.

9. Simplify $8^{\aleph_0}$.

13. Cardinal Arithmetic

10. Prove that every infinite set has a subset which can be placed in one-one correspondence with **N**.

11. Project: Look up the Schroeder-Bernstein Theorem and prove:

$$\aleph_0^{\aleph_0^{\aleph_0}} = 2^{2^{\aleph_0}}.$$

14
Continued Fractions

As Fowler shows in *The Mathematics of Plato's Academy*, continued fractions are implicit in ancient Greek mathematics. As far as we know, they were not explicitly defined before 1618, when Daniel Schwenter rediscovered them.

Continued fractions are implicit in Euclid's Algorithm (Book VII, Proposition 2) dating from 300 BC. This is a procedure for finding the greatest common divisor (gcd) of two positive integers. The *greatest common divisor* of two positive integers a and b is the positive integer d whose divisors are precisely the common divisors of a and b. In other words, d divides a and b and any common divisor of a and b divides d. This notion can easily be extended to natural numbers or even integers, but one has to be careful. For example, $\gcd(0, 17) = 17$, but $\gcd(0, 0) = 0$, even though 17 is also a common divisor of 0 and 0, yet $17 > 0$. For the gcd of two nonzero integers, say 12 and -15, one has two candidates that meet the above definition, 3 and -3 in this case, and one usually chooses the positive one. The algorithm is best described with the help of an example. Suppose we want to find the gcd of 502 and 1604. We perform four divisions as follows.

	3		5		8		6
502	1604	98	502	12	98	2	12
	1506		490		96		12
	98		12		2		0

Since $dividend = (divisor \times quotient) + remainder$, any number which divides dividend and divisor will also divide the remainder. Thus any factors common to the two original integers are carried through the calculations until they surface in the last nonzero remainder. The last nonzero remainder is thus the gcd of the original integers. The above calculations can be rewritten as follows:

$$\frac{1604}{502} = 3 + \frac{1}{\frac{502}{98}} = 3 + \frac{1}{5 + \frac{12}{98}} = 3 + \frac{1}{5 + \frac{1}{\frac{98}{12}}} = 3 + \frac{1}{5 + \frac{1}{8 + \frac{2}{12}}} = 3 + \frac{1}{5 + \frac{1}{8 + \frac{1}{6}}}$$

The final expression is a *(simple) continued fraction.*

Standard abbreviations for continued fractions allow us to write the above fraction as $3 + \frac{1}{5+} \frac{1}{8+} \frac{1}{6}$ or as (3,5,8,6). Thus (a_0, a_1, a_2) is the fraction

$$a_0 + \cfrac{1}{a_1 + \cfrac{1}{a_2}}.$$

When one divides by a positive integer, one always obtains a remainder which is less than the divisor. Thus the sequence of divisors in Euclid's Algorithm steadily decreases. Using Euclid's Algorithm, we can write any positive rational number as a finite continued fraction $(a_0, a_1, a_2, \ldots, a_n)$, where the a_i are natural numbers and, for $i > 0$, $a_i > 0$.

However, even irrational real numbers can be written as continued fractions, and this too was known to the ancient Greeks, according to Fowler. We then obtain infinite continued fractions $(a_0, a_1, a_2, \ldots)$; but these may be approximated by finite continued fractions $c_0 = a_0$, $c_1 = (a_0, a_1)$, $c_2 = (a_0, a_1, a_2)$, etc., called *convergents*. Ultimately we are interested in the case when the a_n are all positive integers, with the possible exception of a_0. For the following argument only, we shall allow the a_n to be positive rational numbers for $n > 0$.

It is easily seen that $c_0 = \frac{a_0}{1}$ and $c_1 = \frac{a_0 a_1 + 1}{a_1}$. To calculate c_n when $n > 1$ we define two sequences of rationals (ultimately integers):

$$p_0 = a_0, \; p_1 = a_0 a_1 + 1, \; p_n = a_n p_{n-1} + p_{n-2} \text{ if } n > 1,$$

$$q_0 = 1, \; q_1 = a_1, \; q_n = a_n q_{n-1} + q_{n-2} \text{ if } n > 1.$$

We claim that $c_n = \frac{p_n}{q_n}$ for all n. This is evidently so when $n = 0$ or 1, so assume the result for n for any admissible $(a_0, a_1, \ldots, a_n)$; we shall show that it also holds for $n + 1$. Now c_{n+1} can be obtained from

$$c_n = \frac{a_n p_{n-1} + p_{n-2}}{a_n q_{n-1} + q_{n-2}}$$

by replacing a_n by $a_n + \frac{1}{a_{n+1}}$, with the help of the induction assumption applied to $(a_0, \ldots, a_n + \frac{1}{a_{n+1}})$. Multiplying top and bottom of the resulting ratio by a_{n+1}, the top becomes

$$\begin{aligned} a_{n+1}\left(\left(a_n + \frac{1}{a_{n+1}}\right)p_{n-1} + p_{n-2}\right) &= a_{n+1}(a_n p_{n-1} + p_{n-2}) + p_{n-1} \\ &= a_{n+1}p_n + p_{n-1} = p_{n+1}. \end{aligned}$$

Similarly, the bottom becomes q_{n+1}, so $c_{n+1} = p_{n+1}/q_{n+1}$, and therefore the result holds by mathematical induction.

The proof involved the rational number $a_n + \frac{1}{a_{n+1}}$ in the induction assumption, but from now on we shall stick to integer a_n. We have proved:

Theorem 14.1. *If $a_0 \in \mathbf{N}$ and $0 < a_n \in \mathbf{N}$ for $n > 0$, the nth convergent of $(a_0, a_1, a_2, \ldots)$ is p_n/q_n, where the p_n and q_n are defined inductively as above.*

Note that, if the a_n are all natural numbers, as we are now supposing, the inductively defined p_n and q_n will be positive integers, in fact, strictly increasing sequences of positive integers.

Theorem 14.2. *Let p_n and q_n be defined inductively as above. For all $n > 0$, $p_n q_{n-1} - p_{n-1} q_n = (-1)^{n-1}$.*

Proof: This is evidently so when $n = 1$ or 2. Assume the result for n. Then

$$\begin{aligned} p_{n+1}q_n - p_n q_{n+1} &= (a_{n+1}p_n + p_{n-1})q_n - p_n(a_{n+1}q_n + q_{n-1}) \\ &= p_{n-1}q_n - p_n q_{n-1} \\ &= -(-1)^{n-1} \\ &= (-1)^n. \end{aligned}$$

An immediate consequence of this theorem is

Corollary 14.3. *For all $n \in \mathbf{N}$, $\gcd(p_n, q_n) = 1$; in other words, p_n/q_n is in lowest terms.*

Another immediate consequence is

Corollary 14.4. *For all $n > 0$, $c_n - c_{n-1} = \frac{p_n}{q_n} - \frac{p_{n-1}}{q_{n-1}} = \frac{(-1)^{n-1}}{q_n q_{n-1}}$.*

It follows that, for odd n,

$$c_n - c_{n-2} = \frac{1}{q_n q_{n-1}} - \frac{1}{q_{n-2}q_{n-1}} < 0,$$

since the q_n are strictly increasing, and similarly, for even n, $c_n - c_{n-2} > 0$. Thus we have a strictly increasing sequence

$$c_1 > c_3 > c_5 > \cdots$$

and a strictly decreasing sequence

$$c_0 < c_2 < c_4 < \cdots.$$

Moreover, by Corollary 14.4, the difference of the two sequences tends to 0, hence they must have a common limit a and we write $a = (a_0, a_1, a_2, \ldots)$.

For example, the infinite continued fraction $(1, 1, 1, \ldots)$ has a limit x, hence $x = 1 + \frac{1}{x}$, so $x^2 - x - 1 = 0$, hence $x = \frac{1}{2}(1 + \sqrt{5})$.

Exercises

1. Find the gcd of 10403 and 2987.

2. Show that $\gcd(a, b) = \gcd(a, a - b)$.

3. Calculate (1,2,1,2,1,2, ...).

4. Show that every positive rational number b/c can be written as a simple continued fraction with an even number of a_i.

5. Let b/c be a positive proper fraction in lowest terms. Using Exercise 4, show that b/c can be written in the form $1/d + e/f$ where $b > e$ and $d > c > f$.

6. Using Exercise 5, show that every proper reduced fraction can be expressed as a sum of distinct *unit fractions* (that is, fractions with numerator 1 and a positive integer denominator).

7. Use the method of Exercises 5 and 6 to express 67/120 as a sum of distinct unit fractions.

8. Associate with each letter the number of its place in the (English) alphabet. Then each word is associated with a sequence $a_0, a_1, \ldots, a_n$. Encode the word into $(a_0, a_1, \ldots, a_n)$, simplified into an ordinary fraction. How would you decipher such a fraction?

9. Show that $(2a, a, 2a, a, 2a, a, \ldots) = a + \sqrt{a^2 + 2}$.

15
The Fundamental Theorem of Arithmetic

One can use Euclid's algorithm to find the gcd of two positive integers a and b. One can also exploit the algorithm to express the gcd d in the form $d = ax + by$ where x and y are integers. Actually, d is the smallest positive integer with this property, and this fact can also be used to describe the gcd of a and b.

Theorem 15.1. *Given positive integers a and b, their gcd is the smallest positive integer d such that $d = ax + by$ with $x, y \in \mathbf{Z}$.*

Proof: Note that the set $\{ax + by | x, y \in \mathbf{Z}\}$ does contain positive integers, e.g., $2ab$. Let d be the smallest positive integer in the set. Clearly any common divisor of a and b divides d. So, to prove that $d = \gcd(a, b)$, we only have to show that d divides a and b. To prove that d divides a, divide a by d to get quotient q and remainder r, so that

$$a = qd + r, \ 0 \leq r < d.$$

Then $r = a - qd = a - q(ax + by) = a(1 - qx) + b(-qy)$ is also of the form $ax' + by'$ with $x', y' \in \mathbf{Z}$. Since d was the smallest positive integer of this form and since $0 \leq r < d$, it follows that $r = 0$, hence d divides a. Similarly, d divides b, so d is a common divisor of a and b.

The following consequence of the theorem is known as the *Fundamental Lemma of Arithmetic*.

Lemma 15.2. *Given positive integers a, b and c, if a divides bc and*

$\gcd(a, b) = 1$, *then a divides c. In particular, if a prime number p divides bc, then p divides b or p divides c.*

Proof: If $\gcd(a, b) = 1$, it follows from the theorem that there are integers x and y such that $ax + by = 1$. Moreover, if a divides bc, there is an integer z such that $bc = az$. Therefore,

$$c = axc + byc = a(xc + yz),$$

and so a divides c.

The Fundamental Theorem of Arithmetic asserts that every positive integer has a factorization into primes; moreover, if we disregard the order of the primes, this factorization is unique. This theorem follows quite easily from the above lemma and we leave it as an exercise. For another proof and a discussion of its history, see Part I, Chapter 3.

Given positive integers a and b, say with $a > b$, we obtain a continued fraction expansion $a/b = (a_0, a_1, \ldots, a_n)$. Even though this continued fraction is *finite*, we can still use the analysis of the previous section to calculate its convergents p_0/q_0 up to p_n/q_n. In particular, $a/b = p_n/q_n$, hence $aq_n = bp_n$. Since $\gcd(p_n, q_n) = 1$, it follows from the Fundamental Lemma of Arithmetic that p_n divides a, say $a = p_n d$, whence also $b = q_n d$. Evidently, d is the gcd of a and b. But it also follows from Theorem 14.2, upon multiplying by d, that $aq_{n-1} - bp_{n-1} = d(-1)^{n-1}$. This allows us to find particular integers x and y such that $d = ax + by$, namely,

$$x = (-1)^{n-1} q_{n-1}, \qquad y = (-1)^n p_{n-1}.$$

Exercises

1. Prove that the smallest divisor > 1 of a positive integer > 1 is a prime number.

2. Deduce that every positive integer > 1 is a product of prime numbers and use the Fundamental Lemma of Arithmetic to show that this factorization into primes is unique.

3. If a, b and c are positive integers and $c > 1$, show that

$$\gcd(c^a - 1, c^b - 1) = c^d - 1,$$

where $d = \gcd(a, b)$. (Hint: use the above theorem.)

16
Linear Diophantine Equations

A *linear Diophantine equation* in two variables is an equation of the form $ax + by = c$ where a, b and c are given integers, and x and y are unknown integers. Sometimes x and y are restricted to the set of *positive* integers.

For example, $4x + 6y = 8$ is a linear Diophantine equation. It has solution $x = 2$ and $y = 0$ (among others). Here we are not interested in noninteger solutions such as $x = \frac{1}{2}$, $y = 1$.

In order to simplify the presentation, we shall allow negative numbers as solutions, but we shall take a, b and c to be positive. By dividing the material into different cases, we could elaborate the whole theory in terms of positive integers. Thus, there is nothing in this chapter which would be inaccessible to the ancient Greeks. On the contrary, it is probable that they used essentially the following method to attack these equations.

The adjective 'Diophantine' comes from the name 'Diophantus'. Diophantus of Alexandria lived about 250 AD. However, in his equations he did not restrict x and y to be integers, but allowed them to be rationals. It was Brahmagupta of India (628 AD) who gave the first complete solution to the linear Diophantine equation (Boyer [1989], pp. 244-47).

Let $d = \gcd(a, b)$. Then there are integers a' and b' such that $a = da'$ and $b = db'$. If $ax + by = c$ then $d(a'x + b'y) = c$ and hence d is a factor of c. Thus, if d is not a factor of c, the Diophantine equation has no solutions. On the other hand, if d is a factor of c, then $c = dc'$ for some integer c', and we can cancel d to get $a'x + b'y = c'$, with $\gcd(a', b') = 1$. In giving a solution of $ax + by = c$, we can thus, without loss of generality, begin by assuming that $\gcd(a, b) = 1$.

To solve the Diophantine equation $ax + by = c$ when $\gcd(a, b) = 1$:

16. Linear Diophantine Equations

1. Use Euclid's algorithm to write $a/b = (a_0, a_1, \ldots, a_n)$.

2. Since a/b is in lowest terms, $a = \pm p_n$ and $b = \pm q_n$. Since $p_n q_{n-1} - p_{n-1} q_n = \pm 1$, it follows that, with some selection of signs, $x_0 = \pm c q_{n-1}$, $y_0 = \pm c p_{n-1}$ is a solution of the Diophantine equation. Thus, using the recursive definitions given in Chapter 14, calculate p_{n-1} and q_{n-1} and pick the signs so that x_0, y_0 is a solution.

3. If t is any integer, $x = x_0 + bt$, $y = y_0 - at$ is another solution of the Diophantine equation.

4. There are no integer solutions other than those mentioned in (3) above. For if $ax + by = c = ax_0 + by_0$, then $a(x - x_0) = b(y_0 - y)$ and b is a factor of $x - x_0$, since $\gcd(a, b) = 1$. Hence, for some integer t, $x = x_0 + bt$.

Example: Solve $25x + 55y = 50$.

First we get the gcd of 25 and 55 by using Euclid's algorithm:

$$25/55 = 0 + \frac{1}{55/25} = 0 + \frac{1}{2 + 5/25} = 0 + \frac{1}{2 + 1/5} = (0, 2, 5).$$

The gcd is the last divisor, namely, 5. (It is merely a coincidence that the last partial quotient is also 5.) We note that 5 divides 50 and so the equation does have integer solutions. We factor out the 5, obtaining $5x + 11y = 10$. We calculate the penultimate convergent $p_1/q_1 = (0, 2) = 1/2$. With the right signs, one solution is $x_0 = \pm 10 \times 2$ and $y_0 = \pm 10 \times 1$. Indeed, we can take $x_0 = -20$ and $y_0 = 10$. The general solution is $x = -20 + 11t$, $y = 10 - 5t$, where t is any integer.

Sometimes we want only positive solutions. In that case we must have $x_0 + bt > 0$ and $y_0 - at > 0$. In the above example this would require an integer t such that $1 < 20/11 < t < 10/5 = 2$, which is impossible.

Exercises

1. Solve the Diophantine equation $101x + 753y = 100,000$. (There are two positive integer solutions.)

2. Solve the Diophantine equation $158x + 57y = 20,000$. (There are two positive integer solutions.)

3. Solve the Diophantine equation $91x + 221y = 1053$. (There are no positive integer solutions.)

4. Show that the following Diophantine equation has a unique solution in positive integers: $17x + 19y = 320$.

5. The Sultana used to divide her maids into two companies, one which would follow her five abreast and the other which would follow her seven abreast – both in rectangular formation. These companies would consist of different numbers of maids on each of nine different days. What is the smallest number of maids the Sultana could have had?

6. Show that, if a and b are positive integers, then $ax + by = c$ has no solutions in positive integers when $[-x_0/b] \geq [y_0/a]$. (Here x_0 and y_0 are the solutions described above, and $[z]$ is the greatest integer not exceeding z.)

17
Quadratic Surds

Let d be a positive nonsquare integer. Using the Fundamental Theorem of Arithmetic, it is not hard to show that $\sqrt{d}$ is irrational. This was first proved by Theaetetus in about 400 BC.

If a and $b \neq 0$ are integers, the expression $(a+\sqrt{d})/b$ is called a *quadratic surd*. Note that if a' and b' are integers, and $(a + \sqrt{d})/b = (a' + \sqrt{d})/b'$ then $a = a'$ and $b = b'$. The proof is left as an exercise.

In this chapter we study continued fraction expansions of quadratic surds. These expansions can be used to solve quadratic Diophantine equations, such as the 'Pell equation' $x^2 - dy^2 = 1$.

To begin with an example, suppose that $x = (1,1,1,\ldots)$. Then $x = 1 + 1/x$ and hence $x^2 - x - 1 = 0$, with the result that $x = \frac{1}{2}(1 + \sqrt{5})$. (We cannot take the other root of the equation since x is positive.) This is a quadratic surd. Indeed, it is a very famous one known as 'the golden ratio'. It is the ratio of the side to the base in the triangles obtained by connecting the five points of the Pythagorean star.

As another example, suppose that $y = (1, 1, 2, 1, 2, 1, \ldots)$. To evaluate this continued fraction, let $x = (1, 2, 1, 2, 1, \ldots)$. Then $y = 1 + 1/x$. Furthermore, $x = 1 + \frac{1}{2+1/x}$, so that $2x^2 - 2x - 1 = 0$ and hence $x = \frac{1}{2}(1+\sqrt{3})$. Thus $y = \sqrt{3}$.

Generalizing from these two examples, it is easily shown that any continued fraction which is ultimately periodic represents a quadratic surd. Less obvious is the converse of this statement, namely, that every quadratic surd has a continued fraction expansion which is ultimately periodic. This was first proved by Lagrange in about 1770. We shall prove the special case:

17. Quadratic Surds

Theorem 17.1. *If d is a positive integer which is not a perfect square, the continued fraction expansion of $\sqrt{d}$ is ultimately periodic.*

Proof: We define sequences of integers a_n and b_n and a sequence of rationals r_n as follows, where $[\rho]$ denotes the greatest integer in ρ:

$$a_0 = 0, \quad r_0 = 1, \quad b_n = \left[\frac{\sqrt{d}+a_n}{r_n}\right], \quad a_{n+1} = b_n r_n - a_n, \quad r_{n+1} = (d_n - a_{n+1}^2)/r_n.$$

Then it is easily verified that

$$\frac{\sqrt{d}+a_n}{r_n} = b_n + \cfrac{1}{b_{n+1} + \cfrac{1}{b_{n+2} + \cdots}},$$

in particular,

$$\sqrt{d} = b_0 + \cfrac{1}{b_1 + \cfrac{1}{b_2 + \cdots}}$$

and hence $b_{n+1} > 0$. It also follows by mathematical induction that the r_n are integers, once it is realized that

$$r_{n+2} = r_n - b_{n+1}^2 r_{n+1} + 2b_{n+1}a_{n+1}.$$

Let $x_n = (\sqrt{d}+a_n)/r_n$ and let t_n be its conjugate $(-\sqrt{d}+a_n)/r_n$. From Chapter 14 we know that

$$\sqrt{d} = \frac{x_n p_{n-1} + p_{n-2}}{x_n q_{n-1} + q_{n-2}}.$$

Taking conjugates, we obtain

$$-\sqrt{d} = \frac{t_n p_{n-1} + p_{n-2}}{t_n q_{n-1} + q_{n-2}}.$$

Solving for t_n, we obtain

$$t_n = \left(-\frac{q_{n-2}}{q_{n-1}}\right)\left(\frac{\sqrt{d}+p_{n-2}/q_{n-2}}{\sqrt{d}+p_{n-1}/q_{n-1}}\right).$$

As n increases, the second factor tends to 1. The q's are positive, so, for sufficiently large n, $t_n < 0$. Now $x_n > 1$ so, for sufficiently large n,

$$\frac{2\sqrt{d}}{r_n} = x_n - t_n > 1$$

and hence $r_n > 0$ and $2\sqrt{d} > r_n$. Since $t_n < 0$, it then follows that $a_n < \sqrt{d}$ while $x_n > 1$ implies $a_n > -\sqrt{d}$. Since

$$2\sqrt{d} > r_n > 0,$$

$$\sqrt{d} > a_n > -\sqrt{d},$$

there are only a finite number of possibilities for $x_n = (\sqrt{d} + a_n)/r_n$ and so the simple continued fraction must repeat.

Exercises

1. Find the continued fraction expansions of $\sqrt{2}$ and $\sqrt{10}$.

2. Find the quadratic surd which is represented by the periodic continued fraction $(a, b, c, a, b, c, \ldots)$.

3. Fill in the details in the proof of the above theorem.

4. Show that for nonsquare d, the continued fraction expansion of $\sqrt{d}$ is of the form $\sqrt{d} = (a_0, \overline{a_1, \ldots, a_{n-1}, 2a_0})$, with the part under the bar periodic.

5. For d as in Exercise 4, show that $p_{n-1}^2 - dq_{n-1}^2 = \pm 1$, where p_{n-1}/q_{n-1} is the $(n-1)$th convergent of $\sqrt{d}$. (Hint: show that

$$\sqrt{d} = \frac{\alpha_n p_{n-1} - p_{n-2}}{\alpha_n q_{n-1} - q_{n-2}}$$

where $\alpha_n = \sqrt{d} + a_0$.)

6. Find a positive integer solution to the equation $x^2 - 61y^2 = \pm 1$.

7. Prove that in $\frac{a + \sqrt{d}}{b}$, the positive integers a and b are uniquely determined.

Note that Exercises 4 and 5 are more difficult.

18
Pythagorean Triangles and Fermat's Last Theorem

A *Pythagorean triangle* is a right triangle all of whose sides have integer lengths. For example, let ABC be a triangle with a right angle at vertex C. Suppose that the sides AC and BC have lengths 5 and 12, respectively. Then the hypotenuse has length $\sqrt{5^2 + 12^2} = 13$. Since all three sides have integer lengths, this is a Pythagorean triangle.

Note that if k is any positive integer, the triangle with sides of lengths $5k, 12k$ and $13k$ is a Pythagorean triangle too. This triangle is just a magnification of the previous one and so it is not very interesting. However, it does suggest that it would be worthwhile to find all the Pythagorean triangles whose gcd is 1. These are called *primitive* Pythagorean triangles. It is easy to show that if x, y and z are the sides of a right triangle, then $\gcd(x, y, z) = \gcd(x, y) = \gcd(y, z)$. In what follows we shall assume that this gcd is equal to 1.

We shall use the following result:

Lemma 18.1. *If m and n are nonnegative integers such that $\gcd(m, n) = 1$ and mn is a square, then both m and n are squares.*

Professor Tournesol discovered an amusing proof of this theorem while he was in jail for having failed the president's son. The prisoners were put in a long row of cells. At first all the doors were unlocked, but then the jailor walked by and locked every second door. He walked by again and stopped at every third door, locking it if it was unlocked, but unlocking it if it was locked. On his next round he stopped at every fourth door, locking it if it was unlocked, unlocking it if it was locked, and so on. Professor Tournesol soon realised that the mth cell would be unlocked in the end just in case

m had an odd number of divisors. Now, if d divides m then so does m/d and it would seem that the divisors of m come in pairs. Unless...'what if $d = m/d$?', thought the professor, 'then the divisor d does not pair off with another, and $d = m/d$ just in case m is a square.'

Let $\tau(x)$ be the number of divisors of a positive integer x. Then $\tau(x)$ is odd just in case x is a square. Now, if $\gcd(m,n) = 1$ then a typical divisor of mn is of the form dg where d divides m and g divides n. Thus $\tau(mn) = \tau(m)\tau(n)$. If mn is a square then $\tau(mn)$ is odd, and hence both $\tau(m)$ and $\tau(n)$ are odd. Thus m and n are both squares. QED.

In order to solve the Diophantine equation $x^2 + y^2 = z^2$ with $\gcd(x,y,z) = 1$, first note that x and y are not both odd. For if $x = 2a+1$ and $y = 2b+1$ then $x^2 + y^2 = 4(a^2 + b^2 + a + b) + 2$ which is not a square, since squares of odd numbers have the form $4(c^2 + c) + 1$ and squares of even numbers have the form $4c^2$. Since $\gcd(x,y) = 1$, it follows that x and y are not both even either. Hence exactly one of x and y is even. Without loss of generality, let us say that $y = 2y'$ and that x is odd. Then x^2 is also odd and y^2 is even. It follows that $z^2 = x^2 + y^2$ is odd and thus z is odd. Since x and y are both odd, $\frac{1}{2}(z+x)$ and $\frac{1}{2}(z-x)$ are both integers. Moreover, their gcd is 1, since any factor which divides them both also divides their sum z and their difference x. But $\gcd(x,z) = 1$.

Since $x^2 + y^2 = z^2$, we have $\frac{1}{2}(z+x)\frac{1}{2}(z-x) = y'^2$. From Professor Tournesol's discovery it follows that there are positive integers u and v with $\gcd(u,v) = 1$ such that $\frac{1}{2}(z+x) = u^2$ and $\frac{1}{2}(z-x) = v^2$. This gives $z = u^2 + v^2$, $x = u^2 - v^2$ and $y = 2y' = 2uv$. Note that u and v are not both odd, since z is not even. The above may be summarized as follows:

Theorem 18.2. *Let x, y and z be positive integers. Then $x^2 + y^2 = z^2$ with y even and $\gcd(x,y,z) = 1$ if and only if, for some positive integers u and v, not both of which are odd, with $u > v$ and $\gcd(u,v) = 1$,*

$$x = u^2 - v^2, \ y = 2uv, \ and \ z = u^2 + v^2.$$

The sufficiency of the above condition (that is, the 'if' part of the theorem) was known to the Mesopotamians about 4000 years ago (Neugebauer).

The eighth problem in the second book of the *Arithmetica* of Diophantus is to express 16 as a sum of two rational squares. Fermat (1601–1665) had a copy of this book and he enjoyed writing notes in its margins. Unlike Diophantus, Fermat was only interested in (positive) integer solutions to the equations in the *Arithmetica*. Fermat knew that the square of an integer can often be expressed as a sum of two positive integer squares. In the margin beside the problem about expressing 16 as a sum of two squares, Fermat wrote:

18. Pythagorean Triangles and Fermat's Last Theorem

On the other hand it is impossible to separate a cube into two cubes, or a biquadratic into two biquadratics, or generally *any power except a square into two powers with the same exponent.* I have discovered a truly marvellous proof of this, which, however, the margin is not large enough to contain' (p. 145, Heath's translation of the *Arithmetica*).

Fermat's assertion is called his 'Last Theorem', although, until quite recently, it would have been safer to call it a 'conjecture'. It is now believed that Fermat only proved the special case when the power $n = 4$, and the 'theorem' remained an open problem for 350 years, though many special cases were proved in that period. The complete theorem was finally proved by Andrew Wiles of Princeton University in 1994. His proof depended on the work of many other mathematicians, notably on a crucial result by K. A. Ribet, as well as ideas from G. Y. Taniyama, G. Shimura, B. Mazur and G. Frey, among others, and the last minute collaboration of Richard Taylor.

Fermat himself showed that $z^4 - x^4 = w^2$ has no solution in positive integers, and from this it follows at once that $x^4 + y^4 = z^4$ has no solution in positive integers (*Oeuvres* I, p. 340 and *Arithmetica*, 2nd edn., p. 293). We shall give a proof of this, due to Euler, which uses Fermat's *method of descent* (from a larger to a smaller solution):

Theorem 18.3. *There are no positive integers x, y and z such that*

$$x^4 + y^4 = z^2.$$

Proof: To obtain a contradiction, suppose there are such integers. Let us take such a triple with the product xy minimized. Then $\gcd(x, y) = 1$. Since x^2, y^2 and z are the sides of a primitive Pythagorean triangle, exactly one of x and y is even. Without loss of generality, let us say that x is even. By our previous theorem, there are positive integers u and v, not both odd, with $\gcd(u, v) = 1$, such that $x^2 = 2uv$ and $y^2 = u^2 - v^2$. Since $v^2 + y^2 = u^2$ and y is odd, v must be even. Since $\gcd(2v, u) = 1$ and $2vu = x^2$, it follows that $2v$ and u are squares (recall Professor Tournesol). Thus $u = c^2$ for some positive integer c.

Again by our above theorem, there are positive integers s and t, not both odd, with $\gcd(s, t) = 1$, such that $v = 2st$ and $u = s^2 + t^2$. Since $2v$ is a square, so is $2v/4 = v/2 = st$. Thus there are positive integers a and b such that $s = a^2$ and $t = b^2$.

The fact that $u = s^2 + t^2$ implies that $a^4 + b^4 = c^2$. Moreover, $(ab)^2 = st = v/2 < 2uv = x^2 \leq (xy)^2$ so that $ab < xy$. But this contradicts the minimality of xy.

Exercises

1. Prove that if a triangle has sides of lengths x, y and z, and $x^2+y^2 = z^2$ then the triangle has a right angle.

2. Show that if x, y and z are integers such that $x^2 + y^2 = z^2$ then $\gcd(x, y, z) = \gcd(x, z)$.

3. How many primitive Pythagorean triangles are there with hypotenuse < 50?

4. Show that if m and n are positive integers with $\gcd(m, n) = 1$ and mn is a cube then m is a cube.

5. Show that if u, v, u' and v' are positive integers such that $u^2 - v^2 = u'^2 - v'^2$ and $2uv = 2u'v'$ then $u = u'$ and $v = v'$.

6. Solve the Diophantine equation $x^{28} + y^{28} = z^{28}$.

7. Find all Pythagorean triangles with perimeter 1716.

19
What Is a Calculation?

At the second International Congress of Mathematicians (Paris, 1900), David Hilbert (1862–1943) presented a list of 23 problems, which he hoped would occupy mathematicians in the 20th century. We shall only talk about three of these problems here, as they concern the foundations of mathematics.

1. Prove or disprove the Continuum Hypothesis (Chapter 12).

2. Show that arithmetic, described as an axiomatic system, is consistent, that is, that it does not admit a proof that $0 = 1$. (As Paul Erdös would say, if such a proof were ever to be discovered, the universe would vanish.)

3. Find an effective method or 'algorithm', as it is now called, for deciding whether a given polynomial Diophantine equation (with integer coefficients) is solvable (in integers). (For the origin of the word 'algorithm', see Part I, Chapter 22.)

This last problem is actually Hilbert's Problem 10. Today we know that these three problems cannot be solved in the way Hilbert had intended. Kurt Gödel showed in 1938 that the Continuum Hypothesis cannot be proved and Paul Cohen showed in 1964 that it cannot be disproved either! The existence of mathematical statements that can be neither proved nor disproved had already been established by Gödel in 1931. It followed from his argument that the consistency of any formal system of arithmetic cannot be proved, unless we allow a method of proof which is more powerful than the method of proof in the given system. We shall return to Gödel's result in a later chapter. After much preliminary work by Martin Davis, Hilary Putnam and Julia Robinson, Problem 10 was ultimately disposed of by Yuri Matijasevič, as we shall see presently. But before tackling this problem, one had to determine what Hilbert meant by an 'effective method'.

By a *numerical function* we mean any function $f : \mathbf{N}^n \to \mathbf{N}$, where $\mathbf{N}$ is the set of natural numbers (including 0) and $n \geq 0$. Among these are the *identity* function $fx = x$, and the successor function $fx = Sx$. These basic functions can surely be calculated. Moreover, the set of calculable functions is evidently closed under the following schemes, where we have written $\bar{x}$ for the string $x_1 x_2 \ldots x_m$, $\bar{z}$ for $z_1 \ldots z_k$, etc:

1. substituting one calculable function $g\bar{y}$ for u in another calculable function $f\bar{x}u\bar{z}$ to obtain a function $h\bar{x}\bar{y}\bar{z} = f\bar{x}g\bar{y}\bar{z}$;

2. interchanging two variables: if $f\bar{u}xy\bar{v}$ can be calculated, so can $g\bar{u}xy\bar{v} = f\bar{u}yx\bar{v}$;

3. contracting two variables: if $f\bar{u}xy\bar{v}$ can be calculated, so can $g\bar{u}x\bar{v} = f\bar{u}xx\bar{v}$;

4. introducing superfluous variables: if $f\bar{u}\bar{v}$ can be calculated, so can $g\bar{u}x\bar{v} = f\bar{u}\bar{v}$;

5. the *recursion scheme*: if $g\bar{x}$ and $h\bar{x}yz$ are calculable, so is $f\bar{x}y$ defined 'recursively' by the equations

$$f\bar{x}0 = g\bar{x}, \qquad f\bar{x}Sy = h\bar{x}yf\bar{x}y.$$

With the help of (5) we can calculate $x + y$, $x \times y$, x^y and many other numerical functions.

The functions which can be calculated so far, namely from the basic functions with the help of schemes (1) to (5), are called *primitive recursive functions*. They were introduced by Gödel in the proof of his famous Incompleteness Theorem, which led to the disposal of Hilbert's second problem and to which we shall return in a later chapter.

How are these primitive recursive functions to be calculated? Before the invention of modern computers and pocket calculators, we could make calculations with pencil and paper; but to explain what this means in precise terms is not easy. The first people to give rigorous answers to the question of what a calculation is were Alan Turing and Emil L. Post, independently in the same year, 1936. The work of Post is not as well-known as that of Turing, who invented a theoretical machine, the *Turing machine*, which may be seen as the ancestor of all modern computers. However, building a Turing machine is not an easy way to compute a given primitive recursive function.

19. What Is a Calculation?   247

A simpler answer would have been available if people had thought about the origin of the word 'calculation', which is derived from the Latin word 'calculus' meaning 'pebble'. Indeed, the ancients performed calculations by moving pebbles from one groove of a table, called an 'abacus', to another. We shall discuss how an abacus may be programmed to calculate any primitive recursive function.

For theoretical purposes we shall assume that an *abacus* consists of a potentially infinite number of *locations* and that we have an inexhaustable supply of *pebbles*. A *program* is made up of two basic instructions:

- put a pebble at location A, then go to ...

- take a pebble from location B, if B is not empty, then go to ..., else go to ...

(the right arrow referring to the 'else' case).

A program is easily illustrated by a *flow diagram*. For example, the following program

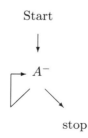

is the program 'empty A' and the next program

19. What Is a Calculation?

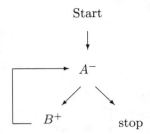

is the program 'transfer the content of A to B'.

To *calculate* a numerical function $z = fxy$ we begin with x pebbles at location X, y pebbles at location Y and no pebbles at any other location. We expect the calculation to stop with x pebbles at X, y pebbles at Y, fxy pebbles at Z and no pebbles at any other location. Similarly we define what it means to calculate $z = fx_1, \ldots, x_n$. The following flow diagram illustrates the programs for calculating the successor function $y = Sx$:

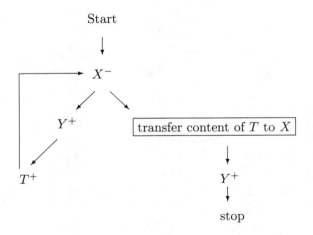

Note the *subroutine* 'transfer content of T to X'. T is a temporary storage location, which is empty at the beginning and at the end of the calculation.

To convince the reader that in fact all primitive recursive functions can be calculated on an abacus, we present a program for calcuating a function fxy obtained by the recursion scheme $fx0 = gx$, $fxSy = hxyfxy$ from functions gx and $hxyz$, which are already known to be calculable:

19. What Is a Calculation?   249

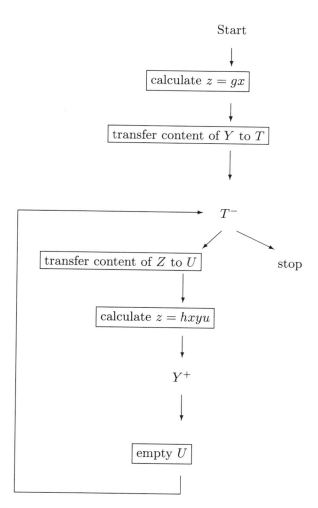

Primitive recursive functions are not the only numerical functions which can be calculated on an abacus. There is one more scheme, the *minimization scheme*:

6. given a calculable function $g\bar{x}y$, $\bar{x} = x_1 x_2 \ldots x_n$, we can calculate $f\bar{x} =$ *the smallest y such that $g\bar{x}y = 0$*.

Here is the flow diagram:

19. What Is a Calculation?

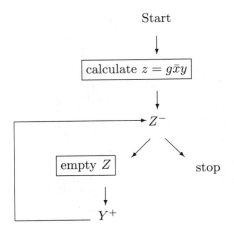

Of course, $f\bar{x}$ is only well-defined if $\forall x \exists y (g\bar{x}y = 0)$, otherwise f is not a function but a *partial function*. The class of (partial) functions obtained by adding the minimization scheme to the schemes discussed earlier is the class of *(partial) recursive functions*. These are in fact the only numerical (partial) functions which can be calculated on an abacus or, for that matter, on a Turing machine or on a modern computer.

Exercises

Construct programs for calculating the following numerical functions on an abacus:

1. $z = x + y$, $x \cdot y$, x^y;
2. $z = y!$;
3. $y = [x/2]$, where $[\alpha]$ is the greatest integer in α;
4. $y = [\sqrt{x}]$.

20

Recursive and Recursively Enumerable Sets

A set A of natural numbers is said to be *recursive* if there is a recursive (or calculable) function $f : \mathbf{N} \to \mathbf{N}$ such that $A = \{x \in \mathbf{N} | f(x) = 0\}$. In other words, A is recursive provided, for any natural number x, we can determine whether $x \in A$ by performing a calculation f on x. (Actually, it suffices here to take for f a primitive recursive function.) For example, the set of primes is recursive, since we can determine whether a given number is prime by dividing it by all natural numbers less than or equal to its square root. With a bit of work, one can find a recursive function f to do the job.

A set of natural numbers is *recursively enumerable* if there is a recursive function $g : \mathbf{N} \to \mathbf{N}$ such that $A = \{g(0), g(1), g(2), \ldots\}$, where the $g(n)$ are not neccessarily arranged in order of magnitude and may repeat. (Again, it actually suffices to take g to be primitive recursive.) In other words, A is recursively enumerable if there is a calculation which will generate all and only the elements of A in some order, possibly with repetitions. Although the empty set does not conform to the definition above, it will be convenient to consider it to be recursively enumerable.

An important observation linking the above two concepts is the following due to Kleene:

Proposition 20.1. *Let A be a set of natural numbers and A^c its complement in $\mathbf{N}$. Then A is recursive if and only if both A and A^c are recursively enumerable.*

Proof sketched: Suppose A is recursive. Then there is a calculable function f such that $x \in A$ if and only if $f(x) = 0$. We may assume that A is not empty, for otherwise the result holds trivially. Let a be the smallest element of A. Define $g(x) = x$ if $f(x) = 0$ and $g(x) = a$ if $f(x) \neq 0$. Then g is surely calculable and A is its range. Thus A is recursively enumerable. One shows similarly that A^c is recursively enumerable.

Conversely, suppose both A and A^c are recursively enumerable, both nonempty. Then A is the range of a calculable function g and A^c is the range of a calculable function h. Now every natural number is in the range of g or the range of h and we can determine whether x is an element of A by calculating

$$g(0), h(0), g(1), h(1), g(2), \ldots.$$

We put $f(x) = 0$ if, for some y, $g(y) = x$ and $f(x) = 1$ if, for some y, $h(y) = x$. Then f is calculable and $A = \{x \in N | f(x) = 0\}$, hence A is a recursive set.

The above considerations can be extended from natural numbers to finite strings of symbols in a formal language, e.g., the language of mathematics. We assume that the set of symbols is finite. For example, if mathematics is based on set theory, we know that the following symbols suffice:

$$x, ' \, (,), \wedge, \Rightarrow, \vee, \neg, \forall, \exists, =, \in, 0, S.$$

Assuming that there are s symbols in the alphabet (e.g., $s = 14$), we can think of a string of symbols as a natural number expressed in the base s. If you prefer, you may say that the number *encodes* the string.

A mathematical *formula* is a finite string of symbols, combined according to some simple rules, and so is a *proof* in mathematics: a finite list of formulas, each of which is an axiom or else derived from earlier formulas in the list by a rule of inference. There is a finite procedure for 'calculating' whether a formula in a given list meets one of these two requirements, hence whether the given list is a proof. It follows that the set of proofs in mathematics, viewed as the set of natural numbers which encode these proofs, is a recursive set. The last formula in a proof is called a *theorem*.

We can program a computer to look at each natural number in turn, convert it to base s, and determine whether it is (encodes) a formula or a proof. In the second case, we can also tell the computer to print out the number which encodes the last formula of the proof. The computer will then generate a list which consists of all and only those base s natural numbers which encode theorems of mathematics. It follows that the set of theorems in mathematics is recursively enumerable.

On the other hand, it was proved by Alonzo Church that there is no mechanical procedure or algorithm for determining, in general, whether a given formula is a theorem of mathematics. In other words, the set of theorems is not recursive.

If we combine this result of Church with the above proposition of Kleene, we may deduce that the complement of the set of theorems is not recursively enumerable. Moreover, we may deduce the following:

20. Recursive and Recursively Enumerable Sets

Theorem 20.2. (Incompleteness Theorem)
There are mathematical formulas p such that neither p nor $\neg p$ are provable.

Proof: Let T be the set of theorems and C the set of contradictions, that is, formulas p such that $\neg p$ is a theorem. The computer can print out the members of C as easily as those of T, hence both sets are recursively enumerable. Suppose that, for every formula p, either p is provable or $\neg p$ is provable. Then p is a nontheorem if and only if $\neg p$ is provable, hence C is the complement of T. Thus both T and its complement are recursively enumerable, which contradicts the result of Church.

We should note that the incompleteness theorem was first proved by Gödel (1906–1978) and that Church used Gödel's result to prove his own. In a subsequent chapter, we shall present Gödel's original proof, which does not depend on Church's result.

21
Hilbert's Tenth Problem

Hilbert's tenth problem asked for an algorithm to determine whether any given polynomial Diophantine equation has a solution in integers. After important preliminary work by Martin Davis, Hilary Putnam (the philosopher) and Julia Robinson, Yuri Matiyasevič showed that no such algorithm exists. The proof is long and we shall give only a few of the highlights here. For a complete treatment, the reader may wish to consult Davis [1973].

First let us reduce the problem of solving a Diophantine equation in *integers* to one of solving it in *positive integers*, using the fact that every positive integer x can be written in the form

$$x = x_0^2 + x_1^2 + x_2^2 + x_3^2 + 1,$$

where the x_i are integers, in view of Lagrange's Theorem (Chapter 9). For example, if we want to know whether

$$x^{17} + y^{17} - z^{17} = 0$$

has a solution in positive integers, we may test whether the following equation has a solution in integers:

$$(x_0^2 + x_1^2 + x_2^2 + x_3^2 + 1)^{17} + (y_0^2 + \cdots + 1)^{17} - (z_0^2 + \cdots + 1)^{17} = 0.$$

A set A of positive integers is said to be *Diophantine* if there is a polynomial $p(t, x_1, \ldots, x_n)$ with integer coefficients such that $t \in A$ if and only if there are positive integers $x_1, x_2, \ldots, x_n$ such that $p(t, x_1, \ldots, x_n) = 0$. We shall write $A = A_p$ to express the relationship between the set A and the polynomial p. For example, the set of *composite* positive integers is Diophantine of the form A_p, where

$$p(t, x_1, x_2) \equiv t - (x_1+1)(x_2+1) \equiv t - x_1 x_2 - x_1 - x_2 - 1.$$

Lemma 21.1. *Every Diophantine set is recursively enumerable.*

Proof: Given the polynomial $p(t, x_1, \ldots, x_n)$ with integer coefficients and any positive integer m, let S_m be the set of all (n+1)-tuples of positive integers, each less than or equal to m. Then S_m is clearly finite. We can enumerate the elements of the Diophantine set A_p by looking at each of $S_1, S_2, \ldots$ in turn and checking whether it contains a solution of the equation $p(t, x_1, x_2, \ldots, x_n) = 0$. Whenever we do find a solution, we list its first member t.

Lemma 21.2. *Suppose there is an algorithm for deciding whether, for any given t, the polynomial equation $p(t, x_1, \ldots, x_n) = 0$ has a solution in positive integers. Then the Diophantine set A_p is recursive.*

Proof: This is so because we can perform a calculation on t to see whether $t \in A_p$.

The next lemma is due to Julia Robinson. It has to do with the notion of *exponential growth*. For example, consider the *Fibonacci sequence* F_m:

$$1, 1, 2, 3, 5, 8, 13, \ldots$$

in which each term, except $F_1 = 1$ and $F_2 = 1$, is the sum of the preceding two terms: $F_{m+2} = F_{m+1} + F_m$. This had been studied by Leonardo of Pisa (1180–1250), also known as 'Fibonacci', in connection with the growth of rabbit populations. As we saw in Part I, Chapter 23,

$$F_m = \frac{(\tfrac{1}{2}(1+\sqrt{5}))^m - (\tfrac{1}{2}(1-\sqrt{5}))^m}{\sqrt{5}}.$$

Since $(\tfrac{1}{2}(1-\sqrt{5}))^m/\sqrt{5}$ is small, F_m is in fact equal to the integer nearest to $(\tfrac{1}{2}(1+\sqrt{5}))^m$. This explains why rabbit populations grow exponentially.

Lemma 21.3. (Julia Robinson)
A sufficient condition for every recursively enumerable set to be Diophantine is that there is a polynomial equation with integer coefficients

$$p(u, v, x_2, \ldots x_n) = 0$$

such that, in its positive integer solutions, v grows exponentially relative to u.

The proof of Lemma 21.3 is too long to be included here.

In 1970, Matiyasevič found a polynomial Diophantine equation whose $(m-1)$th positive integer solution, for $m \geq 2$ has the form

$$(m, F_{2m}, x_2, \ldots, x_n).$$

By Lemma 21.3, he inferred

Theorem 21.4. (Matiyasevič)
Every recursively enumerable set of positive integers is Diophantine.

It follows by Lemma 21.1 that a set of positive integers is Diophantine if and only if it is recursively enumerable. We saw in Chapter 20 that not every recursively enumerable set of positive integers is recursive, e.g., the set of positive integers which encode the theorems of mathematics. Therefore, in view of Lemma 21.2, there is no algorithm for testing whether any given polynomial equation has a solution in positive integers. We conclude that *Hilbert's tenth problem is unsolvable*.

Matiyasevič's result had curious repercussions on the existence of *prime representing polynomials*. Consider the polynomial

$$f(x) = x^2 - x + 41.$$

For $x = 1, 2, 3, \ldots, 40$, $f(x)$ is a prime number. While this might convince a physicist that, for any positive integer x, $f(x)$ is always prime, $x = 41$ is a counterexample.

On the other hand, consider the polynomial

$$g(x) = -x^2 + 3.$$

The set of all $g(x) \geq 0$ such that x is a positive is a *subset* of the set of prime numbers, albeit a very small subset.

These examples suggest the following questions:

1. Is there a polynomial with integer coefficients all of whose values for positive integer arguments are primes?

2. Is there a polynomial with integer coefficients such that the set of its *positive* values for positive integer arguments is just the set of primes?

The answer to the first question is 'no', as the reader will be invited to verify in the Exercises. Surprisingly, the answer to the second question is 'yes'. This uses nothing about the set of prime numbers except that it is recursively enumerable, hence Diophantine by Theorem 21.4.

Theorem 21.5. (Putnam)
For any Diophantine set A_p there is a polynomial

$$q(t, x_1, x_2, \ldots, x_n)$$

with integer coefficients such that A_p is the set of all positive values of $q(t, x_1, \ldots, x_n)$ for positive integers $t, x_1, \ldots, x_n$.

Proof: We recall that $t \in A_p$ if and only if $p(t, x_1, \ldots, x_n) = 0$. Let
$$q(t, x_1, \ldots, x_n) = t(1 - (p(t, x_1, \ldots, x_n))^2).$$
We shall illustrate the argument by taking $n = 1$.

(a) Suppose t and x are positive integers and $q(t, x) > 0$. Then
$$t(1 - (p(t, x))^2) > 0$$
and hence $1 > (p(t, x))^2$, so $p(t, x) = 0$, and hence $t \in A_p$.

(b) Suppose $t \in A_p$. Then there is a positive integer x such that $p(t, x) = 0$, hence $q(t, x) = t$.

It follows from Putnam's Theorem that there is a polynomial
$$q(t, x_1, \ldots, x_n)$$
with integer coefficients, such that the set of its positive values, for positive integer assignments of the $n + 1$ variables, is exactly the set of prime numbers. Such a polynomial can be found in Browder [1976], p. 331.

Exercises

1. Explain why the set of positive integers which are not powers of 2 is Diophantine.

2. Describe an algorithm for solving Diophantine equations in one variable:
$$a_0 x^n + a_1 x^{n-1} + \cdots + a_n = 0,$$
where the a_i are given integers.

3. Let $f(x)$ be a polynomial with integer coefficients. Show that there is a positive integer x such that $f(x)$ is not prime.

4. Let $f(x_0, x_1, \ldots, x_n)$ be a polynomial with integer coefficients. Show that there are positive integers $x_0, \ldots, x_n$ such that $f(x_0, \ldots, x_n)$ is not prime.

5. Prove that the set $\{2, 3, 4\}$ is Diophantine and find a polynomial $q(t, x_1, \ldots, x_n)$ with integer coefficients whose positive values, for positive integer arguments, are just the members of this set.

6. Find a polynomial with integer coefficients whose positive values, for positive integer arguments, are all and only composite numbers.

22
Lambda Calculus

The Lambda Calculus of Alonzo Church represents an attempt to understand mathematical entities as *functions*. Usually, people think of a function $f : A \to B$ as having a domain A and a codomain B. But, in the *untyped* version of the lambda calculus, one makes the implicit assumption that $A = B$ is some kind of universal set and that f is is defined everywhere; it is even possible to apply f to itself.

We write $f`a$ for the value of $f$ at $a$ and we read this as $f$ *of* $a$. For example, if $f$ is the squaring function, we write $f`a$ for a^2. Similarly, if $\phi(x)$ is the expression $x^3 + x + 1$, we can introduce a function g such that $g`x = \phi(x) \equiv x^3 + x + 1$. It is customary to write $g = \lambda_x(x^3 + x + 1)$, where λ_x is the *abstraction operator*. It is sometimes important to distinguish between the expression $\phi(x)$ and the function $\lambda_x \phi(x)$ which sends x to $\phi(x)$. In particular, $\lambda_x \phi(x)`2 = 2^3 + 2 + 1 = 11$ and $\lambda_x \phi(x)`y = \phi(y) \equiv y^3 + y + 1$. In 1937, Church and Turing showed independently that every calculable numerical function can be expressed in terms of the untyped lambda calculus. In particular, the natural numbers and the usual arithmetic operations on natural numbers can be so expressed, as we shall see.

Conversely, every numerical function definable in terms of the untyped lambda calculus is calculable, thus
$$\text{recursive} = \text{calculable} = \lambda\text{-definable}.$$
According to the *Church-Turing Thesis*, any of these three equivalent concepts captures the intuitive notion of what it means to be 'computable'.

We shall now give a rigorous presentation of the *untyped lambda calculus*. We assume, to begin with, that there is a supply of countably many

22. Lambda Calculus

variables $x_1, x_2, \ldots$. Rather than referring to specific variables, such as x_{17}, we shall use letters $x, y, z, f, g, h, \ldots$ as arbitrary variables. We now define *terms* of the λ-calculus:

1. each variable is a term;

2. if ϕ and ψ are terms, so is $(\phi`\psi)$;

3. if ϕ is a term and x is a variable, then $(\lambda_x \phi)$ is a term;

4. all terms are built up according to the above rules.

We have inserted parentheses to avoid ambiguity. However, when there is no danger of ambiguity we may omit them.

We define a *free* occurrence of a variable in a term as follows:

1. a variable x occurs freely in the term x;

2. if x occurs freely in ϕ or ψ, then it occurs freely in $\phi`\psi$;

3. if y is a variable distinct from x and y occurs freely in ϕ, then it also occurs freely in $\lambda_x \phi$; however, x does *not* occur freely in $\lambda_x \phi$.

If the variable x occurs freely in the term ϕ, then it said to be *bound* to λ_x in $\lambda_x \phi$. We often write the term ϕ as $\phi(x)$ to indicate the possible free occurence of x in ϕ. This has the advantage that we can write $\phi(\alpha)$ for the result of substitutung α for x in ϕ. But we are only *allowed to substitute* α for x in $\phi(x)$ if no variable occurring freely in α becomes bound in $\phi(\alpha)$.

In addition to the usual properties of equality — reflexivity, symmetry, transitivity and the rule allowing substitution of equals — we postulate the following three rules:

R1. $(\lambda_x \phi(x))`\alpha = \phi(\alpha)$, if we are allowed to substitute α for x in $\phi(x)$;

R2. $\lambda_x(\phi`x) = \phi$, if x is not free in ϕ (possibly because x does not occur in ϕ at all);

R3. $\lambda_x \phi(x) = \lambda_y \phi(y)$, if we are allowed to substitute y for x in $\phi(x)$.

Here are some examples to illustrate these ideas, which the reader may skip if she has already understood them.

1. $(\lambda_x(y`x))`(x`z)$ is a term with the first occurrence of x not free, it being bound to λ_x, and the second occurrence of x free.

2. Suppose $\phi(x)$ and $\psi(x)$ are terms not containing λ_y and suppose we know that $\phi(x) = \psi(x)$. Then $\lambda_x\phi(x)`y = \lambda_x\psi(x)`y$, by the properties of equality, hence $\phi(y) = \psi(y)$ by R1.

3. $\lambda_f(g`f) = g$ by R2 (assuming $g \neq f$) and $(\lambda_g g)`f = f$ by R1.

4. What can go wrong if we disregard the 'if' clauses of the rules? From 3 we see that
$$(\lambda_g(\lambda_f(g`f)))`f = (\lambda_g g)`f = f.$$
But disregarding the 'if' clause in R1 would yield
$$(\lambda_g(\lambda_f(g`f)))`f = \lambda_f(f`f)$$
(taking $x \equiv g$ and $\phi(x) \equiv \lambda_f(g`f)$). But $f$ and $\lambda_f(f`f)$ are not the same, as is seen by applying both to a, say. $f`a$ depends on $f$, but $(\lambda_f(f`f))`a = a`a$ does not; f in the last term is a *dummy* variable.

We shall now show how to do arithmetic in the lambda calculus. First note that any two functions f and g can be *composed* to form $f \circ g$, where $(f \circ g)`x = f`(g`x)$, hence $f \circ g = \lambda_x(f`(g`x))$ by R2. In particular, $f \circ f$ is usually written $f^2$, the *iterate* of $f$, so $f^2 = \lambda_x(f`(f`x))$. Church had the idea that the number 2 should be defined as the *process of iteration*, as the function which assigns f^2 to f, that is
$$2 = \lambda_f(f^2) = \lambda_f(\lambda_x(f`(f`x))).$$

Since it is customary to think of f^0 as the identity function and of f^1 as f, we may also define
$$0 = \lambda_f(\lambda_x x), \qquad 1 = \lambda_f(\lambda_x(f`x)) = \lambda_f f,$$
hence $0 = \lambda_f 1$ since $\lambda_f f = \lambda_x x$ by R3. The reader is invited to define the number 3 in the lambda calculus. In general, $n = \lambda_f(f^n)$, where $f^n = f \circ f \circ \cdots \circ f$ is the nth iterate of f. More precisely, f^n is defined inductively by $f^{\Sigma`n} = f \circ f^n$, where $\Sigma`n$ is the successor of n. Thus $n`f = f^n$.

Substituting m for f in this equation, we have $n`m = m^n$. But this tells us how to define *exponentiation* in the lambda calculus. In particular, $m^0 = 0`m = 1$. Since $(m \times n)`f = f^{m \times n} = (f^n)^m = m`(n`f) = (m \circ n)`f$, we may define *multiplication* by $m \times n = m \circ n = \lambda_x(m`(n`x))$. For *addition* we have
$$(m+n)`f = f^{m+n} = f^m \circ f^n = (m`f) \circ (n`f),$$
so that $m + n = \lambda_f((m`f) \circ (n`f))$.

We can now prove some of the basic laws of arithmetic. For example, since composition of functions is associative, multiplication of numbers is associative. On the other hand, composition of functions is not commutative, yet multiplication of numbers is; this must be proved by mathematical induction.

22. Lambda Calculus

Of considerable interest in computer science is the following:

Theorem 22.1. (Fixpoint Theorem)
For any term ϕ, there is a term α such that $\phi`\alpha = \alpha$ can be proved.

This implies, in particular, that it is not possible to incorporate a term into the lambda calculus which corresponds to the negation symbol in logic.

Proof: Let $\beta = \lambda_x(\phi`(x`x))$ and $\alpha = \beta`\beta$. Then

$$\alpha = \beta`\beta = (\lambda_x(\phi`(x`x)))`\beta = \phi`(\beta`\beta) = \phi`\alpha.$$

Surprisingly, it is possible to get rid of the λ-abstraction (and all bound variables) in the lambda calculus. This discovery goes back to Schönfinkel (1924). In fact, if we write

$$I = \lambda_x x \qquad \text{(identity)},$$
$$K = \lambda_x \lambda_y x \qquad \text{(constancy operator)},$$
$$S = \lambda_x \lambda_y \lambda_z ((x`z)`(y`z)) \qquad \text{(Schönfinkel operator)},$$

we have

$$I`x = x \qquad \text{(thus } I = 1\text{)},$$
$$(K`x)`y = x \qquad \text{(thus } K`x \text{ is the function with constant value } x\text{)},$$
$$((S`f)`g)`x = (f`x)`(g`x).$$

We define *combinators* as follows:

1. all variables are combinators;
2. I, K, and S are combinators;
3. if ϕ and ψ are combinators, so is $\phi`\psi$;
4. all combinators are obtained from the above rules.

Theorem 22.2. (Schöfinkel)
Every term of the lambda calculus is provably equal to a combinator.

Proof: In view of rules (1) to (3), this will follow if we show that, if ϕ is equal to a combinator, so is $\lambda_x \phi$. We prove this by induction on the length of ϕ.

Since $\lambda_x x = I$, $\lambda_x y = K`y$, $\lambda_x I = K`I$, $\lambda_x K = K`K$ and $\lambda_x S = K`S$, we know that the induction hypothesis holds for all combinators ϕ of length 1.

Suppose now that ϕ has length greater than 1 and that every combinator ψ shorter than ϕ is such that $\lambda_x \psi$ is equal to a combinator. We claim that $\lambda_x \phi$ is equal to a combinator. Indeed, $\phi = \psi' \chi$, where ψ and χ are shorter than ϕ, hence

$$\lambda_x \phi = \lambda_x(\psi'\chi) = (S'(\lambda_x\psi))'(\lambda_x\chi).$$

This is a combinator because $\lambda_x \psi$ and $\lambda_x \chi$ are combinators by our inductional assumption.

Using the methods of this proof, we can express 2 as a combinator:

$$\begin{aligned}
2 &= \lambda_f \lambda_x(f'(f'x)) \\
&= \lambda_f((S'\lambda_x f)'\lambda_x(f'x)) \\
&= \lambda_f((S'\lambda_x f)'f) & \text{by } \mathbf{R2} \\
&= (S'(\lambda_f(S'\lambda_x f)))'\lambda_f f \\
&= (S'(\lambda_f(S'(K'f))))'I \\
&= (S'((S'\lambda_f S)'\lambda_f(K'f)))'I \\
&= (S'((S'(K'S))'K))'I & \text{by } \mathbf{R2}.
\end{aligned}$$

Exercises

1. Assuming that m, p and q are natural numbers expressed in the lambda calculus, show that $(m^p)^q = m^{(q \times p)}$.

2. Prove that $0^{\Sigma' n} = 0$.

3. Prove that $I = (S'K)'K$.

4. Express m^n, $m \times n$, and $m + n$ in terms of I, K, S, m and n.

23
Logic from Aristotle to Russell

Logic was not always regarded as a branch of mathematics, certainly not by Aristotle (384–322 BC), who was the first to write about logic in the West. Among the principles which he recognized are the following:

$\neg\neg p \iff p$ (double negation),
$p \vee \neg p$ (excluded third),
$(p \Rightarrow q) \iff (\neg q \Rightarrow \neg p)$ (contraposition).

He also looked at modal logic and showed how possibility can be defined in terms of necessity.

Aristotle's major concern was with a type of argument called the 'syllogism', which predominated in logical thinking for the next two thousand years. It dealt with four types of basic statements:

SaP   meaning *all* S *are* P,
SeP   meaning *no* S *are* P,
SiP   meaning *some* S *are* P,
SoP   meaning *some* S *are not* P.

He realized that PeS is equivalent to SeP and that PiS is equivalent to SiP and he adhered to a convention that SaP implies SiP. (Today we use

23. Logic from Aristotle to Russell

words differently: we assert that *all* unicorns have horns, but deny that *some* unicorns have horns. Evidently Aristotle did not believe in the empty set.)

A *syllogism* is an argument which infers one such basic statement from two others. Here are the first four 'figures' of the syllogism:

$$\frac{\text{MaP}}{\text{SaM}} \qquad \frac{\text{MeP}}{\text{SaM}} \qquad \frac{\text{MaP}}{\text{SiM}} \qquad \frac{\text{MeP}}{\text{SiM}}$$
$$\overline{\text{SaP}} \qquad \overline{\text{SeP}} \qquad \overline{\text{SiP}} \qquad \overline{\text{SoP}}$$

William of Shyreswood (1250 AD) gave these syllogisms the names

b<u>a</u>rb<u>a</u>ra,   c<u>e</u>l<u>a</u>r<u>e</u>nt,   d<u>a</u>r<u>ii</u>,   f<u>e</u>r<u>io</u>,

— to make them easier to remember. There were more such figures, which we shall not discuss here. Here is a typical argument illustrating the 'ferio':

$$\frac{\text{no minister is prudent}}{\text{some socialists are ministers}}$$
some socialists are not prudent

The Stoics (200 BC), Philo of Megara in particular, essentially introduced truth tables into logic, thus anticipating Ludwig Wittgenstein (1889–1951). They discussed the problem of whether 'p or q' is true when both p and q are true and whether 'if p then q' is always true when p is false. They arrived at the modern conventions, expressed in the following *truth tables*:

$p \vee q$	$p \Rightarrow q$
T T T	T T T
T T F	T F F
F T T	F T T
F F F	F T F

The Stoics were commited to the view that there are only two truth values (T and F). In particular, they believed that a statement like 'there will be a battle tomorrow' is either true or false, although Aristotle seems to have had some second thoughts about this. This belief was associated in their minds with the view that the future is determined by fate (Kneale p. 48).

Today we tend to dismiss the detailed elaboration of Aristotelian logic by the medieval Scholastics and recognize as a major advance only the ideas of Gottfried W. Leibniz (1646–1716). Leibniz conceived of a universal symbolic language which would be adequate not only for mathematics, but for all of science. Unfortunately, he did not get around to publishing the details of his proposal, perhaps because he was preoccupied by his controversy with Newton concerning the invention of 'the' calculus and by his successful diplomatic efforts to put George I on the throne of England.

It was only in 1847 that full-blown *symbolic logic* finally saw the light of day. This was the year in which both George Boole (1815–1864) and Augustus DeMorgan (1806–1871) published their first works in logic. The former saw propositional logic as a branch of algebra, distinguished from the usual algebra of 'quantities' by the 'idempotent' law: $p \times p = p$. De Morgan is remembered for his laws expressing the duality between disjunction ('or') and conjunction ('and'):

$$\neg(p \wedge q) = \neg p \vee \neg q, \qquad \neg(p \vee q) = \neg p \wedge \neg q.$$

The next major step was taken by Gottlob Frege (1848–1925), who was the first to have a modern view of universal and existential quantifiers, although without using the modern notation: $\forall$ and $\exists$. This was surprisingly recent, considering that every student nowadays is familiar with these notions. Frege also attempted to express all of mathematics in terms of logical symbols, in fact, reducing mathematics to logic, thus espousing a philosophical position called 'Logicism'.

Crucial to Frege's project is the assumption that, corresponding to any property expressed by a predicate of his language, there is a uniquely determined set whose elements are just the entities with that property. Frege expressed this assumption by the following *comprehension scheme*:

$$\exists_y \forall_x (x \in y \iff P(x)),$$

where $P(x)$ is any formula, possibly containing the free variable x. (The double arrow means 'if and only if' or 'just in case'.) This scheme is accompanied by the *axiom of extensionality* to ensure the uniqueness of the set y whose existence has been asserted above:

$$\forall_y \forall_z (\forall_x (x \in y \iff x \in z) \implies y = z).$$

The axiom of extensionality implies, in particular, that there can be at most one entity y with no elements, the *empty set*. This axiom implicitly contains the assumption that all entities discussed by Frege's language are sets.

Frege had just written a book propounding these views when he received a letter from Bertrand Russell (1872–1970), pointing out that there was a serious problem when $P(x)$ was the formula $\neg(x \in x)$. Indeed, if y was such that $\forall(x \in y \iff \neg(x \in x))$, one would obtain as a special case

$$y \in y \iff \neg(y \in y),$$

which is a contradiction.

This argument is known as 'Russell's paradox'. One way to avoid the contradiction is to forbid expressions such as $x \in x$. Russell and Whitehead propose a *theory of types*, according to which each symbol denoting an entity should have attached to it a certain natural number, its *type*, and the formula $a \in b$ is permitted only if the type of b is one higher than the type of a. This theory was developed in excruciating detail in the three-volume *Principia Mathematica*, an unnecessarily complicated treatise that is more talked about than studied.

Although up-to-date versions of type theory are now available, most mathematicians prefer other methods for avoiding Russell's and similar paradoxes. On the whole, if mathematicians worry about such problems at all, they subscribe to the set theory of Gödel and Bernays, which distinguishes between *sets* and *classes*: only the former can be elements. Unfortunately, one has to add a number of axioms to specify which classes are sets. Logicians, on the other hand, prefer the set theory of Ernst Zermelo (1871–1953) and Abraham Fraenkel (1891–1965), who modify the comprehension as follows:

$$\forall_z \exists_y \forall_x (x \in y \iff (x \in z \wedge P(x))).$$

They too had to introduce additional axioms, which spoiled the simplicity of Frege's project.

The fact that Frege's simple and natural comprehension scheme led to contradictions startled many mathematicians, and some became sceptical about all but the most basic procedures. Other mathematicians were already sceptical. We shall discuss L. Kronecker (1823–1891) and H. Poincaré (1854–1912) in this chapter, leaving L. E. J. Brouwer to the next chapter.

It was Kronecker who said 'God made the whole numbers, all the rest is the work of man'. He was suspicious of Cantor's infinite cardinals; but most of all he rejected *nonconstructive* arguments such as the following proof that there exist irrational numbers α and β such that α^β is rational.

Consider $\sqrt{2}^{\sqrt{2}}$. If it is rational, we are done, since we know that $\sqrt{2}$ is irrational. Suppose $\sqrt{2}^{\sqrt{2}}$ is irrational. Call it α and take $\beta = \sqrt{2}$. Then

$$\alpha^\beta = \sqrt{2}^{(\sqrt{2} \times \sqrt{2})} = \sqrt{2}^2 = 2,$$

which is surely rational.

This proof depends on the Stoic idea, also endorsed by Aristotle, that every proposition, here the proposition that $\sqrt{2}^{\sqrt{2}}$ is rational, is either true or false. What Kronecker would have objected to is that, at the end of the proof, we don't know whether $\alpha = \sqrt{2}^{\sqrt{2}}$ or $\alpha = \sqrt{2}$.

It is known, using some deep mathematics, that $\sqrt{2}^{\sqrt{2}}$ is irrational, but that is beside the point here. It is also beside the point that the stated theorem has an easy constructive proof: take $\alpha = \sqrt{2}$ and $\beta = 2\log_2 3$. It is possible to exhibit mathematical theorems for which no constructive proof exists, for instance the theorem which asserts that, if the axiom of choice is true, then there exist nonprincipal ultrafilters on the set of natural numbers. (The reader unfamiliar with ultrafilters may ignore this example.)

Poincaré, on the other hand, objected to *impredicative* definitions or constructions, essentially those which define or construct an entity in terms of entities of a higher type. For example, consider the usual proof that every nonempty set of real numbers that is bounded above has a least upper bound. According to Dedekind, a *real number* is a set α of rational numbers such that

1. both α and its complement α^c are nonempty;

2. every element of α is less than every element of α^c;

3. α has no greatest element.

Now let m be a nonempty set of real numbers which is bounded above. Put

$$\alpha \equiv \{x \in \mathbf{Q} \mid \exists_{y \in m} x \in y\}.$$

It is not hard to show that α satisfies 1 to 3 and that it is the least upper bound of m. However, being a *set* of reals, m has a higher type than α. The construction of α is thus impredicative in Poincaré's sense.

Most mathematicians feel that Poincaré was too sceptical here. If we disallowed constructions such as that of the least upper bound of m above, most of analysis would have to be abandoned.

Exercises

1. Suppose the barber in a certain village shaves all and only those men of the village who do not shave themselves. Prove that the barber is a woman.

2. Prove that, in the set theory of Zermelo–Fraenkel, there does not exist a 'universal' set y, such that $\forall_x (x \in y \Leftrightarrow x = x)$.

3. Prove that the impredicatively defined set α above is a real number according to Dedekind's definition as given in the text.

24
Intuitionistic Propositional Calculus

It seems that, over many centuries, no philosopher or mathematician ever seriously questioned Aristotle's *law of the excluded third:* for every proposition p, either p or not p, symbolically $p \vee \neg p$. In retrospect, it appears that Aristotle himself had some doubts about applying this law when talking about events in time, e.g., when p was the proposition: there will be a sea battle tomorrow. But in mathematics, which deals with unchanging entities, the law of the excluded third was accepted as gospel truth, as was the equivalent assertion: for every proposition p, $\neg\neg p \Rightarrow p$; two negations make an affirmation.

It was the topologist Luitzen Egbertus Jan Brouwer (1882–1966) who observed that all nonconstructive arguments in mathematics depend on Aristotle's law and he proposed that we simply drop it, together with all its consequences, at least when talking about infinite collections. Surprisingly, it turns out that, if one follows Brouwer's suggestion, one is still left with a rich logical system adequate for all constructive mathematics. We shall present the outlines of such a system here, starting with the propositional calculus.

We consider the logical symbols $\top$, $\bot$, $\wedge$, $\vee$ and $\Rightarrow$, the first two counting as formulas, the last two being binary connectives between formulas, so that $A \wedge B$, $A \vee B$, and $A \Rightarrow B$ are formulas if A and B are. The usual reading of these formulas is as follows:

$$\begin{aligned} \top &\equiv \text{true}, \\ \bot &\equiv \text{false}, \\ A \wedge B &\equiv \text{(both) } A \text{ and } B, \end{aligned}$$

24. Intuitionistic Propositional Calculus

$$A \vee B \equiv \text{(either) } A \text{ or } B,$$
$$A \Rightarrow B \equiv \text{if } A \text{ then } B.$$

However, *intuitionists*, as the followers of Brouwer are called, understand these words in a way subtly different from that of classical mathematicians, as we hope to make clear in the next chapter.

It is customary to make use of the *entailment* symbol $\vdash$, where

$$A_1, \ldots, A_n \vdash B$$

means that the assumptions $A_1, \ldots, A_n$ entail the conclusion B, or that B may be deduced from $A_1 \ldots, A_n$, n being any natural number. We often denote strings of formulas by capital Greek letters; thus $\Gamma \vdash B$ means that B may be inferred from the formulas in Γ. Note that Γ may be empty ($n = 0$), or consist of a single formula ($n = 1$).

We shall adopt the following *axioms:*

$$\vdash \top; \qquad \bot \vdash A; \qquad A \wedge B \vdash A; \qquad A \wedge B \vdash B;$$

$$A, B \vdash A \wedge B; \qquad A \vdash A \vee B; \qquad B \vdash A \vee B; \qquad A, A \Rightarrow B \vdash B.$$

The last of these has the Latin name 'modus ponens', which we shall abbreviate as MP. We also adopt two *rules of inference:*

$$\frac{\Gamma, A \vdash C, \quad \Gamma, B \vdash C}{\Gamma, A \vee B \vdash C}; \qquad \frac{\Gamma, A \vdash B}{\Gamma \vdash A \Rightarrow B}.$$

These are called 'argument by cases', abbreviated AC, and 'deduction rule', abbreviated DR, respectively.

In classical logic, one would be quite happy to establish these axioms and rules of inference with the help of truth tables. In intuitionistic logic, we are not allowed to use truth tables, as they would also establish Aristotle's $A \vee \neg A$ and $\neg \neg A \Rightarrow A$, which we have discarded, provided we define $\neg A \equiv A \Rightarrow \bot$, as we shall do from now on.

In addition to the above axioms and rules of inference, which describe the logical connectives, we also have the following, which describe the entailment symbol:

$A \vdash A$ (inferring a formula from itself);

$$\frac{\Lambda \vdash A \quad \Gamma, A \vdash B}{\Gamma, \Lambda \vdash B}$$ (replacing an assumption by others which entail it);

$$\frac{\Gamma, A, B, \Delta \vdash C}{\Gamma, B, A, \Delta \vdash C}$$ (interchanging two assumptions);

$$\frac{\Gamma \vdash B}{\Gamma, A \vdash B}$$ (introducing a superfluous assumption);

$$\frac{\Gamma, A, A \vdash B}{\Gamma, A \vdash B}$$ (contracting two identical assumptions into one).

These so-called *structural rules* were formally introduced by Gerhard Gentzen (1919–1945). Except for the axiom $A \vdash A$, they are often tacitly understood and not mentioned in actual arguments.

We will show how to establish deductions of the form $\Gamma \vdash B$ by looking at a few examples.

EXAMPLE 1. To prove that $A \vee B, \neg B \vdash A$.

Informally, we would argue as follows. We are given the assumptions $A \vee B$ and $B \Rightarrow \bot$. Suppose A. Then surely A, by the axiom $A \vdash A$. Suppose B. Then $\bot$ by modus ponens from the second given assumption. Therefore A, by the axiom $\bot \vdash A$. Since A in either case, we invoke the argument by cases and infer that A holds in view of the given assumptions.

It is customary to rewrite such an argument more formally in a vertical fashion:

1	(1)	$A \vee B$	given
2	(2)	$B \Rightarrow \bot$	given
3	(3)	A	assumed
2,3	(4)	A	introducing a superfluous hypothesis
5	(5)	B	assumed
2,5	(6)	$\bot$	MP 2,5
2,5	(7)	A	by axiom $\bot \vdash A$
2,1	(8)	A	AC 4,7 replacing 3 and 5 by 1
1,2	(9)	A	interchanging two arguments

Note that the middle column contains the formulas given, assumed or inferred at different stages of the argument, numbered consecutively; the left column lists the numbers of all the hypotheses, given or assumed, upon which the formula in the middle column depends, and the right column indicates the justification for writing it down. The first two entries in the last line say precisely that $A \vee B, B \Rightarrow \bot \vdash A$, as was to be proved.

EXAMPLE 2. To prove that $A \vdash \neg\neg A$.

Here is the informal argument: we are given A. Suppose $A \Rightarrow \bot$. Then $\bot$ by modus ponens. Therefore, $(A \Rightarrow \bot) \Rightarrow \bot$ by the deduction rule.

Formally:

1	(1)	A	given
2	(2)	$A \Rightarrow \bot$	assumed
1,2	(3)	$\bot$	MP 1,2
1	(4)	$(A \Rightarrow \bot) \Rightarrow \bot$	DR 2,3

EXAMPLE 3. To prove that $\vdash A \Rightarrow (B \Rightarrow A)$.

We shall give the formal argument only.

1	(1)	A	assumed
2	(2)	B	assumed
1,2	(3)	A	introducing a superfluous hypothesis
1	(4)	$B \Rightarrow A$	DR 2,3
	(5)	$A \Rightarrow (B \Rightarrow A)$	DR 1,4

After some practice, the student may stop mentioning the structural rules, such as line (3) above, or lines (4) and (9) in Example 1.

EXAMPLE 4. To prove that $A \Rightarrow B \vdash \neg B \Rightarrow \neg A$.

1	(1)	$A \Rightarrow B$	given
2	(2)	$B \Rightarrow \bot$	assumed
3	(3)	A	assumed
1,3	(4)	B	MP 1,3
1,2,3	(5)	$\bot$	MP 2,4
1,2	(6)	$A \Rightarrow \bot$	DR 3,5
1	(7)	$\neg B \Rightarrow \neg A$	DR 2,6

EXAMPLE 5. To prove that $\vdash \neg\neg(A \vee A)$.

1	(1)	$(A \vee \neg A) \Rightarrow \bot$	assumed
2	(2)	A	assumed
2	(3)	$A \vee \neg A$	axiom $A \vdash A \vee B$
1,2	(4)	$\bot$	MP 1,3
1	(5)	$A \Rightarrow \bot$	DR 2,4
1	(6)	$A \vee \neg A$	axiom $B \vdash A \vee B$
1	(7)	$\bot$	MP 1,6
	(8)	$((A \vee \neg A) \Rightarrow \bot) \Rightarrow \bot$	DR 1,7

Exercises

Prove the following.

1. $A \vee \neg A \vdash \neg\neg A \Rightarrow A$.

2. $((A \Rightarrow B) \Rightarrow A) \vdash \neg\neg A$.

3. $C \Rightarrow (A \wedge B) \vdash (C \Rightarrow A) \wedge (C \Rightarrow B)$.

4. $(C \Rightarrow A) \wedge (C \Rightarrow B) \vdash C \Rightarrow (A \wedge B)$.

5. $A \Rightarrow (B \Rightarrow C) \vdash (A \wedge B) \Rightarrow C$.

6. $(A \wedge B) \Rightarrow C \vdash A \Rightarrow (B \Rightarrow C)$.

Note that the classical result $((A \Rightarrow B) \Rightarrow A) \vdash A$ follows from (2) if $\neg\neg A \vdash A$, but it does not hold intuitionistically.

25
How to Interpret Intuitionistic Logic

To explain the subtle difference between intuitionistic and classical logic, we shall present an intuitionistic interpretation of the logical connectives that goes back to Brouwer, Heyting and Kolmogorov. It involves talking about *reasons* for a formula. A *reason* for A may be thought of as a proof of A from some suitable assumption. We would like to say

- there is exactly one reason for $\top$ (namely, quoting the axiom);
- there is no reason for $\bot$;
- a reason for $A \wedge B$ consists of a reason for A and a reason for B;
- a reason for $A \vee B$ is a reason for A or a reason for B;
- a reason for $B \Rightarrow C$ is a rule for converting a reason for B into a reason for C.

Now compare these statements with the following statements about sets (see Chapter 13, where 0 was defined as the empty set and 1 as $\{0\}$):

- there is exactly one element of 1;
- there is no element of 0;
- an element of $A \times B$ is a pair of elements of A and B, respectively;
- an element of $A + B$ is an element of A or an element of B;

- an element of C^B is a function that converts an element of B into an element of C.

Comparing intuitionistic logic with the arithmetic of sets, we are led to the following analogies:

$$\begin{array}{cc} \top & 1 \\ \bot & 0 \\ A \wedge B & A \times B \\ A \vee B & A + B \\ B \Rightarrow C & C^B \end{array}$$

Moreover, a deduction from A to B, namely, an argument showing that $A \vdash B$, corresponds to a mapping $A \to B$. If there is a deduction from A to B and a deduction from B to A, we shall write $A \vdash\dashv B$. This corresponds to mappings $A \to B$ and $B \to A$ and we may write $A \leftrightarrow B$. Frequently these two mappings are inverse to one another, so we have a one-to-one correspondence between A and B, that is, $A \cong B$.

For example, we can prove intuitionistically that

$$C \Rightarrow (A \wedge B) \vdash\dashv (C \Rightarrow A) \wedge (C \Rightarrow B),$$

$$A \Rightarrow (B \Rightarrow C) \vdash\dashv (A \wedge B) \Rightarrow C,$$

$$(A \vee B) \Rightarrow C \vdash\dashv (A \Rightarrow C) \wedge (B \Rightarrow C).$$

These equivalences correspond to the following one-to-one mappings between sets:

$$(A \times B)^C \cong A^C \times B^C,$$

$$(C^B)^A \cong C^{A \times B},$$

$$C^{A+B} \cong C^A \times C^B.$$

As we saw in Chapter 13, these are just generalizations of the familiar laws of arithmetic:

$$(a \times b)^c = a^c \times b^c,$$

$$(c^b)^a = c^{a \times b},$$

$$c^{a+b} = c^a \times c^b.$$

One cannot but be impressed by the remarkable unity pervading logic, set theory and arithmetic.

A word of warning: $A \leftrightarrow B$ does not always mean $A \cong B$. For example, the intuitionistic equivalence

$$A \Rightarrow (B \Rightarrow A) \vdash\dashv \top$$

translates into
$$(A^B)^A \leftrightarrow 1,$$
with mappings in both directions; but the 1-to-1 correspondence $(A^B)^A \cong 1$ holds only if $A \cong 1$ or $A \cong 0$ or $B \cong 0$.

We should also point out that the interpretation of $A \vee B$ advocated here does not work for classical logic. As we see in Chapter 28, there is a formula G in number theory for which one can prove neither G nor $\neg G$. According to the intended interpretation of $A \vee B$, $G \vee \neg G$ has no proof. Yet, classically, $G \vee \neg G$ is just a case of Aristotle's axiom of the excluded third and thus quoting this axiom would constitute a proof.

Exercises

1. Take any of the one-to-one correspondences of Chapter 13, translate it into a statement of intuitionistic logic and prove the latter.

2. Take any intuitionistic theorem of the form $A \vdash B$ and find the corresponding mapping $A \to B$ between sets.

26
Intuitionistic Predicate Calculus

We shall be dealing with formulas A which contain so-called *free variables* such as x in $x^2+2=0$ or x and y in $xy+x=y$. To indicate which variables may be present we often write $A(x)$ or $A(x,y)$ instead of just A.

From $A(x)$ one may obtain the formulas $\forall_x A(x)$ and $\exists_x A(x)$ in which x is no longer free; it is *bound* to the *universal quantifier* $\forall_x$, meaning *for all* x, or to the *existential quantifier* $\exists_x$, meaning *for some* x. Similarly, one may form $\forall_x \forall_y A(x,y)$, $\forall_x \exists_y A(x,y)$, etc.

From $A(x)$ one may also obtain the formula $A(t)$, the result of substituting the *term* t for x. Here t may be 5 or y or $x+3y$ or whatever — it may even be x. However, if $A(x)$ is $\exists_y B(x,y)$, we are not supposed to substitute y or $x+3y$ for x, because y is no longer free in $\exists_y B(y,y)$ or $\exists_y B(3x+y,y)$. This prohibition is spelled out in the following definition: t is *substitutable* for x in $A(x)$ provided any free variable in t (maybe t itself) remains free in $A(t)$. This definition is needed for stating the following two axioms:

$$\forall_x A(x) \vdash A(t) \quad \text{(universal specification)},$$

$$A(t) \vdash \exists_x A(x) \quad \text{(existential generalization)},$$

— subject to the restriction that t is substitutable for x in $A(x)$.

Were it not for this restriction, we would have as a special case of the first axiom

$$\forall_x \exists_y B(x,y) \vdash \exists_y B(y,y)$$

and from 'everybody blames somebody' we could infer that 'somebody blames himself'.

In addition to the above two axioms, we shall also adopt the following two rules of inference:

$$\frac{\Gamma \vdash A(x)}{\Gamma \vdash \forall_x A(x)} \quad \text{(universal generalization),}$$

provided x is not free in Γ;

$$\frac{\Gamma, A(x) \vdash B}{\Gamma, \exists_x A(x) \vdash B} \quad \text{(existential specification),}$$

provided x is not free in Γ or B.

We abbreviate the names of these axioms and rules of inference by US, EG, UG, and ES, respectively. The reason for the restriction on UG, for example, is to avoid inferring from 'x is afraid' that 'everyone is afraid'.

We shall present some examples to illustrate arguments involving quantifiers.

EXAMPLE 1. To prove $\forall_x(F(x) \wedge G(x)) \vdash \forall_x F(x) \wedge \forall_x G(x)$.

1	(1)	$\forall_x(F(x) \wedge G(x))$	given
1	(2)	$F(x) \wedge G(x)$	US 1
1	(3)	$F(x)$	axiom for $\wedge$, 2
1	(4)	$\forall_x F(x)$	UG 3 (x not free in 1)
1	(5)	$G(x)$	axiom for $\wedge$, 2
1	(6)	$\forall_x G(x)$	UG 5 (x not free in 1)
1	(7)	$\forall_x F(x) \wedge \forall_x G(x)$	axiom for $\wedge$, 4,6

EXAMPLE 2. To prove $\exists_y \forall_x F(x, y) \vdash \forall_x \exists_y F(x, y)$.

1	(1)	$\exists_y \forall_x F(x, y)$	given
2	(2)	$\forall_x F(x, y)$	assumed
2	(3)	$F(x, y)$	US 2
2	(4)	$\exists_y F(x, y)$	EG 3
1	(5)	$\exists_y F(x, y)$	ES 4 (y not free in 4)
1	(6)	$\forall_x \exists_y F(x, y)$	UG 5 (x not free in 1)

EXAMPLE 3. To prove $\neg \exists_x F(x) \vdash \forall_x \neg F(x)$.

$$\begin{array}{r|ll l}
1 & (1) & \exists_x F(x) \Rightarrow \bot & \text{given} \\
2 & (2) & F(x) & \text{assumed} \\
2 & (3) & \exists_x F(x) & \text{EG 2} \\
1,2 & (4) & \bot & \text{MP 1,2} \\
1 & (5) & F(x) \Rightarrow \bot & \text{DR 4,2} \\
1 & (6) & \forall_x (F(x) \Rightarrow \bot) & \text{UG 5 (x not free in 1)}
\end{array}$$

Exercises

1. Prove the converse of Example 1.

2. What goes wrong if you try to prove the converse of Example 2?

3. Can you prove the converse of Example 3?

4. Prove that $\neg \forall_x \neg F(x) \vdash \neg\neg \exists_x F(x)$, using Example 3 above and Example 4 of Chapter 24. Classically, but not intuitionistically, one can infer from this that $\neg \forall_x \neg F(x) \vdash \exists_x F(x)$. If we could prove $\neg \forall_x \neg F(x)$, we would have a nonconstructive proof of $\exists_x F(x)$.

5. Prove that $\exists_x \forall_y F(x, y) \vdash \exists_x F(x, x)$.

27
Intuitionistic Type Theory

This chapter is an adaptation of the appendix in Couture and Lambek [1991], giving a brief overview of a recent formulation of type theory in Lambek and Scott [1986], which is adequate for elementary mathematics, including arithmetic and analysis, when treated constructively. As far as we know, the only proofs in these disciplines which are essentially nonconstructive depend on the *axiom of choice*. One formulation of this axiom asserts that, for any nonempty collection of nonempty sets, there exists a set containing exactly one element from each of the given sets.

From basic types 1, Ω and N one builds others by two processes: if A is a type so is PA; if A and B are types, so is $A \times B$. The intended meaning of these types is as follows:

- 1 is the type of a specified single entity (introduced for convenience);

- Ω is the type of truth values or propositions (here there are more than two truth values);

- N is the type of natural numbers;

- PA is the type of sets of entities of type A;

- $A \times B$ is the type of pairs of entities of types A and B, respectively.

We allow arbitrarily many variables of each type and write $x \in A$ to mean that x is a variable of type A. In addition, we construct terms of different types inductively as follows:

27. Intuitionistic Type Theory

1	Ω	N	PA	$A \times B$
$*$	$a = a'$	0	$\{x \in A | \phi(x)\}$	$<a, b>$
	$a \in \alpha$	Sn		

— it being assumed that a and a' are terms of type A already constructed, α of type PA, n of type N, $\phi(x)$ of type Ω and b of type B.

Logical symbols may be defined as follows:

$$\begin{align} \top &\equiv \; * = *, \\ p \vee q &\equiv \; <p, q> = <\top, \top>, \\ p \Rightarrow q &\equiv \; p \wedge q = p, \\ \forall_{x \in A} \phi(x) &\equiv \; \{x \in A \mid \phi(x)\} = \{x \in A \mid \top\}, \end{align}$$

where it is understood that p, q and $\phi(x)$ are terms of type Ω. From these symbols one may define others, taking care not to make implicit use of De Morgan's rules (Prawitz [1965]):

$$\begin{align} \bot &\equiv \; \forall_{t \in \Omega} t, \\ \neg p &\equiv \; \forall_{t \in \Omega}(p \Rightarrow t), \\ p \wedge q &\equiv \; \forall_{t \in \Omega}(((p \Rightarrow t) \wedge (q \Rightarrow t)) \Rightarrow t), \\ \exists_{x \in A} \phi(x) &\equiv \; \forall_{t \in \Omega}((\forall_{x \in a}(\phi(x) \Rightarrow t)) \Rightarrow t). \end{align}$$

Other symbols, such as appear in $\exists_{x \in A}\phi(x)$, $\{a\}$, $\alpha \subseteq \beta$, $\alpha \times \beta$, etc., are defined in the usual fashion.

Axioms and rules of inference are stated in terms of a deduction symbol $\vdash_X$, where X is a finite set of variables. The permissible deductions take the form

$$p_1, \ldots, p_n \vdash_X p_{n+1},$$

where the p_i are terms of type Ω and X contains all the variables which occur freely in the p_i. (When X is empty, the subscript may be ommitted.) The axioms and rules of inference hold no surprises. For the purpose of illustration, here are a few special cases:

- $p \vdash p$;

- $\dfrac{\phi(x) \vdash_{\{x\}} \psi(x)}{\phi(\alpha) \vdash \psi(\alpha)}$;

- $<a, b> = <c, d> \vdash a = c$;

- $\dfrac{\phi(x) \vdash_{\{x\}} \phi(Sx)}{\phi(0) \vdash_{\{x\}} \phi(x)}$.

The reader may recognize the last mentioned rule of inference as the principle of mathematical induction. Although we have not stated all the axioms and rules of inference here, they will of course imply the usual axioms and rules for the intuitionistic propositional and predicate calculi.

For a less formal treatment of constructive arithmetic and analysis, the reader may consult Goodstein [1970] and Bishop [1967], respectively.

28
Gödel's Theorems

A formal language is (among other things) a system for dealing with strings of symbols. An interpretation of these symbols is called a *model*. For example, the Lambda Calculus tells us how to arrange marks such as ' and λ_x. The interpretation of these marks in terms of functional application and functional abstraction is a model.

The formal languages that interest us here contain the notion of a proof. A *proof* is a finite sequence of formulas, each of which is either an axiom or follows from some previous members of the sequence by a rule of inference. We call a formal language *consistent* if there is no proof in the language whose last line is $\bot$, that is, it does not contain the proof of a contradiction.

In this chapter we consider a formal language L which is adequate for *arithmetic*. We assume that L includes the intuitionistic predicate calculus together with some axioms for arithmetic, and we assume that all the basic laws of arithmetic are provable in L. For example, L might be the type theory considered in the previous chapter.

In 1930, Kurt Gödel (1906–1978) proved a completeness result for the classical predicate calculus. In 1950, this result was extended by Leon Henkin to classical type theory. It was later extended to intuitionistic type theory (Lambek and Scott [1986]). This completeness result may be expressed as follows:

Theorem 28.1. (Completeness)
A formula is provable in L if it is true under all possible interpretations of the nonlogical symbols in L, i.e., in all models of L.

We shall not give a proof here, but we remark that the proof of this theorem depends on the axiom of choice and is thus unacceptable to intuitionists. The converse of this theorem is called the soundness theorem; its proof is straightforward and is acceptable to intuitionists. As a corollary of the completeness-soundness result, L is consistent if and only if it has a model.

If n is a natural number, let $S^n 0$ be the expression in L formed by placing the letter S n times before the symbol 0. This expression will normally be interpreted as the natural number n. We call a model ω-*complete* if, for any formula $A(x)$ of L, x being of type N, whenever $A(S^n 0)$ is true in that model for each natural number n, then $\forall_{x \in N} A(x)$ is also true in that model.

In 1931, Gödel proved his *incompleteness theorem* for arithmetic, which may be expressed in our terminology as follows:

Theorem 28.2. (Incompleteness)
There is a formula in L which is true in any ω-complete model, but not provable in L, assuming L to be consistent.

Combining this with the completeness theorem, we may conclude that some models are not ω-complete.

To an intuitionist, the notion of truth is equivalent to that of knowability, which we shall here interpret as provability. Thus we may conclude that the world in which we live is ω-complete provided, whenever $A(S^n 0)$ has a proof for each n, then $\forall_{x \in N} A(x)$ does also. However, there is no particular reason to believe that this is the case. Even if, for each n, the formula $A(S^n 0)$ showed up as the last line of a proof, it would not guarantee that $\forall_{x \in N} A(x)$ showed up as the last line of a proof. Hence the intuitionist has no particular reason to think of the world we live in as ω-complete.

Platonists, of whom Gödel was one, see truth as the property of an eternal and immutable reality which is independent of finite human minds. A classical Platonist believes that the real world contains an infinite collection of natural numbers. Now if it is true of each of these numbers that it has a property A, then it is true that they all have property A, whence the real world is an ω-complete model of L. Since Gödel was a classical Platonist, he concluded from his incompleteness theorem that there is a formula of arithmetic which is eternally true, but which is not the last line of a proof in L. (Note that for Platonists, the language of arithmetic is about the real world, hence it is consistent.)

29
Proof of Gödel's Incompleteness Theorem

We shall now sketch a proof of the incompleteness result in a manner acceptable to intuitionists. We begin with a lemma.

Lemma 29.1. (Gödel's Lemma)
Suppose $R(m,n)$ is a recursive relation between the natural numbers m and n. That is, assume that, for any two numbers m and n, there is a finite effective procedure for deciding whether they are in the relation R. Then there is a formula $F(x,y)$ in L, with x and y of type N, such that

- *if $R(m,n)$ then $\vdash F(S^m 0, S^n 0)$,*
- *if not $R(m,n)$ then $\vdash \neg F(S^m 0, S^n 0)$.*

(Here $\vdash$ means 'there is a proof in L that'.)

As an example of this lemma, let R be the relation 'is 1 greater than'. Let $F(x,y)$ be the formula $x = Sy$. Then, if m is 1 greater than n, it is provable in L that $S^m 0 = SS^n 0$, and if m is not greater than n, it is provable in L that $S^m 0 \neq SS^n 0$.

We shall not prove this lemma here, but it should seem reasonable, inasmuch as L is meant to capture ordinary number theory.

Theorem 29.2.
(Gödel's Incompleteness Theorem (Semantic Version))
If L is consistent, there is a formula in L which is true in any ω-complete model but not provable in L.

Proof: We begin by enumerating all expressions in L of type PN, that is, of the form $\{x \in N | A(x)\}$, where $A(x)$ is a formula in L, with x of type

N. Since there are only countably many finite strings of symbols, there are only countably many expressions of this form. Call them $\alpha_1, \alpha_2, \ldots$.

We next enumerate all the proofs of L. There are only countably many — assuming that proofs are finite — since there are only countably many finite sequences of formulas. Call the proofs $P_1, P_2, \ldots$.

Let $R(m, n)$ be the relation which is satisfied just in case P_n is a proof of the formula $S^m 0 \in \alpha_m$. $R(m, n)$ is recursive, since we can decide whether it is true for a given m and n by looking to see whether $S^m 0 \in \alpha_m$ is the last line of the proof P_n.

By Gödel's Lemma, there is a formula $F(x, y)$ such that, if $R(m, n)$, then $\vdash F(S^m 0, S^n 0)$ and, if not $R(m, n)$, then $\vdash \neg F(S^m 0, S^n 0)$. Consider $\{x \in N | \neg \exists_{y \in N} F(x, y)\}$. This is one of the $\alpha_1, \alpha_2, \ldots$, say α_g. Then, from the definition of α_g, it is provable in L that

$$S^g 0 \in \alpha_g \iff \neg \exists_{y \in N} F(S^g 0, y). \quad (\clubsuit)$$

In both intuitionistic and ordinary logic, it follows that

$$\vdash \neg S^g 0 \in \alpha_g \iff \vdash \exists_{y \in N} F(S^g 0, y).$$

Consider the formula $S^g 0 \in \alpha_g$. If it is provable in L, then there is some natural number n such that P_n is a proof of $S^g 0 \in \alpha_g$. In that case, $R(g, n)$. But then $\vdash F(S^g 0, S^n 0)$, and hence $\vdash \exists_{y \in N} F(S^g 0, y)$. We thus have $\vdash \neg S^g 0 \in \alpha_g$.

Hence it is not the case that $S^g 0 \in \alpha_g$ (assuming L is consistent). That is, for all n, not $R(g, n)$, and hence, for all n, $\vdash \neg F(S^g 0, S^n 0)$ (by Gödel's Lemma). Call this result $(*)$, for future reference.

From the soundness result, we may conclude that, for all n, $\neg F(S^g 0, S^n 0)$ is true in all models of L. Hence in any ω-complete model of L, the proposition $\forall_{y \in N} \neg F(S^g 0, y)$ is true.

Even intuitionistic logic allows us to infer $\neg \exists_y G(y)$ from $\forall_y \neg G(y)$. Hence we may conclude that $\neg \exists_{y \in N} F(S^g 0, y)$ is true in any ω-complete model, and thus $S^g 0 \in \alpha_g$ is true in any such model by $(\clubsuit)$. Thus, although $S^g 0 \in \alpha_g$ is not provable in L (since L is consistent), $S^g 0 \in \alpha_g$ is nonetheless true in all ω-complete models.

Exercises

1. Prove the two claims made about inferences in intuitionistic logic.

2. If $R(m, n)$ means 'm and n both equal 3', find the corresponding formula $F(x, y)$ whose existence is asserted by Gödel's Lemma.

30
More about Gödel's Theorems

We say that a language L is ω-*consistent* provided, for any formula $A(x)$ of L, if $A(S^n0)$ is provable for each natural number n, then it is not the case that $\vdash \neg\forall_{y \in N} A(y)$. It is not hard to show that ω-consistency implies consistency.

If L is ω-consistent, then it follows from (∗) in the previous chapter that it is not the case that
$$\vdash \neg\forall_{y \in N} \neg F(S^g 0, y).$$

From (♣) in the previous chapter, it now follows that it is not the case that $\vdash \neg S^g 0 \in a_g$. This gives us

Theorem 30.1.
(Gödel's Incompleteness Theorem (Syntactic Version))
If L is ω-consistent, there is a formula G such that neither G nor $\neg G$ is provable in L.

Rosser showed that ω-consistency here can be replaced by plain consistency. His proof is short, but tricky, and we shall skip it.

If neither a statement nor its negation is provable in a language, it is called *undecidable* relative to that language. There is no way to get rid of all undecidable statements in a language by adjoining a finite number of new axioms. If we added $S^g 0 \in a_g$ to the axioms of L, then there would still be a g' such that $S^{g'} 0 \in a_{g'}$ is an undecidable statement relative to this new language.

Hilbert's second problem was to prove the consistency of arithmetic using only formal arithmetic to do so. Gödel's Incompleteness Theorem implies that this is impossible. For let *Cons* be a statement in L which expresses the idea that there is no n such that P_n is a proof of $\perp$. Then, if L is consistent, and *if there is a proof in L of this*, we shall have $\vdash$ *Cons*. If L is not consistent, we can prove anything, so we can have $\vdash \neg Cons$ as well as $\vdash Cons$.

One can formalize the proof of Gödel's Incompleteness Theorem to show that $\vdash Cons \Rightarrow S^g 0 \in a_g$. Hence, if there were a proof in L that L was consistent, namely, if $\vdash Cons$, then we would have $\vdash S^g 0 \in a_g$. But this we have seen is not the case.

Gödel's Incompleteness Theorem, being a metamathematical result, has different implications depending on one's conception of mathematics. A classical *formalist* views arithmetic as nothing more than strings of symbols, manipulated according to certain rules. He does not want to rely on any interpretation of these symbols in order to ensure the consistency of the rules. He would hope that consistency could be established within the system itself. Gödel's Theorem shows, against the hopes of the classical formalist, that the consistency of arithmetic cannot be demonstrated within arithmetic. Moreover, it seems that the idea of truth cannot be captured by the notion of provability.

From an *intuitionist* point of view, Gödel's Incompleteness Theorem is not unwelcome. For intuitionists, 'true' should mean 'provable' (with a finite proof). They maintain that there are some statements which are neither true nor false; Gödel's result merely confirms this belief by showing that there are statements which are neither provable nor disprovable.

For the classical *Platonist*, the Incompleteness Theorem shows that there are statements true in the real world which are not provable. In other words, the realities of mathematics are too profound to be captured by any finite axiom system. There will always be truths in mathematics which cannot be cranked out by a computer, but which must await new philosophical insights for their discovery.

31
Concrete Categories

In the 20th century, we find a great deal of concrete, practical mathematics. Statistics is flourishing, the computer has proved the Four Colour Theorem, and numbers with upwards of two hundred digits can be factored.

Another trend in the 20th century is a degree of abstraction never seen before in mathematics. For example, the study of the Euclidean plane has been replaced by the study of vector spaces and topological spaces that abstract some of its properties. A prominant and influential proponent of this trend in algebra was Emmy Noether (1882–1935). The supreme abstraction is the notion of a *category*, to which we shall turn our attention in this chapter.

Nowadays, when studying vector spaces, we are forced to look also at linear transformations. Similarly, when studying topological spaces, we are led to continuous mappings. When studying groups (themselves an abstraction of permutation groups), we have to look at homomorphisms. To abstract the properties which these examples have in common, we introduce the notion of a 'concrete category'.

A *concrete category* is a class of sets, each endowed with a certain structure, together with the class of all functions which map one set to another while preserving this structure.

EXAMPLE 31.1

The class of sets together with the class of all functions between them is a concrete category. Here there is no structure to preserve, so the condition on the functions is trivially satisfied.

EXAMPLE 31.2

A *monoid* is a set containing a special *identity* element 1 together with a binary operation · between members of that set, such that $(a \cdot b) \cdot c = a \cdot (b \cdot c)$, and $1 \cdot a = a \cdot 1 = a$. A *monoid homomorphism* is a function f from a monoid A to another monoid A' which preserves structure: $f(a \cdot b) = f(a) \cdot f(b)$ and $f(1) = 1$. (When we write '$f(1) = 1$' it is understood that the first 1 is the special element of A and the second 1 is the special element of A'.) For example, taking 0 as the special element and + as the binary operation, the natural numbers form a monoid. As another example, the singleton set $\{1\}$ is a monoid with $1 \cdot 1 = 1$. The mapping from the natural numbers to $\{1\}$ is a monoid homomorphism. Note that group homomorphisms are a special case of monoid homomorphisms. The class of monoids, together with the monoid homomorphisms forms a concrete category.

EXAMPLE 31.3

A *pre-ordered set* is a set, together with a binary relation $\leq$ on that set which is reflexive and transitive: $a \leq a$, and if $a \leq b$ and $b \leq c$ then $a \leq c$. A *monotone mapping* is a function f from a pre-ordered set A to a pre-ordered set A' that preserves the order: if $a \leq b$ then $f(a) \leq f(b)$.

For example, both the natural numbers and the even numbers together with the relation $\leq$ are pre-ordered sets. The doubling function maps the natural numbers into the set of even numbers in an order preserving way, and is thus a monotone mapping. The collection of all pre-ordered sets together with all monotone mappings between them forms a concrete category.

Note that in all cases the identity function on a set will preserve its structure. Also, the composition of two structure preserving functions will preserve structure.

More abstract than a concrete category is a *category*, which we shall define in Chapter 32.

Exercise

1. Verify that groups and group homomorphisms form a concrete category.

32
Graphs and Categories

A *graph* (more precisely: an *oriented multigraph*) consists of a class of *arrows* (or *directed edges*) together with a class of *objects* (or *nodes*), and also two mappings from the class of arrows to the class of objects. The mappings are called S (*source* or *domain*) and T (*target* or *codomain*).

$$\{\text{arrows}\} \xrightarrow[\text{target}]{\text{source}} \{\text{objects}\}$$

If f is an arrow, $S(f) = A$ and $T(f) = B$, we write

$$f : A \to B \quad \text{or} \quad A \xrightarrow{f} B.$$

A *category* is a graph subject to the following conditions:

1. associated with any two arrows $f : A \to B$ and $g : B \to C$ (so that $T(f) = S(g)$) is an arrow $g \circ f : A \to C$;

2. if $f : A \to B$, $g : B \to C$ and $h : C \to D$, then $(h \circ g) \circ f = h \circ (g \circ f)$;

3. associated with each object A, there is an identity arrow 1_A, whose source and target are A;

4. if $f : A \to B$ then $f \circ 1_A = f$; if $g : B \to A$ then $1_A \circ g = g$.

A concrete category is a category whose objects are sets with structure and whose arrows are the structure preserving functions between them. In concrete categories, 1_A is the identity map on A, and ∘ is function composition.

EXAMPLE 32.1
Let A be any set. Let the class of objects be A. Let the class of arrows be A. Let $S(a) = T(a) = a$ for all $a \in A$. Let 1_a be a and let $a \circ a = a$. Then the conditions for a category are satisfied and we have what is called the *discrete* category corresponding to A. So a set may be viewed as a category.

EXAMPLE 32.2
Let $(A, 1, \cdot)$ be a monoid (with special object 1 and binary operation $\cdot$). For the class of objects take the singleton set $\{*\}$. For the class of arrows, take A. Let $S(a) = T(a) = *$ for all $a \in A$. Let 1_* be 1 (the monoid identity) and let $a \circ b$ be $a \cdot b$. The graph we have just constructed is a category, thanks to the structure of the monoid. In this way a monoid may be viewed as a category. If we consider the monoid of the natural numbers with addition, in this way, as a category, then what is the number 2? It is the unique arrow in the natural number monoid, viewed as a category, which can be written as a composition of nonidentity arrows in exactly one way.

EXAMPLE 32.3
Let $(A, \leq)$ be a pre-ordered set. Let the class of objects be A, and let the class of arrows be $\{(a,b) | a \leq b\}$. (Here a and b are assumed to be elements of A.) Let $S((a,b)) = a$ and $T((a,b)) = b$. By the reflexivity of $\leq$, (a, a) is an arrow for all $a \in A$. Let this be the identity arrow associated with a, so that $1_a = (a, a)$. Define the composition of arrows thus: $(b, c) \circ (a, b) = (a, c)$. By transitivity of $\leq$, we know that (a, c) is an arrow if (a, b) and (b, c) are arrows. Again, we have a category.

The examples from Chapter 31 show that *many interesting objects in mathematics congregate in categories*. The examples in this chapter illustrate that *interesting mathematical entities may often be viewed as categories*.

Exercise

1. Show that there is a category with exactly two objects and exactly one nonidentity arrow. (This category is sometimes said to *be* the number 2.)

33
Functors

If A and B are two categories, a *functor* F from A to B is a mapping sending objects of A to objects of B and, at the same time, a mapping sending arrows of A to arrows of B, so that

1. if g is any arrow in A with source a and target a', then $F(g)$ is an arrow in B with source $F(a)$ and target $F(a')$;

2. $F(1_a) = 1_{F(a)}$;

3. $F(g \circ h) = F(g) \circ F(h)$.

We saw in Chapter 32 that sets, monoids and pre-ordered sets may all be viewed as categories. How do the structure preserving mappings between such entities compare with the functors between them when they are viewed as categories? We answer the question as follows.

EXAMPLE 33.1
Suppose A and B are sets, each viewed as a category (Example 32.1). Let F be any map from A to B. If $a \in A$, then F maps a to $F(a)$. But a is just 1_a and $F(a)$ is just $1_{F(a)}$. Hence $F(1_a) = 1_{F(a)}$. If a and b are any elements of A, then they are also arrows of A. If $a \neq b$ then $S(a) = T(a) = a \neq b = S(b) = T(b)$, so they cannot be composed. If $a = b$ then $F(a \circ a) = F(a) = F(a) \circ F(a)$. Hence F is a functor.

EXAMPLE 33.2

Suppose A and B are monoids, each viewed as a category. Let F be a monoid homomorphism from A to B. Without loss of generality, we may suppose that $\{*\}$ is the class of objects for both categories A and B (see Example 32.2). Suppose $F(*) = *$. Since F is a homomorphism, $F(1_*) = 1_{F(*)}$. Moreover, $F(a \circ a') = F(a \cdot a') = F(a) \cdot F(a') = F(a) \circ F(a')$. Hence F is a functor.

EXAMPLE 33.3

A monotone mapping between pre-ordered sets may be viewed as a functor. The details are left as an exercise.

The next three examples illustrate the observation that *many entities of interest in mathematics may be viewed as functors*.

EXAMPLE 33.4

A set may be viewed as a category, as we saw in Example 32.1. It can also be viewed as a functor. Let A be the discrete one-element category and B the category of sets. For any set S there is a unique functor from A to B such that $F(1_A)$ is the identity function on S. This functor can be viewed as the set S.

EXAMPLE 33.5

Let A be a category with two objects, **a** and **o**, and with four arrows: $1_\mathbf{a}$, $1_\mathbf{o}$, $s : \mathbf{a} \to \mathbf{o}$, and $t : \mathbf{a} \to \mathbf{o}$. This category may be pictured thus:

$$\mathbf{a} \underset{\to}{\to} \mathbf{o}$$

Suppose F maps **a** to a set X and **o** to a set Y. Suppose $F(s)$ and $F(t)$ are functions with domain X and codomain Y. Then F is a functor from A to the category of sets. We can think of this functor as a graph with class of objects Y, class of arrows X, source mapping $F(s)$ and target mapping $F(t)$.

EXAMPLE 33.6

Let M be a monoid and X a set. Suppose $m : M \times X \to X$ is a function such that, for elements a and a' of M and b of X, $m(a \cdot a', b) = m(a, m(a', b))$. Suppose also that $m(1, b) = b$. Then (M, X, m) is an *M-set*. For example, M might be the monoid of positive integers with multiplication, and X might be the set of segments constructible in Euclidean geometry, and $m(a, CD)$ might be the function mapping a segment CD to a segment CE (with E on CD produced) which is a times as long as CD. One often writes $m(a,b)$ as ab.

How can we make an M-set into a functor? Let F map the category M to the category of sets so that $F(*)$ is X and, for any $a \in M$. $F(a): X \to X$ is the function which sends $b \in X$ onto $F(a)(b) = ab$. Thus $F(1)$ is the identity function on X and $F(a \cdot a') = F(a) \circ F(a')$. For, if $b \in X$ then $F(a \cdot a')(b) = (aa')b = a(a'b) = F(a)(F(a')(b))$.

Call a category *small* if its class of objects and its class of arrows are both *sets*. (The distinction between 'set' and 'class' is made clear in the set theory of Gödel and Bernays.) *Cat* is the category whose objects are small categories and whose arrows are the functors from one small category to another. The fact that the small categories themselves form a category again illustrates the slogan that 'interesting objects congregate in categories'.

Exercises

1. Complete the details of Example 33.3 on pre-ordered sets.

2. Show in detail that *Cat* is indeed a category.

3. A graph is *small* if its class of objects and its class of arrows are both sets. Show that there is a category *Grph* whose objects are small graphs, and whose arrows are like functors, except that they need not satisfy the equations (2) and (3) in the definition of a functor. (*Grph* will be discussed in some detail in the following section.)

34
Natural Transformations

Category theory began in 1945 with Eilenberg and Mac Lane's article 'General Theory of Natural Equivalences'. In this chapter we will investigate the notion of a 'natural equivalence'.

Let A and B be categories and let F and G be functors from A to B. A *natural transformation* t from F to G is a mapping that assigns to every object a of A an arrow $t(a)$ in B from $F(a)$ to $G(a)$, such that, for any arrow f in A from a to b, $G(f) \circ t(a) = t(b) \circ F(f)$. This can be pictured as follows:

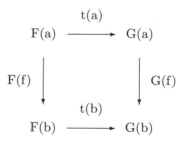

Note that a and b are objects in category A, whereas all of the objects and arrows in the picture are in B.

EXAMPLE 34.1

We saw in Example 33.4 that a set S can be viewed as a functor F from the discrete one-object category to the category of sets, in such a way that F takes its object to S and its arrow to the identity function on S. Let S and S' be two sets, and F and F' the corresponding functors. Given a function f from S to S', let t be a mapping that assigns to the object $*$ of the discrete one-object category the arrow f from S to S'. Then $F'(1_*) \circ t(*) = 1_{S'} \circ f = f = f \circ 1_S = t(*) \circ F(1_*)$. Hence t is a natural transformation from F to F'. Conversely, it is easy to show that every natural transformation from F to F' must have this form.

EXAMPLE 34.2

A *small graph* consists of a set of objects, a set of arrows, and two mappings (*source* and *target*) from the set of arrows to the set of objects. A *morphism* F between small graphs is a mapping which sends the objects of the first graph to the objects of the second, and the arrows of the first graph to the arrows of the second. Moreover, if $f : a \to b$ in the first graph, we require that $F(f) : F(a) \to F(b)$. That is, graph morphisms preserve source and target. In Example 33.5 we saw that a graph may be viewed as a functor. Now we shall show that a graph morphism can be viewed as a natural transformation between functors which represent graphs.

Let A be the two-object category of Example 33.5 and F any functor from A to the category of sets. We saw that $(Y, X, F(s), F(t))$ forms a small graph, call it G.

Suppose h is a graph morphism from G to $G' = (Y', X', F'(s), F'(t))$, where F' is a second functor from A to the category of sets and $F'(\mathbf{a}) = X'$, $F'(\mathbf{o}) = Y'$. Let τ be a function from the set $\{\mathbf{a}, \mathbf{o}\}$ to the class of arrows in Set, which assigns to $\mathbf{a}$ the map $h : X \to X'$ and to $\mathbf{o}$ the map $h : Y \to Y'$. To show that τ is a natural transformation, we must show that, if f is either s or t, then $F'(f) \circ \tau(\mathbf{a}) = \tau(\mathbf{o}) \circ F(f)$. (The equations $F'(1_\mathbf{a}) \circ \tau(\mathbf{a}) = \tau(\mathbf{a}) \circ F'(1_\mathbf{a})$ and $F'(1_\mathbf{o}) \circ \tau(\mathbf{o}) = \tau(\mathbf{o}) \circ F'(1_\mathbf{o})$ follow at once.)

Both $F'(s) \circ \tau(\mathbf{a})$ and $\tau(\mathbf{o}) \circ F(s)$ map X to Y'. If $a \in X$, then $(F'(s) \circ \tau(\mathbf{a}))(a) = F'(s)(h(a))$, which is the source of $h(a)$ in Y'. Since h is a graph morphism, $F'(s)(h(a)) = h(F(s)(a))$. But this equals $\tau(\mathbf{o}) \circ F(s)(a)$. Similarly, one can show that $F'(t) \circ \tau(\mathbf{a}) = \tau(\mathbf{o}) \circ F(t)$. Thus τ is a natural transformation.

EXAMPLE 34.3

We saw in Chapter 33 that an *M-set* can also be regarded as a functor. If (M, X, m) and (M, X', m') are two M-sets, an *M-homomorphism* is a function f from X to X' such that, if $a \in M$ and $b \in X$, $f(m(a,b)) = m'(a, f(b))$. We usually write this equation as $f(ab) = af(b)$. We will let the reader show that such an M-homomorphism may be viewed as a natural transformation between the functors that represent the M-sets.

It may seem from the above examples that category theory is merely a complicated way of expressing simpler ideas. However, one should bear in mind that the abstract definitions of the theory embody the ideas and methods of many branches of mathematics at once, and thus may serve to unify their separate proofs and results. Chapter 35 should help to illustrate this fact.

Exercises

1. Show that every M-homomorphism can be viewed as a natural transformation.

2. Write an essay supporting one of the following views:

    (a) Category theory is a perfect example of useless abstraction. Instead of giving us something new in mathematics, it merely burdens us with a new jargon.

    (b) Category theory is the crown of contemporary mathematics. It combines insights from different branches of mathematics and provides a common language for discussing them.

35
A Natural Transformation between Vector Spaces

We begin with a review of vector spaces over the field of real numbers, although any other field may be substituted for **R**.

A *vector space* V over **R** is an Abelian group together with a mapping from $V \times \mathbf{R}$ to V sending (x, r) to xr, such that, if r and s are real numbers and x, y are in V, then

$$(x+y)r = xr + yr,$$

$$x(r+s) = xr + xs,$$

$$x(rs) = (xr)s,$$

$$x1 = x.$$

Note that **R** is a vector space over itself.

A *linear transformation* from a vector space V to a vector space V' is a mapping $f : V \to V'$ such that $f(x+y) = f(x) + f(y)$ and $f(xr) = (f(x))r$ for any $x, y \in V$ and $r \in \mathbf{R}$.

Taking vector spaces as objects and linear transformations as arrows, it is easy to show that the vector spaces (over the reals) form a concrete category which we shall call *Vect*.

A linear transformation with codomain **R** is called a *linear functional*. If f and g are linear functionals on V, we define $(f+g)(x) = f(x) + g(x)$ and $(fr)x = f(xr)$. Now the set of linear functionals on V forms a vector space over **R**, called the *dual* space V^* of V.

The above procedure may be repeated to obtain the *double dual* of V, namely, $V^{**} = (V^*)^*$. This double dual is closely related to V.

35. A Natural Transformation between Vector Spaces

Let $\sim : V \to V^{**}$ so that, if $x \in V$, then $\tilde{x}$ is the transformation from V^* to $\mathbf{R}$ that maps any linear functional f to $f(x)$, that is, $\tilde{x}(f) = f(x)$. Two things follow immediately:

I. $\tilde{x}$ is a linear transformation from V^* to $\mathbf{R}$, that is, a linear functional on V^*;

II. $\sim$ is a linear transformation from V to V^{**}.

In the case that V has finite dimension, $\sim$ is an isomorphism.

If h is a linear transformation from a vector space V to a vector space V', we define h^{**} as the function from V^{**} to V'^{**} such that, if $p \in V^{**}$, $h^{**}(p)$ is the member of V'^{**} that maps f' in V'^* to $p(f' \circ h)$. Note that

III. $f' \circ h : V \to \mathbf{R}$ and thus $f' \circ h \in V^*$, which is the domain of p;

IV. h^{**} is a linear transformation from V^{**} to V'^{**}.

In proving **IV**, we note that, if $f' \in V'^*$, then

$$h^{**}(pr)(f') = (pr)(f' \circ h) = p((f' \circ h)r)$$

$$= p((f'r) \circ h) = h^{**}(p)(f'r) = (h^{**}(p)r)(f').$$

Now suppose F maps each object V in the category *Vect* to V^{**}, and each arrow $h : V \to V'$ in *Vect* to h^{**}. If h is the identity function on V, then h^{**} is the identity function on V^{**}, since then $h^{**}(p)(f') = p(f' \circ h) = p(f')$.

Moreover, if $h : V \to V'$ and $g : V' \to V''$ are linear transformations, then so is $g \circ h : V \to V''$. $F(g \circ h) = (g \circ h)^{**}$ maps $p \in V^{**}$ to the member of V''^{**} that maps f'' in V''^* to $p(f'' \circ (g \circ h))$. That is, if $f'' : V'' \to \mathbf{R}$,

$$(g \circ h)^{**}(p)(f'') = p(f'' \circ (g \circ h)).$$

(Note that $f'' \circ (g \circ h) : V \to \mathbf{R}$, so that $f'' \circ (g \circ h) \in V^*$, which is the domain of $p \in V^{**}$.)

Since

$$(F(g) \circ F(h))(p)(f'') = g^{**}(h^{**}(p))(f'') = h^{**}(p)(f'' \circ g)$$

$$= p((f'' \circ g) \circ h) = p(f'' \circ (g \circ h)),$$

it follows that F is a functor from *Vect* to *Vect*.

Another functor from *Vect* to *Vect* is the identity functor I.

Let t assign to every vector space V the linear transformation from V to $F(V) = V^{**}$ which we called $\sim$. That is, let $t(V)(x) = \tilde{x}$. Suppose $h : V \to V'$ and let x be any element of V. Then $(F(h) \circ t(V))(x) = h^{**}(\tilde{x})$.

35. A Natural Transformation between Vector Spaces

Also, $(t(V') \circ I(h))(x) = (h(x))\tilde{}$. These two elements of V'^{**} are in fact equal. For let $f' : V' \to \mathbf{R}$, so that $f' \in V'^*$. Then

$$\begin{aligned} h^{**}(\tilde{x})(f') &= \tilde{x}(f' \circ h) \\ &= (f' \circ h)(x) \\ &= f'(h(x)) \\ &= (h(x))\tilde{}\,(f'). \end{aligned}$$

We may conclude that $F(h) \circ t(V) = t(V') \circ I(h)$, and hence t is a natural transformation from the functor I to the functor F.

Examples such as this have led to the slogan *that many objects of interest in mathematics are functors and that the arrows between them are natural transformations*. This and the slogans mentioned earlier were first proposed by F. W. Lawvere.

Exercises

1. Show that *Vect* is a concrete category.

2. Show that the sum of two linear transformations (from V to V') is a linear transformation.

3. Show that V^* is a vector space.

4. Verify **I**, **II**, **III** and **IV** from the text.

5. Generalize the results of this chapter from vector spaces to M-sets. (Things become a little easier if it is assumed that multiplication in M is commutative.)

References

[1] *Dictionary of Scientific Biography*. New York: Charles Scribner's Sons, 1976.

[2] Allen, Reginald E. *Greek Philosophy: Thales to Aristotle*. New York: The Free Press, 1966.

[3] Altmann, Simon L. 'Hamilton, Rodrigues, and the Quaternion Scandal'. Mathematics Magazine, 62 (1989) 291-308.

[4] Anbona, S. 'Un Traité d'Abu Jafar [al Kazin] sur les Triangles Rectangles Numériques'. J. History Arabic Studies 3 (1979), 134-178.

[5] Anglin, W. S. *Mathematics: A Concise History and Philosophy*, New York: Springer-Verlag, 1994.

[6] Anglin, W. S. *The Queen of Mathematics*, Dordrecht: Kluwer, 1995.

[7] Anglin, W. S. 'Using Pythagorean Triangles to Approximate Angles'. American Mathematical Monthly, 95 (1988), 540-41.

[8] Aquinas, Thomas *Summa Contra Gentiles*. Trans. Anton C. Pegis et al. London: University of Notre Dame Press, 1975.

[9] Archibald, R. C. 'Gauss and the Regular Polygon of Seventeen Sides'. American Mathematical Monthly, 27 (1920), 323-26.

[10] Archimedes *The Works of Archimedes*. Trans. T. L. Heath. New York: Dover, 1897.

[11] Aryabhata *Aryabhatiya of Aryabhata*, New Delhi: Indian National Science Academy, 1976.

[12] Aschenbrenner, Karl *The Concept of Value*. Dordrecht: D. Reidel, 1971.

[13] Ascher, M. *Ethnomathematics*. Pacific Grove, California: Brooks/Cole, 1991.

[14] Ayoub, R. 'What is a Napierian Logarithm?'. Amer. Math. Monthly, 100 (1993), 351-364.

[15] L. Bakhtiar *Sufi Expressions of the Mystic Quest*. London: Thamer and Hudson, 1976.

[16] Ball, W. W. Rouse *A Short Account of the History of Mathematics*. New York: Dover, 1960.

[17] Barker, Stephen F. *Philosophy of Mathematics*. Englewood Cliffs: Prentice-Hall, 1964.

[18] Bailey, Cyril *The Greek Atomists and Epicurus*. Oxford: Clarendon, 1928.

[19] Barnes, Jonathan *Early Greek Philosophy*. London: Penguin, 1987.

[20] Barnes, Jonathan *PreSocratic Philosophers*. Vol. 2. London: Routledge and Paul, 1979.

[21] Barnsley, Michael *Fractals Everywhere*. Boston: Academic, 1988.

[22] Bell, E. T. *The Development of Mathematics*. 2nd ed. New York: McGraw-Hill, 1945.

[23] Bell, E. T. *Men of Mathematics*. New York: Simon and Schuster, 1965.

[24] Benacerraf, Paul, and Hilary Putnam, ed. *Philosophy of Mathematics*. 2nd ed. Cambridge: Cambridge University, 1983.

[25] Berggren, J. L. *Episodes in the Mathematics of Medieval Islam*. New York: Springer-Verlag, 1986.

[26] Birkhoff, Garret, and Saunders Mac Lane *A Survey of Modern Algebra*. 4th ed. New York: Macmillan, 1977.

[27] Bishop, Errett, and Douglas Bridges *Constructive Analysis*. Berlin: Springer-Verlag, 1985.

[28] Bleicher, M. N. 'A New Algorithm for the Expansion of Egyptian Fractions'. Journal of Number Theory, 4 (1972), 342-82.

[29] Blumenthal, L. M. *A Modern View of Geometry.* San Francisco: Freeman, 1961.

[30] Boyer, Carl B. and Uta C. Merzbach *A History of Mathematics.* 2nd ed. New York: John Wiley, 1989.

[31] Browder, F. E. (ed.) *Mathematical Developments Arising from Hilbert Problems.* Proc. Symposia Pure Math. 28, Providence: Amer. Math. Soc., 1976.

[32] Bruckheimer, M. and Y. Salomon 'Some Comments on R. J. Gillings' Analysis of the 2/n Table in the Rhind Papyrus'. Historia Mathematica, 4 (1977), 445-52.

[33] Burton, David M. *The History of Mathematics.* Dubuque: Wm. C. Brown, 1985.

[34] Cajori, Florian *A History of Mathematics.* 2nd ed. New York: Macmillan, 1919.

[35] Cantor, Georg *Transfinite Numbers.* New York: Dover, 1955

[36] Cardano, Girolamo *The Great Art.* Trans. T. Richard Witmer Cambridge, Mass.: MIT, 1968.

[37] Cardano, Girolamo *The Book of My Life.* Trans. Jean Stoner, New York: Dover, 1962.

[38] Carruccio, Ettore *Mathematics and Logic in History and in Contemporary Thought.* Trans. I. Quigly. London: Faber and Faber, 1964.

[39] Carslaw, H. S. 'Gauss's Theorem on the Regular Polygons which can be Constructed by Euclid's Method'. Proceedings of the Edinburgh Mathematical Society, 28 (1910), 121-8.

[40] Chace, Arnold Buffum et al. *The Rhind Mathematical Papyrus.* Oberlin, Ohio: Mathematical Association of America, 1927-9.

[41] F. Chung and S. Sternberg ' Mathematics and the Buckyball'. Amer. Scientist 81 (1993).

[42] Church, A. *The Calculi of Lambda Conversion.* Ann. Math. Studies 6, Princeton University, 1941.

[43] Connell, I. *Modern Algebra.* New York: North Holland, 1982.

[44] Couture, J. and J. Lambek 'Philosophical Reflections on the Foundations of Mathematics'. Erkentniss 34 (1991), 187- 209.

[45] Coxeter, H. S. M. and S. L. Greitzer *Geometry Revisited.* Washington: The Mathematical Association of America, 1967.

[46] Dauben, J. W. 'Abraham Robinson and Nonstandard Analysis'. in: W. Asprey and P. Kitcher (eds.), *Studies in the Philosophy of Science* 11 (1988), Minneapolis: University Minneapolis, 1988.

[47] Dauben, J. W. *Georg Cantor: His Mathematics and Philosophy*. Cambridge, Mass: Harvard University, 1979.

[48] Davis, Martin 'Hilbert's Tenth Problem is Unsolvable'. American Mathematical Monthly, 80 (1973), 233 -69.

[49] Descartes, René *The Geometry of Rene Descartes*. Trans. D. E. Smith and M. L. Latham. New York: Dover, 1954.

[50] L. E. Dickson *History of the Theory of Numbers*. New York: Chelsea, 1971.

[51] Diophantus *Arithmetica*. 2nd ed. Trans. Thomas L. Heath. New York: Dover, 1964.

[52] Diophantus *Books IV to VII of Diophantus' Arithmetica*. Trans. Jacques Sesiano. New York: Springer-Verlag, 1982.

[53] Dodgson, Charles L. *Euclid and his Modern Rivals*. London: Macmillan, 1879.

[54] Dummett, M. *Elements of Intuitionism*. Oxford: Oxford University Press, 1977.

[55] Dunham, William *Journey through Genius*. New York: Wiley, 1990.

[56] Eilenberg, S. and S. Mac Lane 'General Theory of Natural Equivalence'. Trans. Amer. Math. Soc., 58 (1945) 231-94.

[57] Ellison, W. J. 'Waring's Problem'. Amer. Math. Monthly 73 (1971), 10-36.

[58] Euclid *Elements*. Trans. T. L. Heath. 2nd ed. New York: Dover, 1956.

[59] Euler, Leonhard *Opera Omnia*. Geneva: 1944.

[60] Eves, Howard *An Introduction to the History of Mathematics*. 5th ed. New York: Holt, Rinehart and Winston, 1969.

[61] Eves, Howard *A Survey of Geometry*. Boston: Allyn and Bacon, 1963.

[62] Ewald, Gunter *Geometry: An Introduction*. Belmont, California: Wadsworth, 1971.

[63] Federico, P. J. *Descartes on Polyhedra*. New York: Springer-Verlag, 1982.

[64] Fermat, Pierre *Oeuvres de Fermat.* Paris: Gauthier-Villar, 1891.

[65] Fibonacci *Le Livre des Nombres Carres.* Trans. Paul ver Eecke. Paris: Albert Blanchard, 1952.

[66] Fowler, D. H. *The Mathematics of Plato's Academy.* Oxford: Clarendon, 1987.

[67] Freed, S. and R. S. Freed 'Origin of the Swastika', Natural History, 1981, 68-75.

[68] Gardner, Martin 'Mathematical Games' Scientific American, 233 (December 1975), 117-8.

[69] Gillings, Richard J. *Mathematics in the Time of the Pharaohs.* New York: Dover, 1972.

[70] Gittleman, Arthur *History of Mathematics.* Columbus: Charles E. Merrill, 1975.

[71] Glass, A. M. W. 'Existence Theorems in Mathematics'. The Mathematical Intelligencer, 11 (1989), 56-62.

[72] Goodstein, R. L. *Recursive Number Theory.* New York: Academic, 1970.

[73] Graves, Robert Perceval *Life of Sir William Rowan Hamilton.* Dublin: Hodges, Friggs Co., 1885.

[74] Guthrie, Kenneth Sylvan *The Pythagorean Sourcebook and Library.* Grand Rapids: Phanes, 1987.

[75] Hall, H. S. and Knight, S. R. *Higher Algebra.* London: Macmillan, 1955.

[76] Hardy, G. H. *A Mathematician's Apology.* Cambridge: Cambridge University, 1992.

[77] Hatcher, William S. *The Logical Foundations of Mathematics.* Oxford: Pergamon, 1982.

[78] Hawkins, G. S. *Stonehenge Decoded.* New York: Doubleday, 1965.

[79] Heaslett, M. A. and J. V. Uspensky *Elementary Number Theory.* New York: McGraw Hill, 1939.

[80] Heath, T. L. *A History of Greek Mathematics.* Oxford: Clarendon, 1921.

[81] Heuer, K. *City of Stargazers.* New York: Charles Scribner's Sons, 1972.

[82] Hilbert, D. *The Foundations of Geometry*. La Salle: Open Court, 1962.

[83] Hintikka, J. *The Philosophy of Mathematics*. Oxford: Oxford University, 1969.

[84] Hofstadter, D. R. *Gödel, Escher, Bach*. New York: Basic Books, 1979.

[85] Hooper, Alfred *Makers of Mathematics*. New York: Random House, 1948.

[86] Hoyle, F. *On Stonehenge*. San Francisco: Freeman, 1977.

[87] Huntley, H. E. *The Divine Proportion*. New York: Dover, 1970.

[88] Jones, P. James 'Diophantine Representations of Fibonacci Numbers over Natural Numbers'. In *Applications of Fibonacci Numbers*. Vol. 3. Ed. G. E. Bergum et al. Dordrecht: Kluwer, 1990.

[89] Jones, Philip S. 'Recent Discoveries in Babylonian Mathematics 1'. The Mathematics Teacher, 50 (1957), 162-5.

[90] Johnson, Roger A. *Advanced Euclidean Geometry*. New York: Dover, 1960.

[91] Jung, Karl *Man and His Symbols*. New York: Doubleday, 1983.

[92] Kershner, R. B., and L. R. Wilson *The Anatomy of Mathematics*. New York: Ronald, 1950.

[93] Kershner, R. B. 'On Paving the Plane'. American Mathematical Monthly, 75 (1968), 839-44.

[94] Khayyam, O. *The Algebra of Omar Khayyam*. Ed. D. S. Kasir New York: AMS, 1972.

[95] Khayyam, O. *Rubaiyat*. New York: Doubleday, 1952.

[96] Khwarizmi, M. *The Algebra*. Trans. Frederic Rosen. London: Oriental Translation Fund, 1831.

[97] Klarner, David A., ed. *The Mathematical Gardner*. Boston: Prindle, Weber and Schmidt, 1980.

[98] Kleene, S. C. *Intoduction to Metamathematics*. New York: Van Nostrand, 1952.

[99] Kline, Morris *Mathematics in Western Culture*. New York: Oxford University, 1953.

[100] Kneale, W. C., and M. Kneale *The Development of Logic*. Oxford: Clarendon, 1984.

[101] Koblitz, Neal *A Course in Number Theory and Cryptography*. New York: Springer-Verlag, 1987.

[102] Koestler, Arthur *The Watershed*. New York: Anchor Books, Doubleday, 1960.

[103] Lagrange, J. *Oeuvres de Lagrange*. Paris: Gauthier-Villars, 1869.

[104] Lam, C. W. H. 'How Reliable is a Computer-Based Proof?' The Mathematical Intelligencer, 12(1) (1990), 8-12.

[105] Lambek, J. 'Are the Traditional Philosophies of Mathematics Really Incompatible?'. The Mathematical Intelligencer, 16 (1994), 56-62.

[106] Lambek, J. 'If Hamilton had Prevailed: Quaternions in Physics'. The Mathematical Intelligencer, to appear.

[107] Lambek, J. 'How to Program an Abacus?'. Can. Math. Bulletin, 4 (1961), 295-302.

[108] Lambek, J. and P. J. Scott *Introduction to Higher Order Categorical Logic*. Cambridge: Cambridge University, 1988.

[109] Legendre, Adrien-Marie *Theorie des Nombres*. 4th ed. Paris: Firmin Didot, 1830.

[110] LeLionnais, Francois, ed. *Great Currents of Mathematical Thought*. Trans. Charles Pinter and Helen Kline, New York: Dover, 1971.

[111] Leonard de Pise. See Fibonacci.

[112] Lipschitz, M. 'Recherches sur la transformation...'. Journal de Mathematiques Pures et Appliquees 4th Ser. Vol. 2 (1886) 373-439, especially p. 404.

[113] Mac Lane, S. *Categories for the Working Mathematician*. New York: Springer-Verlag, 1971.

[114] Marius, Cleyet-Michaud *Le Nombre d'Or*. Que-Sais-je?, 1530.

[115] Martin, Georges E. *Transformation Geometry*. New York: Springer-Verlag, 1982.

[116] Maxwell, J. C. 'Remarks on the Classification of Physical Quantities'. Proc. London Math. Soc. 3 (1869), 224-232.

[117] Maziarz, Edward A. and Thomas Greenwood *Greek Mathematical Philosophy*. New York: Frederick Ungar, 1968.

[118] McClenon, R. B. 'Leonardo of Pisa and his Liber Quadratorum'. American Mathematical Monthly, 26 (1919), 1-8.

[119] Menninger, K. *Number Words and Number Systems; a cultural history of numbers.* Cambridge, Mass.: M.I.T., 1969.

[120] Midonick, Henrietta O., ed. *The Treasury of Mathematics.* New York: Philosophical Library, 1965.

[121] Mohanty, S. P. 'Integer Points of $y^2 = x^3 - 4x + 1$'. Journal of Number Theory, 30 (1988), 86-93.

[122] Mordell, L. J. *Diophantine Equations.* New York: Academic, 1969.

[123] Nagel, E. and J. R. Newman *Gödel's Proof.* London: Routledge and Kegan Paul, 1959.

[124] Nahin, P. J. 'Oliver Heaviside'. Scientific American, 1990, 122-129.

[125] Napier, John *The Construction of the Wonderful Canon of Logarithms.* Trans. William Rae MacDonald. Edinburgh: William Blackwood and Sons, 1889.

[126] Nelson, Harry L. 'A Solution to Archimedes' Cattle Problem'. Journal of Recreational Mathematics, 13 (1980-81), 164-76.

[127] Neugebauer, O. *The Exact Sciences in Antiquity.* 2nd ed. New York: Dover, 1969.

[128] Newman, James R., ed. *The World of Mathematics.* New York: Simon and Schuster, 1956.

[129] Ogilvy, C. Stanley *Excursions in Geometry.* New York: Oxford University, 1969.

[130] Ore, Oystein *Cardano, the Gambling Scholar.* New York: Dover, 1965.

[131] Ore, Oystein *Number Theory and Its History.* New York: McGraw-Hill, 1948.

[132] Pappus of Alexandria *La Collection Mathématique.* Trans. Paul ver Eecke, Paris: Bruges, 1933.

[133] Peano, Ioseph *Arithmetices Principia.* Rome: Fratres Bocca, 1889.

[134] Peitgen, H. O. and P. H. Richter *The Beauty of Fractals.* Berlin: Springer-Verlag, 1986.

[135] Penrose, Roger *The Emperor's New Mind.* New York: Oxford University, 1989.

[136] Plato *The Collected Dialogues.* New Jersey: Princeton University, 1989.

[137] Plutarch, *Makers of Rome*. London: Penguin, 1965.

[138] Poincaré, H. *Science and Hypothesis*. New York: Walter Scott, 1905.

[139] Prawitz, D. *Natural Deduction*. Stockholm: Almquist and Wiskell, 1965.

[140] Robins, Gay, and Charles Shute, ed. *The Rhind Mathematical Papyrus*. London: British Museum Publications, 1987.

[141] Rosen, F. *The Algebra of Mohammed Ben Musa*. London: 1831.

[142] Rosenbloom *The Elements of Mathematical Logic*. New York: Dover, 1950.

[143] Russell, Bertrand *A History of Western Philosophy*. New York: Simon and Schuster, 1972.

[144] Russell, Bertrand *Mysticism and Logic*. London: Penguin Books, 1953.

[145] Shapiro, Harold N. *Introduction to the Theory of Numbers*. New York: John Wiley, 1983.

[146] Sierpinski, W. *Elementary Theory of Numbers*. Warsaw: Polska Akademia Nauk, 1964.

[147] Silberstein, L. *The Theory of Relativity*. 2nd ed. London: Macmillan, 1924.

[148] Smith, David Eugene *History of Mathematics*. New York: Dover, 1958.

[149] Smith, David Eugene *A Source Book in Mathematics*. New York: Dover, 1959.

[150] Smith, T. V., ed. *From Thales to Plato*. Chicago: The University of Chicago, 1965.

[151] Stillwell, John *Mathematics and Its History*. New York: Springer-Verlag, 1989.

[152] Tymoczko, Thomas *New Directions in the Philosophy of Mathematics*. Boston: Birkhauser, 1985.

[153] Tzanakis, N. and B. M. M. de Weger 'On the Practical Solution of the Thue Equation'. Journal of Number Theory, 31 (1989), 99-132.

[154] J.V. Uspensky and M.A. Heaslet *Elementary Number Theory*. McGraw Hill, New York: 1939.

[155] Uspensky, V. A. *Post's Machines*. Moscow: Mir Publications, 1979.

[156] Van der Waerden, B. L. *Geometry and Algebra in Ancient Civilizations*. Berlin: Springer-Verlag, 1983.

[157] Van der Waerden, B. L. *A History of Algebra*. Berlin: Springer-Verlag, 1985.

[158] Van der Waerden, B. L. *Science Awakening*. Vol. 1. 4th ed. Trans. Arnold Dresden, Dordrecht: Kluwer, 1975.

[159] van Nooten, B. 'Binary Numbers in Indian Antiquity'. J. Indian Philosophy 21 (1993), 31-50.

[160] Ver Eeke, P. *Diophante d'Alexandrie*. Paris: Blanchard, 1959.

[161] Viète, Francois *Opera Mathematica*. Hildesheim: Georg Olms Verlag, 1970.

[162] Wantzel, P. L. 'Recherches sur les moyens de reconnaitre si un probleme de geometrie peut se resoudre par la regle et le compas'. Journal de Mathematique, 2 (1837) 366-72.

[163] Weyl, André *Number Theory: An Approach through History*. Basel: Birkhauser, 1984.

[164] Weyl, Herman *Space, Time and Matter*. New York: Dover, 1922.

[165] Zaslavski, C. *Africa Counts*. Boston: Prindle, Weber and Schmidt, 1973.

Index

abacus 247
Abel 136
Abelian 184
Abraham 21
abstraction 259, 295, 305
Abu'l Wafa 165
Academy 63, 109
Achilles 55
Adelhard 121
Agha Khan 119
Ahmose 7, 11
aleph-null 68, 219
Alexander the Great 83
Alexandria 29, 83, 93, 95, 107
algebra 106, 118
algorithm 118, 245, 256
al-Khayyami 118
al-Khazin 106, 122, 196
al-Khwarizmi 117
al-Mamun 117
amicable 34, 119
Amru 109
Anaximander 31
Anaximenes 31
Antiphon 60

Apollonius 94, 130
Aquinas 84, 109
arbelos 101
Archimedes 27, 97, 104, 112, 141, 153
Archytas 63, 84
Aristarchus 93
Aristotle 7, 33, 47, 53, 68, 84, 99, 100, 156, 265, 266, 279
Arius 107
arrow 55, 162, 297
Aryabhata 113
Asia Minor 2
astrology 105, 114, 129, 130, 143
astronomy 1, 105, 141
Aswan 95
Athanasius 107
Athens 59
atom 57, 69
Augustine 39
axiom 84, 289
axiom of Archimedes 92, 99, 146
axiom of choice 285

Babylonian 21

322 Index

Bachet 149
backwards causation 54
Baghdad 115
Barrow 155
Beltrami 90
Ben Ezra 121
Berkeley 56, 160
Bernays 92, 268, 301
Bhaskara 113, 114
Bible 109, 126
binary 11
Binomial Theorem 155
biquaternion 212
Boethius 109
Bolyai 90
Bombelli 130
Book of Changes 11
Boole 267
Bourbaki 83, 84, 87, 175
Brahe 142
Brahmagupta 3, 113, 233
Briggs 141
Brouwer 176, 271, 277
buckminster fullerene 45
buckyball 45

calculation 245
calculus 98, 118, 156, 281
Cantor 68, 219
Cardano 3, 129, 195
cardinal 219
cartesian 90, 170
Casanova 80
category 162, 295, 297
cattle problem 100
Cauchy 191
Cavalieri 153
chemical equation 42
Chen Jing-Run 18, 112
China 3, 11, 18, 111, 125, 149, 151
Ch'in Chiu Shao 112
Chiu Chang Suan Shu 111
Chinese remainder 111
chord 104, 113, 130, 153

Christian 112
Christina 146
Ch'ung Chih 112
Church 147, 176, 252, 259
Chu Shih Chieh 112
cipher key 18
circle 9, 60, 64, 98, 153
circle squaring 60, 71, 79
classical problems 71, 79
Cleopatra 83
Cohen 220, 245
Cole 38
Columbus 95, 126
combinator 262
commensurable 47
complete ordered field 192
completeness 86, 289, 290
complex 3, 122, 129, 130, 176, 195
comprehension scheme 267
concrete category 295
conic section 64, 94, 147
Constantine 107
construction, ruler and compass 71
continous 30, 61, 85, 199
continued fractions 228
continuum hypothesis 245
convergent 165, 228
Copernicus 1, 94, 141
cosines, law of 86
count nouns 30
Couture 285
Coxeter 151, 165
creator 39, 43
Crotone 33
cryptography 18
cubic equation 118, 119, 128, 132
Curry 224
cycloid 154

da Coi 128
d'Alembert 199
Davis 246
Dedekind 3, 269

Index   323

De Gang Ma 112
de la Vallée Poussin 18
Delian problem 62, 71
Democritus 7, 31, 57
De Moivre 73, 124, 164, 165, 197
De Morgan 267, 286
Desargues 145
Descartes 44, 75, 146, 164
Dieudonné 87, 175, 176
Diophantine 100, 233, 255
Diophantus 27, 100, 105, 149, 233
Dirac 216
directrix 95
Dirichlet 149
discrete 30, 61
distributive law 48
divine 34, 126
division ring 204
double negative 265, 271, 275
doubling cube 62, 65, 71, 80, 81, 147

eccentricity 95
eclipse 2, 29
Egypt 7, 16, 41, 159
Eilenberg 303
Einstein 212, 217
electromagnetic 211, 217
electron 54
Elements 16, 61, 83, 121
ellipse 153
elliptic geometry 85, 91
Empedocles 57
ennumerable 251
Eratosthenes 17, 95
Erdös 10, 16, 245
Euclid 16, 34, 37, 48, 83, 156, 196, 228
Eudoxus 3, 61, 63–64, 84, 86, 191
Euler 39, 44, 82, 164, 207
Euler's Formula 44, 165
excluded middle 68, 100, 265, 271
Ezekiel 22

factorization 16, 18
Faltings 150
Fermat 81, 148, 165, 242
Ferrari 129
Ferro 128
Feynman 54
Fibonacci 106, 122, 256
field 80, 87, 162, 184, 189
figurative 35, 169
Fiore 128
Fitzgerald 119
fixpoint 262
flow diagram 247
fluent 157
fluxion 155, 163
focus 95
formalist 176, 294
Four Colour Theorem 295
Four Weight Problem 12
Fowler 227
Fraenkel 268
Francis 122
Frederick II 122, 166
Frege 267
Frey 243
function 259, 262, 296
functor 299
Fundamental Theorem of Algebra 199
Fundamental Theorem of Arithmetic 16, 231

Galileo 141, 147, 156, 320
Galois 136
Gauss 3, 16, 81, 86, 90, 169, 199
generating function 123
Gentzen 273
George I 160
Gerbert 121
Girard 199
God 22, 148, 150, 151, 161, 176, 268
Gödel 220, 245, 253, 268, 289
Goldbach 17, 18, 112

golden ratio 64, 74, 124, 126, 230, 237
Golenishchev 7
graha 1
graph 297, 304
gravity 156
Gregory 154, 155
group 162, 184

Hadamard 18
Hafiz 119
Hamilton 3, 175, 203
Harriot 126
Hasan 119
Heath 106
Heaviside 211
Hegel 53
Henkin 289
Heraclitus 31, 53
Hermite 79, 212
Herodotus 7
Heron 22, 104, 114
Heyting 277
Hilbert 17, 87, 91, 107, 146, 245, 255, 293
Hipparchus 22, 105
Hippasus 43, 47
Hippias 59
Hippocrates 61, 84
Hiram of Tyre 27
Hooke 156
Horner 112
Hoyle 2
Hrotsvitha 121
Huygens 154
Hypatia 108
hyperbolic geometry 90

icosahedron 42, 44
India 3, 112, 155, 233
infinite 31, 68, 114, 118, 219
infinitesimal 56, 160
injective 189
integers 183
integral domain 183

intuitionist 68, 176, 272, 294
intuitionistic logic 271, 281
Ionia 29, 69
Iraq 2, 21
irrationality 35, 47

Jesus 107, 129, 139
Jones 98
Jordanus 122
Joseph 8
Jupiter 1,142
Justinian 109

kebu 1
Kepler 142, 156
Khayyam 119, 128
Kingsley 108
Kleene 251
Klein-Gordon 215
Koerbero 27
Kolmogorov 277
Kronecker 28, 176, 268
Kummer 149

Lagrange 113, 166, 207, 237, 255
Lambda Calcutus 259, 289
Laplace 167
Lawvere 54, 309
Least Action 161
Legendre 17, 86, 89, 149, 167, 170
Leibniz 56, 159, 267
Lenin 54
Leonardo, see Fibonacci
Lewis 109
Lilavati 114
limit 55, 160
Lindemann 79
Lipschitz 208
linear equations 111, 114, 233
linear functional 307
linear transformation 307
Lobachevsky 90, 203
logarithm 17, 139
logic 53, 265

logicist 176, 267, 294
Lorentz transfomation 212
Louis XIV 159
love 57
Lucas 38
lune 61

Mac Lane 303
Maclaurin 163
magic square 111
Magna Grecia 2
Mahavira 114
Marcellus 98
Marie Antoinette 166
Mars 1, 142, 143
Marx 53, 57
mass nouns 30
mathematical induction 61, 151, 179
Matiyasevič 17, 123, 246, 257
matrix 203
Maupertuis 161
Maxwell 211
Mazur 243
means 34
Menaechmus 64
Menelaus 104, 107
Meno 63
Mercator 154
Mercury 1
Mersenne 38, 145, 148
mesopotamian 21, 122, 242
M-homomorphism 304
Miletus 29
Mill 161
Minkowski 212
model 289
monad 162
monoid 296, 298, 300
month 1
moon 1, 104, 166
Moscow Papyrus 7
M-set 300
Museum 83
mysticism 39, 162

Napier 139
natural number 175, 179
natural transformation 303, 307
Nebuchadnezzar 21
negative number 3, 113, 123, 126
Nehemiah 28
Nelson 101
Newton 56, 84, 97, 148, 155
Nicomachus 108
Nim 12
Nizam-ul-Mulk 119
nodes 1, 297
Noether 295
nonconstructive 271, 272
non-Euclidean 89
norm 122, 204
notation 125

Omar I 109
omega-complete 290
omega-consistent 293
Oresme 122

Pacioli 126
Paganini 39
Pappus 107, 146
papyrus 7
paradox 55, 268
Parallel Postulate 85, 89, 118, 203
Parmenides 54
Pascal 112, 150
Pauli 215
Peano 176, 179, 182
Pell 113, 160, 237
pentagon 42
perfect 37, 86, 121, 122, 145, 164
Philo of Alexandria 39
Philo of Megara 266
pi 9, 27, 77, 98, 112, 113, 114, 118, 155
Pingala 112
Planck 215
planet 1

Plato 29, 42, 50, 59, 60, 63, 67, 70, 162
Platonist 68, 176, 290, 294
Playfair 89
Plimpton 27, 33
Plotinus 105
Plutarch 98
Poincaré 90, 213, 269
polyhedron, see regular solid
positron 54
Post 246
postulate 84
prime 15, 18, 81, 86, 257
primitive recursive 246
primitive root 169
probability 129, 151, 164
Proclus 50, 89, 109
program 247
projective geometry 145
proof 30
Ptolemy 22, 105, 108, 130, 141
Ptolemy I 83
Putnam 246, 257
pyramid 7, 31, 57, 113
Pythagoras 27, 31, 33, 43
Pythagoreans 3, 33, 47, 108, 118, 142, 237
Pythagorean triangle 27, 241

Qin Jiushao 112
quadratic equation 25
quadratic reciprocity 168, 169
quadratic surds 237
quadratrix 60
quadrivium 34
quantum mechanics 155, 215
quantum of time 56
quartic equation 129, 135
quaternions 3, 175, 203

rahu 1
Ramanujan 115
rational operation 75
rational 187
real 3, 191

Recorde 126
recursive 251
Regiomontanus 125
regular polygon 41, 60, 73, 81, 86
regular solid 41, 63, 87, 142
relativity 212
Renaissance 125
Rhind Papyrus 7
Ribet 243
Riemann 91
Riese 126
rigour 86, 87
ring 183
Robinson, A. 56, 160, 162
Robinson, J. 246, 255
Rodrigues 3
Rudolff 126
Ruffini 136
rule of signs 147
Russell 162, 176, 267
Rutherford 57

Sacchieri 89
Saturn 1, 142
scale (base) 9, 11
Schanuel 192
Schönfinkel 262
Schrödinger 215
Schroeder-Berstein 225
Schwenter 227
Shimura 243
Shyreswood 266
siddhanta 113
Silberstein 212
sine 104, 113
Six Weight Problem 12
skew field 204
Snell 161
Socrates 59, 62, 162
spherical triangle 126
Spinoza 84
square 47, 63, 67, 86, 207
squaring circle, see circle squaring
Steiner 91

Stevin 141
Stifel 126, 139
Stonehenge 2
string 47, 176, 217, 252, 289, 292
subroutine 248
Sultana 235
Sulvasutras 112
Sumeria 21
Sun 1, 93
Sun Tsu 111
Syene 95
syllogism 266
Sylvester 121
Syracuse 97

Taniyama 243
Tartaglia 128
Taurinus 90
Taylor, B. 163
Taylor, R. 243
Thabit 39, 118
Thales 3, 29, 30, 33
Theaetetus 63, 84, 237
Theodorus 63
Thibault 89
Timaeus 42, 63
tortoise 55
Tournesol 177, 241, 243
trigonometry 60, 104, 118
trisection of angle 60, 71, 130
Trotsky 47
Turing 246, 259
type 268, 285
Tzanakis 36

unique factorisation 16, 169, 231
unit fraction 9, 100, 230
Ur 21

vector 211, 307
Venus 1
Viète 130
Vinogradov 18
Voltaire 156, 161
Von Neumann 176

Wallis 89, 148, 154, 155
Wantzel 60, 65
water 30, 31, 42, 57
week 1
Weierstrass 160, 199
Whitehead 268
Widman 126
Wiles 149, 243
William of Shyreswood 266
Williams 101
Wilson 166
Wittgenstein 266
Wren 156

year 1

Zeno 54, 61
Zermelo 268
zero 105, 112, 114, 121, 185
Zhu Shijie 112
ziggurat 27

Undergraduate Texts in Mathematics

Anglin: Mathematics: A Concise History and Philosophy.
Readings in Mathematics.
Anglin/Lambek: The Heritage of Thales.
Readings in Mathematics.
Apostol: Introduction to Analytic Number Theory. Second edition.
Armstrong: Basic Topology.
Armstrong: Groups and Symmetry.
Bak/Newman: Complex Analysis.
Banchoff/Wermer: Linear Algebra Through Geometry. Second edition.
Berberian: A First Course in Real Analysis.
Brémaud: An Introduction to Probabilistic Modeling.
Bressoud: Factorization and Primality Testing.
Bressoud: Second Year Calculus.
Readings in Mathematics.
Brickman: Mathematical Introduction to Linear Programming and Game Theory.
Cederberg: A Course in Modern Geometries.
Childs: A Concrete Introduction to Higher Algebra. Second edition.
Chung: Elementary Probability Theory with Stochastic Processes. Third edition.
Cox/Little/O'Shea: Ideals, Varieties, and Algorithms.
Croom: Basic Concepts of Algebraic Topology.
Curtis: Linear Algebra: An Introductory Approach. Fourth edition.
Devlin: The Joy of Sets: Fundamentals of Contemporary Set Theory. Second edition.
Dixmier: General Topology.
Driver: Why Math?
Ebbinghaus/Flum/Thomas: Mathematical Logic. Second edition.
Edgar: Measure, Topology, and Fractal Geometry.
Fischer: Intermediate Real Analysis.
Flanigan/Kazdan: Calculus Two: Linear and Nonlinear Functions. Second edition.
Fleming: Functions of Several Variables. Second edition.
Foulds: Combinatorial Optimization for Undergraduates.
Foulds: Optimization Techniques: An Introduction.
Franklin: Methods of Mathematical Economics.
Hairer/Wanner: Analysis by Its History.
Readings in Mathematics.
Halmos: Finite-Dimensional Vector Spaces. Second edition.
Halmos: Naive Set Theory.
Hämmerlin/Hoffmann: Numerical Mathematics.
Readings in Mathematics.
Iooss/Joseph: Elementary Stability and Bifurcation Theory. Second edition.
Isaac: The Pleasures of Probability.
Readings in Mathematics.
James: Topological and Uniform Spaces.
Jänich: Linear Algebra.
Jänich: Topology.
Kemeny/Snell: Finite Markov Chains.
Kinsey: Topology of Surfaces.
Klambauer: Aspects of Calculus.
Lang: A First Course in Calculus. Fifth edition.
Lang: Calculus of Several Variables. Third edition.
Lang: Introduction to Linear Algebra. Second edition.
Lang: Linear Algebra. Third edition.
Lang: Undergraduate Algebra. Second edition.
Lang: Undergraduate Analysis.
Lax/Burstein/Lax: Calculus with Applications and Computing. Volume 1.
LeCuyer: College Mathematics with APL.
Lidl/Pilz: Applied Abstract Algebra.
Macki-Strauss: Introduction to Optimal Control Theory.
Malitz: Introduction to Mathematical Logic.

(continued)

Undergraduate Texts in Mathematics

Marsden/Weinstein: Calculus I, II, III. Second edition.
Martin: The Foundations of Geometry and the Non-Euclidean Plane.
Martin: Transformation Geometry: An Introduction to Symmetry.
Millman/Parker: Geometry: A Metric Approach with Models. Second edition.
Moschovakis: Notes on Set Theory.
Owen: A First Course in the Mathematical Foundations of Thermodynamics.
Palka: An Introduction to Complex Function Theory.
Pedrick: A First Course in Analysis.
Peressini/Sullivan/Uhl: The Mathematics of Nonlinear Programming.
Prenowitz/Jantosciak: Join Geometries.
Priestley: Calculus: An Historical Approach.
Protter/Morrey: A First Course in Real Analysis. Second edition.
Protter/Morrey: Intermediate Calculus. Second edition.
Ross: Elementary Analysis: The Theory of Calculus.
Samuel: Projective Geometry.
Readings in Mathematics.

Scharlau/Opolka: From Fermat to Minkowski.
Sigler: Algebra.
Silverman/Tate: Rational Points on Elliptic Curves.
Simmonds: A Brief on Tensor Analysis. Second edition.
Singer/Thorpe: Lecture Notes on Elementary Topology and Geometry.
Smith: Linear Algebra. Second edition.
Smith: Primer of Modern Analysis. Second edition.
Stanton/White: Constructive Combinatorics.
Stillwell: Elements of Algebra: Geometry, Numbers, Equations.
Stillwell: Mathematics and Its History.
Strayer: Linear Programming and Its Applications.
Thorpe: Elementary Topics in Differential Geometry.
Troutman: Variational Calculus and Optimal Control with Elementary Convexity. Second edition.
Valenza: Linear Algebra: An Introduction to Abstract Mathematics.
Whyburn/Duda: Dynamic Topology.
Wilson: Much Ado About Calculus.